Peace Through Tourism

Peace through Tourism considers the possibilities for tourism to contribute to efforts to unmask conflict and promote peace. This edited volume considers the intersections between tourism, peace, justice, and sustainability through conceptual and empirical works surveying practices, problems, and challenges all around the globe. It presents a complex and critical approach, arguing that peace through tourism is dialogic and not as simple as describing a few "good" niche segments of tourism.

The pedagogies of peace represented here work to analyse structural violence associated with tourism—such as in the dominance of neoliberal market imperatives over local or social economies; colonising, patriarchal and anthropocentric practices in tourism; and tourism's complex role in post-conflict settings. Analyses found here place scholars, industry, and communities in conversation about building shared tourism futures where peace is understood as peace with justice and differences are bridged through dialogues towards understanding. In light of the many challenges in attaining sustainable development in the 21st century, this volume is an important and timely endeavour. Radical practices are explored that support more 'just' tourism futures.

With a new introduction, this book is an insightful resource for scholars and researchers of Tourism and Peace and Conflict Studies. The chapters in this book were originally published in *Journal of Sustainable Tourism.*

Freya Higgins-Desbiolles is Adjunct Senior Lecturer, Business Unit, University of South Australia, and Co-chair of the Peace Tourism Commission of the International Peace Research Association (IPRA).

Lynda-ann Blanchard is Vice-President of the Australian Council for Human Rights Education; Honorary Affiliate, Department of Peace and Conflict Studies at the University of Sydney, Australia and, Co-chair of the Peace Tourism Commission of the International Peace Research Association (IPRA).

Yoko Urbain is affiliated with St. Marianna University Nursing School, Kawasaki City, and Soka University Department of Letters, Tokyo, Japan; and Co-chair of the Peace Tourism Commission of the International Peace Research Association (IPRA).

Peace Through Tourism

Critical Reflections on the Intersections between Peace, Justice and Sustainable Development

Edited by
Freya Higgins-Desbiolles, Lynda-ann Blanchard and Yoko Urbain

Routledge
Taylor & Francis Group
LONDON AND NEW YORK

First published 2023
by Routledge
4 Park Square, Milton Park, Abingdon, Oxon OX14 4RN

and by Routledge
605 Third Avenue, New York, NY 10158

Routledge is an imprint of the Taylor & Francis Group, an informa business

British Library Cataloguing in Publication Data
A catalogue record for this book is available from the British Library

ISBN13: 978-1-032-29014-0 (hbk)
ISBN13: 978-1-032-29018-8 (pbk)
ISBN13: 978-1-003-29957-8 (ebk)

DOI: 10.4324/9781003299578

Typeset in Myriad Pro
by Newgen Publishing UK

Publisher's Note
The publisher accepts responsibility for any inconsistencies that may have arisen during the conversion of this book from journal articles to book chapters, namely the inclusion of journal terminology.

Contents

Part II
Peace Tourism Pedagogies

Part III
Radical Peace Tourism in Practice

Part IV
Postscript

Citation Information

The chapters in this book were originally published in various volumes and issues of the *Journal of Sustainable Tourism*. When citing this material, please use the original page numbering for each article, as follows:

Chapter 1

Fortress tourism: exploring dynamics of tourism, security and peace around the Virunga transboundary conservation area
Lisa Trogisch and Robert Fletcher
Journal of Sustainable Tourism, volume 30, issue 2–3 (2022), pp. 352–371

Chapter 2

Tourism, peace and sustainability in sanctions-ridden destinations
Siamak Seyfi, Colin Michael Hall and Tan Vo-Thanh
Journal of Sustainable Tourism, volume 30, issue 2–3 (2022), pp. 372–391

Chapter 3

Insurgent citizens: mobility (in)justice and international travel
Pooneh Torabian and Heather Mair
Journal of Sustainable Tourism, volume 30, issue 2–3 (2022), pp. 392–407

Chapter 4

The role of dark commemorative and sport events in peaceful coexistence in the Western Balkans
Metod Šuligoj and James Kennell
Journal of Sustainable Tourism, volume 30, issue 2–3 (2022), pp. 408–426

Chapter 5

Peacebuilding and post-conflict tourism: addressing structural violence in Colombia
Mónica Guasca, Dominique Vanneste and Anne Marie Van Broeck
Journal of Sustainable Tourism, volume 30, issue 2–3 (2022), pp. 427–443

Chapter 6

Disrupting structural violence in South Africa through township tourism
Meghan L. Muldoon and Heather L. Mair
Journal of Sustainable Tourism, volume 30, issue 2–3 (2022), pp. 444–460

Chapter 7

"We are reconciliators": when Indigenous tourism begins with agency
Nicole Curtin and Steven Bird
Journal of Sustainable Tourism, volume 30, issue 2–3 (2022), pp. 461–481

Chapter 8

Exploring a unifying approach to peacebuilding through tourism: Abraham and Israel/Palestine
Jack Shepherd
Journal of Sustainable Tourism, volume 30, issue 2–3 (2022), pp. 482–499

Chapter 9

Promoting sustainable tourism futures in Timor-Leste by creating synergies between food, place and people
Tracy Berno, Gobie Rajalingam, Agueda Isolina Miranda and Julia Ximenes
Journal of Sustainable Tourism, volume 30, issue 2–3 (2022), pp. 500–514

Chapter 10

Living in the Wake *of Rural Irish Troubles: building an institution for sustainable peace through emotive out-of-place tourism*
John Erwin and Tristan Sturm
Journal of Sustainable Tourism, volume 30, issue 2–3 (2022), pp. 515–532

Chapter 11

Beyond multicultural 'tolerance': guided tours and guidebooks as transformative tools for civic learning
Meghann Ormond and Francesco Vietti
Journal of Sustainable Tourism, volume 30, issue 2–3 (2022), pp. 533–549

Chapter 12

One stone, two birds: harnessing interfaith tourism for peacebuilding and socio-economic development
Dagnachew Leta Senbeto
Journal of Sustainable Tourism, volume 30, issue 2–3 (2022), pp. 550–564

Chapter 13

"Don't look back in anger". War museums' role in the post conflict tourism-peace nexus
Fabio Carbone
Journal of Sustainable Tourism, volume 30, issue 2–3 (2022), pp. 565–583

Chapter 14

Dances with despots: tourists and the afterlife of statues
Elizabeth Carnegie and Jerzy Kociatkiewicz
Journal of Sustainable Tourism, volume 30, issue 2–3 (2022), pp. 584–599

Chapter 15

Making waves: Peace Boat Japan as a model of sustainable peace through tourism
Lynda-ann Blanchard, Sumiko Hatakeyama and Akira Kawasaki
Journal of Sustainable Tourism, volume 30, issue 2–3 (2022), pp. 600–617

Chapter 16

A diverse economies approach for promoting peace and justice in volunteer tourism
Phoebe Everingham, Tamara N. Young, Stephen L. Wearing and Kevin Lyons
Journal of Sustainable Tourism, volume 30, issue 2–3 (2022), pp. 618–636

Chapter 17

Te Awa Tupua: peace, justice and sustainability through Indigenous tourism
Jason Paul Mika and Regina A. Scheyvens
Journal of Sustainable Tourism, volume 30, issue 2–3 (2022), pp. 637–657

Chapter 18

WWOOFing in Australia: ideas and lessons for a de-commodified sustainability tourism
Adrian Deville, Stephen Wearing and Matthew McDonald
Journal of Sustainable Tourism, volume 24, issue 1 (2016), pp. 91–113

Chapter 19

Gender and sustainability – exploring ways of knowing – an ecohumanities perspective
Kumi Kato
Journal of Sustainable Tourism, volume 27, issue 7 (2019), pp. 939–956

Chapter 20

"This is a holy place of Ama Jomo": buen vivir, *indigenous voices and ecotourism development in a protected area of Bhutan*
Heidi Karst
Journal of Sustainable Tourism, volume 25, issue 6 (2017), pp. 746–762

Chapter 21

The land has voice: understanding the land tenure –sustainable tourism development nexus in Micronesia
T. S. Stumpf and C. L. Cheshire
Journal of Sustainable Tourism, volume 27, issue 7 (2019), pp. 957–973

Chapter 22

Colonizing space and commodifying place: tourism's violent geographies
Jennifer A. Devine
Journal of Sustainable Tourism, volume 25, issue 5 (2017), pp. 634–650

Notes on Contributors

Tracy Berno, Department of Hospitality and Tourism, Auckland University of Technology, Auckland, New Zealand.

Steven Bird, Northern Institute, Charles Darwin University, Australia.

Lynda-ann Blanchard, Department of Peace and Conflict Studies, University of Sydney, Australia; Peace Tourism Commission of International Peace Research Association (IPRA).

Fabio Carbone, Centre for Trust, Peace and Social Relations & Faculty of Business and Law, Coventry University, Coventry, UK.

Elizabeth Carnegie, Newcastle Business School, Northumbria University, Newcastle, UK.

C. L. Cheshire, Pacific Business Center Program, University of Hawaii at Manoa, Honolulu, HI, USA.

Nicole Curtin, Northern Institute, Charles Darwin University, Australia.

Adrian Deville, Clarence Valley Experience P/L, Grafton, New South Wales, Australia.

Jennifer A. Devine, Department of Geography, Texas State University, San Marcos, TX, USA.

John Erwin, Geography, School of Natural and Built Environment, Queen's University Belfast, Belfast, UK.

Phoebe Everingham, The University of Newcastle, Callaghan, New South Wales, Australia.

Robert Fletcher, Wageningen School of Social Sciences, Sociology of Development and Change Group, Wageningen University and Research, Wageningen, The Netherlands.

Mónica Guasca, Division of Geography and Tourism, University of Leuven, KU Leuven, Leuven, Belgium.

Colin Michael Hall, College of Hotel & Tourism Management, Kyung Hee University, Seoul, South Korea; School of Business and Economics, Linnaeus University, Kalmar, Sweden; Department of Management, Marketing and Entrepreneurship, University of Canterbury, Christchurch, New Zealand.

Sumiko Hatakeyama, Peace Boat, Tokyo, Japan.

Freya Higgins-Desbiolles, Business Unit, University of South Australia, Adelaide, Australia; Peace Tourism Commission of the International Peace Research Association (IPRA).

Heidi Karst, Department of Geography and Environmental Management, University of Waterloo, Waterloo, Canada.

Kumi Kato, Center for Tourism Research, Wakayama University, Wakayama, Japan.

Akira Kawasaki, Peace Boat, Tokyo, Japan.

James Kennell, Faculty of Business, Department of Marketing, Events and Tourism, University of Greenwich, London, UK.

Jerzy Kociatkiewicz, Institut Mines-Télécom Business School, Evry, France.

Kevin Lyons, The University of Newcastle, Callaghan, New South Wales, Australia.

Heather L. Mair, Department of Recreation and Leisure Studies, University of Waterloo, Waterloo, Ontario, Canada.

Matthew McDonald, Department of Management, RMIT University Vietnam, Ho Chi Minh City, Vietnam.

Jason Paul Mika, Te Raupapa Waikato Management School, University of Waikato, Waikato, New Zealand.

Agueda Isolina Miranda, The Asia Foundation, Dili, Timor-Leste.

Meghan L. Muldoon, Sustainable Tourism, University of Groningen, Groningen, Netherlands; School of Community Resources and Development, HNU-ASU Joint International Tourism College (HAITC), Arizona State University, Haikou, China.

Meghann Ormond, Cultural Geography Chair Group, Environmental Sciences Department, Wageningen University & Research, Wageningen, The Netherlands.

Gobie Rajalingam, The Asia Foundation, Dili, Timor-Leste.

Regina A. Scheyvens, School of People, Planning and Development, Massey University, Palmerston North, New Zealand.

Jack Shepherd, European Tourism Research Institute (ETOUR), Mid Sweden University, €Ostersund, Sweden.

Dagnachew Leta Senbeto, Institute of International Business and Governance, Lee Shau Kee School of Business and Administration, The Open University of Hong Kong, Ho Man Tin, Kowloon, Hong Kong.

Siamak Seyfi, Geography Research Unit, University of Oulu, Oulu, Finland.

T. S. Stumpf, School of Business, Humboldt State University, Arcata, CA, USA.

Tristan Sturm, Geography, School of Natural and Built Environment, Queen's University Belfast, Belfast, UK.

Metod Šuligoj, Faculty of Tourism Studies – Turistica, University of Primorska, Portorose, Slovenia.

Pooneh Torabian, Department of Tourism, University of Otago, Dunedin, New Zealand.

Lisa Trogisch, Wageningen School of Social Sciences, Sociology of Development and Change Group, Wageningen University and Research, Wageningen, The Netherlands.

Yoko Urbain, Peace Studies, Soka University, Japan; Peace Tourism Commission of the International Peace Research Association (IPRA).

Dominique Vanneste, Division of Geography and Tourism, University of Leuven, KU Leuven, Leuven, Belgium.

Anne Marie Van Broeck, Division of Geography and Tourism, University of Leuven, KU Leuven, Leuven, Belgium.

Francesco Vietti, Riccardo Massa Department of Human Sciences for Education, University of Milan Bicocca, Milan, Italy.

Tan Vo-Thanh, La Rochelle Business School – Excelia Group, CEREGE, La Rochelle, France.

Stephen L. Wearing, UTS Business School, Management Discipline Group, University of Technology, Sydney, Australia.

Julia Ximenes, The Asia Foundation, Dili, Timor-Leste.

Tamara N. Young, The University of Newcastle, Callaghan, New South Wales, Australia.

Peace through Tourism: Critical reflections on the intersections between peace, justice, sustainable development and tourism

ABSTRACT

United Nations Sustainable Development Goal 16 focuses on fostering inclusive, just and peaceful societies to ensure sustainable development. This can be an important catalyst to sustainable tourism and draws needed attention to the value of peace through tourism thinking. We might ask: by what means could we address structural injustices in tourism and additionally harness tourism for greater inclusivity, justice and sustainability? Peace through tourism analyses could benefit from engagement with peace studies thinking on peace with justice, issues of violence, frameworks for fostering positive peace, peace pedagogies and critical literacies in non-violence. Additionally, this book argues tourism must be understood in its wider context and the contribution it may make to both causing and countering structural injustices. We briefly explore emerging thinking on ecocide and decolonisation in order to illuminate the need for greater engagement in multi-logical, multi-layered and critical thinking in tourism studies. By furthering the peace tourism agenda, this work underscores the need to frame tourism in terms of peace and justice concerns. It also supports recent claims that tourism must be assessed and addressed in its wider structural context. This is a critical and timely discussion as global tourism struggles through multiple crises and challenges.

Introduction

> "Tourism can be a platform for overcoming the pandemic. By bringing people together, tourism can promote peace, solidarity and trust." - UN Secretary-General Antonio Guterres (UNWTO, 2020)

It has become a catchphrase that tourism contributes to peace and sustainability. The prologues to important tourism declarations, documents such as the "Global Code of Ethics for Tourism" (UNWTO, 1999) and numerous tourism strategies have presented such assertions. The statement above is a recent example and was pronounced when leaders and scholars debated the possibilities and desirability of a tourism reset in the face of COVID-19 (Lew et al., 2020). However, despite such rhetoric, some analysts have assessed the peace tourism proposition as weak, mistaking correlative and causal relationships (e.g. Litvin, 1998; Pratt & Liu, 2016) and proceeding from flawed and wishful premises (Litvin, 2020). It is thus a timely moment to revisit the topic of peace and tourism and to insert more fully concerns with justice and sustainable development into the discussion.

It is perhaps not intuitive that tourism is in fact a peace matter. The common understanding of tourism is that it is the business of supplying products and experiences for tourists to enjoy while they are on holidays. These are the twin pillars of the tourism business system, the industry and the tourists (Leiper, 1979). Such a narrow focus of course is impossible because tourism is a complex system and is embedded within larger social, economic, environmental and political contexts (Farrell & Twining-Ward, 2005). These facts are what contribute to issues of peace and conflict arising within tourism. For instance, Perkumiene and Pranskuniene (2019) examined the problem of overtourism and demonstrated the importance of the social system in addressing the interactions of tourists and locals, requiring a balancing between the democratic rights of locals to live in their home areas with the rights of tourists to travel. Peace and conflict issues emerge from the tensions that the tourism system exacerbates or creates as tourism processes work to package, commodify and market tourism products and experiences in the marketised tourism system (e.g. Mbaiwa, 2011).

Thus, it is valuable to more fully engage with the peace studies discipline. Although, much like tourism studies, peace studies have been characterised as "still in its infancy" (Webel, 2007, p. 4), there are nonetheless theories and ideas we can draw on from peace studies to help inform and advance our thinking on peace through tourism. The fact that both tourism and peace studies are developing disciplines offers an exciting space for reflection, dynamism, respect for process and creativity.

Peace studies has distinguished between "positive" and "negative" peace (Galtung, 2012). "'Positive' peace denotes the simultaneous presence of many desirable states of mind and society, such as harmony, justice, equity, etc. 'Negative' peace has historically denoted the 'absence of war' and other forms of widescale violent human conflict" (Webel, 2007, p. 6). Some peace analysts have emphasised the importance of "peace with justice" (Rees, 2003). As Rees explained: "A just peace would be marked not only by an end to hostilities between peoples but also by fairness in social, economic and political arrangements" (2003, p. 8). In his most recent work, Rees spoke of "diverse forms of advocacy for a common humanity through literacy about non-violence and for the health-promoting values of creative, non-destructive uses of power" (2020, p. 8). These works emphasise that peace is not simple and cannot be easily packaged as a product the way tourism can. Peace is a process; it is fluid, multi-layered and multi-logical. Such work offers a broader base to both critique tourism and its impacts as well as envision and build more positive futures in and through tourism.

We developed this Special Issue focused on Peace through Tourism and received author abstracts pre-COVID-19. The crisis caused by the COVID-19 pandemic has upended everything, stopping tourism and travel in a way previously unimaginable in our ever-globalising world. Social media memes of the time likened it to a game of Jumanji, referencing a popular film, as COVID-19 and government responses to it inserted unpredictable dynamics in our societies and economies. This will certainly have unexpected consequences to our efforts to shape tourism towards ethics, responsibility and sustainability. While leadership of tourism organisations such as the World Tourism Organisation (UNWTO) have not hesitated to boldly assert that tourism promotes peace, trust and solidarity (as above), this Special Issue takes a more critical approach to the topic. As Boluk, Cavaliere and Higgins-Desbiolles asserted: "The academy must work critically to reflect the realities of global communities, as related to, and impacted by, sustainable tourism development" (2019, p. 848). We are honoured in receiving articles for this Special Issue from noted tourism and peace scholars, emerging scholars in both disciplines and practitioners in the field building peace tourism ventures from which we can learn.

Peace through Tourism and the United Nation Sustainable Development Goals

> "United Nations Sustainable Development Goal 16: Promote peaceful and inclusive societies for sustainable development, provide access to justice for all and build effective, accountable and inclusive institutions at all levels." (United Nations, n.d.a)

Tourism scholars have taken inspiration from the United Nations Agenda for Sustainable Development (2015–2030) and its Sustainable Development Goals (SDGs) (Boluk et al., 2019). The SDGs followed on from the Millennium Development Goals (MDGs) to supposedly address their deficiencies and failures:

> One of the reasons that new goals are necessary and the MDGs failed to meet all of their targets was the absence of a critical consciousness that considered the structural causes of poverty [...] Above all, we have failed to relate the dominant neoliberal economic model to persistent levels of poverty and climate change. (McCloskey, 2015, p. 186)

The global 2030 Agenda for Sustainable Development included 17 SDGs and 169 specific targets (UN, 2019). Tourism authorities, including the UNWTO, presented tourism as a tool to advance this agenda and embraced the rhetoric. Goal 16 makes it clear that sustainability is dependent on peace, justice and sound governance. Despite emphasising tourism's contribution to peace in many of its documents and declarations, the UNWTO has not identified goal 16 as a particular focus in its commitment to the UN SDG agenda. It instead focused on: Goal 8 on tourism's contributions to economic growth and employment: Goal 12 on tourism's role in securing sustainable production and consumption; and Goal 14 on tourism in marine environments (UNWTO, n.d.). This would seem a lost opportunity considering that tourism is argued to be a catalyst to some of the world's largest peaceful movements of people (Budeanu, 2005) and that it is a key facet for intercultural interactions today.

As analysts build a body of knowledge on tourism and the SDGs, the development of critical thinking and holistic analyses are imperative (Boluk et al., 2019). Such critical works may help explain why bodies such as the UNWTO assumed the particular focus it did in engaging with the SDG agenda. For instance, Hall applied a managerial ecology lens to the engagement of tourism authorities with the SDGS, asking:

> Why... are the SDGs predominantly framed in a particular growth and business way in a tourism context that emphasise the importance of market oriented approaches and managerialism, when the SDGs are also concerned with broader social and environmental concerns as well as economic change? (2019, p. 1045)

Hall's analysis leads us to understand that there is an unwillingness on the part of some tourism proponents to engage with tourism in its broader context and address the relationship between tourism and wider injustices, inequities and exclusions as well as dynamics of unsustainable growth and development. One important aspect to SDG 16 specifically is that it uses important terms including "peaceful", "inclusive" and "justice" which are imperative to ensuring that the SDGs are qualitatively better than the MDGs that preceded them in terms of addressing the structural injustices to which McCloskey (2015) drew attention. SDG 16 inspires crucial questions of tourism, including: who is included in the benefits of tourism and who is excluded; how does tourism impact the enjoyment of peaceful futures in the places where tourism occurs (often called "destinations"); and how does tourism support or undermine justice and sustainability at all levels. It is precisely for these reasons that the intersections between tourism and peace offer potentiality far more rich, multi-layered and multi-logical considerations that expand beyond the narrow confines of a tourism-centric focus.

The Peace through Tourism Proposition

The topic of peace through tourism has been contentious and debated (Salazar, 2006). Some analysts have considered it a fruitful domain to consider the ways that tourism and peace intersect (e.g. Blanchard & Higgins-Desbiolles, 2013a; Moufakkir & Kelly, 2010); while others have claimed it is a confusion of correlative and causal relationships (e.g. Litvin, 1998). Critiques of the studies offered so far have characterised the work as wishful, anecdotal, naïve or driven by limited case study analysis (e.g. Pratt & Liu, 2016; Pedersen, 2017; Wolmuther & Wintersteiner, 2014, p. 52). To date, peace through tourism analyses are haunted by these endless debates on whether the peace through tourism proposition has merit and how we might overcome this.

In understanding the dynamics of peace tourism considerations, it is important to understand its early origins. In a context of concerns with the impacts of terrorism on tourism in the 1980s, Louis D'Amore promoted the idea that peace through tourism had a role to play in securing a more peaceful world. He argued tourism could serve as a form of "citizen diplomacy" (2010). He explained: "This human journey toward a collective will of humankind can only begin by first building bridges of understanding between civilizations, cultures, and religions. Travel and tourism – the world's largest industry – has a key role to play toward this end" (D'Amore, 2010, p. 562). He was instrumental in forming the non-governmental organisation the International Institute for Peace through Tourism (IIPT) in 1986 as an organisation to promote the idea that tourism fosters peace. Since that time, IIPT has held many conferences and summits, released the Amman Declaration on Peace through Tourism and supported the development of Peace Parks around the world (IIPT, n.d.; Becken & Carmignani, 2016, p. 64).

Academics soon took up the study of the relationship between tourism and peace. One early theoretical approach was what has been called the "contact theory", which hypothesised that inter-cultural contact through tourism could lead to greater understanding and reduce intercultural tensions (see Tomlejnovic, 2010 for a review). Particularly prevalent were case studies that examined the ways in which tourism, conflict and peace might interrelate in a specific context where conflicting parties were brought together through tourism. For instance, Kim, Prideaux and Prideaux (2007) examined peace through tourism in the Korean Peninsula, finding a form of "track two" diplomacy resulting from the travel of South Korean citizens to selected sites in North Korea. In another example, Higgins-Desbiolles (2003) considered how Indigenous tourism experiences might be used to foster reconciliation in the settler-colonial context of Australia and offered a case study analysis of Camp Coorong in South Australia. Numerous case studies have been offered addressing peace tourism possibilities in places of conflict and post-conflict, including Northern Ireland (e.g. Simone-Charteris & Boyd, 2010), Cyprus (Scott, 2012) and Bosnia and Herzegovina (Causevic, 2010). Others have been highly sceptical of the claim that contact through tourism can foster peace, arguing that the opposite might occur, that is, contact can lead to reinforced stereotyping and less understanding (e.g. Farmaki, 2019; Pizam, Jafari & Milman, 1991).

The edited volume on peace through tourism by Blanchard and Higgins-Desbiolles (2013a) showcased case studies drafted by leaders from non-government organisations (NGOs) and communities that were at the cutting-edge of engaging with tourism in terms of human rights, peace and justice. This included Equations in India (2013), Tourism Concern in the UK (Stancliffe, 2013), the Ngarrindjeri Aboriginal community in South Australia (Bennett & Ngarrindjeri Being Heard, 2013), Peace Boat (Tatsuya, 2013) and the Alternative Tourism Group in Palestine (Kassis, 2013); as well as a chapter from a peace tourism business founder, Touchdown Tours (Preketes, 2013). Collectively, these case studies demonstrated that tourism involves more than just for-profit businesses that are the typical object of focus. It also demonstrated that NGOs and communities were active and effective in demanding that human rights and justice feature in considerations. Blanchard and Higgins-Desbiolles gathered these case studies in their collection as part of their promotion of "a pedagogy of peace" in peace tourism work (2013a). The book was developed following a conference convened in Sydney in 2010 supported by a grant from the Toda Institute for Peace Policy and Research for an ideas exchange and dialogue. The intention was to create opportunities for dialogue and co-learning on peace pedagogy and peace tourism practices. In a review of the book, Steve Pan noted it helped promote better understanding of "tourism as a social force" and he called for more case studies (2015, p. 317).

Nonetheless, later scholars of peace through tourism have characterised such efforts to offer case studies as inconclusive in supporting the peace tourism proposition. For instance, Wolmuther

& Wintersteiner assessed the deficiencies in peace tourism analyses to that point and alleged "...we sometimes find confusion between normative goals and empirical facts" (2014, p. 52). Such criticisms resulted in concerted efforts to quantify and more rigorously measure the relationship between tourism and peace and thereby arguably more adequately assess the peace through tourism proposition. For instance, Becken and Carmignani (2016) developed a probit model based on UNWTO data from 126 countries over a 15-year period. Their findings provided clear quantitative evidence demonstrating that greater tourism arrivals correspond to a reduction in the risk of conflict. They found:

> In the context of tourism, the importance of strong institutional arrangements for the country's ability to leverage peace-contributing potentials from tourism was highlighted earlier ... If unsupported by broader peace building strategies ... the private tourism industry, tends to pursue a (passive) principle of "doing no harm", rather than actively promoting transformational change or peace-building initiatives. (2016, p. 77)

Utilising data from the Global Peace Index and international tourism arrivals in 111 countries, Pratt and Liu (2016) attempted to answer the question of whether tourism contributes to world peace; seemingly the opposite of more "micro" level case studies. From their findings, they concluded that tourism is more a beneficiary of peace (thus supporting Litvin, 1998) and also peace is most likely to positively impact tourism development in middle income countries (Pratt & Liu, 2016, p. 89). Importantly, Pratt and Liu (2016) acknowledged their study was focused on the indicators of negative peace (as per Galtung) rather than positive peace; they suggested future studies should seek to also study indicators of positive peace using a tool such as the positive peace index.

Another angle of criticism for peace through tourism was offered by Blanchard and Higgins-Desbiolles (2013c). In this work, they explained concerns that certain proponents were opportunistically seizing the rhetoric: "Too often an uncritical approach is taken in harnessing the philosophy, language and practice of 'peace tourism'. The terms may be hijacked by industry perspectives to justify business agendas" (Blanchard & Higgins-Desbiolles, 2013c, p. 20). As Salazar noted: "It seems contradictory ... that an industry which is so laden with conflict claims to be in a privileged position to foster world peace" (Salazar, 2006, p. 330). This dynamic has much in common with industry usage of the malleable term "sustainability", which has led to accusations of "greenwashing" and usurpation of agendas (see Weaver, 2009).

As a way to overcome the impasse in peace tourism analysis regarding conceptualisation of the relationship between tourism and peace, some analyses focused on "post-conflict tourism". For instance, Famaki (2017) proposed to better theorise the tripartite relationship between conflict, peace and tourism, offering a continuum that was premised on the argument that the "peace facilitating role of tourism should be placed in a context of conflict and consider the influence of tourism to the conflict context and the moderating impact of the context on tourism's peace effects" (p. 536). Farmaki argued this focused attention on the important questions of which aspects of tourism could be targeted to support peace and "how the conflict context can restrain tourism's positive effects on reconciliation" (2017, p. 537).

Causevic and Lynch (2011) introduced the concept of "phoenix tourism" to tourism studies. This describes a situation when tourism supports destination recovery from conflict by fostering normalisation of social relationships. Based on research conducted in Bosnia and Herzegovina and using the concept of double burial which highlights rebirth, Causevic and Lynch (2011) proposed phoenix tourism not as a type of tourism but as a process. This process occurs when places that have experienced intense conflict might use tourism to improve social relations in a post-conflict situation to eventually arrive at a more normal form of tourism for economic development. In such a scenario, war memorabilia with intense meaning might evolve to become heritage to be interpreted through tourism. This analysis is a very sophisticated argument combining rich theorisation with detailed empirical work. One clear message in this work is that in understanding tourism's peace capacities, engaging with local context is essential.

Other analysts have examined the possibilities for tourism to contribute to peace-building efforts, often in situations of post-conflict or intercommunal tensions. In one example, Wolmuther and Wintersteiner proposed "peace-sensitive tourism":

> Instead of speaking of tourism as an 'instrument for peace' we prefer the term 'peace-sensitive tourism'. Peace-sensitive tourism – both from the supplier's and from the consumer's side – is conceived not as an isolated peace strategy, but as part of a huge and historical social transformation towards a culture of peace. Peace-sensitive tourism is aware of its economic and social power and thus of its social responsibility. It aims to contribute – within the limits of its opportunities but aware of the power of alliances – to the democratisation of the society concerned, of international relations and of the world system as such. (2014, p. 20)

In another case, Blanchard and Higgins-Desbiolles proposed a "pedagogy of peace" to shape peace through tourism practices arguing:

> Dialogue based on the goals of peace education would join the notions of tourism as an industry and tourism as a social force ... It involves collaborative critical thinking and action between the academy and the industry to include the voices of people who are seldom heard coupled to analyses that expose injustice. (2013c, pp. 29–30)

This review indicates that peace tourism thinking is still evolving. One key issue is whether to place the emphasis on tourism or on peace and the associated disciplines that purport knowledge about these phenomena. Many of the analysts of peace through tourism are based in the tourism discipline and therefore, arguably, their emphasis is on tourism's role in supporting peace or conflict (e.g. Farmaki, 2017; Causevic & Lynch, 2011; Becken & Carmignani, 2016). Greater transdisciplinary approaches are needed to inform our understanding of the dynamics operative in peace through tourism and to move beyond tourism-centric framing. One example comes from Scott in her study on Cyprus when she posed some critical questions: "where do governments and the tourism industry position themselves in terms of tourism and peace priorities? What are the consequences of adopting a 'peace first' or a 'tourism first' approach? What are the mechanisms at work?" (2012, p. 2116). The collaboration between Higgins-Desbiolles, Blanchard and Urbain, the former a tourism scholar and the latter two peace scholars, spanning some two decades has been an attempt to open a dialogue across the disciplines and to bridge these disciplinary knowledges (see Blanchard & Higgins-Desbiolles, 2013a). It is important that peace and conflict studies perspectives are fully utilised to further our understanding but equally important is a dialogue between these disciplines so that we can better understand the possibilities for peace tourism to develop and support positive outcomes for people and places.

Understanding Peace and Violence in Tourism

As noted above, a number of scholars have focused on issues of conflict in their discussions of peace tourism (e.g. Causevic, 2010; Farmaki, 2017; Simone-Charteris & Boyd, 2010). However, drawing on peace studies perspectives, Blanchard and Higgins-Desbiolles argued that conflict is not the problem; it is violence (2013b). From this perspective, conflict, diverse views and tensions are sites of creativity and engagement; they offer opportunities for dialogue and building understanding. Tourism authorities on the other hand are less tolerant of conflict because it represents a threat to the conduct of tourism, the stories that are packaged and told, and the comfort of tourists who are often risk adverse.

Peace scholar Johan Galtung's (2012) analysis of forms of violence offers some valuable insights for peace tourism thinking. Considered one of the founders of peace studies, his theorisation of negative and positive peace is helpful in strengthening our understanding of the interface between peace and tourism. He distinguished direct violence, structural violence and cultural violence (Galtung, 2012). Direct violence is intended (often visible) violence, often between states and engaged in by military and other forces; it is what we conventionally call to mind when we think of violence.

Galtung distinguished this from structural (often invisible) violence which he characterises as "the nonintended slow, massive suffering caused by economic and political structures of exploitation and repression" (2012, p. 1). Cultural violence is that which legitimises these forms of direct and structural violence (Galtung, 2012, p. 1).

With this conceptual framework, we come to a full peace agenda:

> This gives us six peace tasks: eliminating the direct violence that causes suffering, eliminating the structures that cause suffering through economic inequity – or, say, walls once placing Jews, now Palestinians, in ghettos – and eliminating cultural themes that justify one or the other. The task known as ceasefire is only one - sixth of a complete peace process. But then come the three tasks of building direct, structural, and cultural peace. The parties exchange goods, not "bads", not violence. The structural version of that builds cooperation and sustainability into the structure, with equity for the economy, and equality for the polity. The goal is to build a structure based on reciprocity, equal rights, benefits, and dignity … and a culture of peace, confirming and stimulating an equitable economy and an equal polity. (Galtung, 2012, p. 1)

If the peace through tourism agenda was to be harnessed to such a peace-building vision, then peace tourism proponents would have to commit to a much broader agenda than just harnessing tourism opportunities in post-conflict contexts or insuring conflict is suppressed so that tourists are not put off. This is what Blanchard and Higgins-Desbiolles (2013a) meant when they discussed peace through tourism in terms of human rights and justice and presented a vision of "human security through international citizenship". However, there is a wide chasm between such an approach and the current conventional understanding of "tourism as an industry" (Higgins-Desbiolles, 2006, 2021) with no further obligation beyond commercial imperatives and abiding by relevant regulations. In fact, "tourism as an industry" today is underpinned by profound inequity or structural violence and this is an important fact to confront.

Tourism's relationship to violence is not necessarily well understood nor how violence plays a significant factor in the unsustainability of tourism. Andrews edited a volume on tourism and violence that assisted in addressing this gap (2014, p. 6): "Of central importance to this work … is the way in which the violence that moulds tourism products, as well as the violence within them, inform the social world and power relations and moral order …". This work offered detailed conceptual and case study analyses to understand the myriad ways that tourism is linked to various forms of violence. Andrews charged that: "violence is manifest in many aspects of tourism practices and encounters …" (2014, p. 5). In Andrews' view, this includes: interpersonal violence in tourism, violence as a tourism attraction, the structural inequities of tourism that result in violence and the ways that military technologies and practices have catalysed tourism developments. Additionally, important are the ways in which inappropriate tourism development can result in a violent deterioration in the quality of life for resident communities (see da Cal Seixas et al., 2014). More recently, Devine and Ojeda (2017) edited a special issue of the *Journal of Sustainable Tourism* which addressed and theorised violence and dispossession in tourism.

In this Special Issue, we have a number of articles that address issues of structural violence and the obstacles for securing peace in a number of contexts. The article offered by Trogisch and Fletcher (this issue) importantly reveals how "peace tourism" to the Virunga transboundary protected area located on the borders of Uganda, Rwanda and the Democratic Republic of Congo actually maintains and exacerbates ongoing conflict in the region. Their findings reinforce the need for critical attention as peace tourism in this case was found to be largely rhetorical due to the dynamics under neoliberal development and conservation dynamics which sees further militarisation in the interests of selling security to potential tourists.

Similarly, Seyfi and Hall (this issue) provide an analysis of sanctions against the Islamic Republic of Iran and demonstrate that contrary to proponents, sanctions do not serve as a tool for advancing peace. Their work supports arguments that peace tourism studies must be much more engaged in issues of positive peace and peace building rather than being myopically focused on the absence of overt violence. The article by Guasca, Vanneste and Broeck (this issue) presents the case of post-conflict Colombia and analyses possibilities for rural agency and self-determination against ongoing structural violence and the possible use of tourism as a tool for peace building. Similarly, Muldoon and Mair (this issue) consider the possibilities for township tours in South Africa to disrupt the structural violence (e.g. silencing and invisibilisation) infringing the welfare, well-being and quality of life for Black South Africans.

Peace as process versus tourism as packaged

In the continuing debates in tourism concerning whether the peace tourism proposition is supported with evidence and empirical facts, a difference between peace studies and tourism studies becomes apparent. Drawing insights from peace studies, we identify that peace studies often emphasises process whereas tourism studies relies on packaging peace for tourists who are often very risk adverse and so the majority of works focus on post-conflict cases. As peace studies scholar Webel explained: "On the contrary, peace in its progressive or dialectical mode denotes active individual and collective self-determination and emancipatory empowerment. Peace entails continuous peacekeeping and peacemaking. And peacemaking requires active and continual personal and collective transformation ..." (2007, p. 8).

There are works that demonstrate insights into the concept of peace through tourism as a complex process. For instance, Blanchard and Higgins-Desbiolles (2013a) built their theorising of a justice approach to peace through tourism on the "peace with justice" concept developed by Rees (2003). Engaging with multiple dimensions of peace, the creative possibilities in conflict and a commitment to processes, their work asked challenging questions of what we might ask of tourism. They demonstrated how this might be enacted from the micro to the global, with such goals as challenging the taken-as-given relationships between "hosts" and "guests" and envisioning a role for tourism in international relations and global citizenship (Blanchard & Higgins-Desbiolles, 2013b, p. 2).

Scott's (2012) study of tourism processes in Cyprus acting as a "confidence building measure" is another study that explains peace tourism as a process. In this work she uncovered complex outcomes from efforts to resolve the long-standing conflict in Cyprus and what role tourism played in advocating for peace; a dual divide for action was revealed. Scott explained how government actions worked to secure the interests of large, foreign owned tourism corporations "... producing benefits for the tourism industry, rather than furthering the interests of peace" (Scott, 2012, p. 2128). However, Scott also described a civil society initiative to use village tourism cooperation across the greenline (which demarcates the conflict) as a more grassroots initiative to revive trust, reciprocity and build confidence. The latter project did not succeed; Scott draws conclusions from that, including consideration that the wider structural contexts may limit the ability to revive trust and reciprocity in places such as divided Cyprus. In terms of this analysis, though, Scott's work is a good example of understanding peace as a process and emphasising the need for peace concerns, particularly positive peace, to not be overshadowed by tourism concerns.

We also can better understand the possibilities for peace through tourism by examining recent research by Fry et al. (2021) that offered a comparative anthropological perspective on societies within peace systems that were found to not only avoid war but also to actively foster intergroup connection. This robust research uncovered five significant factors: overarching common identity; positive social interconnectedness; interdependence; non-warring values and norms; non-warring myths, rituals, and symbols; and peace leadership. These, they argued, can be used in policy development to promote and sustain peace, cohesion and cooperation that could assist in addressing even our wickedest threats and interwoven global challenges such as global pandemics, oceanic pollution, loss of biodiversity, nuclear proliferation, and climate change (Fry et al., 2021).

This perspective of peace as process encourages us to think in terms of peace and peace through tourism pedagogies and what tools, infrastructures and processes we might have to support such peacemaking, peacebuilding and peace sustaining work. As Jenkins and Reardon noted: "Peace education is concerned with developing pedagogies that enable learners to think in terms of complexities beyond the standard curricula on controversial issues that usually teach students to consider little more than the two major opposing positions ..." (2007, p. 216). As an example, the importance of peace museums in educating and providing an infrastructure for peace has been emphasised in the work of van den Dungen (2009). War museums and monuments proliferated in the aftermath of the World Wars, but peace museums have been multiplying from the 1980s and drawing attention to the need for education against war and violence, particularly in the nuclear age (van den Dungen, 2009; Urbain, 2013).

Melotti, Ruspini and Marra (2018) offered a peace tourism example from the Nobel prize nominated island of Lampedusa, which has experienced the direct impacts of the refugee crisis in Europe. In this work, they analysed the Museum of Trust and Mediterranean Dialogue as a tool for enhancing awareness and mutual understanding. They found this "preparatory exhibition tried to develop a complex cultural discourse, interweaving local community, tourists, and migrants in a relatively small space, where residents, tourists, and migrants coexist in a fluid though asymmetrical way" (2018, p. 222). While not fully successful, it is an example of peace tourism as an imperfect and evolving practice. In their work, they envisioned developing a new "diffused eco-museum" which would "enhance a shared local memory starting from three main socio-cultural dimensions: the original fishing culture, the tourist function, and the cultures of migration" (Melotti et al., 2018, p. 223).

In this volume, we have a number of articles which suggest possibilities for peace tourism pedagogies through diverse mechanisms, operating at many levels and using a variety of creative and effective means and processes. Blanchard, Hatakeyama and Kawasaki (this issue) explain the use of a mass tourism facility, a cruise ship, as a floating venue for peace making, anti-nuclear education and education for global citizenship. Carbone (this issue) has expanded work examining peace museums to consider the possibilities of harnessing war museums and war monuments to the task of educating for peace. Berno, Rajalingam, Miranda and Ximenes (this issue) discuss development of Timor Leste food tourism as a pathway for dialogue in a post-conflict situation, reconciliation and healing from violent occupation. Ormond and Vietti (this issue) through engaged research on Migrantour guided walking tours and the Roots Guide guidebook explain how tours and guidebooks can move past approaches of multicultural tolerance to engage diverse communities in visits with the Other that accept frictions and undermine facile stereotypes. Erwin and Strum (this issue), like Ormond and Vietti, report from engaged research that analyses the use of an interactive performance to engage local participants in rural Northern Ireland in understanding diverse perspectives on the conflict and its aftermath.

Peace tourism can also be viewed as a dialogic space for communicating across different worldviews and dividing lines to have conversations about differences and how we might bridge these (Blanchard & Higgins-Desbiolles, 2013a; see also Shelley, Ooi & Denny, 2021). The rich possibilities for such dialogues can be found in the work of Shepherd (this issue) which examines the Abrahamic myth as a basis for shared dialogue across the Israel-Palestine divide. In the context of Ethiopia, Senbeto (this issue) explains how religious pilgrimage sites and practices can support interfaith understanding. In a different vein, Kociatkiewicz and Carnegie (this issue) explain how problematic historical monuments and statues in Eastern Europe have been treated differently in their process of heritagisation and they consider the ways in which tourists engage in "dances with despots" in their new tourism settings. In the settler-colonial context of Australia, Curtin and Bird (this issue) listened to Aboriginal tourism operators who indicated clear commitment to use tourism as a tool for reconciliation and dialogue with tourists in order to assert their agency through hosting, connecting and sharing. This is a use of tourism as a dialogic tool and indicates how it can contribute to conflict reduction and peace building; by engaging conflict in tourism spaces, such as contested histories over identity and belonging to place. The values of Ubuntu featured in the article by Muldoon and Mair (this issue) on South African township tours reveal this worldview of relatedness through shared humanity as a dialogic approach conducive to overcoming stark differences.

The importance of the wider structural context in peace through tourism thinking

An additional insight that peace and conflict studies can provide is the imperative to engage with the wider structural context in which either peace or conflict is supported and sustained. This is one of the key points to Rees' (2003) articulation of "peace with justice" which requires defining a just peace as one that features not only the absence of violent conflict but also the presence of fairness, justice and equity in society, economy and politics. This is a similar point in the emphasis placed on the concept of "positive peace" in the work of Galtung (e.g. 2012) and others.

This attention to the wider context and concerns with issues of structural justice is something that many tourism analysts are hesitant to engage with, arguing that tourism is just an industry with no obligations beyond that (see Butcher, 2020). There are important exceptions and work in this vein has been evolving (e.g. Pezzullo, 2009; Jamal, 2019; Higgins-Desbiolles, 2018). In terms of peace through tourism analysis, this was the very challenge that Higgins-Desbiolles and Blanchard (2010) presented to peace tourism proponents, arguing we must engage with human rights, justice and peace issues in our work on tourism.

We can no longer ignore serious crises that threaten both humanity, other species and whole ecologies. In recent years, extraordinary warnings have been issued by scientists that the world is on a dangerous pathway. For instance, in 2018 scientists explored the possibilities that human-induced climate change is leading us on a pathway to "hothouse earth": "If the threshold is crossed, the resulting trajectory would likely cause serious disruptions to ecosystems, society, and economies" (Steffen et al., 2018, p. 8252). In addition to the grave environmental challenges we face, which may be of existential proportions, we also face grave social crises. We have witnessed "civilisation" disputes and jockeying for global power. This has been expressed through such diverse phenomena as the "war on terror", expressions of white supremacy, Brexit, the Belt and Road project, among numerous others. Tourism occurs in such contexts, contributes to such phenomena and holds some responsibilities for supporting pathways to peace and justice.

In this Special Issue, Torabian and Mair employ a critical mobilities lens to explain how racial hierarchies exist in access to international mobility and state bordering practices wreak violence and injustices that should call into question notions of peace and justice in tourism. Additionally, Suligoj and Kennell (this issue) explore dark commemorative events in the Western Balkans through a media analysis and demonstrate that such events occur in a social system of ongoing nationalism and intolerance which impairs the ability to develop sustainable events and tourism pathways in the region. For this introductory article, we will offer two brief illustrative examples to demonstrate how these issues of wider structural justice are of vital concern to tourism studies. These are: tourism's relationship to ecocide and tourism's decolonisation imperative.

Tourism and Ecocide: a crime against peace

> "Climate change is the biggest security threat that modern humans have ever faced." – Sir David Attenborough speaking before the UN Security Council, 24 February 2021 (Lynch, 2021)

The role of tourism in environmental damage and the need for environmental ethics in tourism have been well studied (e.g. Holden, 2003). Today things are arguably qualitatively different as the global scale of environmental change has increased exponentially, with the threat posed by global climate change now top of mind. In such a context, it is now possible to consider tourism's contributions to ecocide (White, 2019). Ecocide refers to "'the extensive damage, destruction to or loss of ecosystems of a given territory, whether by human agency or by other causes, to such an extent that peaceful enjoyment by the inhabitants of that territory has been severely diminished" (Higgins, 2012, p. 3). Higgins argued that such actions caused by humans amounted to a crime and her work has been instrumental in efforts at incorporating this into some legal codes. Recently, legal experts have advocated to have the worst offences of the crime of ecocide prosecutable before the International Criminal Court (Siddique, 2021).

We would argue here that ecocidal forms of tourism also represent a crime against peace. The continuous expansion and development of tourism globally has had momentous impacts on natural environments. This is based not only on the widespread dissemination of the mass tourism model but also niche market forms of tourism including ecotourism. Ecotourism may even be worse as ecotourists seek out more pristine environments thereby spreading out the impacts of tourism even further afield (see Cater, 1993). Added to this is the newer niche of "last chance tourism" which features seeing places, cultures, fauna, flora and phenomena before they disappear; ironically sometimes due to tourism impacts (see Lemelin et al., 2010). White noted: "These transformations not only change nature but simultaneously transform human relationships and the habitats of human and nonhuman alike. This fundamentally undermines environmental health and wellbeing globally" (2019, p. 285). In addition to the work to encode ecocide as a crime, legal experts are working to enshrine the rights of nature as another mechanism to overturn human devastation of nature.

While tourism development plays a serious role in these damaging dynamics, it also can be used as a tool to counter them. Tourism can contribute to changes in ecological consciousness with considerable transformative capacities. Recent work by Bawaka Country et al. (2017) explained how Indigenous tourism experiences on Country can lead to transformations:

> … more-than-humans have agency that may shape or enable possibilities for transformative experiences… transformation can be grounded in the notion of continual co-becoming of Country. Here, rather than a passive backdrop to experience or place, Country is an active participant in learning ... Indeed, visitors become part of Country, understanding themselves as connected in deep ways, both to each other and to Country itself. (p. 446)

Higgins-Desbiolles wrote about this in regard to Indigenous ecotourism through which "indigenous communities us[e] the opportunity that ecotourism provides to educate non-indigenous people about indigenous values and lifeways in the hopes of overturning the destructive nature of the Western environmental paradigm" (2009, p. 144). Guasca, Vanneste and Broeck (this issue) illustrate how ecological concern was one of the motivations for their informants from Montes de María, Colombia to use tourism as a tool of "environmental and social pedagogy" in order to protect traditional territories and lifeways. Mika and Scheyvens (this issue) offer insights into the conduct of Indigenous-controlled tourism along the Whanganui River in Aotearoa/New Zealand, a place where the rights of the river (through the granting of the status of legal personhood) have been recognised and enacted.

The need for decolonising tourism

As recent news reports of a resurgence in white supremacy, right-wing violence in a number of countries and the injustices committed against oppressed and marginalised populations living around the world (see UN CTED, 2020; Spence, 2020), the issues of ongoing imperialism and the need to decolonise are clearly an important priority. Recent work has engaged with decolonisation of tourism practices and studies (e.g. Chambers & Buzinde, 2015; Grimwood, 2019; Wijesinghe, 2020). There is also an impressive decolonial movement in tourism that is carrying out praxis that should inform studies in tourism sustainability and peace. One example is the DeTours in Hawai'i (also see book review, this issue):

> Within the militouristic landscape of O'ahu, they draw upon the practice of Indigenous remapping, "a key strategy for decolonizing space and place in the neo-colonial present, thereby revalorizing . . . Indigenous ontologies" while challenging neocolonial authority. DeTours mark the collisions that have defined the islands for well over a century, that have delineated Hawai'i as occupied territory—and envisioning otherwise. (Mei-Singh & Vicuña Gonzalez, p. 175)

The work of Engmann (2020) offered insights into the formation of heritage tourism in Ghana catering to African diasporic communities generated by the Atlantic slave trade. This case of African diaspora travel shows a multi-layered engagement and re-empowerment as Ghana led decolonisation efforts in the Global South. Through its slave heritage tourism offerings, Ghana connected with African diasporic communities to help them in the reviving their identities and African connections to overcome the legacies of Western enslavement. Through invaluable interviews with revolutionary leader and former President of Ghana Jerry Rawlings, Engmann (2020) captured a time when tourism could be seen as part of the global movement for decolonisation. Rawlings stated:

> Ghana was seen as a place where the black man had every reason to feel proud and was not exploited by neo-colonialism, so that was something in and of itself...In our case... [it] is a connection to blackness, to "Africanness". We were one country that exhibited defiance to exploitative relations. Ghana had experienced some serious setbacks. But the 1979 revolution restored a sense of justice that led to a very energetic society. (Engmann, 2020, p. 9)

From Australia, Aboriginal scholar Emma Lee and her collaborators used the term "love-bombing" to describe their strategy in decolonising state agencies' relationships with her community in Tasmania:

> Our engagement strategy saw our peoples' position as: culture, not politics; family, not organisations; and relationships before agreements. We have used the language and strategies of family violence prevention and applied it to our TWWHA [Tasmania Wilderness World Heritage Area] country. We had been in an abusive relationship with the governing agencies, but we needed to make our families and communities healthy and functional again. We introduced love back into our negotiations (because this reflects our love of TWWHA country) and termed our strategy "love-bombing", because we gave the government cultural gifts, strategic advice, and invitations to share in country and kinship. Even when they did not want it and rejected us, we came back the next day and demonstrated how our peoples can contribute to the Tasmanian economy through our joy, happiness, sustainable business and conservation plans, and willingness to work together. (Lee & Hamilton, 2016)

Lee's works suggest that using Indigenous epistemologies, values and practices can reset longstanding violent and negative relationships in settler-colonial contexts. This "love-bombing", with its paradoxical juxtapositions, describes overturning power relationships by inviting non-Indigenous others to transform their relationships to place and First Nations, to the benefit of all. Such a vision moves on from reconciliation as mutual tolerance to reconciliation as mutual thriving through relatedness and reciprocity. The work of Devine and Ojeda (2017) was also indicative of decolonising thinking for sustainability in tourism. Their recommendations included: "... redefining 'sustainability' as Indigenous and Native sovereignty, advocating for grassroots and collective forms of tourism, reducing tourism's role in climate change by traveling locally, and contesting the reproduction of colonial itineraries and practices of 'Othering'" (2017, p. 605).

Decolonisation thinking in tourism studies could also be used in a broader way. Serge Latouche (2009) has called for the "decolonization of the imaginary" from capitalistic thinking through developing a degrowth agenda. Kallis explained this as: "an active process of liberating thought, desires and institutions from the logic of growth, productivism and accumulation for accumulation's sake" (2011,

p. 877). These efforts at adding decolonisation to the practices of peace through tourism denote an approach to "positive peace" that also challenges the structural contexts in which tourism operates. It is this more challenging but comprehensive approach that enables the capacity to truly transform tourism towards more just, sustainable and peaceful futures. In this issue, Everingham et al. make a contribution to such work in their analysis of the value of a diverse economies approach through their case study of a volunteer tourism organisation operating in Ecuador. Mika and Scheyvens (this issue) offer a case study of Māori tourism enterprises on Te Awa Tupua, the Whanganui River, and derive a model they entitle "kaupapa tāpoi" which encompasses Māori worldviews of justice, peace and sustainability which could shape the future conduct of Indigenous tourism.

Conclusion

> The first principle is peace. The second principle equity, justice for the people. And third, the power of the good minds, of the collective powers to be of one mind: unity. And health. All of these were involved in his basic principles. And the process of discussion, putting aside warfare, as a method of reaching decisions, and now, using intellect. (Oren Lyons speaking on the peacekeeping traditions of the Iroquois Confederacy on Radio KUNM, 2005)

As we discuss concerns with peace, tourism and peace through tourism, it would be remiss to not recognise that as of 2016, "more countries experienced violent conflict than at any point in almost 30 years" and that violence and conflict are on the rise (UN, n.d.b). This is the wider structural context in which tourism operates and of which peace through tourism analyses must be cognisant. In recent years, analyses of the intersections of peace and tourism have taken on some sophistication and diversity. This work could be further advanced by assuming a critical and questioning approach focused on processes and questioning whose voices and ideas are enabled (Boluk et al., 2019). It might ask: what are peace building systems in tourism and how do we further foster these?

As the discussion in this article has demonstrated, links to the wider structural contexts—within which tourism is situated—are essential. Drawing on Rees' (2003) discussion of peace with justice, we can surmise that peace tourism with justice requires attention to the attainment of fairness in social, political, economic and ecological arrangements. It is clear from this discussion that linking tourism processes to wider development processes and securing structural justice (as per Rees, 2003; Galtung, 2012) are key to realising peace through tourism. As Becken and Carmignani noted: "the evidence presented here hints at a broader set of relationships linking tourism and development. Tourism dynamically interacts with other macroeconomic, political, and institutional factors to determine development outcomes" (2016, p. 78).

The industry is making inroads on this as it confronts multiple and intimidating crises. In speaking on VisitScotland's decision to be the first national tourism board to join the Tourism Declares initiative (on the climate emergency), Malcolm Roughead explained:

> We want to set out our vision for the future of responsible tourism and how sustainability, inclusivity and communities will be major focuses for us in the years to come. Our declaration is a reaffirmation of our commitment to place responsible tourism firmly at the heart of everything we do. (Tuppen, 2021)

But this does not go far enough as it currently incrementally moves from a commitment to promoting tourism to instead promoting more "responsible tourism". It is the argument of this article that peace through tourism analyses press us to understand tourism in its larger context and to address its deep embroilment in serious structural injustices. Too often a key failure of tourism industry and tourism academy efforts is the narrow focus on tourism interests, tourism dynamics and a tourism first philosophy (see Higgins-Desbiolles, 2021). Additionally, with a peace with justice approach we might consider reparations, restitution and restoration rather than merely responsibility and reconciliation in our efforts to set tourism on a pathway to address structural justice concerns in tourism. The earlier sections focused on "ecocide" and "decolonisation" impress upon us that we must reset tourism on a pathway for securing social and ecological justice, and considerations of peace with justice are conducive for such endeavours.

Looking forward

Such an agenda of peace through tourism invites new horizons for research, teaching and engagement. Specifically, we need more collaborations between all stakeholders that help empower the disempowered – with NGOs (as seen here with Peace Boat), with communities (as seen here with Indigenous communities), with overlooked players (such as refugees as seen here) and with ecologies (like the Whanganui River in this Special Issue). Instituting critical and radical practices also means fostering research collaborations that build supportive collegial research environments; rather than competition, we could foster a culture of sharing research whilst building on existing work with constructive critique.

We need to theorise "critical" perhaps even "radical" pedagogies (Blanchard & Nix, 2019; Boluk & Carnicelli, 2019) in the classroom, with the industry and with the public in order to reset tourism's trajectory towards social and ecological justice. However, learning from human rights educator Audrey Osler (2015, pp. 30–32), pedagogy can be "an emancipatory or transformatory form…where [we] get to study justice/injustice and equality /inequality… to support greater justice and equality in [our] lives and to promote solidarity with others to achieve these ends in the wider society"; this involves a pedagogy that addresses "asymmetrical power relations" and recognises that human rights can "become part of discourse used to control, if rights and principles are applied without dialogue and without consideration of people's specific needs". Thus, we also need critical, emancipatory methodologies (as seen here). This means embracing qualitative research and taking seriously case study analysis as well as more creative forms of evidence such as "storytelling as research"—finding ways to seek the human faces and environmental spaces behind the statistical and marketing claims. In seeking minoritised or marginalised voices and creating ethical research partnerships with communities in tourism spaces, we are engaging the process of peace through tourism. Additionally, we need to strengthen the engagement with all stakeholders in order to foster greater dialogue on the meaning and value of tourism and peace, including further deliberations considering peace with justice approaches. We cannot offer a toolkit for such a programme of work, because respect for specific contexts, cultures and communities is essential for just outcomes; the heart of this work is in the process and in the becoming.

As we can see from the contributions to this Special Issue such an approach *is* being taken seriously and in multi-faceted ways, with the result that empowering and impactful work is being achieved. Of course, there is also work to be realised, as foreshadowed in the sections here on "ecocide" and "decolonising tourism". In a context where crises are multiplying and compounding, peace through tourism approaches—based on peace with justice foundations—provide creative, empowering and progressive responses through critical, multi-logical and multilayered dialogues amongst those concerned with tourism and its possibilities.

Postscript

As well as looking forward, there is also value in looking back. When we were asked to add an additional five chapters to this volume as a book publication, we embraced the chance to look back at the *Journal of Sustainable Tourism* to see where peace through tourism in theory and practice might be found in existing documented research. This is invaluable, to remind us that although 'peace tourism' might be a new and developing field in the academic literature, tourism scholars and practitioners have long been concerned with understanding ethics, equity and justice (or the 'positive peace') dimensions of tourism. These additional studies engage a critical, constructive and sometimes radical lens.

In Chapter 18, Adrian Deville, Stephen Wearing & Matthew McDonald (2016) investigate a particular practice in Australia (WWOOFing) to analyse the tourism context in terms of relations of power, authenticity and sustainability. At the time of this research in 2013, the authors argued this illustration represents a type of "de-commodified sustainability tourism" that was as yet "unexplored in the sustainable tourism literature". In Chapter 19, Kumi Kato (2019) explores the traditional practice of *ama* in Japan and its application to tourism:

> "The study identifies the power of women's knowledge ... [and illustrates] that while tourism can be an important social and economic force, it can also devalue these core qualities as a result of the negative impacts caused by gender stereotypes".

Similarly, in Chapter 20, Canadian scholar Heidi Karst (2017) is concerned with 'voice' as her research with communities in Bhutan explores "perceptions of ecotourism and well-being" and contributes "an analysis that embraces an indigenous ontology" in critique of claims about ecotourism as "a tool for increasing the economic well-being of local communities while conserving natural resources". Chapter 21, is even more 'radical' in an analysis that gives a 'voice' to nature, with the authors' work—"The Land has a voice: understanding the land tenure-sustainable tourism development nexus in Micronesia" (2019). The research findings here "elucidate how the current SDG targets lack sufficient provision for how Micronesian societies function, what is valued by the people that comprise these societies, and how Indigenous land holding rights form the bedrock of the sustainability of Micronesian livelihoods". Finally, in Chapter 22 Jennifer Devine (2017) unmasks how the structural violence hidden in the economic and political dimensions of tourism enables "[neoliberalism] combined with tourism's violent practices of spatial colonisation and the commodification of place" to generate resistance from disenfranchised communities, as she explains: "This violence often engenders diverse forms of resistance by 'hosts' living in other people's playgrounds around the world".

From these additional studies notions of 'a gender and ecohumanities perspective', 'Indigenous voices and ecotourism', 'decommodification and sustainability' and 'decolonisation of tourism spaces' all suggest linking 'positive peace' dimensions to tourism. Instituting critical and radical research agendas within academic studies in tourism needs to be viewed as a means of fostering research collaborations including building supportive collegial and interdisciplinary research environments—rather than competitive 'ownership' over ideas and concepts. Scholarship concerned with 'peace tourism' as 'sustainable tourism' would foster a culture of sharing research whilst building on existing work with constructive critique. Of course, there is also more work to be realised, as foreshadowed above with regard to "ecocide" and "decolonising tourism". This is not obscure thinking of 'radical' academics. If we reflect on the 'new' war that has been initiated while this collection was in publication, we can see the traditional tropes of tourism playing out. As tourism academics and the tourism industry rush to condemn the unjust war in Ukraine and highlight the effects on global and local tourism spaces (e.g., Dolnicar & McCabe, 2022), one might ask where is this same industry solidarity when it comes to Syria, Somalia and Palestine? In a global context where crises are multiplying and compounding, understanding tourism based on peace with justice foundations provides opportunities to spotlight 'injustice' in tourism policy and practice, to create interdisciplinary dialogue amongst those concerned with the work for sustainable peace and sustainable tourism.

Acknowledgements

We are grateful to Professor James Higham for his support. We are also grateful to all of our authors and the large number of reviewers who helped us create this Special Issue in the uniquely challenging times of the pandemic.

References

Andrews, H. (ed.). (2014). *Tourism and violence*. London: Routledge.

Bawaka Country, Wright, S., Lloyd, K., Suchet-Pearson, S., et al. (2017). Meaningful tourist transformations with Country at Bawaka, North East Arnhem Land, northern Australia. *Tourist Studies*, 17 (4), 443–467.

Becken, S. & Carmignani, F. (2016). Does tourism lead to peace? *Annals of Tourism Research*, 61, 63–79.

Bennett, J. & Ngarrindjeri Being Heard (2013). An experiment with tourism. In L. Blanchard and F. Higgins-Desbiolles (Eds.). *Peace through tourism: Promoting human security through international citizenship* (pp. 156–170). Abingdon: Routledge.

Blanchard, L. & Higgins-Desbiolles, F. (Eds.) (2013a). *Peace through tourism: Promoting human security through international citizenship*. Abingdon: Routledge.

Blanchard, L. & Higgins-Desbiolles, F. (2013b). Introduction: Peace matters, tourism matters. In L. Blanchard and F. Higgins-Desbiolles (Eds.). *Peace through tourism: Promoting human security through international citizenship* (pp. 1–16). Abingdon: Routledge.

Blanchard, L. & Higgins-Desbiolles, F. (2013c). A pedagogy of peace: the tourism potential. In L. Blanchard and F. Higgins-Desbiolles (Eds.). *Peace through tourism: Promoting human security through international citizenship* (pp. 19–33). Abingdon: Routledge.

Blanchard, L. & Nix, M. (2019). Creating spaces for radical pedagogy in higher education. *Human Rights Education Review*, 2(2), 65–83. doi:http://doi.org/10.7577/hrer.3363

Boluk, K. & Carnicelli, S. (2019). Tourism for the emancipation of the oppressed. *Annals of Tourism Research*, 76, 168–179. https://doi.org/10.1016/j.annals.2019.04.002

Boluk, K.A., Cavaliere, C.T. & Higgins-Desbiolles, F. (2019). A critical framework for interrogating the United Nations Sustainable Development Goals 2030 Agenda in tourism. *Journal of Sustainable Tourism*, 27(7), 847–864. doi:10.1080/09669582.2019.1619748

Budeanu, A. (2005). Impacts and responsibilities for sustainable tourism: a tour operator's perspective. *Journal of Cleaner Production*, 13(2), 89–97.

Butcher, J. (2020). The war on tourism. *Spiked* Online. Retrieved 5 May 2020. Available at: www.spiked-online.com/2020/05/04/the-war-on-tourism/

Cater, E. (1993). Ecotourism in the third world: Problems for sustainable tourism development. *Tourism Management*, 14(2), 85–90.

Causevic, S. (2010). Tourism which erases borders: An introspection into Bosnia and Herzegovina. Moufakkir, O. & Kelly, I. (2010). *Tourism, progress and peace* (pp. 48–64). Wallingford: CABI.

Causevic, S. & Lynch, P. (2011). Phoenix tourism: Post-conflict tourism role. *Annals of Tourism Research*, 38(3), 780–800.

Chambers, D. & Buzinde, C. (2015) Tourism and decolonisation: locating research and self. *Annals of Tourism Research*, 5 (1), 1–16.

Da Cal Seixas, S.R., de Moraes Hoeffel, J.L., Botterill, D., Carnevale Vianna, P.V. & Renk, M. (2014). Violence, tourism, crime and the subjective. In H. Andrews (ed.). *Tourism and violence* (pp. 145–164). London: Routledge.

D'Amore, L. (2010). Peace through tourism: The birthing of a new socio-economic order. *Journal of Business Ethics*, 89, 559–568.

Devine, J. & Ojeda, D. (2017). Violence and dispossession in tourism development: a critical geographical approach. *Journal of Sustainable Tourism*, 25(5), 605–617.

Dolnicar, S. & McCabe, S. (2022). Solidarity tourism how can tourism help the Ukraine and other war-torn countries? *Annals of Tourism Research*, 94. https://doi.org/10.1016/j.annals.2022.103386

Engmann, R.A.A. (2020). Coups, castles, and cultural heritage: Conversations with Flight Lieutenant Jerry John Rawlings, former President of Ghana. *Journal of Heritage Tourism*, 16 (6): 722–737. doi: 10.1080/1743873X.2020.1817929

Equations (2013). Of peoples and places: Tourism and zones of conflict in India. In L. Blanchard and F. Higgins-Desbiolles (Eds.). *Peace through tourism: Promoting human security through international citizenship* (pp. 116–134). Abingdon: Routledge.

Farmaki, A. (2017). The tourism and peace nexus. *Tourism Management*, 59, 528–540.

Farmaki, A. (2019). Visitor-host encounters in post-conflict destinations. The case of Cyprus. In R. K. Isaac, E. Çakmak & R. Butler (Eds.). *Tourism and hospitality in conflict-ridden destinations* (pp. 240–252). New York, NY: Routledge.

Farrell, B. & Twining-Ward, L. (2005). Seven steps towards sustainability: Tourism in the context of new knowledge. *Journal of Sustainable Tourism*, 13(2), 109–122. doi: 10.1080/09669580508668481

Fry, D.P., Souillac, G., Liebovitch, L., et al. (2021). Societies within peace systems avoid war and build positive intergroup relationships. *Humanities and Social Sciences Communications*, 8(17). https://doi.org/10.1057/s41599-020-00692-8

Galtung, J. (2012). Peace, positive and negative. In D.J. Christie (Ed). *The encyclopedia of peace psychology* (pp. 1–7). Blackwell. doi:10.1002/9780470672532

Grimwood, B.S.R., Stinson, M.J. & King, L.J. (2019). A decolonizing settler story. *Annals of Tourism Research*, 79, 102763. https://doi.org/10.1016/j.annals.2019.102763

Hall, C.M. (2019). Constructing sustainable tourism development: The 2030 agenda and the managerial ecology of sustainable tourism. *Journal of Sustainable Tourism*, 27(7), 1044–1060. doi:10.1080/09669582.2018.1560456

Higgins, P. (2012). *Earth is our business: Changing the rules of the game.* London: Shepheard-Walwyn Publishers.

Higgins-Desbiolles, F. (2003). Reconciliation tourism: Tourism healing divided societies. *Tourism Recreation Research*, 28(3), 35–44.

Higgins-Desbiolles, F. (2006). More than an Industry: Tourism as a social force. *Tourism Management*, 27(6), pp. 1192–1208.

Higgins-Desbiolles, F. (2009). Indigenous ecotourism's role in transforming ecological consciousness. *Journal of Ecotourism*, 8(2), 144–160. doi:10.1080/14724040802696031

Higgins-Desbiolles, F. (2018). The potential of justice through tourism. *Via@ - Tourism review*. https://journals.openedition.org/viatourism/2469

Higgins-Desbiolles, F. (2021). The "war over tourism": Challenges to sustainable tourism in the tourism academy after COVID-19. *Journal of Sustainable Tourism*, 29(4), 551–569. doi: 10.1080/09669582.2020.1803334

Higgins-Desbiolles, F. & Blanchard, L. (2010). Challenging peace through tourism: placing tourism in the context of human rights, justice and peace. In O. Moufakkir & I. Kelly (Eds.). *Tourism, progress and peace* (pp. 35–47). Wallingford: CABI.

Holden, A. (2003). In need of new environmental ethics for tourism? *Annals of Tourism Research*, 30(1), 94–108. doi:10.1016/S0160-7383(02)00030-0

IIPT (n.d.). History. Retrieved 4 February 2021. Available at https://peacetourism.org/history/

Jamal, T. (2019). *Justice and ethics in tourism*. Abingdon: Routledge.

Jenkins, T. & Reardon, B.A. (2007). Gender and peace: Towards a gender inclusive, holistic perspective. In C. Webel & J. Galtung (Eds.). *Handbook of Peace and Conflict Studies* (pp. 209–231). London: Routledge.

Kallis, G. (2011). In defence of degrowth. *Ecological Economics*, 70(5), 873–880.

Kassis, R. (2013). The struggle for justice in Palestine. In L. Blanchard and F. Higgins-Desbiolles (Eds.). *Peace through tourism: Promoting human security through international citizenship* (pp. 225–240). Abingdon: Routledge.

Kim, S.S., Prideaux, B. & Prideaux, S. (2007). Using tourism to promote peace on the Korean Peninsula. *Annals of Tourism Research*, 34(2), 291–309.

Latouche, S. (2009). *Farewell to growth*. Oxford: Polity Press.

Lee, E. & Hamilton, F. (2016). Tasmania – After a long journey, World Heritage Area delivers Indigenous rights. Retrieved 10 August 2020. Available at: www.iccaconsortium.org/index.php/2016/09/25/tasmania-after-a-long-journey-world-heritage-area-delivers-indigenous-rights/

Leiper, B. (1979). The framework of tourism: Towards a definition of tourism, tourist, and the tourist industry. *Annals of Tourism Research*, 6(4), 390–407.

Lemelin, H., Dawson, J., Stewart, E.J., Maher, P. & Lueck, M. (2010). Last-chance tourism: the boom, doom, and gloom of visiting vanishing destinations. *Current Issues in Tourism*, 13(5), 477–493, DOI: 10.1080/13683500903406367.

Lew, A.A., Cheer, J.C., Haywood, M., Brouder, P. & Salazar, N.B. (2020). Visions of travel and tourism after the global COVID-19 transformation of 2020. *Tourism Geographies*, 22(3), 455–466. doi:10.1080/14616688.2020.1770326

Litvin S. W. (1998). Tourism: the world's peace industry? *Journal of Travel Research*, 37(1), 63–66.

Litvin, S.W. (2020). Tourism and peace: A review and commentary. *Tourism Review International*, 23, 173–181. doi: https://doi.org/10.3727/154427219X15741004672675

Lynch, S. (2021). Climate change is biggest threat to security, Attenborough tells UN. Irish Times (Online). Retrieved 21 March 2021. Available at: www.irishtimes.com/news/world/us/climate-change-is-biggest-threat-to-security-attenborough-tells-un-1.4493232

Lyons, O. (2005). Peace talks: Peacekeeping traditions of the Iroquois Confederacy – with Oren Lyons on Radio KUNM. Retrieved 20 December 2020. Available at www.goodradioshows.org/peaceTalksL33.html

Mbaiwa, J.E. (2011). Cultural Commodification and tourism: The Goo-Moreni community, Central Botswana. *Journal of economic and Human Geography*, 102(3), 290–301.

McCloskey, S. (2015). Viewpoint: From MDGs to SDGs: We need a critical awakening to succeed. In S. McCloskey (Ed.), *Policy and practice: A development education review* (pp. 186–194). Carson, CA: Center for Global Education.

Mei-Singh, L. & Vicuña Gonzalez, V. (2017). DeTours: Mapping decolonial genealogies in Hawai'i. *Critical Ethnic Studies*, 3(2), 173–192.

Melotti, M., Ruspini, E. & Marra, E. (2018). Migration, tourism and peace: Lampedusa as a social laboratory. *Anatolia*, 29(2), 215–224. doi:10.1080/13032917.2017.1414441

Moufakkir, O. & Kelly, I. (2010). *Tourism, progress and peace*. Wallingford: CABI.

Osler, A. (2015). Human rights education, postcolonial scholarship, and action for social justice. *Theory and Research in Social Education*, 43(2), 244–274. https://doi.org/10.1080/00933104.2015.1034393

Pan, S. (2015). Peace through tourism. *Anatolia*, 26(2), 315–317. doi: 10.1080/13032917.2014.1003748

Pedersen, S.B. (2017). Peace through tourism: A brief history of a popular catchphrase. In M. Andren (ed.). *Cultural Borders and European integration* (pp. 29–38). Goteborg: Centrum för Europaforskning.

Perkumiene, D. & Pranskuniene, R. (2019). Overtourism: Between the right to travel and residents' rights. *Sustainability*, 11(7), 2138. doi:10.3390/su11072138

Pezzullo, P. (2007). *Toxic tourism: Rhetorics of pollution, travel and environmental justice.* Tuscaloosa: University of Alabama Press.

Pizam, A., Jafari, J. & Milman, A. (1991). Influence of tourism on attitudes: US students visiting USSR. *Tourism Management*, 12(1), 47–54.

Pratt, S. & Liu, A. (2016). Does tourism really lead to peace? A global view. *International Journal of Tourism Research*, 18, 82–90. doi:10.1002/jtr.2035

Preketes, J. (2013). Touchdown Tours: The business of peace tourism. In L. Blanchard and F. Higgins-Desbiolles (Eds.). *Peace through tourism: Promoting human security through international citizenship* (pp. 217–222). Abingdon: Routledge.

Rees, S. (2003). *A passion for peace: Exercising power creatively*. Sydney: University of New South Wales Press.

Rees, S. (2020). *Cruelty or humanity*. Bristol: Policy Press.

Salazar, N.B. (2006, July–December). Building a "Culture of Peace" through tourism: Reflexive and analytical notes and queries. *Universitas Humanística*, 62, 319–333.

Scott, J. (2012). Tourism, civil society and peace in Cyprus. *Annals of Tourism Research*, 39(4), 2114–2132. https://doi.org/10.1016/j.annals.2012.07.007

Shelley, B., Ooi, C-S. & Denny, L. (2021). The dialogic negotiation of justice. *Journal of Sustainable Tourism*, 29(2–3), 488–502. doi:10.1080/09669582.2020.1727487

Siddique, H. (2021, 23 June). Legal experts worldwide draw up 'historic' definition of ecocide. *The Guardian* Online. Retrieved 23 June 2021. Available at www.theguardian.com/environment/2021/jun/22/legal-experts-worldwide-draw-up-historic-definition-of-ecocide

Simone-Charteris, M.T. & Boyd, S. (2010). Northern Ireland re-emerges from the ashes. In O. Moufakkir & I. Kelly (Eds.). *Tourism, progress and peace* (pp. 179–199). Wallingford: CABI.

Spence, S. (2020). Right-wing extremism: The new wave of global terrorism. *The Conversation* (Online). Retrieved 22 March 2021. Available at https://theconversation.com/right-wing-extremism-the-new-wave-of-global-terrorism-147975

Stancliffe, A. (2013). Tourism Concern: Putting human rights principles into practice. In L. Blanchard and F. Higgins-Desbiolles (Eds.). *Peace through tourism: Promoting human security through international citizenship* (pp. 75–88). Abingdon: Routledge.

Steffen, et al. (2018). Trajectories of the earth system in the anthropocene. *Proceedings of the National Academy of Sciences of the United States of America*, 115(33), 8252–8259. https://doi.org/10.1073/pnas.1810141115

Tatsuya, Y. (2013). The floating peace village: An experiment in nonviolence. In L. Blanchard and F. Higgins-Desbiolles (Eds.). *Peace through tourism: Promoting human security through international citizenship* (pp. 135–143). Abingdon: Routledge.

Tomljenovic, R. (2010). Tourism and intercultural understanding or contact hypothesis revisited. In O. Moufakkir & I. Kelly (Eds.). *Tourism, Progress and Peace* (pp. 17–34). Wallingford: CABI.

Tuppen, H. (2021, 25 January). First national tourist board to declare a climate emergency, VisitScotland is putting sustainable tourism and the environment first. *South China Morning Post*, Online. Retrieved 27 January 2021. Available at www.scmp.com/lifestyle/travel-leisure/article/3119087/first-national-tourist-board-declare-climate-emergency

UN (n.d.a). Goal 16. Retrieved 31 January 2021. Available at: https://sdgs.un.org/goals/goal16

UN (n.d.b). A new era of conflict and violence. Retrieved 21 March 2021. Available at: www.un.org/en/un75/new-era-conflict-and-violence

UN (2019). Sustainable development goals. Retrieved 2 October 2020. Available at: www.undp.org/content/undp/en/home/sustainable-development-goals.html

UN CTED (2020). Member states concerned by the growing and increasing threat of extreme right-wing terrorism. CTED Trends Alert, April. Retrieved 22 March 2020. Available at: www.un.org/sc/ctc/wp-content/uploads/2020/04/CTED_Trends_Alert_Extreme_Right-Wing_Terrorism.pdf

UNWTO (1999). Global code of ethics for tourism. Retrieved 31 January. Available at: www.unwto.org/global-code-of-ethics-for-tourism

UNWTO (n.d.). Tourism and the SDGs. Retrieved 3 March 2019. Available at: http://icr.unwto.org/content/tourism-and-sdgs

UNWTO (2020, 9 June). Tourism can be a platform for overcoming the pandemic. Retrieved 20 July 2020. Available at: www.unwto.org/news/tourism-can-promote-solidarity-un-secretary-general-antonio-guterres

Urbain, Y. (2013). Awareness-Raising and Global Citizenship Through Peace Tourism: Case Studies of Hiroshima and Nagasaki. In L. Blanchard and F. Higgins-Desbiolles (Eds.). *Peace through tourism: Promoting human security through international citizenship* (pp. 144–155). Abingdon: Routledge.

van den Dungen, P. (2009). Towards a Global Peace Museum Movement: A progress report (1986–2010). *Peace Forum*, 24(34), 63–74.

Weaver, D.B. (2009). Reflections on sustainable tourism and paradigm change. In S. Gossling, C.M. Hall & D.B. Weaver (Eds.). *Sustainable tourism futures* (pp. 33–42). New York: Routledge.

Webel, C. (2007). Introduction: Toward a philosophy and metapsychology of peace. In C. Webel & J. Galtung (Eds.). *Handbook of Peace and Conflict Studies* (pp. 3–13). London: Routledge.

White R. (2019). Loving the planet to death: Tourism and ecocide. In T. Raymen & O. Smith (Eds.) *Deviant Leisure* (pp. 285–304). Cham: Palgrave Macmillan. https://doi.org/10.1007/978-3-030-17736-2_13

Wijesinghe, S. (2020). Researching coloniality: A reflection on identity. *Annals of Tourism Research*, 82, 102891. https://doi.org/10.1016/j.annals.2020.102901

Wohlmuther, C. & Wintersteiner, W. (Eds.) (2014). *International Handbook on Tourism and Peace*. Klangenfurt, Austria: Drava.

Part I

Addressing structural violence

OPEN ACCESS

Fortress tourism: exploring dynamics of tourism, security and peace around the Virunga transboundary conservation area

Lisa Trogisch and Robert Fletcher

ABSTRACT

This paper examines the role of tourism in promoting peace around transboundary protected areas (TBPA) in conflict-affected regions through a case study of the Virunga TBPA straddling the Democratic Republic of Congo, Rwanda and Uganda. Also praised as 'peace parks', TBPAs embody the rationale that stimulating a tourism economy in shared conservation spaces will provide incentives for formerly antagonistic states and actors to cooperate. Virunga TBPA exemplifies this strategy by promoting high-end gorilla tourism in a region scarred by longstanding conflict. Drawing on twelve months of fieldwork, we found that these optimistic aspirations are contradicted by militarisation of the three National Parks constituting the TBPA, as well as prevalence of violence and political segregation among the countries. As park authorities sell 'feelings of security' by framing their neighbours as threats in order to attract visitors, intrastate competition and conflict intensifies. Peace tourism rhetoric and revenues allow the governments to justify and finance the militarisation as necessary to protect 'their' tourists while concealing their security interests in protecting national borders. This contradiction between tourism's ostensive 'peace dividend' and the violence it generates within transboundary conservation efforts poses the question: what kind of 'peace' can be generated through tourism, and for whom?

Introduction: 'peace tourism' to transboundary protected areas

Before the Corona pandemic abruptly halted international travel in 2020, tourism was the fastest growing industry in the world and thus increasingly promoted as 'peace tourism': a key strategy for peacebuilding in conflict-affected regions (UNWTO, 2016, WTTC, 2016). Given that natural sites, and particularly those containing charismatic wildlife, are popular travel destinations, the United Nations World Tourism Organisation (UNWTO) advocates tourism promotion within protected areas as a substantial opportunity for low-income countries to achieve a "triple win" combining sustainable economic development, nature conservation and peacebuilding. Within this broader promotion, one proposed but deeply contested modality to pursue this triple win is the creation of Transboundary Protected Areas (TBPA). Since TBPAs straddle the borders of at least two countries, they also became praised under the compelling name of "Peace Parks" to encourage cooperation among states characterized by historical and/or current conflict. Proponents argue that the mutual interest to generate revenue from nature-based tourism within the parks incentivizes states to collaborate over conservation of the flora and fauna they share (Ali, 2007; Hanks, 2003). In this respect, TBPAs have been incorporated into broader environmental peacebuilding initiatives by the United Nations positing a relationship between overpopulation and exploitation of natural environments in stimulating violent conflict in low-income countries (UNEP, 2009, UNEP, 2016). This "recognition that environmental issues can contribute to violent conflict underscores [TBPAs'] potential significance as pathways for cooperation, transformation and the consolidation of peace in war-torn societies" (UNEP, 2009, p. 5). 'Peace parks' thus offer the potential to harness 'peace tourism' as a market-based, business-oriented solution to these challenges (D'Amore, 2009).

The prospects and pitfalls of peace tourism more generally have been subject to substantial scholarly analysis and discussion (see e.g. Blanchard & Higgins-Desbiolles, 2013; Kelly, 2006; Pizam, 1996). Yet research concerning promotion of tourism for peace within TBPAs in conflict-affected regions remains scarce. The few scholars engaging with tourism in protected areas in conflict contexts either focus on the negative impact of warfare in reduced economic revenues for conservation (Baral & Heinen, 2005; de Merode et al., 2007) or on tourists' concerns and travel choices (see e.g. Alluri, 2009), but rarely on socio-economic and political impacts of 'peace tourism' on local stakeholders in relation to this particular mode of conservation. This paper aims to fill this gap by analysing the role of tourism within a TBPA in relation to its triple-win framing as a simultaneous sustainable development, conservation and peacebuilding strategy.

We focus on the transboundary *Virunga Conservation Area* (VCA)[1] composed of *Virunga National Park* (PNVi) in the DRC, *Volcanoes National Park* (VNP) in Rwanda and *Mgahinga Gorilla National Park* (MGNP) in Uganda. VCA is an ideal case to interrogate the peace tourism narrative for two main reasons. First, the Virungas are the last remaining habitat of endangered mountain gorillas and are therefore deemed both a biodiversity hotspot by conservationists and a 'unique selling proposition' for one of the most expensive wildlife tourism attractions in the world. Secondly, VCA is situated within the Great Lakes Region, an area infamously described as the epicentre of "Africa's World War" (Prunier, 2011) due to its history of intra- and interstate warfare including the Rwandan genocide and its aftermath characterized by persistent violent conflict in the DRC. Since 2007, the Dutch government has financed the intergovernmental *Greater Virunga Transboundary Collaboration* (GVTC) to institutionalise transboundary conservation within VCA on the model of a 'Peace Park'. One focus of the trilateral GVTC Secretariat is to bring together protected area authorities, governmental representatives, researchers and conservation non-governmental organizations (NGOs) to promote gorilla tourism as a peacebuilding strategy founded in socio-economic and political cooperation among the three countries.

Yet the promising official agreements among the DR Congo, Rwanda and Uganda are contradicted by the prevalence of various kinds of violence, from reported killings of mountain gorillas (Jenkins, 2008) to tourists (Maekawa et al., 2013) to park rangers (Marijnen & Verweijen, 2016). Other forms of violence, ranging from social exclusion to imprisonment and murder of civilians living adjacent to the TBPA, are also frequent but rarely documented by mass media, official reports or academic researchers. Despite these highly problematic developments, there has been no research on the effects of 'peace tourism' inspired measures within the Virungas in relation to this violence. In bringing together current debates over 'peace tourism' with critical conservation studies, this paper contributes to both discussions by offering a pioneering analysis of the role of tourism in peacebuilding efforts within TBPAs in conflict-affected regions.

In relation to these issues, we argue that the transboundary collaboration efforts centred on gorilla tourism development have paradoxically spurred increased economic competition among the three states that has in turn precipitated political distancing as well as militarisation within the three National Parks constituting the VCA. Focusing mainly on the relationship between the DRC and Rwanda, our analysis demonstrates how gorilla tourism has merged with the selling of 'security' as a luxury, high-end tourism product in its own right. The revenue generated from these endeavours finances the militarisation of conservation actors, areas and national borders, thereby functioning as a geopolitical strategy to make the National Parks a military bulwark against the neighbouring countries. The resultant segregation among the three states undermines collaboration and trust among national conservation authorities and tourism actors and thereby contradicts the promoted 'peace tourism' narrative in the TBPA treaty. Hence, 'gorilla tourism for peace' seems to remain largely an appealing piece of rhetoric that attracts international funding for gorilla conservation, private investments into tourism-related businesses and tourists that is used to finance military expenditures in the name of securitising the great apes and their visitors.

After a brief description of our methods, we situate our study within the debate on 'peace tourism' intersecting with critical conservation studies. Thereafter, we explore the functional role of gorilla tourism in peacebuilding and collaboration efforts within the VCA by focussing specifically on narratives and practices used by different governmental and non-governmental conservation actors in Rwanda and the DRC on national and transnational levels. We then shift our attention to the local context to analyse the perceived impacts of gorilla tourism on 'peace' for local tourism stakeholders as well as people not directly involved in tourism living adjacent to the VCA. We conclude by advocating more critical engagement with the concealed pitfalls of high-end wildlife tourism in conflict-affected regions packaged within the 'Peace Parks' brand, particularly as we continue to struggle with the ongoing fallout of the COVID-19 pandemic.

Methods

This paper draws on twelve months of qualitative field research in communities adjacent to the the three National Parks that constitute the mountain gorilla habitat in the Virungas conducted by the first author from January to December 2018. These communities were chosen since they comprise the greatest part of the population evicted from the Parks and therefore are viewed as the main stakeholders to benefit from tourism interventions (GVTC, 2014; UNWTO (United Nations World Tourism Organization), 2015). Long-standing exploratory research within the area by the first author since 2015 provided the needed trust for beginning the formal participatory research period within the park-adjacent villages of Kinigi (Rwanda) and Ntebeko (Uganda) as well as access to key informants. Proceeding from these personal connections, snowball sampling enabled data collection with other local stakeholders ranging from owners and employees of the park-adjacent tourism facilities, tour and ranger guides as well as community members not directly engaged in tourism or conservation practices. With permission of local chiefs and park authorities, partaking in daily life as well as conservation and tourism related activities within the National Parks afforded the researcher insights into day-to-day events and conversations, which were then substantiated with 60 semi-structured interviews.

Due to the volatile political situation, the outbreak of Ebola and unpredictable violence and kidnappings in the region of the Congolese park in 2018, participatory methods with local communities in the Eastern Congo were restricted to an interactive collaboration with a Congolese NGO in the regional capital of Goma, which enabled 28 open and semi-structured interviews during field trips to seven park-adjacent villages. In this regard, we want to emphasize that all informants are anonymized to honour and protect those who agreed to participate despite the danger that the ongoing interregional conflict, mistrust and surveillance present for them. Additional participant observation was conducted with park authorities from the *Institut Congolais pour la Conservation de la Nature (ICCN)*, the *Rwandan Development Board* (RDB) and the *Ugandan Wildlife Authority* (UWA) during overnight stays at the park's headquarters and patrol outposts, during which the researcher accompanied anti-poaching patrols, gorilla monitoring and touristic activities within the parks.

These insights concerning location-specific dynamics were complemented by 12 focus group interviews followed up by 75 semi-structured interviews with protected area authorities, 36 with representatives of (international) conservation NGOs working in the conservation-tourism sector and 47 with people directly working for the tourism sector to gain an overarching understanding of conservation and tourism management in the VCA generally.[2] In addition, frequent attendance of GVTC workshops, meetings and conferences throughout the year provided insight into discussions of 'peace tourism' among representatives of national research institutions, security services, government officials, the GVTC and the United Nations international development and peacebuilding initiatives in the Great Lakes Region.

The results of these interactive methods were recorded with extensive fieldwork notes in a two-journal design, one holding descriptive field notes and the other used for reflective notes in order to foster self-awareness concerning how the researcher's presence influenced the situation, interaction with and responses by informants. This primary data collection was triangulated by iterative "member-checking" discussing the accuracy of findings with participants (Creswell & Plano Clark, 2011, p. 211) and a comprehensive review of policy briefs, treaties and reports discussing transboundary peace, tourism and conservation as well as popular media representations of the Virungas. Our analysis of these various data sources employed a grounded theory approach (Strauss & Corbin, 1997) in pursuit of a holistic understanding of the complex interrelations among different actors on multiple scales. This analysis relied on an iterative process of inductive pattern building from specific data points rather than formal content analysis in line with the qualitative, experiential focus of the overarching project.

Peace as security through the militarisation for tourism?

Since the establishment of *Waterton-Glacier International Peace Park* in 1932 to symbolize peace and friendship between the United States and Canada, the number of Transboundary Protected Areas (TBPA) in the model of "Parks for Peace" has increased steadily throughout the world (IUCN, 2020; Sandwith et al., 2001). Based on the assumption that environmental problems correlate with violent conflicts (see e.g. Ali, 2007), proponents argue that transboundary conservation offers a solution to these conflicts "to promote and strengthen peace and co-operation between neighboring countries [...] as a prerequisite for sustainable economic development and foreign investment" (Hanks, 2003, p. 136). In particular, scholars and practitioners focused on environmental governance promote TBPAs as an environmental peacebuilding strategy in its own right (UNEP, 2009).

In light of persistent environmental problems and social conflicts around conservation areas worldwide, the compelling concept of TBPAs has been questioned by a growing body of critics in the past decades. Some argue that TBPAs produce new conflicting interests among a multitude of actors as conservation strategies are aligned to the profit-maximising logic of capitalism (see e.g. Duffy, 2006). Fairhead (2005) pointed specifically to the creation of TBPAs as a covert tactic of neo-colonial governance by industrialised states to ensure the exploitation of resource-rich environments in sub-Sahara Africa. Büscher and Ratmusindela (2015) call attention to the increasingly violent methods that contradict the promoted intentions of 'Peace Parks' within the *Great Limpopo Transfrontier Park*. They refer to the wider debate in the critical conservation literature that scrutinizes the militarisation of all types of protected areas and actors (Duffy et al. 2019; Lunstrum, 2014, Ybarra, 2016) and the entanglement between violent and militaristic conservation practices and political violence, conflict or warfare within the larger surroundings (see e.g. Lombard & Tubiana, 2020; Minarchek, 2020). Others point to conservation violence and its destructive consequences for local residents (Bocajero & Ojeda, 2016; Büscher & Fletcher, 2017; Duffy, 2014). Specifically in the enclosure of protected areas, the forceful eviction, dispossession and exclusion of local residents from their means of survival, traditions and beliefs continues to be problematized in reference to the persistence of "fortress conservation" as a central environmental protection strategy (Brockington, 2002; Neumann, 2004; Pemunta, 2019).

As one response to this increasing critique, international development institutions have praised 'eco-tourism' as a new panacea to fund conservation and simultaneously benefit communities in low-income countries in proximity to protected areas (Honey & Gilpin, 2009). The 'peace parks' narrative expanded this notion of eco-tourism into a triple-win scenario by promoting TBPAs as tourism destinations that can support both environments and people through the economic prosperity generated from tourism revenues that will ultimately also contribute to regional cooperation and peace (see e.g. Hanks, 2003; Mearns, 2012). While tourism was already promoted as a "passport to peace" in 1967,[3] it gained increasing attention as a peacebuilding approach after the "Amman Declaration on Peace through Tourism" was formally adopted by the UN in 2000 (UNWTO, 2016). In 1988, D'Amore proposed 'peace tourism' as 'the world's first global peace industry' in founding the *International Institute for Peace through Tourism* (IIPT, 2017). In outlining its approach, IIPT refers to traditional conflict studies defining 'negative' peace as the absence of the 'direct use of physical force' (WHO, 2002) and 'positive' peace as the absence of 'structural' violence reflected in the unequal socio-economic and political organization of society (Galtung, 1969, 1996). In relation to this distinction, IIPT and other 'peace tourism' proponents stress tourism's potential not only to reduce direct violence but also to enhance positive peace indicators, in particular intercultural understanding, ecological sustainability and human welfare through "attitudes, institutions and structures that create and sustain peaceful societies" (IEP, 2017, p. 4; see also Haessly, 2010).

Based on the widespread assumption that violent conflict correlates with a lack of development (Collier & Hoeffler, 2000; Nassani et al., 2017), the ostensible 'virtuous cycle' of peace tourism to TBPAs in particular is aligned with the conservative neoliberal paradigm asserting that economic growth contributes to the Gross Domestic Product (GDP) of a nation state, which ultimately 'trickles down' to create development amongst the local population (Williamson, 1989). In consequence it is argued that citizens depending on the tourism industry for their well-being have an incentive to remain or become peaceful to help create a positive image of their country and attract more tourists (Haessly, 2010). Tourism's contribution to economic growth is thus seen as key to a form of development within low-income societies that will bring 'liberal peace' after the Western model: expanding free markets, democracy, human rights and the rule of law (Doyle, 2005; Levy & Hawkins, 2009). An extensive literature emphasizes this putative causal relationship between tourism and peace (e.g. D'Amore, 2009, Honey & Gilpin, 2009; Kelly, 2006) and even stresses the potential for tourism to create a global 'culture of peace' (Moufakkir & Kelly, 2010). By contrast, a growing body of critical tourism scholarship problematizes this posited relationship between tourism and peace (Litvin, 1998) as well as the focus on economic growth through neoliberal market engagement underlying this peace rhetoric (Blanchard & Higgins-Desbiolles, 2013).

Discussion of the potential for tourism to facilitate peacebuilding in conflict areas is nowadays complemented by attention to "human security". The UNWTO asserted in 1996 that "the success or failure of a tourism destination depends on being able to provide a safe and secure environment for visitors" (UNWTO, 1996). While the United Nations has more generally depicted 'human security' as an outcome of expanding military investments by the state (Haq, 1999), a significant shift to increased militarisation for ensuring security of 'the tourist' occurred in the aftermath of the 11 September 2001 attacks in the United States, after which terrorism was declared a global threat influencing travel advisories (Hall et al., 2004). While tourism scholars have thereby increasingly incorporated 'human security' into the debate concerning tourism's peace-generating potential (e.g. Blanchard & Higgins-Desbiolles, 2013), a few have also pointed to the consequent problematic entanglement between the 'securitisation of travel' and the military-industrial complex in inventing counter-terror measures to protect wealthy consumers (Lisle, 2016). In the same vein, Weaver shows that tourism strategies become increasingly militarized to serve the tourists' demand for 'security' leading to a normalization of military presence in tourist destinations, and thereby in the everyday reality of local people (2011, p. 674). Scholars have also highlighted the separation between locals and tourists through enclosed facilities and services in "enclave tourism" practiced in all-inclusive resorts in particularly remote areas (Mbaiwa, 2005; Nunkoo & Ramkissoon, 2016). Our study builds on all of this previous research to analyze "fortress tourism" (Fletcher, 2011) as a particular dynamic in which militarization is used to securitize tourism activities in (post-)conflict areas outside of enclave resorts.

Within critical conservation research, the relationship between tourism, peace and security in the context of conflict-affected conservation areas has been touched on in broader debates on violence within peace parks but has not yet been systematically investigated. While some conservation literature on tourism in conflict-affected contexts asserts tourism's potential to finance nature conservation in such areas (Hanks, 2003; Joras et al., 2011), others have questioned whether tourism's propensity to commodify natural and cultural 'products' may instead augment violence and conflicts among people (Brockington et al., 2008; Büscher & Fletcher, 2017), as well as among states in competing to attract more visitors within the global tourism market (Tripp, 2010). In addition, the eviction and dispossession of communities in the establishment of protected areas (Andersson et al., 2013), the lack of compensation and employment opportunities often attending these displacements (Adams & Infield, 2003) and the marginal revenue-sharing from tourism benefits such initiatives frequently entail (Sandbrook, 2010) speak to the presence of problematic, conflict-aggregating effects of tourism in conservation areas on the micro level.

Building on these initial explorations, the following analysis brings together discussions of peace tourism, peace parks, and the militarization of conservation to pursue a novel investigation of the implications of tourism as a peacebuilding strategy via a case study of the VCA. While the endangered mountain gorilla became the face of commercial tourism and led to increased international attention to the region, the impact on increased socio-economic development and peacefulness are expected to vary greatly among the DRC, Rwanda and Uganda. While acknowledging the undeniable ecological violence of stress, diseases and habitat destruction from tourism on mountain gorillas (Robbins, 2014) that problematizes the environmental component of 'peace tourism' narratives, we focus here on the socio-economic and political consequences for human stakeholders related to tourism. We argue that gorilla tracking became a form of 'fortress tourism' that sells a provision of security to wealthy visitors that simultaneously funds the military build-up of national borders within the three National Parks of the VCA, while excluding local residents. We begin by briefly outlining the historical context within which the regional tourism efforts originated.

Historical context of the Virungas - which 'conflict' should be brought to peace?

In violent quarrels over colonial boundaries between German East Africa and Belgium in 1902, two mountain gorillas shot by Robert von Beringe brought the great apes to the attention of Western scientists who assigned them their zoological name (*gorilla gorilla beringei*). In 1925, the Belgium king Albert I established *Parc National Albert* in the Eastern DR Congo. This space continues to constitute the main area of *Virunga National Park* (PNVi) today but previously incorporated Rwandan and Ugandan parts of the Virunga volcano chain to protect the broader mountain gorilla habitat. With independence attained by the DRC (then Zaire) [4] in 1960 and Rwanda and Uganda in 1962, management of the forests became separated into independent state-owned entities that gained international prominence through Dian Fossey's work in Rwanda from 1967 to 1985. PNVi in Zaire became a popular tourist destination in the economically vibrant first years of Mobutu Seso Seko's dictatorship (1969–1997). Inspired by neoliberal Structural Adjustment Programmes introduced by the *World Bank* and other international financial institutions (IFIs), Western conservationists promoted mountain gorilla tourism as an 'incentive' for the Rwandan government to commit to conservation beginning in 1979 (Nielsen & Spenceley, 2010; Vedder & Weber, 1990).

With regional conflict placing the Virungas at centre stage of the warfare in the 1990s, conservation and tourism came to an abrupt halt (Kanyamibwa, 1998). Supported by the Ugandan government of Yoweri Museveni, the rebel group *Rwandan Patriotic Front* (RPF) under now-president Paul Kagame began to fight the Hutu regime invading Rwanda from Mgahinga forest through *Volcanoes National Park* (African Rights, 1998). The following years of Rwandan civil war cumulated in the genocide in 1994. Justified as a conservation measure, the Ugandan government declared Mgahinga forest a National Park, precipitating eviction of the local population and prohibition of any entrance to the gazetted area to erect a buffer zone between Uganda and Rwanda. Acknowledging the devastating impact of the genocide on the conservation area, PNVi was declared as 'UNESCO World Heritage site in danger' in 1994.

The aftermath of the genocide led to an estimated 1.2 million refugees into Zaire with the largest camps around PNVi (UNHCR, 2000). Factions of the former Rwandan army and the *Interahamwe*[5] that executed the genocide re-established a military stronghold in PNVi forming the *Forces démocratiques de libération du Rwanda* (FDLR) in 2000. This armed group is held responsible for crimes against humanity and continued violent attacks both within Eastern Congo and in Rwanda via *Volcanoes National Park* (USNCC, 2009). Concerted counterstrike of Rwandan and Ugandan backed military to eliminate associated Rwandan Hutu refugees in the region of Eastern DRC have officially been labelled the two Congo Wars (1996–1997; 1998–2003). However, humanitarian organisations report ongoing interference by armed groups supported by the Rwandan and Ugandan governments, who are suspected to incite insecurity and chaos in Eastern Congo to facilitate the exploitation and trafficking of natural resources (HRW, 2005; Lopez et al., 2017; UNEP-MONUSCO-OSESG, 2015; UNSC, 2000, 2003). While conservation and tourism activities of PNVi have been revamped under a public-private partnership between ICCN and the *Virunga Foundation* with international funding since 2008, the park and its surrounding population are affected by the violent activities of more than approximately 140 armed groups within Eastern Congo (HRW, 2019).

Peace and security – through conservation and tourism?

Considering these complex entanglements and power struggles in the conflictual history between the DRC, Rwanda and Uganda, international peacebuilding and conservation actors stressed the unifying qualities of joint conservation and tourism measures to establish collaboration, understanding and trust between the three states during the United Nations Great Lakes conference for "Peace and security in Greater Virunga Landscape: silencing the guns" in 2018.[6] A representative of the GVTC asked:

> *„Can we create a peace haven in the Greater Virunga Landscape by collaborating over transboundary resources, including mountain gorillas, for benefitting all three countries?"*

What sounded like a committed agenda to work towards generating peace through gorilla conservation together was, in fact, an appeal to establish a partners coalition for the institutional sustainability of GVTC, to replace the in political and financial support withdrawn by the Dutch government following their termination of funding at the end of 2018. In the first formal Memorandum of Understanding in 2004, the three Protected Area authorities and respective Ministries of the DRC, Rwanda and Uganda agreed to "build trust, understanding and cooperation among wildlife authorities, nongovernmental organisations, users and other stakeholders to achieve sustainable conservation and thereby contribute to peace" (GVTC, 2006 [2015]: 6). The treaty emphasized specifically the role of 'gorilla eco-tourism' comprising over 80% of all visitors in the three countries to achieve the objectives in the same spirit as the 'peace through tourism' vision of IIPT. Moreover, governmental officials of each country declared that they would take over the financing of GVTC as a coordinating institution for the concerted efforts (GVTC, 2006 [2015]: 31).

In stark contrast to these written agreements and lip service proclamations, however, all transboundary meetings and projects by GVTC have expired, incapacitating the institution due to its liquidation at the time of writing (November 2020). Witnessing the preceding developments in 2018, the questions arose for us what impact the 'peace tourism' rhetoric has on the region and for the Virunga Conservation Area? What are the practices and outcomes of implementing 'peace tourism' propositions in each country and in how far do they lead to more peaceful relations between the three states? And how do tourism-related state and non-state actors, conservation organisations as well as people living around the parks without direct involvement in gorilla tourism, perceive and experience the impact of tourism on 'peace'?

Transboundary gorillas offer no comparative advantage

Mountain gorilla tracking appears to be an ideal 'product' to advocate for the benefits of tourism for peace in the relationship between the DRC, Rwanda and Uganda. The most evident commonality between neoliberal development policies promoted by the UNWTO and World Bank and 'peace tourism' aspects of the GVTC treaties is the primary focus on gorilla tourism as a foreign exchange earner to stimulate economic growth in each country. In pursuit of this revenue, the protected area authorities and governmental representatives of DR Congo, Rwanda and Uganda agreed to "[d]evelop a regional tourism plan that covers infrastructure development, marketing, visitor services and allows coordinated movement of tourists in the region" (GVTC, 2006 [2014]: 11). The mountain gorilla attraction indeed made tourism the most important export earner of both Uganda (contributing 7.7% to GDP in 2018) and Rwanda (14.9% to GDP in 2018). Yet in interviews with representatives of the main gorilla conservation NGOs in the region, many expressed concern about the deterioration of political and economic relations between the DR Congo, Rwanda and Uganda and the negative impact of this on transboundary conservation efforts. As a representative of the *International Gorilla Conservation Programme* (IGCP), a long-standing transnational conservation NGO, highlighted:

> *"We started to talk to all of them [park authorities] to create relationships between Rwanda, Uganda and the DRC on the conservation level at least, while the governments had a tense relationship. We organized joint trainings and patrols between park officials and rangers from ORTPN,[7] UWA and ICCN since 1991. We were allowed to ignore the borders. But with GVTC everything became political and conservation NGOs have less to say. Now rangers cannot share conservation data or walk together across the borders anymore."*

Another IGCP employee made the explicit link between increasing conflict and tourism:

> *"In my opinion, I think tourism has negative effects on the relations between Rwanda, Uganda and the DRC because of increased competition over attracting tourists and in many cases it becomes more politicized. Gorillas are a golden asset and now each government wants to have an advantage over the other."*

Based on Ricardo's (1817) well-rehearsed economic model of comparative advantage, international development organizations argue that due to lack of industrial capacity and/or 'valuable natural resources' to extract, many developing countries hold their only 'unique selling proposition' in the nature and wildlife they harbor, which they can vend as a competitive tourism product on the world market (UNEP, 2009). Since the forest habitat of mountain gorillas is split between the DR Congo, Rwanda and Uganda, the three states embrace the same 'natural asset' in selling gorilla tracking as a high-end activity to international tourists. Hence, mountain gorillas framed as 'consumer products' confer no comparative advantage – no 'unique product' – to any one of the three countries involved in this activity. Due to the need to compete to attract the biggest share of the limited supply of visitors and tourism-related investors, therefore, the three countries are each encouraged to develop new consumer products to differentiate themselves from one another. As a result, GVTC's attempts to homogenize gorilla tourism across the VCA and share both costs and benefits regionally produced the opposite effect of increased political segregation and militarization of borders between the DRC, Rwanda and Uganda rather than the increased collaboration over management of the shared ecosystem the initiative envisioned. In the next section we analyse the different strategies each country developed to maintain sovereignty and create a 'unique', competitive product to showcase on the global tourism market.

Regional dynamics triggered by Rwanda's gorilla price increase

Gorilla tracking in all three National Parks comprising the VCA includes a guided hike with a military escort through dense forest at 2,200 to 4,300 metres altitude to visit one of the habituated gorilla families for one hour of observation on site. The extreme variation between prices for gorilla tracking – from 350 USD in the DR Congo to 650 USD in Uganda to initially 750 USD in Rwanda – creates the impression that the three countries offer highly differentiated touristic experiences. This image is reproduced by park rangers of RDB who are trained to promote the 'uniqueness' of Rwanda's mountain gorillas compared to the Ugandan and Congolese specimens. Despite the fact that mountain gorillas move freely across political borders in direct accordance with the aims of the transboundary conservation approach, the rangers persistently emphasized the unique Rwandan identity and superiority of 'their' apes. In a focus group interview with park rangers at a patrol outpost, one ranger expressed:

> *"Uganda has less [sic] gorillas and they are not cute. Look at our new gorilla babies in Agashya,[8] they have much, much fur and look so cute."*

Another ranger added:

> *"Tourists cannot see the gorillas in Congo because they hide from the war. When gorillas have fear they are not funny."*

These statements represent a common sentiment among RDB rangers that the Congolese and Ugandan gorilla families are 'less': *'less visible', 'less funny', 'less cute'.*

In this context, GVTC presented its first substantial transboundary achievement in the "transfrontier gorilla tourism" agreement by declaring that when a gorilla group crosses borders from the country in which it has been habituated, the revenues generated from tracking this family has to be shared equally between the two countries (GVTC, 2006, p. 13). In response, every day from 6 in the morning to 6 at night, park rangers and their assistant trackers are assigned to locate each habituated gorilla family and follow them closely throughout the day until they build their nests to go to sleep in the evening, in order to ensure the animals do not leave the national park area. By taking part in monitoring patrols we witnessed these 'safeguarding' measures that are conducted similarly in Rwanda and Uganda, enabled by the small sizes of VNP (160 km^2) and MGNP (33,9 km^2) compared to PNVi in the DRC (7769 km^2). In contrast to the national commitments under GVTC, these measures exemplify the unwillingness by each state to collaborate over transnational revenue-sharing with respect to border-crossing gorillas.

As a second proposition to overcome regional differences in the 'gorilla tourism product', the three states added to the transboundary strategic plan in 2015 the intention to "[h]armonise visitor charges for common tourism products and visitor handling services standards" (GVTC, 2014, p. 11). Hence, RDB rangers and tourism operators claimed to be taken by surprise when Rwanda doubled its gorilla tracking fees to 1500 USD overnight on May 11, 2017 (personal interviews, Musanze, Goma, Kigali, 2018). The Chief Executive Officer of RDB, Claire Akamanzi, officially justified the unprecedented price spike on grounds of ecological sustainability as intended to decrease the number of tourists (and associated impact) without decreasing the foreign exchange Rwanda needs to finance development projects and "empower communities living near the park".[9] In response, an employee of a regional gorilla conservation organisations complained:

> *"The Rwandan government is always doing what it wants. We monitor and work with the gorilla families for four decades but we would not be consulted for any conservation-related advices."*

In a meeting with the trilateral secretariat of GVTC, one representative admitted feeling subverted:

"We were not informed about the price increase. We did not get an official letter or call after the decision. We heard the information on public news."

The unilateral price change made a strong statement against the promoted transboundary collaboration and undermined any political mandate and authority the GVTC retained at that point. Rwanda's official explanation of the price increase was intended to please conservationists, international donors and further serves as part of Rwanda's image campaign to overcome persistent associations with the genocide in 1994 (Grosspietsch, 2006). The tourism management literature accordingly advices, "developing countries must constantly [...] protect their image and maintain their position in the global tourism arena" (Avraham & Ketter, 2016, p. 3) by 'selling' their destination that faces "stereotypes of high risk, hostility and underdevelopment" (Avraham & Ketter, 2016, p. 110). This dominant conceit of the Western tourism literature already points to the imbalance of the global tourism market that is intended to please Western consumer demands.

Since gorilla tracking was initiated again in 1999, a growing group of Rwandan scholars has praised the Rwandan government for adapting to this 'world standard of tourism' to rapidly re-establish the devastated national image and economy creating 'wealth and peace in post-conflict Rwanda' (Maekawa et al., 2013; Spencer & Rurangwa, 2012). The World Bank reinforces this perception in framing Rwanda as the "African success story" pointing to the peace-generating link between mountain gorilla conservation and economic development as a key component of this success (Nielsen & Spenceley, 2010).

Selling security

Tourism exemplifies the capitalist principles of permanent (re-)invention of 'new' consumer goods for offer to the global market. Rwanda's gorilla tracking fee increase was intended to promote a novel, unique tourism product for the country combining the endangered gorillas with high-end security. It is this singular combination that underlines Rwanda's success in ostensibly providing a 'peace haven' for the great apes in the conflict-affected region. Moreover, since tourist arrivals are highly dependent on the stability and security of a destination region (Neumayer, 2004), globally fanned fears of terrorism led to new tourism marketing strategies offering 'security' for purchase more widely (Sönmez, 1998). Rwanda's strategy falls right in line with this approach. In accordance with Hochschild's argument that the 'management of emotions' became an 'instrument' of the capitalist system to invent new products turning 'feeling into a commodity' for sale (Hochschild, 1983 [2012], p. 186), Rwanda promotes an inimitable gorilla tourism experience combined with a luxurious feeling of security for the tourist in the otherwise 'war-torn' region (World Bank, 2013). Since the commercialisation of security is a fruitful strategy to overcome the distorted international crisis narratives surrounding African countries in general, the Rwandan government devotes a significant part of its marketing campaigns as well as national expenditures to the security sector. As the official website of VNP presents:

"Today the people of Rwanda are living in a very stable and peaceful country as the RPF[10] have continued to ensure the safety of all Rwandan borders to ensure that there are no external attacks from any rebel group that might be formed in one of the neighboring countries. The internal security within the country is also very tight with forces always moving around communities to ensure that all the citizens are safe." (VNP, 2018)

In this respect, Rwanda's 'unique selling proposition' in the tourism sector has never been solely the mountain gorilla but rather this combined with the "African success story" that seemingly transformed the country "from a land of zombies and misery to become an economic model on the African continent" (VNP, 2018). Given the importance of this differential framing for capturing tourism revenue, increasing the GVTC's interference within national policies and practices as part of its transboundary peace efforts appear to pose a particular threat to Rwanda's specific tourism asset. Hence, Rwanda's increase of gorilla tracking fees was partially a political measure to reclaim sovereignty and re-create a unique tourism product. Presenting

Rwanda as "more developed", "more progressive" and a "safe haven" (African Wildlife Fund, 2020) relative to its neighbours further justifies the high price as a necessary investment to provide security and high standards for visitors and investors. A Rwandan tour operator explained in the same spirit:

> *"Rwanda has no natural resources, the only thing we have is our security that attracts foreigners: investors and tourists. That is why our government invests everything into security. [...] I get concerned if I don't see a soldier somewhere."*

Tourist 'security' to justify militarisation

This quote underscores how official promotion of Rwanda's tourism seems to be aligned to the security demands of the tourism sector to militarize the destination (Weaver, 2011). Yet Rwanda's government rather appears to make use of the normalised militarization for tourism in Rwanda's everyday reality in a reverse manner. In order to keep the current power structures under the authoritarian regime of Paul Kagame unchallenged, the deployment of military, security services and surveillance infiltrates and controls every space within the country's borders, *regardless* of the presence of tourists. The official explanation of focussing security on high-end tourism generates an internationally accepted and even appreciated legitimization of militarising Rwanda in the name of tourists' safety. More generally, tourism scholars have found a strong positive correlation between military expenditures and tourism growth, thus recommending states to create 'security' through the militarisation of tourism destinations. Such prescriptions offer states further legitimization to devote larger shares of the fiscal base to military expenditure and attract international funding and supply for the national armament (Nassani et al., 2017).

While international and national media commonly portrays Rwanda as an island of security in the conflict-ridden region, the frequent violent skirmishes between armed forces from the DRC and the Rwandan military within *Volcanoes National Park* on Rwandan territory are for the most part successfully concealed by Rwandan censorship of the media, under the imperative not to scare away potential tourists. On 4 October 2019 an exceptionally violent incident reached the international media, when militias entered Rwanda though VNP and killed 14 people in the adjacent communities of Kinigi district where the National Park headquarters are located. The British travel advice service stated in the same moment that "the Government of Rwanda reports that the area is now secure following the deployment of additional security forces and tourism services continue to operate as normal in the Park"[11] .

After the end of the genocide, the Rwandan government established a dense network of the national military, the *Rwanda Defence Force*[12], in and around VNP, to protect the national borders with the DRC from expected and actual attacks by fighters from the deposed former Hutu government and nowadays its remnants within the armed group of the FDLR. With the reintroduction of tourism in 1999, the paramilitary training and armament of park rangers added another militarised set of actors to the park and the national borders. In the name of conservation, the 'green' militarisation (Lunstrum, 2014) of the area is justified to the international public and visitors through two main narratives. First, the training and armament of rangers is justified as necessary for the protection of mountain gorillas against 'poachers'. The 'law enforcement' warden of VNP explained this in the following way:

> *"The biggest problem of the park are poachers, those Historically Marginalized People[13] who are addicted to hunting. First, we have routine patrols every day inside park to protect the animals. They report on illegal park activities. Second, we get information from the communities that there is a poacher who enters the park. We respond with an ambush patrol."*

However, while partaking in patrols close to the Congolese border, park rangers explained that would-be 'poachers' are too poor to afford the fire weapons they would need to present a substantial threat to the animals or to rangers themselves. The leading ranger of the patrol further explained:

> *"The forest is steep and has too many trees, that is why poachers can only use bamboo traps to catch small duikers.[14] But it is our duty to destroy all traps because a gorilla can step into it and get injured."*

This conviction that local hunters could not use firearms in the dense forest on the steep volcano slopes calls into question whether militarization of park patrols for protection of wildlife is either effective or necessary.

The second narrative deployed by protected area authorities and ranger guides from RDB to justify militarization to gorilla tourists is that the military joining every tracking group is needed to protect visitors from wild buffalos. In contrast to the rifles carried by park rangers, RPF soldiers wear bullet-proof security vests and heavy arms, join tourist groups just before entering the National Park and leave again just after the group leaves the forest again to not compromise the welcoming atmosphere of the park headquarters. Yet even on routine patrols with rangers far away from any tourist activity, up to ten armed soldiers would join the armed RDB staff as well, calling into question explanations of their necessity for 'protection from buffalo'.

The Virunga National Park strategy – embracing security through militarization

Rwanda's price spike resulted in an increase of tourism flows from Rwanda to the DRC and Uganda. In 2018, *Volcanoes National Park* stated that they lost an estimated 60% of gorilla tourists to Uganda (VNP, 2018). In 2018, interviews with Rwandan tourism stakeholders and protected area authorities were increasingly coloured by derogatory anecdotes and a hostile attitude towards the neighbouring gorilla destinations. In particular, depictions of 'the Congo' as a place of barbarity, violence and war came through in conversations with park rangers, tourism workers and local residents throughout both Rwanda and Uganda (first author, forthcoming). Western media, international organisations and international travel advisories contribute to undifferentiated and distorted narratives of crisis throughout the Eastern DR Congo (see e.g. Autessere, 2012).

In contrast to Rwanda's reputation as the "most secure African country",[15] the Eastern Congo stands on the opposite end of the safety spectrum as the "world's deadliest spot"[16] and Western governments advice against all travel to Eastern Congo.[17] In an effort to transform this image from curse to blessing, marketing strategies implemented by the new management of PNVi comprising a public-private partnership between ICCN and the British charity *Virunga Foundation* established in 2008. This strategy aims to embrace the DRC's negative image in describing "the fight to save one of the world's most dangerous parks" (National Geographic, 2016). Funding by the Howard Buffet Foundation and the European Union enabled large-scale projects "to foster peace and prosperity through the responsible economic development of natural resources" (PNVi 2PNV020). A central aspect of this economic plan is PNVi's own peace tourism model. As a PNVi staff member explained in an interview:

> *"GVTC extracts money from our investors to finance projects that are completely inappropriate for the Congo. These people [from GVTC] have no understanding of our situation, they just give money to 'Water4Virungas'[18] and wonder that it disappears and nothing happens."*

This statement reflects a common perception among PNVi staff and rangers that external institutions such as GVTC lack an understanding of the Congolese context and of PNVi as characterized by both severe corruption and extraordinary security requirements.

Consequently, PNVi representatives rarely attended transboundary meetings organized by GVTC in 2018, resulting in growing resentment on the part of protected area authorities from RDB and UWA, as exemplified in focus group interviews:

> *"De Merode[19] does not want to listen to other people. To be fair, he faces incredible challenges that most, not even Congolese, could manage. But he thinks he is a better leader and this arrogance makes him work on an island."*

Such claims to autonomous sovereignty by conservation authorities have prompted researchers to argue that PNVi's adoption of securitization and militarization measures in the name of conservation that can be equated with the governance strategy under Belgium colonial rule to create 'a state within the state' (Marijnen, 2018, p. 2).

In interviews, the PNVi park management in the park headquarters in Rumangabo (DRC) gave two official reasons for their withdrawal from participation under GVTC. First, permanent intimidation and blackmailing by Congolese state authorities and armed groups would not allow managers to spend time and money on GVTC efforts that have no political mandate to create any positive change. Second, the ongoing exploitation of the Eastern Congo's natural resources by Rwanda would render any collaboration merely superficial window-dressing. These explanations point to a history of mistrust and anger against Rwanda's overt interventions and backing of armed militias in the Eastern DRC, as exemplified by this statement from one ICCN employee:

> *"Please stress when you go back to your country that it is Rwanda that keeps the insecurity in Eastern Congo. Rwanda controls our mines, our trade, our roads, our borders. Kagame has its own control even in the Congolese military. Our instability is created by Rwanda, they take advantage of our weakness. Rwanda has an interest to maintain the instability in North Kivu. For their own benefits."*

Parallel to Rwanda's self-portrayal as an island of security in sub-Sahara Africa, PNVi's park managers espouse negative narratives about the Eastern Congo in claiming to offer visitors a contrasting safe experience with military escorts in the dangerous environment. In the same vein, tour operators advertise PNVi for adventurous travellers as a reasonably safe, unique experience in "the heart of instability" on the internet. [20] In an interview, a Rwandan tour guide who presented himself as a the founder of a transboundary tourism agency aligned to Virunga's transnational travel sphere, emphasized a kind of 'dark touristic' interest in the Congo:

> *"These people would pay anything just to set one foot on Congolese soil to say they have been in the war zone."*

In this respect, the kidnapping of tourists on the road between Goma and the park headquarters in Rumangabo in May 2018 was unprecedented in the park's history and caused a major shock to the park and tourism stakeholders within the DRC. In contrast to the frequent kidnappings of local people living in proximity to the park barely gaining media attention, the abduction of two British visitors attracted international publicity leading to the closure of all tourism activities in and around PNVi until February 2019. The incident further undermined the relationship between Congolese and Rwandan park authorities and brought all diplomatic discussions concerning transboundary conservation measures under GVTC to a halt. In reaction to Rwanda's sudden price increase, ICCN staff accused Rwanda of having initiated the kidnappings out of jealously that Rwanda would lose its tourists to PNVi. In a focus group interview, one employee of PNVi's tourist lodge at the park's headquarters in Rumangabo stated this clearly:

> *"Rwanda needed to stop it [tourism], to get more tourists. We know all the armed groups in and around the park, they have never kidnapped tourists."*

Investigations from the United Nations mission *United Nations Organization Stabilization Mission in the Democratic Republic of the Congo* (MONUSCO) corroborated other sources in explaining that the kidnappers were probably linked to the FDLR in response to the latest law enforcement activities of PNVI's park management entailing confiscation of large amounts of illegal charcoal from several FDLR cartels.

To respond to the panic of international tourism operators and Western travel advisories that discouraged any travel into the Eastern Congo, PNVi's management hired security service operators and risk analysts from France and the United States to restructure the National Park's whole security system, creating an electronic-fenced, highly protected stronghold out of the park's headquarters and the patrol outposts. While the park management stated that they cannot establish a Virunga specific army, the paramilitary training, equipment and salary of the 700

ICCN rangers extends the resources of the regular Congolese army, *Forces Armées de la Republique Democratique du Congo* (FARDC). As an attempt to overcome a history of violent collision between park rangers and military personnel in the past and to strengthen the regional security in the future, PNVi re-established joint patrols of park rangers and Congolese soldiers, called *Corps Protection Parque de Naturel* (CorPPN), in the aftermath of the kidnapping incident. Furthermore, an arrangement with the MONUSCO was made to safeguard the road from Goma to Rumangabo.

Based on the new security protocol, the park management promoted security for travellers, including warning that safety in relation to transport, accommodation and tourism activities in and around the park can only be guaranteed when provided by PNVi park authorities themselves. The access to a special visa for the DRC that can only be issued via *Virunga National Park* and the monopoly on military security asserted as a necessary tourism service undermines any potential for local tourism operators who are not able to compete with the 'security standards' demanded by international tourism operators and Western governments' travel advice. A number of Congolese tour operators and driver guides confirmed this; as one succinctly summarized:

> *"The park is de Merode's private business. He does not want to share."*

When discussing this perceived exclusion with the PNVi management, an employee explained in response:

> *"There is a political agenda behind all the negative rumours about the park. The Congolese government is blackmailing us because they cannot exploit the oil in Virunga: 'If you want to protect the park, pay us the revenue!' If not, they attack. We have the responsibility that our staff and visitors are safe, we cannot delegate that task to externals."*

This statement points to the complex entanglement of conservation practices with the violent conflict and warfare in the Eastern Congo that informed establishment of the tourism industry for decades. These measures aimed towards reopening the park for tourism illustrate how the understanding of 'peace' in tourism became increasingly linked to the concept of human security, yet the 'humans' deserving of (military) protection within this context appears to be primarily the tourists rather than local citizens. In the context of PNVi in the Eastern Congo and VNP in Rwandan, 'security' seems to be a luxurious privilege for the humans in transit through, not the humans actually inhabiting the destination region. Military escorts, fenced park headquarters and tourism lodges as well as the gazetted, patrolled borders of the National Park create strong boundaries between visitors and locals rather than the intercultural understanding proposed by 'peace tourism' visions.

Local perceptions of 'peace' through tourism

UNWTO proposes that local communities adjacent to protected areas should receive the greatest benefits from tourism as they comprise the greatest part of the population evicted in the establishment of National Parks and further need an incentive to support conservation efforts (UNWTO (United Nations World Tourism Organization), 2015, p. 22). In this spirit, the GVTC treaty between the three states emphasized "revenue-sharing schemes from tourism revenues with local communities to establish positive attitudes towards tourism and the TBPA" (GVTC, 2006 [2014]: 13). Yet, the 'peace tourism' agenda does not consider the violence done to local people and environments *by* tourists, private businesses and the nation states as among the biggest tourism providers. National reports still identify the highest poverty levels in all three countries in direct vicinity to the National Parks (Bush et al., 2010; Kalulu et al., 2016, Kujirakwinja et al., 2010; Twinamatsiko et al., 2014). In addition, and partly in consequence of this rampant poverty, illegal park activities such as logging and hunting in Rwanda[21] and the DRC, and charcoal production and fishing in PNVi specifically, persist (Marijnen & Verweijen, 2016; Sabuhoro et al.,

2020). In response, park managers have introduced stricter law enforcement measures by conservation staff, from imprisonment to the legalization of shoot-on-sight in Uganda and the DR Congo. Studies concerning local perceptions of gorilla tourism confirmed our interviews with park-adjacently living community members in all three countries reflecting the persistent conflict between park staff and local residents who complain about the loss of their harvests to animals exiting the park, the inequitable distribution of revenues, and labor exploitation by tourism businesses (Adams & Infield, 2003; Munanura et al., 2016; Nkurunungi & Ampumuza, 2013). In this regards the promotion of national revenue-sharing models appears to be largely a marketing strategy to appeal to tourists' conscience, enabling them to 'offset' their luxurious consumer footprint by offering them the conviction that their presence and money are serving "help ending poverty" (PNVi, 2020), "empower local communities" (VNP, 2018) and "contribute to improving the livelihoods of households in frontline villages" (UWA, 2020). GVTC's transboundary strategic plan further asserts that

> *[t]he tourism industry creates employment opportunities in both urban and rural areas, thereby reducing pressure on wildlife resources. At the current level of tourism, the available facilities, especially for the up market visitor seem to be adequate in some areas. Focus is therefore required for the lower market visitor in these places.* (GVTC, 2006 [2014]: 10-11)

In direct opposition to this mandate to improve the 'lower-end' market, however, Rwanda's 2017 increase of gorilla tracking fees led to a drastic decrease in the numbers of tourist arrivals with devastating impacts on small-scale tourism businesses including hostels and smaller hotels, local tour guides and restaurants. In order to keep the flow of tourism's foreign exchange for the Rwandan government, Rwanda implemented a range of new tourism rules and regulations to homogenise the tourism landscape around *Volcanoes National Park* to make it "one of the world's most remarkable luxury ecotourism destinations" (RDB, 2020).

Consequently, high-end hotels and services became the face of Rwanda's tourism marketing strategy to attract the wealthiest tourists and celebrities. Five-star hotel projects thus began to mushroom around VNP in 2018, offering helicopter flights to overcome the inaccessibility of stony dirt roads up the volcano slopes.

In the course of 2018, in reaction to this shift, we witnessed the closure of numerous tourist accommodations. While many smaller facilities simply lost out from the lack of international tourist arrivals, one particular narrative circulated around the main tourist areas of VNP, as a hostel owner described:

> *"The big hotels complain to the government of Rwanda that they see no guests anymore and demand to shut us down. Now the government makes up new law every day, to bring licenses and documents that cost a lot of money. We cannot speak to the government and we cannot pay. We obey the law and shut down."*

Many other small tourism businesses similarly claimed that the Rwandan government invented arbitrary policies demanding unfulfillable requirements to ostensibly improve business standards. For example, one day in 2018 an official note sent to local guesthouses that remained in business commanded them to provide a marquise covering the front 'porch' of each house or face prohibition of any tourism services until further notice. After three weeks, the resulting closure cut any possible profit leading to bankruptcy for most owners. In this politically legitimized way the government squeezed remaining competitors for the high-end tourism facilities out of the market.

The promises made by the public-private partnership managing PNVi to "deliver large-scale opportunities for the local community by reducing poverty rates, stabilizing security, and strengthening local infrastructure" (PNVi, 2020) leave many local people living in the vicinity feeling disappointment and rage. In focus group interviews in park-adjacent villages, many expressed their resentment against PNVi personified in the park director de Merode. Despite implementation of two hydroelectric power plants and numerous community development projects funded largely by the Buffet Foundation, many local communities complained that they are not and have never been included in the benefits from these projects, particularly pointing at the electric wires running through their village to the city of Goma, without an electric connection to their houses (first author, forthcoming). Moreover, they explained how they mostly suffer from the persistent insecurity around PNVi, a situation that appears exceptionally unjust to them when large military convoys safeguard a few tourists in and out of the park. One local chief summarized this view:

> *"The park is for the communities, ICCN just manages it. But ICCN takes the crops and the land from our communities, so they [community members] turn against the park. Now the park is for white people and there is no peace between the park and the communities."*

Conclusion

This paper contributes to ongoing debates concerning the role of 'peace tourism' initiatives in the politics and lived realities around transboundary protected areas in conflict-affected settings through a case study of the *Virunga Conservation Area*. While the VCA is a single case study embedded in the complex historical context peculiar to the DR Congo, Rwanda and Uganda, it serves to problematize dynamics of 'peace tourism' rhetoric and practice more generally that will hopefully inspire future interdisciplinary research at the intersection of tourism, conservation and conflict studies.

In particular, this paper questions the dominant conceptualizations of 'peace through tourism' in the context of transboundary protected areas drawing on neoliberal, macro-economic development theories. Protagonists of so-called peace parks suggest that augmenting the fiscal base of the national economy will improve the well-being of citizens as well as collaboration between governments in management of the shared ecosystem. In the case of the VCA, however, our analysis has demonstrated that intergovernmental 'peace tourism' propositions under the umbrella institutions GVTC have instead led to increased competition, mistrust and mutual 'defaming' among the DRC, Rwanda and Uganda to assert sovereignty and capture the lion's share of gorilla tourism revenue. In this context, the historical warfare between the three countries provides an opportunity to maintain and exacerbate accusations and threat discourses legitimizing militarization within the National Parks comprising the VCA.

Simultaneously, the national park authorities exploit this same militarization as an indicator of the 'security' that is sold to tourists as a luxurious (and lucrative) product in paradoxical combination with the opportunity to encounter *endangered* mountain gorilla. In a circular logic, these tourism revenues are used to justify and finance the militarisation of national and park borders to ultimately create a military bulwark against the neighbouring states under the cover of providing security for tourists. Our framing of this dynamic as 'fortress tourism' points to the contradictions that the 'peace tourism' label conceals in safeguarding wealthy visitors using military means that simultaneously exclude and alienate local residents. This provision of 'human security' for tourists thus underscores the structural violence and inequality local inhabitants experience in not being considered as 'human' and therefore barely profiting from the expanding infrastructure projects, risk assessments and security measures implemented around them. In this way, the association of peace with 'human security' promised within peace parks discourse becomes yet another consumer product for the people who can afford to pay for this 'peacefulness'. These dynamics suggest that the improvement of park-to-people relationships the peace park concept intended to facilitate reverses in the case of the Virungas as local people instead experience an increase in inequality and sense of abandonment by conservation, state and international agents due to the provision of security itself.

Ultimately, 'peace tourism' in the VCA thus seems to remain a largely rhetorical strategy intended to please international conservation and development donors and international tourists so as to secure funding for other stakeholders and projects having little or nothing to do with environmental peacebuilding itself. This contradiction between tourism's ostensive 'peace dividend' and the different forms of violence it actually generates within the context of transboundary conservation efforts poses the question of what kind of 'peace' can actually be generated through this type of tourism, and for whom? Who defines what this 'peace' means and entails?

To address these questions became even more urgent since the COVID-19 pandemic drastically impacted the international travel industry, in the process accentuating the highly volatile entanglement between tourism and conservation in relation to both local people's livelihoods *and* biodiversity protection (Fletcher et al., 2020). In VCA, dependency on gorilla tourism led to closure of the three National Parks, leaving local tourism stakeholders without business and diminishing the park's revenue-sharing with local communities even further. This has aggravated an already tense situation as further militaristic securitization of the protected areas confronts a growing number of people who enter the parks unauthorized in pursuit of basic resources for survival as other livelihood options recede. Yet COVID-19 also presents a moment of global uncertainty that could offer potential for new aspirations and innovations to alter these long-standing patterns of inequality entrenched by the flawed concept of tourism as a panacea for peace, conservation and economic development. This moment demands engaged, critical and constructive scholarship and activism to pursue transformative change, in the contexts of conservation, tourism and beyond.

Notes

1. The geographically transboundary nature of the forest around the Virunga volcanoes is intended to become a TBPA named 'Virunga Conservation Area', yet, these branding initiatives are not generally utilised. We refer to the shared forest as 'Virunga Conservation Area' (VCA) , 'Virunga massif' or 'the Virungas'.
2. For protective purposes we anonymized participating organisations if they not explicitly expressed their will to be mentioned by name.
3. United Nations declaration 1967 of the International Tourist Year under the slogan "Tourism: Passport to Peace"
4. The Democratic Republic of Congo was named 'Republic of Zaire' from 1971 to 1997.
5. The *Interahamwe* is a paramilitary organization formed under Rwanda's former president Juvénal Habyarimana who played a central role in the atrocities before and during the Rwandan genocide in 1994 (Gourevitch, 1998).
6. Personal statements by Ramasamy Jayakumar, Head of Natural Sciences Sector, UNESCO, and Gabor Beszterczey, Senior Program Officer, O/SESG during *United Nations Great Lakes* conference, November 26, 2018.
7. Rwanda's conservation authority has been renamed from *Office Rwandais du Tourisme et des Parcs Nationaux* (ORTPN) to Rwandan Development Board in 2009.
8. Mountain gorillas are distinguished by their families with assigned names.
9. *The EastAfrican* 10th May 2017: Rwanda targets wealthy tourists with new gorilla permit fees, Wednesday May 10 2017; https://www.independent.co.uk/travel/news-and-advice/rwanda-gorilla-encounter-fees-doubled-uganda-a7727821.html
10. *Rwandan Patriotic Front* (explanation by the authors)
11. Website: https://www.gov.uk/foreign-travel-advice/rwanda/safety-and-security, accessed 27-2-2020.
12. The *Rwandan Defence Force* was formerly known by the name *Rwandan Patriotic Army* created by the military division of the former rebel group of the *Rwandan Patriotic Front*.
13. After 1994, Rwanda invented the problematic label "Historically Marginalized People" (HMPs) to refer to the indigenous Batwa population to avoid categorisation into ethnicities.
14. Duikers are a type of small antelope.
15. Most travel advisors for Rwanda quote the Gallup report 2017 ranking Rwanda as second safest country in Africa, as leading example the Lonely Planet: https://www.lonelyplanet.com/rwanda/safety, accessed 18-4-2020.

16. Website: https://www.brookings.edu/opinions/democratic-republic-of-congo-worlds-deadliest-spot/, accessed 18-4-2020.
17. Website: https://www.gov.uk/foreign-travel-advice/democratic-republic-of-the-congo/safety-and-security, accessed 18-4-2020
18. The funding by GVTC was based on "the overall objective of Water4Virunga [...] of contributing to the regional stability through an improved human security and inclusive growth" (GVTC n.d.). Yet, the embezzlement of project funds and power struggles between different stakeholders on the Congolese side has torpedoed any implementation on Congolese ground – in contrast to the compelling depictions of success on W4V official website (https://www.water4virungas.org/).
19. The Belgium Emmanuel de Merode is PNVi's park director since 2008.
20. Website: https://www.virungaparkcongo.com/information/safety-virunga-safe/, accessed 27-2-2020.
21. Georgia Tech Center for GIS and the Dian Fossey Gorilla Fund International 2018: 'Poaching in Rwanda's Volcanoes National Park', http://hub.arcgis.com/datasets/6a9e68db05704b119adb5fc3e4708291, accessed 17-1-2020.

Disclosure statement

No potential conflict of interest was reported by the authors.

Funding

This work was supported by Studienstiftung des Deutschen Volkes10.13039/501100004350.

References

Adams, W. M., & Infield, M. (2003). Who is on the gorilla's payroll? Claims on tourist revenue from a Ugandan National Park. *World Development*, *31*(1), 177–190. https://doi.org/10.1016/S0305-750X(02)00149-3
African Rights. (1998). *Rwanda: The insurgency in the Northwest*. African Rights.
African Wildlife Fund. (2020). Endangered mountain gorillas find a safe haven in Africa's oldest park, https://www.awf.org/blog/endangered-mountain-gorillas-find-safe-haven-africas-oldest-park
Ali, S. H. (2007). *Peace parks. Conservation and conflict resolution*. MIT Press.
Alluri, R. M. (2009). The role of tourism in post-conflict peacebuilding in Rwanda. Working Paper, Swiss Peace Foundation.
Andersson, J. A., de Garine-Wichatitsky, M., Cumming, D. H. M., Dzingirai, V., & Giller, K. E. (2013). People at wildlife frontiers in Southern Africa. In J. A. Andersson, M. de Garine-Wichatitsky, D. H. M. Cumming, V. Dzingirai, & K. E. Giller (Eds.), *Transfrontier conservation areas. People living on the edge* (pp. 1–11). Routledge.
Autessere, S. (2012). Dangerous tales: Dominant narratives on the Congo and their unintended consequences. *African Affairs*, *111*(443), 202–222.
Avraham, E., & Ketter, E. (2016). *Tourism marketing for developing countries: battling stereotypes and crises in Asia, Africa and the Middle East*. Palgrave Macmillan.

Baral, N., & Heinen, J. T. (2005). The Maoist people's war and conservation in Nepal. *Political and the Life Sciences, 24*(1–2), 2–11. https://doi.org/10.2990/1471-5457(2005)24 < 2:TMPWAC > 2.0.CO;2

Blanchard, L., & Higgins-Desbiolles, F. (2013). *Peace through tourism: promoting human security through international citizenship*. Routledge.

Bocajero, D., & Ojeda, D. (2016). Violence and conservation: beyond unintended consequences and unfortunate coincidences. *Geoforum, 69*, 176–183.

Brockington, D. (2002). *Fortress conservation: The preservation of the Mkomazi Game Reserve*. James Currey.

Brockington, D., Duffy, R., & Igoe, J. (2008). *Nature unbound. Conservation, capitalism and the future of protected areas*. Earthscan.

Büscher, B., & Fletcher, R. (2017). Destructive creation: capital accumulation and the structural violence of tourism. *Journal of Sustainable Tourism, 25*(5), 651–1858. https://doi.org/10.1080/09669582.2016.1159214

Bush, G. K., Ikirezi, M., Daconto, G., Gray, M., & Fawcett, K. (2010). *Assessing impacts from community conservation interventions around Parc National des Volcans, Rwanda*. Technical Report. International Gorilla Conservation Program.

Collier, P., & Hoeffler, A. (2000). *Greed and grievance in civil war*. Policy Research Working Paper 2355. World Bank.

Creswell, J., & Plano Clark, V. L. (2011). *Designing and conducting mixed methods research*. Sage Publications.

D'Amore, L. (2009). Peace through tourism: the birthing of a new socio-economic order. *Journal of Business Ethics, 89*(S4), 559–568. https://doi.org/10.1007/s10551-010-0407-3

de Merode, E., Smith, K. H., Homewood, K., Pettifor, R., Rowcliffe, M., & Cowlishaw, G. (2007). The impact of armed conflict on protected-area efficacy in Central Africa. *Biology Letters, 3*(3), 299–301. https://doi.org/10.1098/rsbl.2007.0010

Doyle, M. W. (2005). Three pillars of the liberal peace. *American Political Science Review, 99*(3), 463–466. https://doi.org/10.1017/S0003055405051798

Duffy, R. (2006). The potential and pitfalls of global environmental governance: The politics of transfrontier conservation areas in Southern Africa. *Political Geography, 25*(1), 89–112. https://doi.org/10.1016/j.polgeo.2005.08.001

Duffy, R. (2014). Waging a war to save biodiversity: The rise of militarised conservation. *International Affairs, 4*, 384–819.

Fairhead, J. (2005). Transnational dimensions to environmental resource dynamics: Modes of governance and local resource management in eastern DRC. In Q. Gausset, M. A. Whyte, & T. Birch-Thomsen (Eds.), *Beyond territory and scarcity: Exploring conflicts over natural resource management* (pp. 195–215). Nordiska Afrikainstitutet.

Fletcher, R. (2011). 'The only risk is wanting to stay': Mediating risk in Colombian tourism development. *Recreation and Society in Africa, Asia and Latin America, 1*(2), 7–30.

Fletcher, R., Büscher, B., Massarella, K., & Koot, S. (2020). Ecotourism and conservation under COVID-19 and beyond. *ATLAS Tourism and Leisure Review, 2*, 42–50.

Galtung, J. (1969). Violence, peace and peace research. *Journal of Peace Research, 6*(3), 167–191. https://doi.org/10.1177/002234336900600301

Galtung, J. (1996). *Peace by peaceful means: Peace and conflict, development and civilization*. Sage Publications.

Gourevitch, P. (1998). *We wish to inform you that tomorrow we will be killed with our families: stories from Rwanda*. Farrar, Straus and Giroux.

Greater Virunga Transboundary Collaboration. (2006 [2014]). *Ten Year Strategic Plan, Rubavu*.

Grosspietsch, M. (2006). Perceived and projected image of Rwanda: visitor and international tour operator perspectives. *Tourism Management, 27*(2), 225–234. https://doi.org/10.1016/j.tourman.2004.08.005

Haessly, J. (2010). Tourism and a culture of peace. In O. Moufakkir & I. Kelly (Eds.), *Tourism, progress, and peace* (pp. 1–16). CABI Publisher.

Hall, M. C., Dallen, T. J., & Duval, T. D. (2004). Security and tourism. *Journal of Travel & Tourism Marketing, 15*(2–3), 1–18. https://doi.org/10.1300/J073v15n02_01

Hanks, J. (2003). Transfrontier Conservation Areas (TFCAs) in Southern Africa: Their role in conserving biodiversity, socioeconomic development and promoting a culture of peace. *Journal of Sustainable Forestry, 17*(1–2), 127–148. https://doi.org/10.1300/J091v17n01_08

Haq, M. (1999). *Reflections on human development* (2nd ed.). Oxford University Press.

Hochschild, A. R. (1983 [2012]). *The managed heart*. University of California Press.

Honey, M., & Gilpin, R. (2009). *Special report on 'tourism in the developing world: Promoting peace and reducing poverty*. United State Institute of Peace.

Human Rights Watch. (2005). *Democratic Republic of Congo: The curse of gold*, https://www.hrw.org/reports/2005/drc0505/12.htm

Human Rights Watch. (2019). World Report 2019: Democratic Republic of Congo. https://www.hrw.org/world-report/2019/country-chapters/democratic-republic-congo

Institute for Economics and Peace (IEP). (2017). *Global Peace Index 2017*. IEP Report 48.

International Institute of Peace through Tourism (IIPT). (2017). About us - International Institute for Peace through Tourism. http://www.iipt.org/AboutUs.html

International Union for the Conservation of Nature (IUCN). (2020). Transboundary Conservation. https://www.iucn.org/commissions/world-commission-protected-areas/our-work/transboundary-conservation
Jenkins, M. (2008). *Who murdered the Virunga Gorillas?* http://ngm.nationalgeographic.com/2008/07/virunga/jenkins-text
Joras, U., Palme, K., Alluri, R., & Leicher, M. (2011). *Tourismus, Frieden und Konflikt: Effekte, Strategien und das privatwirtschaftliche Engagement in der Friedensförderung*. Deutsche Stiftung Friedensforschung.
Kalulu, R., Tushabe, E., & Nsabimana, E. (2016). Local community attitude and perceptions towards tourism conservation policies in Rwanda case of Volcanoes National Park. *Asian Journal of Science and Technology*, *7*(11), 3811–3819.
Kanyamibwa, S. (1998). Impact of war on conservation: Rwandan environment and wildlife in agony. *Biodiversity and Conservation*, *7*(11), 1399–1406. https://doi.org/10.1023/A:1008880113990
Kelly, I. (2006). Introduction to peace through tourism. IIPT Occasional Paper No. 1, Global Educators Network.
Kujirakwinja, D., Shamavu, P., Hammill, A., Crawford, A., Bamba, A., & Plumptre, A. J. (2010). *Healing the rift: Peacebuilding in and around the protected areas in the Democratic Republic of Congo's Albertine Rift*. Unpublished Report to USAID.
Maekawa, M., Lanjouw, A., Rutagarama, E., & Sharp, D. (2013). Mountain gorilla tourism generating wealth and peace in post-conflict Rwanda. *Natural Resources Forum*, *37*(2), 127–137. https://doi.org/10.1111/1477-8947.12020
Mbaiwa, J. E. (2005). Enclave tourism and its socio-economic impacts in the Okavango Delta, Botswana. *Tourism Management*, *26*(2), 157–172. https://doi.org/10.1016/j.tourman.2003.11.005
Moufakkir, O., & Kelly, I. (2010). *Tourism, progress, and peace*. CABI Publisher.
Levy, S. E., & Hawkins, D. E. (2009). Peace through tourism: Commerce-based principles and practices. *Journal of Business Ethics*, *89*(S4), 569–585. https://doi.org/10.1007/s10551-010-0408-2
Lisle, D. (2016). *Holidays in the danger zone: Entanglements of war and tourism*. University of Minnesota Press.
Litvin, S. (1998). Tourism: The world's peace industry? *Journal of Travel Research*, *37*(1), 63–66. https://doi.org/10.1177/004728759803700108
Lombard, L., & Tubiana, J. (2020). Bringing the tracker guards back in: Arms-carrying markets and quests for status in conservation at war. *Political Geography*, *79*, 102131–102111. https://doi.org/10.1016/j.polgeo.2019.102131
Lopez, E., Awawi, A., & Salcedo-Albarán, E. (2017). *Trafficking of coltan in the Democratic Republic of the Congo. The Global Observatory of Transnational Criminal Networks*. Research Paper No. 6.
Lunstrum, E. (2014). Green militarization: anti-poaching efforts and the spatial contours of Kruger National Park. *Annals of the Association of American Geographers*, *104*(4), 816–832. https://doi.org/10.1080/00045608.2014.912545
Marijnen, E. (2018). Public authority and conservation in areas of armed conflict: Virunga National Park as a 'state within a state' in Eastern Congo. *Development and Change*, *49*(3), 790–814. https://doi.org/10.1111/dech.12380
Marijnen, E., & Verweijen, J. (2016). Selling green militarization: The discursive (re)production of militarized conservation in the Virunga National Park, Democratic Republic of the Congo. *Geoforum*, *75*, 274–285. https://doi.org/10.1016/j.geoforum.2016.08.003
Mearns, K. (2012). Community-based tourism and peace parks benefit local communities through conservation in Southern Africa. *Acta Academica*, *44*, 70–87.
Minarchek, M. (2020). Creating environmental subjects: Conservation as counter-insurgency in Aceh, Indonesia, 1925–1940. *Political Geography*, *81*, 102189–102113. https://doi.org/10.1016/j.polgeo.2020.102189
Munanura, I. E., Backman, K. F., Hallo, J. C., & Powell, R. B. (2016). Perceptions of tourism revenue sharing impacts on Volcanoes National Park, Rwanda: a Sustainable Livelihoods framework. *Journal of Sustainable Tourism*, *24*(12), 1709–1726. https://doi.org/10.1080/09669582.2016.1145228
Nassani, A. A., Zaman, K., Aldakhil, A. M., & Qazi, A. M. (2017). War economy and pleasure: assessing the effects of military expenditure on tourism growth. *Quality & Quantity*, *51*(4), 1733–1754. https://doi.org/10.1007/s11135-016-0362-x
National Geographic. (2016). Inside the fight to save one of the world's most dangerous parks. https://www.nationalgeographic.com/magazine/2016/07/virunga-national-parks-africa-congo-rangers/
Neumann, R. P. (2004). Moral and discursive geographies in the war for biodiversity in Africa. *Political Geography*, *23*(7), 813–837. https://doi.org/10.1016/j.polgeo.2004.05.011
Neumayer, E. (2004). The impact of political violence on tourism. *Journal of Conflict Resolution*, *48*(2), 259–281. https://doi.org/10.1177/0022002703262358
Nielsen, H., & Spenceley, A. (2010). The success of tourism in Rwanda. Gorillas and more. Background Paper for the African Success Stories Study. World Bank.
Nkurunungi, J. B., & Ampumuza, C. (2013). *Assessment of the impacts of mountain gorilla habituation and tourism on their sustainable conservation*. Consultancy Report for the International Gorilla Conservation Programme.
Nunkoo, R., & Ramkissoon, H. (2016). Stakeholders' views of enclave tourism: A grounded theory approach. *Journal of Hospitality & Tourism Research*, *40*(5), 557–588. https://doi.org/10.1177/1096348013503997

Pemunta, N. V. (2019). Fortress conservation, wildlife legislation and the Baka Pygmies of southeast Cameroon. *GeoJournal*, *84*(4), 1035–1055. https://doi.org/10.1007/s10708-018-9906-z

Pizam, A. (1996). Does tourism promote peace and understanding between unfriendly nations? In A. Pizam & Y. Mansfeld (Eds.), *Tourism, crime, and international security issues*. John Wiley & Sons.

Prunier, G. (2011). *Africa's world war: Congo, the Rwandan genocide, and the making of a continental catastrophe*. Oxford University Press.

Ricardo, D. (1817 [1992]). *The principles of political economy and taxation*. Rutland.

Robbins, M. (2014). *Gorilla Journal 48, June, Berggorilla & Regenwald Direkthilfe*. www.berggorilla.org/fileadmin/gorilla-journal/gorillajournal-48-english.pdf

Rwanda Development Board. (2020). Rwanda wins emerging destination award as top 30 place to travel in 2020, and top overall in Africa. https://rdb.rw/rwanda-wins-emerging-destination-award-as-top-30-place-to-travel-in-2020-and-top-overall-in-africa/

Sabuhoro, E., Wright, B. A., Powell, R. B., Hallo, J. C., Layton, P. A., & Munanura, I. E. (2020). Perceptions and behaviors of indigenous populations regarding illegal use of protected area resources in East Africa's Mountain Gorilla Landscape. *Environmental Management*, *65*(3), 410–419. https://doi.org/10.1007/s00267-020-01254-z

Sandbrook, C. (2010). Putting leakage in its place: the significance of retained tourism revenue in the local context in rural Uganda. *Journal of International Development*, *22*(1), 124–136. https://doi.org/10.1002/jid.1507

Sandwith, T., Shine, C., Hamilton, L. S., & Sheppard, D. (2001). *Transboundary protected areas for peace and co-operation*. Best Practice Protected Area Guideline Series No. 7. IUCN.

Sönmez, S. F. (1998). Tourism, terrorism, and political instability. *Annals of Tourism Research*, *25*(2), 416–456. https://doi.org/10.1016/S0160-7383(97)00093-5

Spencer, J. P., & Rurangwa, M. (2012). Tourism as a route for the economic development of rural areas of Rwanda: Vibrant hope or impossible dreams? *African Journal for Physical, Health Education, Recreation and Dance*, *18* (4:2), 1023–1042.

Strauss, A., & Corbin, J. M. (1997). *Grounded theory in practice*. Sage.

Twinamatsiko, M., Baker, J., Harrison, M., Shirkhorshidi, M., Bitariho, R., Wieland, M., Asuma, S., Milner-Gulland, E. J., Franks, P., & Roe, D. (2014). *Linking Conservation, Equity and Poverty Alleviation: Understanding profiles and motivations of resource users and local perceptions of governance at Bwindi Impenetrable National Park, Uganda*. IIED Research Report.

Tripp, A. M. (2010). *Museveni's Uganda. Paradoxes of power in a hybrid regime*. Lynne Rienner Publishers.

Uganda Wildlife Authority (UWA). (2020). *Community and conservation*. https://www.ugandawildlife.org/wildlife-a-conservation-2/conservation/communities-a-conservation

United Nations High Commissioner for Refugees (UNHCR). (2000). The Rwandan Genocide and Its Aftermath. In *The State if the World's Refugees 2000: Fifty Years of Humanitarian Action*. UN Report, pp. 245–275.

UNEP-MONUSCO-OSESG. (2015, April 15). Experts' background report on illegal exploitation and trade in natural resources benefitting organized criminal groups and recommendations on MONUSCO's role in fostering stability and peace in eastern DR Congo. Final report. www.unep.org, 12-12-2019.

UNEP. (2009). *From conflict to peacebuilding: The role of natural resources and the environment*. United Nations Environment Programme. http://www.unep.org.

UNEP. (2016). *Environmental cooperation for peacebuilding programme*. Final Report 2016. : UNEP.

UN Security Council Presidential Statement (S/PRST/2000/20), June 2, 2000.

UN Security Council Resolution 1457 (S/RES/1457/2003), January 24, 2003.

US National Counterterrorism Center (USNCC). (2009). *Report on terrorism*. Office of the Director of National Intelligence, April 30, 2010.

UNWTO (United Nations World Tourism Organization). (2016). *Towards measuring the economic value of wildlife watching tourism in Africa*. Briefing Paper. UNWTO.

UNWTO. (2016). *Tourism, a catalyst for peace and development*. http://media.unwto.org/press-release/2016-07-14/tourism-catalyst-peace-and-development

Vedder, A., & Weber, B. (1990). The Mountain Gorilla Project. In A. Kiss (Ed.), *Living with wildlife: Wildlife resource management with local participation in Africa*. World Bank, 195–211.

Virunga National Park (PNVi). (2020). *Virunga Alliance*. https://virunga.org/alliance

Volcanoes National Park (VNP). (2018). *Rwanda's rise after the genocide*. https://www.volcanoesnationalparkrwanda.com/rwanda/24-years-rwanda-after-genocide.html

Volcanoes National Park (VNP). (2018). Rwanda gorilla safaris decline after price hike. https://www.volcanoesnationalparkrwanda.com/blog.html/page/5/

Weaver, A. (2011). Tourism and the military. Pleasure and the war economy. *Annals of Tourism Research*, *38*(2), 672–689. https://doi.org/10.1016/j.annals.2010.12.005

Williamson, J. (1989). What Washington means by policy reform. In J. Williamson (Ed.), *Latin American readjustment: how much has happened*. Peterson Institute for International Economics, Chapter 2 , https://www.piie.com/commentary/speeches-papers/what-washington-means-policy-reform.

World Bank. (2013). Addressing violence against women in war-torn regions, https://www.worldbank.org/en/news/feature/2013/12/09/addressing-violence-against-women-in-war-torn-regions

World Health Organization. (2002). *World report on violence and health*.

World Travel and Tourism Council. (2016). *Tourism as a driver of peace*. Policy Paper.

Ybarra, M. (2016). Blind passes' and the production of green security through violence on the Guatemalan border. *Geoforum*, *69*, 194–206. https://doi.org/10.1016/j.geoforum.2015.06.004

Tourism, peace and sustainability in sanctions-ridden destinations

Siamak Seyfi , Colin Michael Hall and Tan Vo-Thanh

ABSTRACT

Despite the widespread use of sanctions as a foreign policy tool in the absence of armed intervention and as a means to promote peace, there is notable absence of research on the effects of sanctions on the peacebuilding capacity of tourism and their relationship to the Sustainable Development Goals (SDGs). This situation is surprising given that both sanctions and tourism are promoted as a force for peace and reconciliation processes. Drawing upon international relations and political science and via semi-structured interviews with key stakeholders in the Iranian tourism and hospitality industry, this study investigates whether sanctions contribute to peace and create an environment suitable for tourism development. The findings indicate how the imposition, relaxation and then re-imposition of sanctions by international state actors as a means of peace have paralyzed the Iranian tourism industry through its psychological, sectoral, and societal effects and mobility restrictions. The consequences of sanctions and their sharp contrast with the SDGs are also explored. This study fills a significant gap in tourism research by examining the implications of the application of a widely used coercive geopolitical tool of statecraft in relation to the peace and tourism nexus.

"People talk about smart sanctions and crippling sanctions. I've never seen smart sanctions, and crippling sanctions cripple everyone, including innocent civilians, and make the government more popular." Mohamed ElBaradei, Director General (1997–2009) of the International Atomic Energy Agency (IAEA) and joint awardee of the 2005 Nobel Prize for Peace with the IAEA for their efforts to prevent the military use of nuclear energy (quoted in Warrick, 2009).

Introduction

Tourism has long been heralded as a force and vital source for peace (D'Amore, 1988), a contributor to reconciliation processes (Farmaki, 2017), and redevelopment in post-conflict contexts (Causevic & Lynch, 2011; Hall, 2017). The nexus between tourism and peace has been recognized by major tourism industry bodies such as the World Tourism Organization (UNWTO, 2018), which considers tourism as "a catalyst for development, peace and reconciliation," and the World Travel and Tourism Council (WTTC) (2016) which views tourism as a "driver of peace." Inspired by Goal 16 of the Global Goals (officially known as the Sustainable Development Goals [SDGs]) which focus on promoting just, peaceful and inclusive societies, the UNWTO (2018, p. 123) suggests that "as tourism revolves around billions of encounters between people of diverse cultural backgrounds, the sector can foster multicultural and inter-faith tolerance and understanding, laying the foundation for more peaceful societies. Tourism, which benefits and engages local communities, can also consolidate peace in post-conflict societies." Nonetheless, the economic and cultural significance of the sector alone cannot bring about peace and, clearly, peace, justice and security all contribute to tourism (Blanchard & Higgins-Desbiolles, 2013b; Higgins-Desbiolles & Blanchard, 2010). Pratt and Liu (2016) argue that peace has a far greater impact on tourism than tourism has on peace. Indeed, Khalilzadeh (2018) recommends the tourism and peace discourse be shifted from tourism as a "peacemaker" to tourism as a "peacekeeper."

This study highlights the issue of sanctions, an important yet rarely explored phenomenon in tourism and peace literature. Sanctions are a widely used instrument by international actors, including the UN Security Council (UNSC) and nation-states, to address threats to international peace and security or to preserve or restore peace (Lorenz, 2019; Pape, 1997). However, sanctions can have substantial negative consequences for tourism destinations, from reductions in tourist arrivals through to investment bans (Pratt & Alizadeh, 2018; Seyfi & Hall, 2019a, 2020). Exploring how tourism is both a target of sanctions as well as a response is of significant value, but interestingly major international tourism policy actors, such as the UNWTO, fail to criticise sanctions regimes. More recently, in the case of the COVID-19 pandemic, the unilateral US sanctions regime on countries such as Cuba, Iran and Venezuela was criticized for its hindrance of the capacity to respond to COVID-19 with calls for the Trump administration to temporarily suspend sanctions against these countries for peaceful and humanitarian objectives (Takian et al., 2020).

Despite the widespread use of sanctions in the post-Cold War era that directly and indirectly affect tourism (Pratt & Alizadeh, 2018; Seyfi & Hall, 2019a, 2019b, 2019c, 2020), there is a notable absence of research on sanctions in the tourism and geopolitics literature (Gillen & Mostafanezhad, 2019; Hall, 2017, Timothy, 2017; Weaver, 2010). Although there are major research streams focusing on the paradigms of peace, tourism, conflict reconciliation and post-conflict, the use of sanctions as a means for promoting peace is largely overlooked, despite sanctions being promoted as contributing to peace (Lorenz, 2019), as well as their wider implication for tourism and the SDGs. This study therefore aims to fill these gaps. To the best of the authors' knowledge, no study has previously examined the consequences of economic sanctions on tourism peace-building capacity.

Drawing upon the sanctions and peace literature in political science and international relations, and via a series of semi-structured interviews conducted with key stakeholders in the Iranian tourism industry, this study aims to understand the impacts of sanctions on Iranian tourism and to explore how sanctions affect the SDGs. Iran, which has been subject to sanctions regimes since the 1970s (Dizaji, 2019; Pratt & Alizadeh, 2018; Takeyh & Maloney, 2011) provides an interesting case to explore such interrelationships. This study is also timely given the reimposition of sanctions by the USA against Iran after a short spell of relaxation of the sanctions (2015–2018) imposed in response to Iran's nuclear program and the potential implications for the county's tourism industry (Ghaderi et al., 2018; Khodadadi, 2016).

The paper begins with an overview of sanctions as a means for bringing peace and meeting the SDGs and a discussion of theoretical approaches to sanctions, peace and tourism. Following detailing of the research context, the research design is outlined, and the empirical results reported. The findings of this study help to construct a composite picture of the sanctions and peace nexus within a tourism context and their wider implications for SDGs.

Literature review

Peace through sanctions

Although there is no standard definition of an international sanction, they are usually defined in terms of state related foreign policy actions applied in relation to a perceived wrongdoing by a state, organization, individual or other identified actor (Cortright & Lopez, 2018; Drezner, 2011). Sanctions take such a variety of forms and generate various effects in different sectors of the targeted state's economy. Rossignol (1996) explains such effects:

> Sanctions usually consist of a ban on the sale and shipment of products to a country and on the purchase of its exports. Other elements not directly linked to trade and commerce, such as culture and sports, are now often included. Thus, we see bans on economic, sporting, and cultural contacts between countries, as well as "person-to-person" sanctions such as withholding visas or other diplomatic documents from citizens of another state (Rossignol, 1996).

The economic restrictions imposed by sanctions occur at different scales, ranging from the economy as a whole to individual business operations. Pratt and Alizadeh (2018) categorised the economy-wide effects of sanctions as macroeconomic, direct, indirect, and induced. At the macro level, both the supply and demand side of the economy are adversely affected impacting economic growth and leading to inflationary pressures. Direct impacts include prohibition on foreign investment and imported goods as well as access to export markets. With a decrease in investment and employment cuts, corporate and market trust may also be weakened. Reputation and image are also affected by the negative perceptions often surrounding the sanctions target. Overall, sanctioned countries become more isolated from the global economic community (Ghaderi et al., 2018).

Sanctions are generally viewed as a lower-cost, lower-risk, middle ground between diplomacy and war (Masters, 2017) and often represent "the least unpalatable of the coercive alternatives available to the UN Security Council when faced with the task of taking action to maintain or restore international peace and security" (Farrall, 2007, p. 3). The UN Security Council has increasingly resorted to sanctions as part of its efforts to build peace and prevent conflict (Brzoska, 2015; Mack & Khan, 2000). Based on Chapter VII of the UN charter, if the UN Security Council determines under Article 39 the "existence of any threat to the peace, breach of the peace, or act of aggression," then under Article 41 it can authorize "complete or partial interruption of economic relations" and under Article 42 it can enforce these through "such action by air, sea or land forces as may be necessary" to maintain or restore international peace and security. Indeed, peacebuilding theory originated with the UN efforts to rebuild the international world order following World War Two and, since the end of the Cold War, sanctions have rapidly become the tool of choice for conflict prevention and peacebuilding adopted by the UN and other international organizations (Cortright & Lopez, 2018).

Prior to the post-Cold War era, there was only two notable examples of UN sanctions: comprehensive sanctions against Rhodesia in 1966 and an arms embargo against the apartheid regime in South Africa in 1979. Yet, since the fall of the Berlin Wall in 1989, there has been a dramatic increase in the imposition of sanctions with the USA, UN and European Union having employed sanctions on other nations more than 500 times (Cashen, 2017). There are two main factors that contributed to the rise of sanctions in the aftermath of the Cold War (Farrall, 2007). First, from a political perspective, within the UN Security Council it is extremely difficult to garner the support necessary to authorize collective military action under Article 42 of the UN Charter (Seyfi & Hall, 2020). In contrast, the imposition of non-military sanctions is perceived as entailing fewer costs than using military action. Nevertheless, by authorizing sanctions, the Security Council has still taken symbolic action, without having to assume the costs and responsibilities for using force. Second, given the increased interdependence in the global economy since the end of the Cold War the potential effectiveness of sanctions in achieving foreign policy objectives is perceived to have increased (Farrall, 2007).

UN sanction measures encompass a broad range of economic, financial, and diplomatic options that do not involve the use of armed force (UN Security Council, 2020). The majority of these measures target state actors and have been imposed for various objectives, including compelling an occupying state to withdraw (e.g. Iraq's invasion of Kuwait); preventing a state from developing or acquiring weapons of mass destruction (e.g. North Korea, Iran); countering international terrorism (e.g. Libya, Al Qaida, Hariri sanctions regimes); stemming human rights violations (e.g. South Africa, Iran); and promoting the implementation of a peace process (e.g. Liberia) (Farrall, 2007). In the case of the EU, key objectives when adopting sanctions are: safeguarding EU's values, fundamental interests, security, preserving peace, consolidating and supporting democracy, rule of law, human rights and the principles of international law, preventing conflict, and strengthening international security (European Council, 2020). Unilateral sanctions have also largely been adopted by individual states as a part of their foreign policy to change the behaviour of the target state instead of being involved in direct military conflict (i.e. US sanctions).

Sanctions are usually presented as a peaceful means for achieving foreign policy objectives and, as such, "are inherently part of a purely diplomatic approach geared toward avoiding a military confrontation" (Fathollah-Nejad, 2014, p. 49). Nevertheless, Pape (1997) argued that sanctions are more often a prelude to war, rather than an alternative to it. More fundamentally with respect to peace dividends, sanctions often have highly detrimental effects on the population of targeted states. For example, Peksen (2009) suggests that economic sanctions worsen government respect for human rights, including freedom from disappearance, extra-judicial killing, torture, and political imprisonment. As a result, there has been a shift from comprehensive sanctions regimes (e.g. against Iraq in the 1990s) to more "targeted" or "smart" sanctions (e.g. Russia), which are directed at specific individuals or entities (e.g. through asset freezes and travel bans) or by prohibiting particular activities (e.g. arms embargoes and export bans) (Cortright & Lopez, 2018; Farrall, 2007). The "qualitative" shift toward smart and targeted sanctions aims to put maximum pressure on those directly involved in the sanctioned behaviour and those who benefit from it, rather than shutting down entire economies (Brooks, 2002). Nevertheless, their effectiveness remains debatable, as Fathollah-Nejad (2014, p. 49) commented "smart bombs may well succeed smart sanctions." Agbonifo (2017, p. 66) classifies the effectiveness of sanctions under three main goals: *coercing* (the extent to which sanctions modify the target's proscribed behaviour); *constraining* (the extent to which sanctions limit the target's ability to continue its proscribed behaviour); and *signalling* (the extent to which the target and others are stigmatized about violation of international norms).

There is considerable empirical evidence that sanctions disproportionately affect politically weak and vulnerable groups (e.g. workers, women, children) and benefit supporters of the target regime (Drury & Peksen, 2014; Kaempfer et al., 2004). For instance, in research that highlights the potential impacts of sanctions on the SDGs, Neuenkirch and Neumeier (2015) reported that sanctions restrict the trade and economic development of targeted countries, negatively affect their GDP growth rate, bilateral trade and financial services which, in turn, negatively affects vulnerable groups in the sanctioned countries through increased inflation, poverty and income inequality and shortage of food and medicine.

Although the sanctioning of countries is seemingly conceived and promoted as a gender neutral act, empirical studies suggest that the most vulnerable populations, including women, are negatively affected by sanctions regimes and that they bear the burden of sanctions (Buck et al., 1998; Drury & Peksen, 2014; Gutmann et al., 2020). Drury and Peksen (2014) argue that the costs of trade sanctions are disproportionately imposed on women, who are often already one of the most marginalized and political and economic actors in the target country, with women's social, political, and economic rights substantially affected (Gutmann et al., 2020). Despite this, surprisingly, very few works have been focused on the gendered implications of sanctions.

To the best of our knowledge, there is no study that has previously examined the consequences of economic sanction on tourism peace-building capacity. Furthermore, the literature fails to examine how sanctions contribute to sustainability given that they disproportionately affect the wider population. This will be discussed in the following sections.

Sanctions and SDGs

Economic sanctions place severe constraints on the economic development of sanctioned countries which are also mainly developing countries (Cortright & Lopez, 2018). These barriers potentially negatively impact the targeted countries' efforts to meet the SDGs. Achieving the SDGs requires finance and economic development, multilateral trade and international investment. Goal 17 emphasizes the integration of poor and developing countries in the global economy and strengthening the multilateral trading system (UN, 2015). Although the conceptualisation of the SDGs has been criticised for their reinforcement of contemporary neoliberal capitalism (Hall, 2019), sanctions are clearly barriers for state-industry partnerships and the SDGs as they are constituted. Ironically, even the UN which is a leading actor in the imposition of sanctions, states that the imposition of economic sanctions "negatively affects socio-economic development of the target countries and weaken the contribution of target countries to international economic development "(Afshari, 2019).

In addition to economic impacts, sanctions have longstanding ramifications for the environment, which is an often-overlooked consequence of sanctions. While sanctions are not the main cause of sanctioned countries' environmental problems, they intensify the situation (Rustler, 2019). The latter is in contrast with the Goal 15 of the SDGs which calls for "protecting, restore and promote sustainable use of terrestrial ecosystems, sustainably manage forests, combat desertification, and halt and reverse land degradation and halt biodiversity loss" (UN, 2015). For instance, because of their impact on the transfer of funds and the movement of people, sanctions against Iran reduced capacity for biodiversity conservation (Madani, 2014).

Under sanctions the cooperation, coordination and sharing of responsibilities by the international community as enshrined in Goal 17 of the SDGs becomes restricted. This can lead, for example, to limited access to international scientific and educational expertise and, even, publishers (Mozafari, 2016; Seyfi et al., 2019). Similarly, the US government has put pressure on European or foreign entities providing technological support for cultural conservation in Iran (Seyfi et al., 2019). This is again in contrast with target 11.4 of Goal 11 of the SDGs which calls for strengthening efforts to protect and safeguard the world's cultural and natural heritage. The financial restrictions caused by sanctions (e.g. foreign investment ban, transfer of technology) affect numerous sectors in target countries. Thus, sanctions potentially lock countries into unsustainable practices that cut across many of the SDGs and which can adversely affect the socio-economic and environmental quality of life, including health, for generations (Rustler, 2019).

Peace through tourism and sanctions

Tourism is "as much a political terrain as a cultural practice" (Germann Molz, 2010). The mobility of people across national boundaries has political implications (Baerenholdt, 2013), which sometimes can be manipulated by government for geopolitical purposes as an instrument of political leverage between nations. One notable example of mobility restrictions was the trade embargo against the apartheid regime in South Africa which included restrictions on travel, notably on overflying and landing rights for South African airlines (Pirie, 1990). The longstanding US sanctions and embargo against Cuba is another noteworthy example of sanctions restriction on mobility which significantly affects US citizens and residents wishing to visit Cuba as well as sanctions against cruise lines (Gordon, 2016). More recently, the executive order signed by President Trump known as the Trump travel ban placed stringent restrictions on travel to the US for citizens of Iran, Iraq, Libya, Somalia, Sudan, Syria, and Yemen and placed restrictions on visa renewals for an additional 38 countries (Arafa, 2018). In addition, there are also specific travel bans on individuals and institutions in target countries (Seyfi & Hall, 2020, 2019a, 2019c).

International and domestic tourism flows require peace, safety and security (WTTC, 2016). There has been much rhetoric about tourism's role in promoting world peace (Blanchard & Higgins-Desbiolles, 2013b; Hall, 2005). One view advocates tourism's potential capability in fostering and sustaining peace. This can be seen in the global tourism-related institutions discourse of the WTTC (2016) and the UNWTO (2018) that highlight the transformative power of tourism to contribute to peace through the improvement of human relations. The role of tourism in establishing peace is reported in numerous studies (Becken & Carmignani, 2016; Blanchard & Higgins-Desbiolles, 2013b; D'Amore, 1988; Farmaki, 2017; Gelbman, 2019; Moufakkir & Kelly, 2010). The contribution of tourism in reconciliation efforts in post-conflict settings has also been highlighted in cases including the Former Yugoslavia (Causevic & Lynch, 2011), Cambodia (Winter, 2008), Vietnam (Gillen, 2014), Burundi (Novelli et al., 2012), Cyprus (Farmaki, 2017), Korean Peninsula (Cho, 2007; Kim & Prideaux, 2003), Palestine (Isaac, 2009), Afghanistan (Durko & Petrick, 2016) and China and Taiwan (Rowen, 2014). In contrast, there is another perspective that questions the validity of the causal relationship between tourism and peace (Hall, 1994, 2005; Khalilzadeh, 2018; Pratt & Liu, 2016; Salazar, 2006). For instance, in their empirical investigation, Pratt and Liu (2016) reported that tourism is a beneficiary of peace rather than a cause of peace. Indeed, Rowen (2014) argued that tourism actually aggravates political contradictions and cross-strait relations between China and Taiwan. As Blanchard and Higgins-Desbiolles (2013a, p. 2) observe, "the essential tension between the multiple dimensions of peace (as a process) and the confining dimensions of tourism (as a product) remains the invisible elephant in the room."

While existing tourism literature has opened up significant areas of inquiry pertaining to the peace and tourism nexus, knowledge is limited as to how peace through tourism is affected in countries subjected to sanctions even if tourism is not a direct object of sanctions. The following study examines such issue in the context of Iran.

Methods

Study context

Iran has been the subject of one of the largest, longest and toughest sanctions regime in history for almost four decades (Dizaji, 2019; Pratt & Alizadeh, 2018; Takeyh & Maloney, 2011), which makes it an interesting case study of sanctions and wider politics in relation to peace and tourism. The country has experienced sanctions by individual countries and international bodies since the 1979 revolution in repose to Iran's nuclear program, human rights abuses, and Iran's wider role in the region (Dizaji, 2019; Takeyh & Maloney, 2011). As of March 2020, the US had nearly 8,000 sanctions in place worldwide, with Iran by far the largest state target (Gilsinan, 2019). Four decades of sanctions have triggered a collapse in the country's economic growth, pushed the country into a deep recession with high inflation, and severed Iran's access to international markets (Pratt & Alizadeh, 2018). In this section, we detail the three main stages of Iranian tourism development which show how imposition, lifting and then re-establishment of nuclear-related sanctions have paralyzed Iran's tourism and hospitality industry.

Tourism in Iran before the lifting of sanctions

The position of Iran at the crossroads of major Asian, Middle Eastern and European countries and trading routes (e.g. the Silk Road) has provide a rich heritage resource for tourism (Seyfi & Hall, 2018). Archaeology, cultural heritage, traditions, and natural heritage are among the key reasons attracting inbound visitors (Khodadadi, 2016). Despite this wealth of tourism resources, Iran has failed to gain a significant share of international tourism arrivals, especially when compared with similar countries in the region. While Iran was considered the top tourist destination in the Middle East during 1967–1977 (Morakabati, 2011), in the post-1979 era following Iran's Islamic Revolution, the aftermath of the Iran-Iraqi war (1980–1988) and conflict over Iran's nuclear developments which led to economic sanctions, international travel to Iran became significantly limited. However, Iran's negative image in many major tourist generating markets disguises its long history of tourism and rich heritage. This is especially significant as strengthening inbound tourism and its economic and cultural roles has been a common goal favored by various administrations in modern Iran (Mozaffari et al., 2017). Several other factors have inhibited Iran's tourism development comparing with its neighbors, these include political tensions with the United States, over-reliance on oil revenues, political instability and conflicts in the Middle East, and poor management (Khodadadi, 2016; Mozaffari et al., 2017; Seyfi & Hall, 2018). However, sanctions have had the greatest direct and indirect impacts of Iranian tourism industry, including bans on foreign investment in tourism and hospitality, bans on financial transactions with Iranian banks, bans on purchase of any aircraft and spare parts, and the indirect negative psychological effects of sanctions in the form of perceived insecurity and risk (Pratt & Alizadeh, 2018; Seyfi & Hall, 2018, 2019a).

Tourism in Iran in the post-sanctions era

The partial lifting of international nuclear-related sanctions in the aftermath of the Iran's nuclear agreement in 2015 (officially known as the Joint Comprehensive Plan of Action) had a strong positive impact on the Iranian tourism sector (Pratt & Alizadeh, 2018) and substantially improved the country's image from being a "pariah state" to a "booming destination" (Ghaderi et al., 2018; Khodadadi, 2016). More than five million foreign tourists visited Iran in 2017, almost three times as many as in 2009 (UNWTO, 2019). In the light of the relaxation of sanctions and the release of pent-up international demand, European airlines such as Air France, British Airways, and Lufthansa resumed direct flights to the country. The Iranian authorities also relaxed visa requirements. Iran ordered nearly 200 aircraft from Airbus, Boeing and ATR, worth $36 billion (Khodadadi, 2018). There was also considerable foreign investment in the hotel industry. For example, in 2016, the French group Accor Hotels became the first foreign-branded hotel to open in Iran since the 1979 Islamic Revolution and Spain's Melia Hotels International opened a beach resort hotel on the Caspian Sea. Iran also made significant investments in high-speed rail in collaboration with the Italian company, Ferrovie dello Stato Italiane (FS). In their analysis of economic impact of the lifting of sanctions to tourism and the wider economy in Iran, Pratt and Alizadeh (2018) concluded that tourism development provided a boost to the overall economy with the extra income stimulating investment. Iran's tourism administration hoped to attract more than twenty million tourists by 2025 thereby creating jobs for millions of unemployed young Iranians (Ghaderi et al., 2018). The increase in the number of arrivals and the subsequent employment generation and increases income, also had many positive effects on the empowerment of women who could work in tourism and hospitality related businesses in Iran's religio-patriarchal society (Farahani & Dabbaghi, 2018).

Tourism in Iran and the re-imposition of sanctions

The US withdrawal of the agreement and the reestablishment of sanctions in May 2018 halted Iran's booming tourist industry. This latest round of sanctions imposed by the Trump administration are part of a "maximum pressure" campaign aimed at reducing oil exports (which account for over 80% of the state budget) to zero in order to curb Iran's regional influence (Dizaji, 2019). Following the return of American sanctions, several airlines suspended their flights to Iran, and foreign investors were obliged to leave Iran due to being blacklisted by US sanctions and the fear of losing access to the US market. Although getting access to new commercial aircraft and spare parts for existing old aircraft was one of Iran's key reasons to reach an agreement with the U.S. and Europe to lift sanctions, Iran only received five airplanes out of the nearly 200 ordered. US sanctions also prevented Iranian carriers from purchasing spare parts which further worsened the airworthiness of Iran's ageing air fleet.

The renewed sanctions also affected the mobility of individuals visiting Iran. US sanctions restricts visa-free entry to the US for travelers who have visited Iran. Travel warnings for Iran were reestablished by European countries and major online travel platforms halted their services in Iran. Overall, Iranian tourism is plagued by sanctions and growing tensions between Iran and the United States which shows how the country's tourism industry is directly and indirectly impacted by geopolitical issues which severely affect any attempts for peace through tourism.

Data collection

The aim of the study was to gain an in-depth understanding on the effects of sanctions on Iran's tourism industry as perceived by various stakeholders and their relationship to peace and the SDGs. An exploratory qualitative research approach was adopted to collect and analyse the data (Creswell & Poth, 2018). Given the exploratory nature of this study, the semi-structured interview was deemed appropriate for data collection. This method allowed the researchers to delve into what each research participant had in mind (Creswell & Poth, 2018) and provided opportunities for each interviewee to explore the issues they felt were important about the topics given to the participants.

Purposive sampling was used as it allows researchers to use their judgement and their a-priori understanding of the topic, to select people who better help them to answer their research questions and achieve their goals (Onwuegbuzie & Collins, 2007) in a sensitive political environment (Hall, 2011). After analyzing existing stakeholder and sanctions research (Perks, 2013; Seyfi & Hall, 2019a, 2019b), participants in this study were selected from the public sector and national authorities, the private sector, non-profit organization and academia, the latter having a close relationship to the tourism industry in Iran. Further interviewees were identified by snowball sampling when those first contacted were asked to identify others who may be interested in doing an interview. The informants were carefully chosen to represent different sectors of the industry and allow for enriched views.

The views of the following informants are presented in the study: five tour operators/agent, the owners of four hotels, the managers of two hotels, two airline managers/staff, four tour guides, three persons from the national tourism administration, three persons from provincial tourism organizations, two persons from the national hotelier association, two persons from the national travel agent's association, three persons from a local handicraft association, two persons from a NGO, and four tourism academics. A pilot study was conducted with four interviewees to test the interview questions, along with interview style and approach (Kim, 2011). The interviews took place from June to August 2019. The recorded interviews were thirty to sixty minutes in duration with the questions being framed according to the objectives of the research and included questions related to the effects of the sanctions on the tourism industry prior to the conclusion of the nuclear agreement, how the lifting of sanctions helped to flourish the tourism-related businesses, the effects of the US withdrawal from the agreement and the subsequent reestablishment of sanctions and inhibiting factors, and how tourism continues to operate under a sanctions environment. The general aim in the interviews was to see through the participants' eyes by having them explain their experiences.

The interview questions were open-ended in order to obtain more spontaneous opinions and to avoid potential bias limiting responses to the researcher's fixed categories (Rowley, 2012). Because Iranian officials are often reluctant to talk openly about political issues, interviews were conducted anonymously, we also sought to ensure the confidentiality and identity of our respodents (Farasatkhah et al., 2008). Those who did not wish to be interviewed in person were offered the option of a telephone or email interview. Where possible, unstructured follow-up probes were used to further discuss points emerged during the interview. According to Patton (2002) the sample size in a qualitative survey depends mainly on data saturation, the possibility of reaching the target sample, resources and available time. Interviews were conducted up to theoretical saturation. After interviewing 36 participants, the theoretical level of data saturation achieved as the last interviews did not yield any new information.

Data analysis

As the vast majority of participants spoke only Persian, the interviews were conducted in that language. The data were analysed in Persian and only the final results were translated into English. All interviews were transcribed verbatim and were back translated into English by the lead author to ensure consistency in meaning (Creswell & Poth, 2018) and were eventually verified by the second researcher. Following every interview, a preliminary analysis was conducted until all the interviews had ended. To secure trustworthiness, respondents were provided a copy of their interview transcript, and feedback was sought to enhance the dependability of the findings (Lincoln & Guba, 1985). Frame analysis was used to analyze the data to identify themes in the text utilizing a coding scheme to organize emerging topics into interrelated themes (Talja, 1999). The analytical process included primarily the identification, analysis and reporting of themes within the data set (Braun & Clarke, 2006). The transcripts and notes from the interviews were read and re-read a number of times to further familiarise with the data as well as identifying key themes and concurrent patterns both within and among interviews (Miles & Huberman, 1994). Verbatim spoken words from the transcripts were then copied, re-organised, and cross-referenced to classify the thematic groups. After several rounds of coding, topics emerging from the data were clustered into interrelated themes, followed by subcategories that allowed greater consistency in structure and development on key issues (Hennink et al., 2020).

Findings

This section deals with the constraints that the sanctions have imposed on Iranian tourism actors, particularly on the private sector. The different stakeholders interviewed discussed the constraints of imposed economic sanctions on the tourism sector and the challenges they have faced so far. As noted above, sanctions have multiple effects on the tourism sector. Overall, the findings show how the imposition, relaxation and then re-imposition of sanctions have paralyzed the tourism industry and related businesses in Iran.

Psychological effects

Destination risk perception

In recent decades, due to political instability and Iran's geopolitical ambitions, international media coverage of Iran has had a negative impact on foreign tourist arrivals. A majority of respondents agreed that the country's negative image has made it a perceived high-risk destination for foreign travelers. Respondents believed that many factors shaped this negative image including "the 1979 revolution," and "the American hostage crisis and subsequent sanctions regime" which gave birth to a "doctrine of anti-American sentiment and Americanophobia" among the Iranian ruling elite. They also mentioned "the Iranian nuclear programme" which has been the main reason for many of sanctions as well as the source of many tensions between Iran and the west. As one respondent commented "the negative image of Iran is widespread in the international media … many countries warn their citizens not to visit Iran (interview #11)."

This was echoed by another respondent who explained the mental effects of sanctions on destination image and commented that:

> ... the negative and psychological dimension of the sanctions [perception] has created a negative image of Iran in the minds of foreign travellers ... this image would make them look for another destination than Iran ... Security is one of the most important factors in the formulation of tourism development strategies in the world. Security and tourism are parameters of an equation that are directly and closely linked ... as long as Iran's security is threatened by sanctions, we will suffer the consequences (interview #14)

This was echoed by a tour operator:

> ... a tourist usually goes to a country where security is not threatened. It is true that Iran is a safe country in the Middle East, but we only feel the situation ourselves. The Western media and even Asian media refer to Iran as a country in crisis that will soon experience internal conflict or war (interview #16)

Respondents all agree that the nuclear agreement of 2015 and the subsequent lifting of sanctions largely contributed to changing Iran's image, leading to a substantial increase in the number of arrivals, especially European tourists, and an increase in foreign investment in tourism. A tour operator involved with inbound tours noted that:

> ... the nuclear deal has resulted in the restoration of Iran's image against negative coverage in the foreign media. This could be observed in international newspapers dealing with Iran and on online travel sites such as Tripadvisor, Lonely Planet, etc. (Interview #19)

Looking at the media headlines at that time, it is clear how important a nuclear deal and the lifting of sanction were in changing Iran's image. For example, just as the nuclear deal was announced, CNN published a report entitled "why your next vacation could be in Iran" (Pleitgen, 2015), citing the Iranian government's attention to the importance of the tourism sector by opening the country to more tourists.

The impact of this positive image cannot be underestimated, given that, for example, the number of German tourists increased by 54% and Italian tourists by 75% (Seyfi & Hall, 2018). But following the withdrawal of the United States from the nuclear agreement and the ensuing tensions, Iran was again castigated by the media, damaging its international image. This was noted by a respondent:

> ... the Vienna Agreement was very promising and positive in the first year, but the withdrawal of the US from this agreement has again increased risk and fear for tourists wishing to visit Iran ... This has created new challenges for the Iranian tourism industry ... Although the cost of travel to Iran has decreased by 50% [mainly due to the devaluation of the currency as a result of sanctions], the number of European and American tourists arriving in the country since Trump's exit has largely decreased (interview #8).

Cancellation and reduction of visits

Most responders stated that they have experienced the cancellation of visits and the decline in arrivals of foreign tourists because of sanctions. Tour operators were the main industry group adversely affected by sanctions. For most respondents, trip cancellation is the first reaction of foreign travelers to the sanctions imposed on Iran. A tour guide explained:

> Soon after the reinstatement of sanctions and rising tensions between Iran and the US, austerity measures were stepped up. International companies, such as Booking.com, had just started working with Iran, but immediately left the country when Trump launched the first wave of sanctions on Iran, other popular travel agencies did the same and they no longer sell Iranian travel tours in their packages ... (interview #5).

This echoes the words of another tour guide:

> At present, only a few small companies sell limited tours but are not interested in expanding the market and will soon eliminate all Iranian tours from their websites. Immediately after the US withdrawal from the agreement, several tours from France, China and Germany to Iran were cancelled ... We're just guides and we live off the money we make from tourists ... it means that my life and the lives of my colleagues are seriously affected (interview #6).

A closer look at international websites offering tours shows that Iran's name has been removed. One hotel owner noted:

> A few years ago, Iranian tour operators could still operate on [TourRadar], but this site also removed all Iranian packages following the imposition of sanctions. With the imposition of sanctions, booking requests from foreign agencies have been drastically reduced and our business is no longer profitable (interview #9).

Reduction in foreign investment

With the lifting of international sanctions in 2015, Iran sought to convince foreign investors to invest in various service sectors through the use of incentives such as "a five-year tax exemption" for those investing in tourism-related infrastructure. The respondents from the hotel sector stated that the country has a great potential for hotel sector. One of the respondents suggested that:

> The lifting of sanctions had some major benefits for our tourism industry. we had a foreign brand hotel chain for the first time after the revolution. ... But with the return of American sanctions against our country, the Accor group's contract faced with new constraints and new plans were eventually canceled (interview #4).

He further stated that international hotel chains such as Rotana (an Emirati hotel management company), and Marriott International had expressed their desire to enter the Iranian market and build new hotels. But the return of sanctions changed meant that these contracts were never signed. This was also echoed by a five-star hotel manager:

> ... with US sanctions against Iran, it is not legally possible for foreign hotel chains to enter the Iranian market at the moment. ... the sanctions prevented investors from taking advantage of investor confidence in the hotel sector and the construction of new Iranian hotels (interview #13).

Sectoral effects

Aviation

The Iranian aviation industry has experienced serious problems with regard to airplane components and airplane purchases as sanctions have targeted Iran's purchase of any aircraft as well as parts. Thus, the sector is clearly experiencing operational difficulties directly and indirectly related to the sanctions. The respondents drew attention to various points concerning the impact of sanctions on the Iranian aviation industry. For instance, an airline official explains that: "The average age of the Iranian airline fleet is over 30 years. After decades of sanctions Americans have blocked any possibility of modernization of the air fleet" ... Iranian airlines have suffered grave deterioration in their safety record due to sanctions" (interview #25). This echoed a colleague's view who stated: "Given the sanctions on Iran, it is only possible to import second-hand aircraft into the country." (interview #26).

Iranian civil aviation has long been subject to U.S, UN, and European operational restrictions and sanctions that bar the provision of parts and technical services to Iranian aircraft. One of the respondents commented:

> ... some airports even refused to refuel Iranian planes, forcing them to travel with heavier fuel loads. Sanctions target the provision of fuel to all Iranian airlines at international airports.... many flights experience delays or cancelations" (interview #25).

A tour operator involved with outbound travel mentioned that Iranian airlines are banned from flying to major European countries such as France and Germany:

> In our tour package, we have to use flights by regional airlines such as Turkish Airlines or Qatar Airways. We have no choice of having direct flight to main European capitals, have to use short alternative routes. Some of our customers who are old, mainly refuse such indirect flights" (interview #33).

Online booking system ban

Booking.com, the Dutch online travel booking platform giant, is another company that entered the Iranian market after the sanctions were lifted but has since left, with adverse effects on hotels (interview #34). One tour operator noted that: "The listing of Iranian hotels on Booking.com had many advantages, as tourists could easily book their hotels and this could encourage them to travel to Iran and make their package themselves" (interview #36). Another hotelier noted: "The accommodation on this website is in itself a good marketing tool and this opportunity could be an advantage for the country's tourism industry" (interview #35). But following the US withdrawal from the agreement, Booking.com announced that it was no longer possible for them to continue with Iranian partners because of US sanctions. As a result, there is no reservation system or platform for tourists who decide to travel to Iran. A travel agency manager explains:

> ... almost all foreign companies involved in tourism marketing and promotion are not interested in working with Iranian partners. It's simple, they can't afford to miss the opportunity to work with the huge American tourism market" (interview #31).

Financial services restrictions

One of the notable financial effects of sanctions that was repeatedly commented by the interviewees is the ban and restriction on international payments systems and the banking industry. The county has no connection to international payment networks such as Visa and MasterCard. An interviewee commented:

> Because of the sanctions, it's impossible to transfer any money to Iran or transfer money from Iran to other countries from official banking channels ... we have to go through the exchanges offices which have bureaus in Dubai, Istanbul for any international transactions. they always ask for a high commission rate for any transaction (interview #18).

A tour guide also explains

> Tourists have to carry cash for duration of stay if they want to visit Iran. They can't use their bank card in Iran ... imagine how hard and risky it is for senior tourists which make up the largest market for our inbound tourism ... I have had many problems with this, and our customers always complain about this (interview #15).

Mobility restrictions

Tightening of the visa procedure for Iranians

Iranian passports are at the bottom of the world ranking of the most powerful passports. According to the 2019 Henley Passport Index (HPI), Iran ranks 101st out of 109 countries, together with North Korea, Lebanon, Bangladesh and Eritrea for a total of only 39 easily accessible destinations (Henley & Partners, 2019), making it difficult to travel abroad with an Iranian passport. As a result, Iranians have to undertake visa application procedures through embassies to travel abroad well in advance. One tour operator explains

> ... we have had many problems with the visa application process over the last years because of sanctions. Visa applicants have faced delays in payment, unilateral and unjustified cancellations, or a lack of responsiveness and availability of embassy officials in Iran. For example, applying for a visa to Germany from Iran currently takes about nine months ... embassies, however, do not refund the visa fee if the visa is not granted (interview #32).

Respondents complained about the lengthy visa process. A tour operator involved in outbound tourism to Europe stated that:

> … with the return of sanctions, the number of applications dropped and the process became even more difficult. So many visa applications are rejected, services require applicants to come back to the embassy several times, others want to add fingerprints to the files (interview #34).

Most importantly, this process is not transparent (interview #29). This situation has led Iranian civilian activists to launch campaigns for the "right to travel", as well as a "visa stories" site (*visastory.ir*), so that everyone can write and publish their experiences of visa applications

Halting direct and charter flights

Soon after the US return of sanctions on Iran, major Europeans airlines including British Airways, and Air France-KLM along with other airlines such as Etihad Airways and Thai Airways all withdrew from Iran. The new round of sanctions targeted global business links with Iran, and companies doing business with Iran are subject to being barred from the US market. This has substantially affected the airline capacity to and from Iran. A tour operator noted that:

> In the first phase, direct flights have become indirect flights … airlines are slowly leaving Iran, and European airlines have been replaced by regional airlines such as Qatar Airways or Turkish Airlines (interview #32).

Another respondent also noted:

> With the rise in currency prices, flight prices have been calculated based on the free market exchange rate which are always fluctuating … sometimes we have to ask customers to pay back the differences in price (interview #30).

Societal effects

Decline in purchasing power

Economic sanctions have substantially weakened the purchasing power of Iranians. This has led to dramatic increase in inflation and unemployment. Several respondents commented that sanctions have greatly changed travel habits among Iranians, including to friends and relatives. An accommodation manager stated:

> … the sanctions have made Iranians avoid long stays in destinations. They are no longer able to travel. Given the increasing cost of travel in Iran, Iranians tend to make day trips, not more (interview #12).

This was echoed by another respondent:

> Tourists who had travelled abroad before the economic crisis and currency fluctuations are now turning to local destinations. As for the middle classes, they were already accustomed to travelling locally, and today's rising costs have almost completely eliminated this option from the household basket (interview #15).

Gender disempowerment

Because of their already vulnerable position in the Iran's male-dominated socio-political culture, women faced more socio-economic hardships from the reimposition of sanctions, and job security and opportunities for women have decreased (Farahani & Dabbaghi, 2018). Respondents believe that the effects of sanctions on Iranian women have been substantial. This was observed by a female tour guide:

> By the return of sanctions, women are the first to lose their jobs and have less job security as compared to men … sanctions endangered women's livelihoods and deprive women of their job security and I've seen many cases of layoffs among my friends over the last year (interview #20).

This was echoed by another respondent:

> Sanctions have created a situation that women accept work without any social security insurance by the employer … this economic pain can be felt in a number of ways such as less well paid job in the hotel and

> other lodging sector, less job security, no support from the government without no safeguards... all of these worsen women's economic status in Iran (interview #21).

Discussion

Sanctions are promoted as a means to achieve peace (Brzoska, 2015; Lorenz, 2019; Mack & Khan, 2000). However, the history of sanctions clearly shows that economic sanctions have not been able to maintain peace in the long-term. The analysis of Hufbauer et al. (2008) reveal that only 34% of economic sanctions imposed between 1914 and 1990 achieved their goals. This was also observed by Smeets (2019) who notes that sanctions episodes continue to be a popular foreign policy tool "despite the fact that the literature does not present conclusive evidence that economic sanctions are an effective policy instrument."

With respect to Iran, while it is possible that sanctions may have restricted some of the geopolitical ambitions of Iran's conservative theocrats, the war by proxy between Iran and Saudi Arabia, as well as other Iranian actions in the Middle East suggest that sanctions have not led to peace by proxy in the region (Akbarzadeh & Ibrahimi, 2020; Saad, 2019). Therefore, any notion of peace that such sanctions have served should be regarded as highly restrictive. As noted by Ali (2019), the longstanding sanctions have caused great difficulties for Iranian civilians, that calling into question the legitimacy and efficacy of sanctions regimes.

The removal of many sanctions following the 2015 nuclear agreement and the boost this gave tourism provided massive impetus for employment generation, especially for women and younger people; helped justify environmental and heritage conservation measures; and improve the quality of life in many destinations. In such a situation peace without sanctions was extremely conducive to meeting many of the SDGs, particularly in areas, such as gender and women's rights, where the conservative elements within Iranian state have extremely narrow perspectives (Shahidian, 2002; Tahmasebi, 2018). The comments of respondents as to the effects of the reimposition of sanctions on tourism to and within Iran provides significant insights into the peace and tourism relationship. Peace, in the sense of the absence of all-out war, by itself is insufficient to encourage tourism and tourism developments that meet broader sustainability goals. Instead, sanctions that are usually justified in the name of peace although often reflecting national political agendas, can have awful consequences on the wellbeing of the general population and especially the most vulnerable (Drury & Peksen, 2014; Gottstein, 1999; Kim, 2019; Koc et al., 2007). As the findings of this study shows sanctions have negatively affected the welfare and populace and life of ordinary people and those working in the tourism and hospitality industry. This is important as the empowerment of women through tourism has been made more fragile with the return of sanctions. This is in line with other findings that economic decline caused by the sanctions leads to greater economic grievance often met by further government repression (Buck et al., 1998; Drury & Peksen, 2014; Gutmann et al.2020; Kim, 2019). This may undermine civil society's capacity to criticize the target government and contradict the very notion of sanctions as a non-violent tool of statecraft.

Rather than punish or restrict ruling elites, sanctions affect society as a whole, and often with results at odds with the stated intentions of the sanctions (Nossal, 1989; Peksen, 2009; Seyfi & Hall, 2020). Such barriers can have a detrimental effect on the efforts of the target developing countries to achieve the SDGs, and demonstrates a strong contrast with the goals laid out in the SDGs. As the findings of the study shows sanctions have negatively affected the middle class who are arguably the main consumers of tourism services. The results here support Peksen's (2009) observation that economic coercion via the impositions of sanctions remains a counterproductive policy tool. As such, a peace through sanctions approach clearly does not appear to contribute to the SDGs. As noted earlier, achieving SDGs requires finance and economic development, multilateral trade, and international investment, which all can be undermined with the imposition of sanctions. While, Goal 17 stresses the inclusion of developing countries into the world economy and the expansion of the multilateral trading mechanism in order to achieve the SDGs, arguably as the result of this study revealed, economic sanctions hindered such capacities by putting ban on financial transactions and foreign investment in the target countries, in a way that the countries subjected to sanction have left isolated from the global community (Ali, 2019; Ghaderi et al., 2018; Pratt & Alizadeh, 2018; Seyfi & Hall, 2020). This can also be seen in the case of Cuba sanctions regime where the US has strongly restricted US business and third country investment in different sectors of Cuban economy including tourism and hospitality (Gordon, 2016).

Instead, what is required is an understanding of peace in which the vital role of social, economic and environmental justice is recognised along with human rights (Higgins-Desbiolles, 2006, 2008; Higgins-Desbiolles et al., 2019; Isaac & Hodge, 2011; Jamal & Camargo, 2014). Importantly, this observation extends to both sanctions and the role of tourism as tourism itself can be used to support authoritarian and despotic regimes with little real commitment to development endeavours such as the SDGs (Boluk et al., 2019; Cortright & Lopez, 2018; Hall, 1994). Instead, the notion of empowerment though tourism is replaced by disempowerment via the imposition of sanctions. In Iran this is particularly significant for women's economic empowerment which is central to realizing women's rights and gender equality and to the achievement of the SDGs (UN Women, 2020). While, tourism has been viewed as a sector capable of advancing gender equality and women's empowerment (Alarcón & Cole, 2019; Cole, 2006; Tucker & Boonabaana, 2012) as it is also enshrined in SDG5 which focuses on gender equality and women's empowerment (Boluk et al., 2019), nonetheless, as the findings of this study also showed, the economic disruption ensuing from sanctions are detrimental to such empowerment capacity of tourism.

In the right circumstances, tourism can contribute to peace and peace to tourism. However, it is vital that the role of geopolitical measures, such as sanctions, and their impacts on SDGs indicators stop being ignored by international institutions and organisations, such as the UNWTO, as well as by tourism researchers. The growing popularity of sanctions as a coercive geopolitical choice of foreign policy clearly has implications with respect to the discourse on peace through tourism. As the findings of this study indicated, economic sanctions disrupt financial investment and supply chains in destinations exposed to sanctions and highlight how any tourism development is related to politics and geopolitics (Farmaki et al., 2015; Hall, 1994, 2017; Kim & Prideaux, 2003; Rowen, 2014). As such, this study shows that sanctions imposed by major powers on, usually developing countries negatively affect the peace-building capacity of tourism and in turn generate more depression and conflict. Furthermore, if we take another step back, we can observe that not only is the impositions of sanctions on Iran and its subsequent implications for the Iranian people steeped in politics but so too is the very framing of the notion of tourism-through-peace (Blanchard & Higgins-Desbiolles, 2013b; Cho, 2007) and, we would also suggest, the SDGs (Boluk et al., 2019; Hall, 2019).

Conclusion

In the post-Cold War period, economic sanctions have become a popular response to myriad threats to international peace and security, yet the peacebuilding or peacekeeping capacity of this foreign policy tool is highly questionable. In his review of the sanctions' regimes on Cuba, former Yugoslavia and Iraq, Gottstein (1999) concluded that sanctions inexorably end up hurting the people, not the ruling elite. This research on Iran suggests the same outcome. Sanctions neither produce nor restore peace.

The current research would also suggest that any contribution of tourism to reconciliationneeds to be understood in the specific context of the initial conflict that requires resolution.While government relations remained restrained the short period of the lifting of sanctions didsuggest something of tourism's transformative economic potential and the opportunity for internationalvisitors to see Iran and the Iranian people first-hand. Indeed, the contribution of sanctionsto the impacts of COVID-19 in Iran may only serve to exacerbate tensions.

Since the establishment of an Islamic republic in 1979, tourism in Iran has been plagued by political instability and decades of sanctions. Although sanctions have long been used as a foreign policy instrument by states and international actors such as the United Nations Security Council and the European Union, there is relatively little research that examines the tourism consequences of sanctions on a given destination and in relation to peace and sustainability. This study showed how the imposition, lifting and reestablishment of sanctions has affected the Iranian tourism industry, creating a business environment in which though peaceful, fails to fulfill tourism's potential promise as a mechanism to facilitate the SDGs.

The study's value lies in two distinctive contributions. First, it expands the peace-through-tourism discourse by highlights the role of sanctions as an important yet rarely explored phenomenon in tourism and peace literature. Despite the increasing growth of sanctions, very limited research has examined peace through sanctions and its relation to tourism and wider implication for the SDGs. Second, it reinforces the notion that a far more nuanced idea of peace is required for peace and tourism research, one that focuses on concerns over human rights, socio-economic and environmental justice and security, rather than the relative absence of violence. It is in both these areas that future research is urgently needed.

Disclosure statement

No potential conflict of interest was reported by the authors.

ORCID

Siamak Seyfi http://orcid.org/0000-0002-2427-7958
Colin Michael Hall http://orcid.org/0000-0002-7734-4587
Tan Vo-Thanh http://orcid.org/0000-0001-9964-3724

References

Afshari, L. (2019). *International sanctions and sustainable development goals*. https://www.sdwatch.org/post/international-sanctions-and-sustainable-development-goals

Agbonifo, J. (2017). Sanctions, conflict prevention and peacebuilding: Coercing non-state armed actors in Africa. *Journal of Peacebuilding & Development*, *12*(1), 65–78. https://doi.org/10.1080/15423166.2017.1281679

Akbarzadeh, S., & Ibrahimi, N. (2020). The Taliban: a new proxy for Iran in Afghanistan? *Third World Quarterly*, *41*(5), 764–782. https://doi.org/10.1080/01436597.2019.1702460

Alarcón, D. M., & Cole, S. (2019). No sustainability for tourism without gender equality. *Journal of Sustainable Tourism*, *27*(7), 903–919. https://doi.org/10.1080/09669582.2019.1588283

Ali, O. O. (2019). From critical to comprehensive dialogue: the effectiveness of the EU's policy towards Iran (1992–1998). *British Journal of Middle Eastern Studies*, 1–16. https://doi.org/10.1080/13530194.2019.1701408.

Arafa, M. (2018). A question to the President of the United States, Donald Trump: is it a travel ban, or a Muslim ban, or a travel Muslim ban?. Revista de Investigações Constitucionais, 5(2), 9–33.

Baerenholdt, J. O. (2013). Governmobility: The powers of mobility. *Mobilities*, *8*(1), 20–34. https://doi.org/10.1080/17450101.2012.747754

Becken, S., & Carmignani, F. (2016). Does tourism lead to peace? *Annals of Tourism Research*, *61*, 63–79. https://doi.org/10.1016/j.annals.2016.09.002

Blanchard, L., & Higgins-Desbiolles, F. (2013a). Introduction Peace matters, tourism matters. In L. Blanchard & F. Higgins-Desbiolles (Eds.), *Peace through tourism: Promoting human security through international citizenship* (pp. 1–16). Routledge.

Blanchard, L. A., & Higgins-Desbiolles, F. (Eds.). (2013b). *Peace through tourism: Promoting human security through international citizenship*. Routledge.

Boluk, K. A., Cavaliere, C. T., & Higgins-Desbiolles, F. (2019). A critical framework for interrogating the United Nations Sustainable Development Goals 2030 Agenda in tourism. *Journal of Sustainable Tourism*, *27*(7), 847–864. https://doi.org/10.1080/09669582.2019.1619748

Braun, V., & Clarke, V. (2006). Using thematic analysis in psychology. *Qualitative Research in Psychology*, *3*(2), 77–101. [Database] https://doi.org/10.1191/1478088706qp063oa

Brooks, R. A. (2002). Sanctions and regime type: What works, and when? *Security Studies*, *11*(4), 1–50. https://doi.org/10.1080/714005349

Brzoska, M. (2015). International sanctions before and beyond UN sanctions. *International Affairs*, *91*(6), 1339–1349. https://doi.org/10.1111/1468-2346.12449

Buck, L., Gallant, N., & Nossal, K. R. (1998). Sanctions as a gendered instrument of statecraft: The case of Iraq. *Review of International Studies*, *24*(1), 69–84. https://doi.org/10.1017/S0260210598000692

Cashen, E. (2017). *The impact of economic sanctions*. https://www.worldfinance.com/special-reports/the-impact-of-economic-sanctions

Causevic, S., & Lynch, P. (2011). Phoenix tourism: Post-conflict tourism role. *Annals of Tourism Research*, *38*(3), 780–800. https://doi.org/10.1016/j.annals.2010.12.004

Cho, M. (2007). A re-examination of tourism and peace: The case of the Mt. Gumgang tourism development on the Korean Peninsula. *Tourism Management*, *28*(2), 556–569. https://doi.org/10.1016/j.tourman.2006.04.019

Cole, S. (2006). Information and empowerment: The keys to achieving sustainable tourism. *Journal of Sustainable Tourism*, *14*(6), 629–644. https://doi.org/10.2167/jost607.0

Cortright, D., & Lopez, G. A. (2018). *Economic sanctions: Panacea or peacebuilding in a post-cold war world?* Routledge.

Creswell, J. W., & Poth, C. N. (2018). *Qualitative inquiry and research design: Choosing among five approaches* (4th ed.). Sage.

D'Amore, L. J. (1988). Tourism—A vital force for peace. *Tourism Management*, *9*(2), 151–154. https://doi.org/10.1016/0261-5177(88)90025-8

Dizaji, S. F. (2019). Trade openness, political institutions, and military spending (evidence from lifting Iran's sanctions). *Empirical Economics*, *57*(6), 2013–2041.

Drezner, D. W. (2011). Sanctions sometimes smart: Targeted sanctions in theory and practice. *International Studies Review*, *13*(1), 96–108. https://doi.org/10.1111/j.1468-2486.2010.01001.x

Drury, A. C., & Peksen, D. (2014). Women and economic statecraft: The negative impact international economic sanctions visit on women. *European Journal of International Relations*, *20*(2), 463–490. https://doi.org/10.1177/1354066112448200

Durko, A., & Petrick, J. (2016). The Nutella project: An education initiative to suggest tourism as a means to peace between the United States and Afghanistan. *Journal of Travel Research*, *55*(8), 1081–1093. https://doi.org/10.1177/0047287515617300

European Council (2020). Sanctions: how and when the EU adopts restrictive measures. Retrieved from https://www.consilium.europa.eu/en/policies/sanctions/.

Farahani, B., & Dabbaghi, H. (2018). Tourism and the empowerment of women in Iran. In S. Seyfi & C.M. Hall (Eds.), *Tourism in Iran: Challenges, development and issues* (pp. 177–192). Routledge.

Farasatkhah, M., Ghazi, M., & Bazargan, A. (2008). Quality challenge in Iran's Higher Education: A historical review. *Iranian Studies*, *41*(2), 115–138. https://doi.org/10.1080/00210860801913321

Farmaki, A. (2017). The tourism and peace nexus. *Tourism Management*, *59*, 528–540. https://doi.org/10.1016/j.tourman.2016.09.012

Farmaki, A., Altinay, L., Botterill, D., & Hilke, S. (2015). Politics and sustainable tourism: The case of Cyprus. *Tourism Management*, *47*, 178–190. https://doi.org/10.1016/j.tourman.2014.09.019

Farrall, J. M. (2007). *United Nations sanctions and the rule of law*. Cambridge University Press.

Fathollah-Nejad, A. (2014). Why sanctions against Iran are counterproductive: Conflict resolution and state–society relations. *International Journal: Canada's Journal of Global Policy Analysis*, *69*(1), 48–65. https://doi.org/10.1177/0020702014521561

Gelbman, A. (2019). Sport tourism and peace: Crossing the contested wall. *Tourism Geographies*, *21*(1), 163–189. https://doi.org/10.1080/14616688.2018.1519718

Germann Molz, J. (2010). Tourism, global culture and transnational diplomacy. World Politics Review. https://www.worldpoliticsreview.com/articles/7144/tourism-global-culture-and-transnational-diplomacy

Ghaderi, Z., Soltani, S., Henderson, J., & Zareei, A. (2018). Sanctions, the 2015 agreement and Iran's tourism industry. In S. Seyfi & C.M. Hall (Eds.). *Tourism in Iran: Challenges, development and issues* (pp. 55–65). Routledge.

Gillen, J. (2014). Tourism and nation building at the war remnants museum in Ho Chi Minh city, Vietnam. *Annals of the Association of American Geographers*, *104*(6), 1307–1321. https://doi.org/10.1080/00045608.2014.944459

Gillen, J., & Mostafanezhad, M. (2019). Geopolitical encounters of tourism: A conceptual approach. *Annals of Tourism Research*, *75*, 70–78. https://doi.org/10.1016/j.annals.2018.12.015

Gilsinan, K. (2019). *A boom time for U.S. sanctions*. https://www.theatlantic.com/politics/archive/2019/05/why-united-states-uses-sanctions-somuch/588625/

Gordon, J. (2016). Economic sanctions as 'negative development': The case of Cuba. *Journal of International Development*, *28*(4), 473–484. https://doi.org/10.1002/jid.3061

Gottstein, U. (1999). Peace through sanctions? Lessons from Cuba, former Yugoslavia and Iraq. *Medicine, Conflict and Survival*, *15*(3), 271–285. https://doi.org/10.1080/13623699908409463

Gutmann, J., Neuenkirch, M., & Neumeier, F. (2020). Precision-guided or blunt? The effects of US economic sanctions on human rights. *Public Choice*. *185*, 161–182. https://doi.org/10.1007/s11127-019-00746-9.

Hall, C. M. (1994). *Tourism and politics*. John Wiley.

Hall, C. M. (2005). *Tourism: Rethinking the social science of mobility*. Pearson.

Hall, C. M. (2011). Researching the political in tourism: Where knowledge meets power. In C. M. Hall (Ed.), *Fieldwork in tourism* (pp. 53–68). Routledge.

Hall, C. M. (2017). Tourism and geopolitics: The political imaginary of territory, tourism and space. In D. Hall (Ed.), *Tourism and geopolitics: Issues from Central and Eastern Europe* (pp. 15–24). CABI.

Hall, C. M. (2019). Constructing sustainable tourism development: The 2030 agenda and the managerial ecology of sustainable tourism. *Journal of Sustainable Tourism*, *27*(7), 1044–1060. https://doi.org/10.1080/09669582.2018.1560456

Henley & Partners. (2019). *2019 Henley passport index*. https://www.henleypassportindex.com/assets/2019/Q3/HPI%20Report%20190701.pdf

Hennink, M., Hutter, I., & Bailey, A. (2020). *Qualitative research methods*. Sage.

Higgins-Desbiolles, F. (2006). More than an "industry": The forgotten power of tourism as a social force. *Tourism Management*, *27*(6), 1192–1208. https://doi.org/10.1016/j.tourman.2005.05.020

Higgins-Desbiolles, F., & Blanchard, L. (2010). Challenging peace through tourism. Placing tourism in the context of human rights, justice and peace. In O. Moufakkir & I. Kelly (Eds.), *Tourism, progress and peace* (pp. 35–47). CABI.

Higgins-Desbiolles, F., Carnicelli, S., Krolikowski, C., Wijesinghe, G., & Boluk, K. (2019). Degrowing tourism: Rethinking tourism. *Journal of Sustainable Tourism*, *27*(12), 1926–1944. https://doi.org/10.1080/09669582.2019.1601732

Hufbauer, G. C., Schott, J. J., Elliott, K. A., & Oeggm, B. (2008). *Economic sanctions reconsidered* (3rd ed.). The Peterson Institute for International Economics.

Isaac, R. K. (2009). Alternative tourism: Can the segregation wall in Bethlehem be a tourist attraction? *Tourism and Hospitality Planning & Development*, *6*(3), 247–254. https://doi.org/10.1080/14790530903363381

Isaac, R. K., & Hodge, D. (2011). An exploratory study: Justice tourism in controversial areas. The case of Palestine. *Tourism Planning & Development*, *8*(1), 101–108. https://doi.org/10.1080/21568316.2011.554048

Jamal, T., & Camargo, B. A. (2014). Sustainable tourism, justice and an ethic of care: Toward the just destination. *Journal of Sustainable Tourism*, *22*(1), 11–30. https://doi.org/10.1080/09669582.2013.786084

Kaempfer, W. H., Lowenberg, A. D., & Mertens, W. (2004). International economic sanctions against a dictator. *Economics and Politics*, *16*(1), 29–51. https://doi.org/10.1111/j.1468-0343.2004.00130.x

Khalilzadeh, J. (2018). Demonstration of exponential random graph models in tourism studies: Is tourism a means of global peace or the bottom line? *Annals of Tourism Research*, *69*, 31–41. https://doi.org/10.1016/j.annals.2017.12.007

Khodadadi, M. (2016). A new dawn? The Iran nuclear deal and the future of the Iranian tourism industry. *Tourism Management Perspectives*, *18*, 6–9. https://doi.org/10.1016/j.tmp.2015.12.019

Khodadadi, M. (2018). Donald Trump, US foreign policy and potential impacts on Iran's tourism industry: Post-nuclear deal. *Tourism Management Perspectives*, *26*, 28–30. https://doi.org/10.1016/j.tmp.2018.01.007
Kim, S. S., & Prideaux, B. (2003). Tourism, peace, politics and ideology: Impacts of the Mt. Gumgang tour project in the Korean Peninsula. *Tourism Management*, *24*(6), 675–685. https://doi.org/10.1016/S0261-5177(03)00047-5
Kim, Y. (2011). The pilot study in qualitative inquiry: Identifying issues and learning lessons for culturally competent research. *Qualitative Social Work: Research and Practice*, *10*(2), 190–206. https://doi.org/10.1177/1473325010362001
Kim, Y. (2019). Economic sanctions and HIV/AIDS in women. *Journal of Public Health Policy*, *40*(3), 351–366. https://doi.org/10.1057/s41271-019-00173-6
Koc, M., Jernigan, C., & Das, R. (2007). Food security and food sovereignty in Iraq: The impact of war and sanctions on the civilian population. *Food, Culture & Society*, *10*(2), 317–348. https://doi.org/10.2752/155280107X211467
Lincoln, Y., & Guba, E. G. (1985). *Naturalistic inquiry*. Sage.
Lorenz, J. P. (2019). *Peace, power, and the United Nations: A security system for the twenty-first century*. Routledge.
Mack, A., & Khan, A. (2000). The efficacy of UN sanctions. *Security Dialogue*, *31*(3), 279–292. https://doi.org/10.1177/0967010600031003003
Madani, K. (2014). Water management in Iran: What is causing the looming crisis? *Journal of Environmental Studies and Sciences*, *4*(4), 315–328. https://doi.org/10.1007/s13412-014-0182-z
Masters, J. (2017). *What are economic sanctions?* Retrieved May 17, 2018, from https://www.cfr.org/backgrounder/what-are-economic-sanctions
Miles, M. B., & Huberman, A. M. (1994). *Qualitative data analysis: An expanded sourcebook*. Sage.
Morakabati, Y. (2011). Deterrents to tourism development in Iran. International Journal of Tourism Research, 13(2), 103–123.
Moufakkir, O., & Kelly, I. (2010). Introduction peace and tourism: friends not foes. In O. Moufakkir & I. Kelly (Eds.), *Tourism, progress and peace* (pp. xvi–xxxii). CABI.
Mozafari, M. (2016). Iran and science publishing in the post-sanctions era. *The Lancet*, *387*(10029), 1721–1722. https://doi.org/10.1016/S0140-6736(16)30260-4
Mozaffari, A., Karimian, R., & Mousavi, S. (2017). The return of the 'Idea of Iran' (2005-2015). In R. Butler & W. Suntikul (Eds.), *Tourism and political change* (2nd ed., pp. 186–199). Goodfellow Publishers.
Neuenkirch, M., & Neumeier, F. (2015). The impact of UN and US economic sanctions on GDP growth. *European Journal of Political Economy*, *40*, 110–125. https://doi.org/10.1016/j.ejpoleco.2015.09.001
Nossal, K. R. (1989). International sanctions as international punishment. *International Organization*, *43*(2), 301–322. https://doi.org/10.1017/S0020818300032926
Novelli, M., Morgan, N., & Nibigira, C. (2012). Tourism in a post-conflict situation of fragility. *Annals of Tourism Research*, *39*(3), 1446–1469. https://doi.org/10.1016/j.annals.2012.03.003
Onwuegbuzie, A. J., & Collins, K. M. (2007). A typology of mixed methods sampling designs in social science research. *Qualitative Report*, *12*(2), 281–316.
Pape, R. A. (1997). Why economic sanctions do not work. *International Security*, *22*(2), 90–136. https://doi.org/10.1162/isec.22.2.90
Patton, M. Q. (2002). Qualitative research and evaluation methods. Sage Publications.
Peksen, D. (2009). Better or worse? The effect of economic sanctions on human rights. *Journal of Peace Research*, *46*(1), 59–77. https://doi.org/10.1177/0022343308098404
Perks, R. (2013). Digging into the past: Critical reflections on Rwanda's pursuit for a domestic mineral economy. *Journal of Eastern African Studies*, *7*(4), 732–750. https://doi.org/10.1080/17531055.2013.841025
Pirie, G. H. (1990). Aviation, apartheid and sanctions: Air transport to and from South Africa, 1945–1989. *GeoJournal*, *22*(3), 231–240. https://doi.org/10.1007/BF00711334
Pleitgen, F. (2015). Why your next vacation could be in Iran. *CNN travel*. ttps://edition.cnn.com/travel/article/iran-tourism-persepolis/index.html
Pratt, S., & Alizadeh, V. (2018). The economic impact of the lifting of sanctions on tourism in Iran: A computable general equilibrium analysis. *Current Issues in Tourism*, *21*(11), 1221–1238. https://doi.org/10.1080/13683500.2017.1307329
Pratt, S., & Liu, A. (2016). Does tourism really lead to peace? A global view. *International Journal of Tourism Research*, *18*(1), 82–90. https://doi.org/10.1002/jtr.2035
Rossignol, M. (1996). *Sanctions, the economic weapon in the new world order* (Rev. edn.). Library of Parliament BP-346E. Government of Canada.
Rowen, I. (2014). Tourism as a territorial strategy: The case of China and Taiwan. *Annals of Tourism Research*, *46*, 62–74. https://doi.org/10.1016/j.annals.2014.02.006
Rowley, J. (2012). Conducting research interviews. *Management Research Review*, *35* (3/4), 260–271. https://doi.org/10.1108/01409171211210154
Rustler, A. (2019). Are sanctions stopping us from achieving global climate goals? *Journal of International Affairs*. https://jia.sipa.columbia.edu/online-articles/are-sanctions-stopping-us-achieving-global-climate-goals

Saad, A. (2019). Challenging the sponsor-proxy model: The Iran–Hizbullah relationship. *Global Discourse, 9*(4), 627–650. https://doi.org/10.1332/204378919X15718898344883

Salazar, N. B. (2006). Building a 'culture of peace' through tourism: Reflexive and analytical notes and queries. *Universitas Humanística, 62*, 319–336.

Seyfi, S., & Hall, C. M. (2018). Tourism in Iran: An introduction. In S. Seyfi & C. M. Hall (Eds.), *Tourism in Iran: Challenges, development and issues* (pp. 3–37). Routledge.

Seyfi, S., & Hall, C. M. (2019a). Sanctions and tourism: Effects, complexities and research. *Tourism Geographies*, 1–19. https://doi.org/10.1080/14616688.2019.1663911

Seyfi, S., & Hall, C. M. (2019b). International sanctions, tourism destinations and resistive economy. *Journal of Policy Research in Tourism, Leisure and Events, 11*(1), 159–169. https://doi.org/10.1080/19407963.2018.1482305

Seyfi, S., & Hall, C. M. (2019c). Sanctions and tourism: Conceptualisation and implications for destination marketing and management. *Journal of Destination Marketing & Management, 15*, 100381. https://doi.org/10.1016/j.jdmm.2019.100381

Seyfi, S., & Hall, C. M. (2020). *Tourism, sanctions and boycotts*. Routledge.

Seyfi, S., Hall, C. M., & Fagnoni, E. (2019). Managing world heritage site stakeholders: A grounded theory paradigm model approach. *Journal of Heritage Tourism, 14*(4), 308–324. https://doi.org/10.1080/1743873X.2018.1527340

Seyfi, S., Hall, C. M., & Kuhzady, S. (2019). Tourism and hospitality research on Iran: Current state and perspectives. *Tourism Geographies, 21*(1), 143–162. https://doi.org/10.1080/14616688.2018.1454506

Shahidian, H. (2002). *Women in Iran: Gender politics in the Islamic Republic*. Greenwood.

Smeets, M. (2019). The theory and practice of economic sanctions. In S. Sutyrin, O. Y. Trofimenko, & A. Koval (Eds.), *Russian trade policy: Achievements, challenges and prospects* (pp. 66–78). Routledge.

Tahmasebi, S. (2018). *How US sanctions impede the women's movement in Iran*. https://www.atlanticcouncil.org/blogs/iransource/how-us-sanctions-impede-the-women-s-movement-in-iran/

Takeyh, R., & Maloney, S. (2011). The self-limiting success of Iran sanctions. *International Affairs, 87*(6), 1297–1312. https://doi.org/10.1111/j.1468-2346.2011.01037.x

Takian, A., Raoofi, A., & Kazempour-Ardebili, S. (2020). COVID-19 battle during the toughest sanctions against Iran. *Lancet (London, England), 395*(10229), 1035–1036. https://doi.org/10.1016/S0140-6736(20)30668-1

Talja, S. (1999). Analyzing qualitative interview data: The discourse analytic method. *Library & Information Science Research, 21*(4), 459–477. https://doi.org/10.1016/S0740-8188(99)00024-9

Timothy, D. J. (2017). Tourism and geopolitics in the GCC region. In M.L. Stephenson & A.al-Hamarneh (Eds), *International tourism development and the gulf cooperation council states* (pp. 45–60). Routledge.

Tucker, H., & Boonabaana, B. (2012). A critical analysis of tourism, gender and poverty reduction. *Journal of Sustainable Tourism, 20*(3), 437–455. https://doi.org/10.1080/09669582.2011.622769

UN. (2015). *Transforming our world: The 2030 agenda for sustainable development*. https://sustainabledevelopment.un.org/post2015/transformingourworld

UN Women. (2020). *SDG 5: Achieve gender equality and empower all women and girls*. https://www.unwomen.org/en/news/in-focus/women-and-the-sdgs/sdg-5-gender-equality.

United Nation Security Council (2019). Sanctions. Retrieved from https://www.un.org/securitycouncil/sanctions/information

United Nations Security Council. (2020). *Sanctions*. https://www.un.org/securitycouncil/sanctions/information

UNWTO. (2018). *Tourism in the 2030 agenda*. https://www.unwto.org/tourism-in-2030-agenda

UNWTO. (2019). *UNWTO tourism highlights, 2016 edition*. UNWTO. http://mkt.unwto.org/publication/unwto-tourism-highlights-2016-edition

Warrick, J. (2009). Joby Warrick interviews Mohamed ElBaradei on Iran's nuclear ambitions. *The Washington Post*. https://www.washingtonpost.com/wp-dyn/content/article/2009/12/04/AR2009120402607.html

Weaver, D. B. (2010). Geopolitical dimensions of sustainable tourism. *Tourism Recreation Research, 35*(1), 47–53. https://doi.org/10.1080/02508281.2010.11081618

Winter, T. (2008). Post-conflict heritage and tourism in Cambodia: The burden of Angkor. *International Journal of Heritage Studies, 14*(6), 524–539. https://doi.org/10.1080/13527250802503274

WTTC. (2016). *Tourism as a driver of peace*. https://sp.wttc.org/-/media/files/reports/special-and-periodic-reports/tourism-as-a-driver-of-peace_report-summary—web.pdf?la = es

Insurgent citizens: mobility (in)justice and international travel

Pooneh Torabian and Heather Mair

ABSTRACT

Goal 16 of the Sustainable Development Goals (SGDs) emphasizes the importance of providing just and inclusive societies for all and target 16.10 underscores the importance of 'protect[ing] fundamental freedoms, in accordance with national legislation and international agreements'. The United Nations (UN) and World Tourism Organization (WTO) frequently describe tourism as a phenomenon that promotes peace and dialogue between different cultures. Nonetheless, the UN rarely sees tourism as political. Despite the ongoing struggles of the members of marginalized communities throughout the world, social justice in the context of international travel remains unattainable for some and neoliberal understandings of freedom of movement and right to travel replicate and contribute to the existing systems of injustice and inequality. Guided by a critical mobilities framework, the aim of this research is to explore how dual citizenship shapes the border crossing experiences of Canadians who have travelled internationally in the post 9/11 era. To that end, we had in-depth interviews with 11 Canadian dual citizens to learn how citizenship shape their border crossing experiences. Findings revealed that some dual citizens are treated as insurgent citizens as they are legally citizens but are not treated as such and are deprived of their rights. Moreover, race plays a crucial role in how dual citizens are hierarchically categorized and international mobilities of some citizens are favoured over the others.

Introduction

Article 13 of the Universal Declaration of Human Rights (United Nations, 1948) states that: (1) Everyone has the right to freedom of movement and residence within the borders of each state. (2) Everyone has the right to leave any country, including his own, and to return to his country. The Manila Declaration on World Tourism (1980) has emphasized access and freedom to travel and tourism as a societal and human right (Edgell, 1990) and gave human rights a new meaning by making it universal (Teeple, 2005). However, there is no universally accepted definition of the term human rights or the word freedom to date. Human rights can only be meaningful if exerted by states and if tourism is known to be a social need and a globalized right, then the idea that people should travel and be tourists must be at the core of what it is mean to be a citizen in today's world (Bianchi & Stephenson, 2013). Goal 16 of Sustainable Development Goals (SGDs) focuses on providing just and inclusive societies for all and target 16.10 underscores the importance of 'protect[ing] fundamental freedoms, in accordance with national legislation and international agreements' (United Nations Department of Economic and Social Affairs, 2015, p. 28). Critical thinking, which is essential for exploring dynamics of power, is explicitly absent from the United Nations SDGs (Boluk et al., 2019). Critical thinking is vital in revealing the social, economic, and environmental injustices that are present and perpetuated by and within the tourism industry. Thus, without employing a critical lens that identifies impacts of power, inequality, privilege, and poverty in tourism, sustainability issues cannot be resolved (Boluk et al., 2019).

In the 1960s, the United Nations described tourism as a phenomenon that promotes peace and dialogue between different cultures (Tesfahuney & Schough, 2016) and the World Tourism Organisation portrays tourism the same way today. The United Nations rarely views tourism as political. Nonetheless, international travel is increasingly becoming 'one among many benchmarks of a "civilized life"' (Bianchi & Stephenson, 2014, p. 11) and is seen as a marker of privilege and power (Castells, 1996), as well as citizenship (Bianchi & Stephenson, 2014; Coles, 2008; Urry, 1995, 2000). Freedom of movement and the right to travel are still far from being globally recognized and access to international travel is seen as a symbol of privileged status and citizenship rights (Adler, 1985; Bianchi & Stephenson, 2013, 2014). Although SDGs are meant to '[...] foster [...] an ethic of global citizenship' (United Nations Department of Economic and Social Affairs, 2015, p. 11) and despite the fact that more people travel freely in the modern era compared to the past, tourism is not recognised as a citizen's right until our freedom to travel is jeopardized (Rojek, 1998).

Although it might seem that dual citizenship would lead to the ease of access when it comes to international travel, it is a form of hierarchical citizenship, from which only some can benefit (Stasiulis & Ross, 2006). Moreover, events like the September 11 attacks and more recently COVID-19 have highly affected international travel and tourism practices worldwide. Travelling across or even within borders has become an unpleasant experience for members of minority groups as they may be constantly at risk of facing discriminatory actions in the form of xenophobia, racism, and more recently Islamophobia (Stephenson, 2006; Stephenson & Ali, 2010). Despite the ongoing struggles of members of marginalized communities throughout the world, social justice in the context of international travel remains unattainable for some diverse groups and populations and access to tourism is not universally seen as a right. To that end, insurgent citizens are legally citizens but are not treated as such and are deprived of their citizenship rights (Cresswell, 2013). In this article, we discuss how dual citizens might be considered as insurgent citizens when crossing international borders. Indeed, as Higgins-Desbiolles (2007) contended, the right to travel is 'an ideological act which is predicated on a system of inequity' (p. 324).

Guided by a critical mobilities framework, the purpose of this research is to explore how dual citizenship shapes the border crossing experiences of Canadians who have travelled internationally in the post 9/11 era. Critical mobilities accentuates mobilities as a right while highlighting the interplay of (im)mobilities and inequalities. In this research, we challenge the idea of freedom of movement as a human right and illustrate the inequalities in terms of accessing international mobilities. Further, we illustrate how international travel and bordering practices can act as highly political indicators of social justice and that tourism is an 'idealized commodity' that only some can access (Hall, 2010, p. 34). Furthermore, we challenge SDG 16 in terms of growing inequalities that target certain citizens and portrays them as insurgent citizens that are stripped of their citizenship rights.

Literature review

Tourism mobilities: a right for everyone?

Although for some, taking part in tourism means participation in leisurely activities, a number of tourism scholars have a different view; they see tourism as a social force that can have transformative power (*cf.* Ateljevic et al., 2007; Higgins-Desbiolles, 2006, 2008; Minnaert, 2014;

Minnaert et al., 2006, 2009; Pritchard et al., 2011). However, tourism is still generally viewed 'as an "industry" [which] has overshadowed other conceptualisations of the tourism phenomenon' (Higgins-Desbiolles, 2006, p. 1192). Therefore, the debate about *who* can travel is underdeveloped in tourism scholarship because the issue of access in tourism is frequently perceived as a market issue and is less frequently situated in a socio-political context.

Mobility is 'at the heart of our understanding of tourism' (Hall, 2005, p. 134) and scholars are coming to argue that limiting international movement leads to sustaining inequalities across the globe. Justice tourism is not only seeking to address the inequalities of contemporary tourism, but it also delineates a path towards a more just global order (Higgins-Desbiolles, 2008). If international travel is a right that all people should have access to, it first needs to be separated from the neoliberal notion of freedom and unequal flows of mobility (Bianchi & Stephenson, 2013). Neoliberalism is frequently 'discussed as an economic doctrine with a negative relation to state power, a market ideology that seeks to limit the scope and activity of governing' whereas, it can also be viewed as 'as a new relationship between government and knowledge through which governing activities are recast as non-political and non-ideological problems that need technical solutions' (Ong, 2006, p. 3).

The movement, or temporary mobility, of transnational actors can be a means by which tourism can be studied (Hall & Tucker, 2004). Globalization and advances in technologies have enabled the movement of capital, labour, and resources transnationally (Meethan, 2001). Thus, the landscape of tourism has changed significantly. Despite the importance of access to international mobilities in tourism, with some notable exceptions (*cf.* Bianchi & Stephenson, 2013, 2014; Stasiulis & Ross, 2006; Stephenson, 2006; Stephenson & Ali, 2010; Torabian & Mair, 2017), the politics of international movement and citizenship are understudied in tourism studies. Tourism scholarship benefits from what has been described as the 'mobilities turn' or paradigm as it provides an opportunity to challenge the predominant ways of knowing and thinking about tourism (Coles & Timothy, 2004; Cresswell, 2006; Hall & Tucker, 2004; Sheller & Urry, 2006). Tourism needs to be understood as a process within the wider ontological context of mobilities (Hannam, 2009).

The mobilities paradigm does not differentiate between places and people; instead, it recognizes the materialities of mobilities and re-centres the human body through which we make sense of places and movements (Bondi et al., 2005; Hannam, 2009). When it comes to tourism mobilities, nation states have undeniable impacts on who, when, and for what reason individuals are able to cross international borders freely. Nation states regulate their cross-border flows by controlling tourism mobilities. Today borders are boundaries of inclusion as much as they can be boundaries of exclusion (Sofield, 2006). 'Tourism is used as a metaphor for conveying aspects of citizenship to public audiences and, in turn, has become a mechanism by which to regulate further short-term movements' (Coles, 2008, p. 69). This has become even more evident during the COVID-19 pandemic where national borders have closed, and people of Asian descent are reporting discrimination in travel (Teixeira da Silva, 2020; Yu & Keralis, 2020).

Dominant representations of mobility portray it as liberty or progress (Cresswell, 2006). However, a global mobility regime is emerging which is '[...] oriented to closure and to the blocking of access' (Shamir, 2005, p. 199). Of special concern to critical mobilities is how mobilities are made possible and how they produce inequalities. Critical mobilities develops a foundation for social theory that puts mobilities, immobilities, and their power relations at its centre; it offers 'a research agenda around the study of various complex systems, assemblages, regimes and practices of (im)mobilities; and at times involves a normative emphasis of addressing the future of mobility in relation to ecological sustainability and mobility justice' (Sheller, 2013, p. 2). As Bianchi and Stephenson (2019) argued, the right to tourism cannot be rightfully attained when the international mobilities of many citizens is 'deemed problematic or "illegitimate"' (p. 133).

Citizenship and international travel

The notion of citizenship contains considerations of management and the negotiation of mobilities (Hindess, 2002; Walters, 2006a, 2006b). Several scholars (see for example Agamben, 2005; Bauman, 1998; Salter, 2004, 2006; Sparke, 2006; Walters, 2006b) have discussed how differences are highlighted at border crossings and how citizenship rights might be accepted, declined, or even scrutinized. Globalization has given new meaning to citizenship and extended its meaning beyond the nation state (Bianchi & Stephenson, 2013). Turner (1993) called for a redefinition of citizenship in response to globalization and what he viewed as the decline of nation states. Citizenship is both a process and a status, which is constantly redefined and debated over time in the context of ever changing political, historical, and social situations and it is not solely a sociological concept nor a legal concept but is defined by the relationship between these two (Isin & Wood, 1999).

Citizenship is becoming a globalizing regime that regulates bodies and their mobilities (Rygiel, 2010). Foucault (1979) invited us to consider disciplinary power and to contemplate how citizenship is dependant on different forms of power at the micro-level, such as, at the level of the body. He called for asking how these forms of power affect macro-level forms of power. Rather than viewing citizens and non-citizens based on their legal status and having established identities, citizenship should be viewed as performative (Rygiel, 2010). Therefore, although the more conventional view of citizenship associates it with rights and responsibilities, this perspective implies that the power of citizenship lies within individual's agency and the 'potential to move individuals to act and to disrupt the order of things' (p. 41). Elaborating on citizenship and rights, Ranciere (2004) mentioned that the moment individuals are detained or denied entry because they are seen as a threat is also the moment that they undermine the power of the established system and practice of citizenship by claiming rights that they do not actually have.

The rise of the citizen as a national figure was associated with the emergence of the tourist, the wanderer and still today 'the citizen continues to be twinned with threatening other' (Cresswell, 2013, p. 88). Therefore, there is a binary of citizen and other within the notion of othering. Tourism mobilities are regulated by a global regime that is devoted to tourism's continuous growth and expansion, while other mobilities, deemed to be less desirable, are halted, regulated, and controlled (Bianchi & Stephenson, 2019). While citizenship seeks to regulate and police certain bodies:

> [I]t constructs some identities as being authentic citizen subjects and others as being non-citizen subjects. Thus, in developing a view of citizenship as government, a key area upon which to reflect is the emerging citizenship identities and political subjectivities that are made possible by and through a globalizing regime of governing mobility. (Rygiel, 2010, p. 37)

Borrowing the concept of 'insurgent citizen' from James Holston (1995, 2007, 2008, 2009), Cresswell (2013) argued that insurgent citizens are legally citizens but are not treated as such and are deprived of their rights. The term 'insurgent citizenship' first emerged in the 1990s. Insurgent citizenship accentuates the spatial dimension of citizenship movements as well as citizens' everyday practices in urban spaces (Lebuhn, 2019). Although the concept has spatial dimensions and is initially used to discuss the inequality and marginalization of citizens in urban areas, it denotes that 'citizenship everywhere is defined by formulations of equality and inequality, as even most democracies experience conflict over the terms of national membership and the distribution of rights' (Hagopian, 2011, p. 219). For the purpose of this study, we use the notion of insurgent citizenship to draw attention to the inequalities of dual citizenship.

Several scholars have emphasized the complexity of their multiple citizenship status when they travel internationally. Although it might seem that having multiple citizenships would lead to the enhancement of rights, Stasiulis and Ross (2006) also challenged the liberal view of dual citizenship as always expanding rights and suggested dual citizenship is a form of hierarchical citizenship that only benefits some. For instance, Khosravi (2007) indicated the complexity of his own citizenship status as a Middle Eastern person who is also an EU citizen, 'My legal status as an EU citizen is situational, conditional and unconfirmed. I am a quasi-citizen whose rights can be suspended in the state of emergency. I am included and at the same time excluded" (p. 332). Similarly, Rahimieh (2007) pointed out the complexity of holding multiple citizenship and crossing international borders after 9/11: "When new Homeland Security measures were adopted and when Iran's newly acquired status as part of the "axis of evil" coalesced, I found myself among the ranks of individuals of Middle Eastern ancestry who were routinely stopped and interrogated on the border between Canada and United States' (p. 229). The discriminatory practices at borders are reinforced by certain categorization of citizens' mobilities (Bianchi et al., 2020). Following Foucault (1979, 1988) and Rygiel (2010), in this research, we consider citizenship as a concept that is associated with practices, technologies, policies, discourses, forms of power utilized to govern the mobilities of individuals and groups of people. In the next part, we discuss the Foucauldian notion of biopower and bodies at borders.

Biopower and biometric borders

Border security uses the body, or information about the body, to make bodies known (See Adey, 2009; Daase & Kessler, 2007; O'Malley, 2004). The assignation of risk profiles to bodies at borders occurs based on different fragments of identity such as race, gender, class, religion, etc. Exclusion is fundamental to bordering and together with risk management and biometric technologies, generates strategies of 'pre-emptive immobilization' (Wilson & Weber, 2008, p. 127; see also Adey, 2009) that target certain categories of travellers. As Cresswell (2013) argued, the concept of freedom of movement presumes that a normal, abled body has the means and access to the resources to be freely mobile. As he further contended, mobility of the citizen (i.e. wanted mobility) is constructed in relation to the (im)mobilities that are other (i.e. unwanted mobility).

Border crossing is a 'racialised ontological practice' (Bhandar, 2008, p. 405) through which individuals are subject to classification and interrogation based on their ontological status. Racialised bodies can weaken the national identity and make the bodies known by their racialised difference (Bhandar, 2008). To further clarify how racial hierarchies are connected to power, Foucault is helpful:

> The discourse of race struggle will be re-centered and will become the discourse of power itself. It will become the discourse of a centered, centralized, and centralizing power. It will become the discourse of a battle that has to be waged not between races, but by a race that is portrayed as the one true race, the race that holds power and is entitled to define the norm, and against those who deviate from that norm, against those who pose a threat to the biological heritage. (Foucault, 2003, p. 61)

In addition to racialised experience at borders, gender is constantly negotiated and performed and intersects with marginalization and the act of gathering information at borders. Considering gender dynamics helps us come to terms with other forms of power and exclusion at border crossings. Disciplinary power and control are not only associated with race but with the intersections of race and gender (Koskela, 2012).

Borders seem to be conceptual lines on a map or structures that appeared as an effort to access the resources in colonized lands, but more than anything they are lines written on human bodies (Chattapodhyay, 2018). On the same note, Muller (2010, p. 95) denoted:

> Unlike conventional notions of citizenship that evolve over time and rely on dynamic notions of rights and responsibilities over time, the biometric body never decays, changes or alters but simply lingers in the database, like some boxed cyborg citizen. The risk assessment is ubiquitous, and the border is (in) your body.

Epistemology and methodology

Critical constructionism

This research is positioned within the critical constructionism paradigm. Critiques of power and/ or discourse(s) are at the heart of critical constructionism (Griffin, 2018). Critical constructionism views power as an ongoing, relational construction and does not assume that power is 'one way and uncontested' (Hosking, 2008, p. 670). Critical constructionism invites reflexive recognition of its own constructive potential and participation in power relations (Foucault, 1979). Critical constructionists acknowledge that social structures can shape opportunities and these structures are reproduced or resisted in situations (Pennell & Maher, 2015). Using this paradigm, we have been participants along with those who chose to take part in this research project (Lincoln et al., 2011). We strived to centre the voices of participants along with our own voice. In addition, we adopted a post-structural lens to approach notions of identity and power in this research. Therefore, we considered identity as fluid and something that can be negotiated. We examined power as discursive and considered the dynamics of power that existed among individuals and structures.

Narrative inquiry

The methodological approach underpinning this research is narrative inquiry, which adopts narrative as both the methodology and the phenomena of the study (Pinnegar & Daynes, 2007), and focuses on the uniqueness of each individual as well as finding commonality between them (Polkinghorne, 1995).

The purpose of narrative inquiry is not only to explore the experience of individuals but also to search for the familiar, linguistic, social, cultural, and institutional narratives through which individual experiences are shaped, articulated, and depicted (Clandinin, 2013). Narrative inquiry is not about reconstructing the Truth about the reality of the experiences of the participants; rather it is about understanding how individuals talk about their experiences and conceptualize them (Josselson & Lieblich, 2001). Therefore, the stories participants shared reflect their own reality and even when those narratives are co-created, they are partial. Narrative shifts the emphasis of research away from the validity of the stories that are told and focuses on how the stories impact the narrator, society, and audience (Daly, 2007). People's stories can be constructed and interpreted in many ways to make sense of their reality (Daly, 2007). Stories can be told in a variety of ways and narratives are co-constructed in conversations (Ochs & Capps, 2001). As narrative researchers, we 'highlight the versions of self, reality, and experience that the storyteller produces through telling' (Chase, 2005, p. 657).

Tourism scholars have used narrative inquiry approaches to reveal how identities are formed and how information is socially constructed about destinations and culture (see for example, Bosangit et al., 2009; Chow-White, 2006; Griffin, 2013; Noy, 2004; Noy & Cohen, 2012; Santos et al., 2008). As Adey (2017) claimed, 'Narratives are essential to methods for mobilities' (p. 280). We employed narrative in this study to illustrate the complexity of the border crossing experiences for dual citizens. In so doing, a critical mobilities framework enabled us to critically examine how narratives of mobilities are produced, performed, resisted, or transformed in relation to the multilayered identities of participants.

Narrative provided the participants with an opportunity to tell their own stories and be at the centre of this work. It has been crucial to make sure we do not speak for the participants. Therefore, as a researchers, we strived to ensure that our research objectives and epistemological stance did not silence the participants (Clandinin & Connelly, 2000). In so doing, we engaged with conversational narrative to centre the voices of the participants. Narratives were co-constructed between us and participants in conversations that were audio recorded.

Method and analysis

The first author recruited participants through Facebook using purposeful sampling (Patton, 2002) as we wanted to recruit participants who met certain criteria. First, they needed to be

Canadian dual citizens and second, they had to have travelled internationally in the post-9/11 travel era. In total, 11 participants participated in this research. Four of the 11 participants were men and seven were women. They were all Canadian dual citizens at the time of interview and had travelled internationally as a dual citizen post 9/11. Participants were citizens of the following countries: United States, Trinidad, the Philippines, Iran, the United Kingdom, Egypt, Netherlands, and Israel. In order to explore how dual citizenship shapes the border crossing experiences of Canadians who have travelled internationally in the post 9/11 era, we conducted in-depth, unstructured interviews with participants.

Regardless of the method used, researchers bring their interpretation and preconceptions to the phenomenon they study (Patton, 2002). Since beginning to think about this research, the first author had engaged in reflexive journaling. Engaging in reflexivity helped us with finding our voices and positionality in this research. The first author is an Iranian – Canadian woman and the second author is a Settler Canadian of Scottish heritage. Moreover, keeping a reflexive journal helped with re-storying narratives as some details were not necessarily present in the audio-recorded files of the interviews.

To analyse the interviews, following Polkinghorne's (1995) processes of narrative analysis and analysis of narratives, we engaged in the re-storying process described by Clandinin and Connelly (2000) as a collaborative process of renegotiating or retelling the story between the researchers and the participants of the study. Re-storying narratives is a methodology that employs narrative analysis as well as analysis of narratives (McCormack, 2004). Re-storying narratives helped us better understand the similarities and the differences among participants' experiences while also attending to their complexity.

Findings

Race holding power

Conversations with participants revealed that their bodies and physical features are crucial to border crossing experiences. Sam is an Iranian - Canadian man. Although he enjoys the advantages of being a dual citizen and his Canadian passport allows him to travel more freely, he notes that:

> They don't really accept you in their country and in their community. So, [at] the border, it is the same scenario. Like you may have one officer whose parents are not U.S. Citizens and one that is Native American. They might have different perspective and now they must deal with the person who wants to cross the border. So, for example, if I go to an officer who has black hair, I will be less nervous because I think like okay he can understand. I mean because they have no idea that who could be a terrorist but the point is you can't judge someone upon their appearance. But they don't have that in their mind and I think when they see you as an Iranian, that comes to their minds you know.

He further mentions Middle Eastern men have far more difficulties crossing borders because of both their gender and the colour of their skin:

> I think for men it is worse. Because the ones who are involved in terrorist attacks are men with black hair. So… there are more questions about the men than women and from my experiences with my wife crossing the borders, it's easier for her. She has a Canadian passport but doesn't have any of the questionable features… And I definitely expect to answer more questions than her.

When crossing international borders, Sam is treated as an Iranian citizen rather than Canadian: 'Well, I mean the black hair is one thing, and the Middle Eastern man, um, like the colour of your skin, that's another thing. I mean all the features that can distinguish you from Canadians at the border'. As he explains, Sam's racial and gendered difference strengthens his Iranian citizenship over his Canadian citizenship:

As a non-white Muslim woman, Lutfiya feels that each time she crossed the Israeli border, she was not treated as a Canadian or American citizen:

> [The border officer] immediately points to a room in the corner that she knows I'm familiar with because every time I've come through there, that's what I've had to go through! [. . .] she didn't look at my name, she didn't look at my passport. She looked at me. I would've had better luck if I was wearing my scarf [loosely] but that time I wasn't, it was more traditional. More easily identifiable Muslim, and I'm obviously not Israeli so she sent me immediately. I remember putting my passport a little bit in her face and she looked at it and she's like, 'I told you!' But it doesn't matter, right? In their racialized ethnic whatever logic stratification'.

The special room Lutfiya refers to is an interrogation room where she has been kept for hours each time she crossed the Israeli border.

Anne is a Dutch-Canadian woman who shared her border crossing experience in Germany:

> They were checking passports and tickets just randomly at a checkpoint. It was just the Germans and German officials. Some people in front of me were other ethnic backgrounds and . . . . I interrupt her by asking what she means by the ethnic background and she replies: I think she was, the lady directly in front of me had dark hair and she had darker skin. I think she might have been Indian, but I don't know. And her English wasn't so good. And he was asking her questions and stuff. And then he's like, 'Okay, you can go'. And then I came and I have my passport. And I was like, 'Do you need to see this?' And he's like, 'No, I just need to look at your eyes. Thank you. Go'. And then I was like, 'Okay'. That's why I'm like, 'See? I'm just not a threat! [. . .]He was, um, from my perspective, he was implying that like my eyes looked truthful and that he understood me in some way.

Anne believes 'Anyone who is not coming or is not going from a Western country to a Western country. I mean, if their ethnic background is not from Western countries'.

Jehan is an Iranian – Canadian man and believes 'the colour of your skin plays an important role at borders regardless of the passport you hold'. However, he never had issues crossing borders. Jehan and his wife first travelled to the United States with their Canadian passport to attend a conference on human rights:

> Because of the theme of the conference, the officer went easy on us and I think because this has been reflected on our profiles, we were never questioned, stopped, or interviewed at the US border. Therefore, we have never been worried when crossing the US-Canada border.

John, a Canadian – British citizen, shares his border crossing experience from Vancouver to Mexico with his brother and a friend who held British passport and was of Iraqi descent:

> [W]hen we came to Canadian-American border, the three of us got out of the car and walked through, went through the whole process and they asked me and my brother what is your name and show your passport and all, and that's it! And this friend of ours, even though he has a British passport, was stopped and you know questioned because of his name. His name was very Arab, and it rings bells or whatever! [W]e were hanging out there for a good half an hour and waiting for him to come through and of course there was nothing wrong and it was just because of his name. We didn't have any other problems.

John explains that his brother's friend was 'perceived as a risk although he was a British citizen'. John himself was never stopped at the borders to be further questioned or interrogated because, in his words, '[he] was sort of the white guy! And was fine to go'.

Insurgent citizens

One of the issues discussed by participants was that although they were travelling with their Canadian passports, they were not treated as Canadian citizens at border crossings. Rosa is an Iranian-Canadian woman who had travelled frequently to the United States before 9/11 and used to feel welcomed at the Canada–US border. The first time she wanted to cross the Canada–US border two months after 9/11 attacks, she was stopped for more than three hours with no explanation. This is how she recalled the experience:

> I remember that it was at least five or six other couples. We were the only family . . . I mean, with the kids. But it was different cultural backgrounds, because I can see that they are talking [in] their own language. Basically, it was all non-, hmmm, I mean, non-white people, or non-American people! Yes! People not from

> the North America. One of the girls [officers] that's sitting back there, she didn't stand over or anything just said 'We are processing you!' I asked 'What kind of processing?' and she quickly replied 'Oh we are sending documents back and forth with the government.' [...] and All this time she was sitting in the back, didn't stand up or come forward. Just raising her voice for me to hear. It was after an hour or so that they approached us and gave our passports back. They said, 'Oh, you're okay to go, but you have to report and say we are going home now when you come back.' I did not feel welcomed there at all!

Lutfiya is an American–Canadian woman. She has a Palestinian – Kashmiri heritage. and had travelled to Palestine a number of times to visit her grandmother. To do so she had to cross the Israeli border first and each time was detained at the border. Lutfiya's detention at the Israeli border despite her American – Canadian citizenship illustrates the complexity of her dual citizenship status. Lutfiya believes she is neither perceived as an American nor a Canadian at the Israeli border:

> I was so excited [to travel to Palestine]. I didn't think that anything could stand in the way. And I really was invested in the power of my passport! I'm very naïve! I really thought at the end of the day, I have an American passport, and Americans do not need a visa to enter Israel. You get it there. There's no reason that they can deny me entry! They cannot do it. It's just not possible. It's not in the realm of possibilities. I still have this naivety. I thought that my passport was the end of the story.

When Lutfiya talks about her experience of detention she notes, '[I was] treated differently than "ordinary" American citizens because of my Palestinian heritage, religion, and political activism'. Lutfiya was 'very invested' in the power of her American citizenship to enter Israel but the last time she wanted to visit Palestine she got detained and deported because '[Israeli] military identified [her] as a security threat'. Based on her documentation, she has the 'right' as an American or Canadian citizen to enter Israel, but she could not benefit from either of her passports.

When describing what her Canadian citizenship means to her, Lutfiya says she believes that the notion of Canadian citizenship is still associated with whiteness, therefore she is a not a 'real' Canadian. What Lutfiya mentioned in terms of hierarchy of Canadian citizenship is also evident in how Sophia pictured her Canadian citizenship, '[I]n Canada, I am white enough that people don't know that I'm an immigrant most of the time'. Therefore, she can benefit from the privilege of being a white Canadian citizen'. Sophia is an Israeli-Canadian woman. Like Sam, Sophia enjoys the advantages her Canadian passport brings to her, 'I have an Israeli and a Canadian passport. I'm at the top of the totem pole of rocks'. However, she does not see herself as a Canadian citizen. Instead, she views herself as an immigrant of a 'far lower economic class' that has access to fewer opportunities.

(Un)wanted mobilities

Some participants indicated that dual citizenship and possessing two passports eased access to international travel and provided them with the opportunity to be freely mobile. However, not all participants could benefit from their dual citizenship status since their mobilities were perceived to be unwanted and undesired.

Despite holding a Canadian passport, Rosa was stopped, racially profiled, and treated poorly when she crossed the Canadian border to the United States after 9/11, 'It was just few more questions to not categorize us in a category, a category that we don't belong to!' Rosa could benefit from her Canadian passport only to some extent; to the point that her trips to the United States were wanted mobilities prior to 9/11. Lutfiya's detention is another example of unwanted mobilities where her dual citizenship status could not grant her access to visiting Palestine.

Sophia refers to her experience of crossing checkpoints from Israel to Palestine:

> And there's so many cameras everywhere. And like this soldier yells at you through a megaphone, and you barely understand what she's saying. And she yells in Hebrew, right? So, if you're Palestinian, you don't speak Hebrew, it's your own fucking fault. They don't care. So, she yells at you in Hebrew, and you're like, 'What?' And she yells, 'Show me your visa!' So, you put your passport to the window. And then if she likes it or she doesn't like it, you either spend the rest of the day being interrogated, or you go to jail, or you go back home and you have to turn around, or you get through. So, it's so nerve wracking. Even though I have the right passport. I have an Israeli passport. So, the worst thing that can happen to me is that I'll be interrogated and potentially like reprimanded. If they really want to make a showcase out of me, they would put me in jail, but chances are that won't happen. If I'm Palestinian, I could be dead as a result of this interaction.

Sophia's description speaks to the inequalities that exist at the Israeli borders and how Palestinians and Israelis experience border crossing differently because of this.

Damian is a Trinidadian – Canadian man. For Damian, his Canadian passport, brings him the 'ability' to access the freedom of movement, 'When I'm traveling in most cases I am a Canadian, because the passport gets me, plain and simple, the ability to freely move and you are viewed differently just being a Canadian'. Nonetheless, Damian acknowledges that his Trinidadian heritage might sometimes come first when he wants to cross borders especially at the Canada-US border where he is usually interrogated. He further explains how the interrogation makes him feel:

> I understand again the ridiculousness… and I also understand the plight of the people at the front line because I've worked in an industry, a front-line industry, at the end of the day they are doing their job. Some might be worse than the others, you just hope that it doesn't happen to you, but they may also need to do it because they have hard data saying that okay… from certain countries…yes… it is sad because not everybody from certain countries are bad human beings, but It's a blanket for all! You know what I mean? If somebody is from a certain country, we have to do something, randomly checking or something. You know what I mean?

Anne shares her experience of crossing the border from Romania to Moldova and explained how her experience was different from her Romanian friends:

> When we were getting to the border, the Romanians had a hard time, they were like….the border people didn't so much like it that the Romanians were coming in! […] I give them my Dutch passport and they're like, 'Oh yeah, come on in'. Like, 'Come on in, no problem'. And the Romanians are like, 'How do they like you and they don't like us? They don't like us because we're regular people. They like you because you're a Western European […]whereas we're just nobodies. We're just Romanians.

Anne's experience of crossing the border from Romania to Moldova also demonstrates the hierarchy of citizenship that exists among the citizens of different countries in the European Union.

Maya, a Canadian–Egyptian woman, believes she has never had any negative experiences crossing international borders and mentions she only has 'one significant story'. Maya's family and two other family friends decided to cross the border to the States over one weekend to go shopping and come back to Canada:

> [the border officers] asked where are you going? So, we said we are going shopping because really we didn't have any plans. And then they asked how long are you going to stay and we said for few hours. […] So, we crossed the border and then we decided to…[pause] there was kind of a park there […] and my kids were very young and we thought that they can have fun. And it was the opposite direction from an from the shopping center. So, we went to this place and we were just basically having fun! And then we realized police officers were coming and they stopped. You couldn't really say they were talking in a bad way or there was something unpleasant, but it was just an awkward feeling! Like 'WHAT?!' and my husband said what's the problem of deciding to just stop and have a little fun here and they were like but 'because you were very specific… and you even said the name of the store, and the guy said you seem to be very specific knowing where you are going, and you completely changed so we were concerned about you and if anything went wrong' […] I think this happened because of our friends and their nationality…maybe…One was from Palestine and one from Algeria, from countries with question mark! Maybe that's why.

Naomi thinks media plays a crucial role in how border officers perceive people who cross borders:

> I think media manipulates the truth and it affects the perception of the officers at the border. If you do not educate yourself, you will easily get into the media's trap. For example, Donald Trump! If you do not know anything about Muslims and hear the things he says, you would believe his non-sense! My sister once went to New York on a bus with schoolmates and she was so worried that they would stop the whole bus because of her and she was so panicked. Of course, nothing happened, and she was not stopped but she was so very worried.

Naomi's explanation shows some mobilities are perceived as 'risky' or 'unwanted'. In the next section, I further discuss the findings of this study.

Discussion

SDG 16 underscores the significance of promoting just, peaceful and inclusive societies for all citizens and target 16.10 puts emphasis on 'protect[ing] fundamental freedoms, in accordance with national legislation and international agreements'. Today, 73 years after the Universal Declaration of Human Rights, the right to travel is still far from being reached for most of the world's population. Dual citizenship is perceived as a means of accessing international travel and be freely mobile. However, as Rahimieh's quotation at the beginning of this paper reminds us, dual citizenship does not necessarily benefit the individual who holds the status and the border 'is not a metaphor of subversive transgression and radical hybridity … but rather a site of policing and discipline, control and violence' (Rahimieh, 2007, p. 2290). The results elucidate that border crossing experiences are associated with profiling and segregating certain groups or individuals based on their race, nationality, and origin. In addition, the assignation of risk profiles to bodies at borders occurs based on race, gender, class, religion, etc. Risk profiling systems are constructed based on exclusion and low-risk bodies are the ones with higher economic status, that are often white and male, whose mobilities is encouraged (Rygiel, 2010).

Recent human rights violations of dual citizens are situated in the current geopolitical realities such as the actions governments take in terms of security after 9/11 (Stasilius & Ross, 2006). Since 9/11, as part of the ongoing 'war on terror', mobilities of individuals across borders have been subjected to racial profiling and other discriminatory practices. The ontological status of the person is what is being problematized and results in further control and categorization, often through the use of bio-technologies at the borders (Bhandar, 2008). It is crucial to note that identities are not only constructed at the borders and the time of border crossings but are reshaped and reconstructed beyond and within borders. As the narratives of participants illuminated, the formation of otherness at border crossings and 'the production of racial difference that easily collapses national identity with the racialised body' (Bhandar, 2008, p. 414).

Border crossing is an act of performance that challenges and reinforces the political and social status of the bodies that cross borders. Nation states regulate their cross-border movements by controlling mobilities, which turns borders into boundaries of inclusion as much as they can be boundaries of exclusion (Sofield, 2006). Participants noted how their physical appearance strengthens one of their citizenships over the other when it comes to crossing international borders. As the findings of this study illustrates, bordering practices identify and filter risky and desirable bodies from the undesirable ones (Walters, 2002, 2006a, 2006b). Therefore, the way this system functions is to single out the risky/destructive bodies and let in the productive/normal bodies. As results show, racial profiling at border crossings is integral to scrutinizing bodies and mark them with dualities such as risky/non-risky. Moreover, as a result of discrimination at border crossings, individuals might internalise the gaze (Rygiel, 2010) and consider themselves a risky/unwanted/undesirable/unqualified bodies.

As the results illustrate, Participants did not only feel surveilled at border crossings and regulated their behaviour accordingly, but constantly surveilled themselves and other people. Borders are simultaneously points of exclusion and inclusion where the movement of abled/low-risk/wanted bodies are encouraged while the movement of disabled/high-risk/unwanted bodies are halted. While dual citizenship seems to ease the access to international travel, it is a form of hierarchical citizenship where only some can benefit from it. Cresswell's concept of insurgent citizen reminds us that in spite of being a legal citizen, some citizens are deprived of their citizenship rights and therefore might not have access to freedom of movement within and beyond borders. Indeed, the notion of hierarchical citizenship illustrates that access to freedom of movement and international travel is only available to some dual citizens whereas insurgent citizens cannot easily access them.

Participants noted how their physical appearance strengthens one of their citizenships over the other when it comes to crossing international borders. Foucault's notion of biopower and biopolitics is helpful in making sense of this issue. Despite holding the Canadian passport, some participants could not claim their citizenship rights (Ranciere, 2004) as the notion of Canadian citizenship is still associated with whiteness.

As the border crossing narratives illustrate, borders are sketched on human bodies and bodies carry borders and 'become' borders (Chattapodhyay, 2018), which reaffirms that border crossing is 'less and less about the line in the sand and more and more about the bodies that cross it – each and every body' (Muller 2010, 61).

Conclusion

As was made clear in our discussion, SDG 16 and 'protect[ing] fundamental freedoms' is far from being reached when it comes to accessing international travel and enjoying citizenship rights. To address SDG 16, it is crucial to give attention to the inequalities that exist in terms of citizenship and access to the freedom of movement (Cresswell, 2001). The aim of this research was to explore how dual citizenship shapes the border crossing experiences of Canadians who have travelled internationally in the post 9/11 era. The theoretical framework underpinning this research, critical mobilities, allowed us to accentuate mobilities as a right while highlighting the interplay of (im)mobilities and inequalities. Taking a Foucauldian lens, we approached citizenship as a concept that is associated with practices, technologies, policies, discourses, forms of power utilized to govern the mobilities of individuals and groups of people and we discussed how citizenship is dependant on different forms of power at the micro-level, such as at the level of the body.

Utilizing the concept of insurgent citizenships in the context of border crossings helped us to gain a deeper understanding of the hierarchy of citizenship and mobilities, which denotes that some citizens are higher in the hierarchies of citizenship and, as such, may have easier access to spaces than other citizens (Adey, 2017). Our work leads us to challenge the idea of freedom of movement as a human right and through the narratives of participants, we illustrated the inequalities in terms of accessing international mobilities. We also discussed how border practices are political indicators of justice and how International control and mobility management of bodies at borders occurs based on dualities such as normal/risky (Salter, 2006) or wanted/unwanted (Cresswell, 2013).

This research contributes to the body of knowledge on tourism mobilities and citizenship and advances the discourse of tourism mobilities by focusing on dual citizenship and its implications for border crossing of individuals. Despite the growing body of literature on tourism mobilities and borders, little attention has been given to citizenship and how it shapes tourism mobilities. In terms of citizenship and international travel, future studies should further explore the sense of identity of those with dual citizenship and how they experience their duality within their adopted homeland. To that end, the following questions need to be asked. What are the experiences of dual citizens in their adopted homeland and how does this resonate with their experiences at border crossings? How has the increased securitization, post 9/11, affected these dual citizens *within* borders and how different is the experience at the border crossings? Moreover, and in light of the current context of the COVID-19 pandemic, what will be the implications of border re-openings as vaccinations become more widely available and, as we can expect a growth in surveillance technologies of the body in an effort to protect health and safety, which bodies will be deemed safe in the post-pandemic context? Moreover, due to the COVID-19 pandemic, as we are experiencing a profound pause in travel – when travel 'restart' – what will the new reality be? In this regard, the following questions should be asked by the tourism mobilities researchers, what questions could we ask now that might help make that new reality more equitable? If travel is always available to all – what are the environmental implications?

Tourism scholars should problematize bordering practice and further examine the connections between the processes of citizenship, racialisation, discrimination, and bordering which are fundamental to study (in)justices and tourism (im)mobilities. In addition, to study tourism (im)mobilities, tourism scholars should consider borders beyond essentialist formulations of lines drawn on maps and investigate the embodied experiences of border crossings and how bodies become borders.

Disclosure statement

No potential conflict of interest was reported by the authors.

ORCID

Pooneh Torabian http://orcid.org/0000-0001-7048-0976

References

Adey, P. (2009). Facing airport security: Affect, biopolitics, and the preemptive securitisation of the mobile body. *Environmentand Planning D: Society and Space*, *27*(2), 274–295. https://doi.org/10.1068/d0208

Adey, P. (2017). *Mobility* (2nd ed.). Routledge.

Adler, J. (1985). Youth on the road: Reflections on the history of tramping. *Annals of Tourism Research*, *12*(3), 335–354. https://doi.org/10.1016/0160-7383(85)90003-9

Agamben, G. (2005). *State of exception*, Translated by K. Attell. University of Chicago Press.

Ateljevic, I., Pritchard, A., Morgan, N., (2007). Introduction: Promoting an academy of hope in tourism enquiry. In: Ateljevic, Pritchard, & Morgan, (Eds.). *The critical turn in tourism studies: Innovative research methodologies* (pp. 1–11). Elsevier.

Bauman, Z. (1998). *Globalization: The human consequences*. Columbia University Press.

Bhandar, D. (2008). Resistance, detainment, asylum: The onto-political limits of border crossing in North America. In: D. Cowen & E. Gilbert (Eds.), *War, citizenship, territory* (pp. 404–425). Routledge.

Bianchi, R. V., & Stephenson, M. L. (2013). Deciphering tourism and citizenship in a globalized world. *Tourism Management*, *39*, 10–20. https://doi.org/10.1016/j.tourman.2013.03.006

Bianchi, R. V., & Stephenson, M. L. (2014). *Tourism and citizenship: Rights, freedoms and responsibilities in the global order*. Routledge.

Bianchi, R., & Stephenson, M. L. (2019). Tourism, border politics, and the fault lines of mobility. In: A. Paasi, E-K. Prokkola, J. Saarinen & K. Zimmerbauer (Eds.), *Borderless worlds for whom? Ethics, moralities and mobilities* (pp. 123–138), Routledge.

Bianchi, R., Stephenson, M. L., & Hannam, K. (2020). The contradictory politics of the right to travel: mobilities, borders & tourism. *Mobilities, 15*(2), 290–306. https://doi.org/10.1080/17450101.2020.1723251

Boluk, K. A., Cavaliere, C. T., & Duffy, L. N. (2019). A pedagogical framework for the development of the critical tourism citizen. *Journal of Sustainable Tourism, 27*(7), 865–881. https://doi.org/10.1080/09669582.2019.1615928

Boluk, K. A., Cavaliere, C. T., & Higgins-Desbiolles, E. (2019). A critical framework for interrogating the United Nations Sustainable Development Goals 2030 Agenda in tourism. *Journal of Sustainable Tourism, 27*(7), 847–864. https://doi.org/10.1080/09669582.2019.1619748

Bondi, L., Davidson, J., & Smith, M. (2005). Introduction: Geography's 'emotional turn. In: J. Davidson, L. Bondi & M. Smith, (Eds.) *Emotional geographies* (pp. 1–16). Ashgate.

Bosangit, C., McCabe, S., & Hibbert, S. (2009). What is told in travel blogs? Exploring travel blogs for consumer narrative analysis. In: W. Hoepken, U. Gretzel & R. Law (Eds.), *Information and communication technologies in tourism* (pp. 61–71). Springer.

Bruner, J. (1985). Narrative and paradigmatic modes of thought. In: J. Bruner (Ed.). In *search of pedagogy. Theselected works of Jerome Bruner* (Vol. 2, pp. 116–128). Routledge.

Castells, M. (1996). *The rise of the network society*. Blackwell.

Chase, S. E. (2005). Narrative inquiry: Multiple lenses, approaches, voices. In: N. K. Denzin & Y. S. Lincoln (Eds.), *The sage handbook of qualitative research* (3rd ed., pp. 651–680). Sage Publications.

Chattopadhyay, S. (2018). Borders re/make Bodies and Bodies are Made to Make Borders: Storying Migrant Trajectories. *ACME: An International Journal for Critical Geographies*, 18(1), 149–172. Retrieved from https://acme-journal.org/index.php/acme/article/view/1428.

Chow-White, P. A. (2006). Race, gender and sex on the net: Semantic networks of selling and storytelling sex tourism. *Media, Culture & Society, 28*(6), 883–905. https://doi.org/10.1177/0163443706068922

Clandinin, D. J. (2013). *Engaging in narrative inquiry*. Left Coast Press, Inc.

Clandinin, D. J., & Connelly, M. (2000). *Narrative inquiry: Experience and story in qualitative research*. Jossey-Bass.

Coles, T. (2008). Telling the tales of tourism: Mobility, media, and citizenship in the 2004 EU enlargement. In P. M. Burns & M. Novelli (Eds.). *Tourism and mobilities: Local-global connections* (pp. 65–81). CABI.

Coles, T. E., & Timothy, D. J. (2004). *Tourism, diasporas and space*.

Cresswell, T. (2001). The production of mobilities. *New Formations, 43*, 11–25.

Cresswell, T. (2006). *On the move: Mobility in the modern western world*. Routledge.

Cresswell, T. (2013). Citizenship in worlds of mobility. In: O. Söderström, S. Randeria, D. Ruedin, G. D'Amato, & F. Panese (Eds.), *Critical mobilities* (pp. 81–100). EPFL Press.

Daase, C., & Kessler, O. (2007). Knowns and unknowns in the 'war on terror': Uncertainty and the political construction of danger. *Security Dialogue, 38*(4), 411–434. https://doi.org/10.1177/0967010607084994

Daly, K. (2007). *Qualitative methods for family studies and human development*. Sage Publications.

Edgell, D. L. (1990). *International tourism policy*. Van Nostrand Reinhold.

Foucault, M. (1979). *Discipline and punish: The birth of the prison*. Vintage Books.

Foucault, M. (1988). *Technologies of the self*, L. H. Martin, H. Gutman, & P. H. Huthon (Eds.), University of Massachusetts Press.

Foucault, M. (2003). *Society must be defended: Lectures at College du France, 1975-1976*. M. Bertani & A. Fontana (Eds.). Translated by D. Macey. Picador.

Griffin, T. (2013). Gap year volunteer tourism stories: sharing more than memories. *Journal of Hospitality Marketing & Management, 22*(8), 851–874. https://doi.org/10.1080/19368623.2013.734223

Griffin, T. (2018). Constructionism: The making of meaning in tourism and hospitality research. In: R. Nunkoo (Ed.), *Handbook of research methods for tourism and hospitality management* (pp. 166–175). Edward Elgar Publishing.

Hagopian, F. (2011). Paradoxes of democracy and citizenship in Brazil. *Latin American Research Review, 46*(3), 216–234. https://doi.org/10.1353/lar.2011.0056

Hall, C. M. (2005). *Tourism: Rethinking the social science of mobility*. Pearson Prentice Hall.

Hall, C. M. (2010). Equal access for all? Regulative mechanisms, inequality and tourism mobility. In: S. Cole, & N. Morgan (Eds.). *Tourism and inequality: Problems and prospects* (pp. 34–49). CABI.

Hall, C. M., & Tucker, H. (2004). Tourism and postcolonialism: An introduction. In: C. M. Hall & H. Tucker (Eds.), *Tourism and postcolonialism: Contested discourses, identities and representations* (pp. 1–24). Routledge.

Hannam, K. (2009). The end of tourism? Nomadology and the mobilities paradigm. In: J. Tribe (Ed.). *Philosophical issues in tourism* (pp. 101–114). Channel View Publications.

Higgins-Desbiolles, F. (2006). More than an "industry": The forgotten power of tourism as a social force. *Tourism Management, 27*(6), 1192–1208. https://doi.org/10.1016/j.tourman.2005.05.020

Higgins-Desbiolles, F. (2007). Hostile meeting grounds: Encounters with the wretched of the earth and the tourist through tourism and terrorism in the 21st century. In: P. M. Burns & M. Novelli (Eds.), *Tourism and politics: Global frameworks and local realities* (pp.309–332). Elsevier.

Higgins-Desbiolles, F. (2008). Justice tourism and alternative globalisation. *Journal of Sustainable Tourism*, *16*(3), 345–364. https://doi.org/10.1080/09669580802154132

Hindess, B. (2002). Neo-liberal citizenship. *Citizenship Studies*, *6*(2), 127–143. https://doi.org/10.1080/13621020220142932

Holston, J. (1995). Spaces of insurgent citizenship. *Planning Theory*, *13*, 35–51.

Holston, J. (2007). *Insurgent citizenship*. Princeton University Press.

Holston, J. (2008). *Insurgent citizenship: Disjunctions of democracy and modernity in Brazil*. Princeton University Press.

Holston, J. (2009). Insurgent citizenship in an era of global urban peripheries. *City & Society*, *21*(2), 245–267. https://doi.org/10.1111/j.1548-744X.2009.01024.x

Hosking, D. M. (2008). Can constructionism be critical? In: J. Holstein & J. Gubrium (Eds.), *Handbook of constructionist research* (pp. 669–686). Guilford Publications.

Isin, E. F., & Wood, K. (1999). *Citizenship and identity*. Sage Publications.

Josselson, R., & Lieblich, A. (2001). Narrative research and humanism. In: K. J. Schneider, J. F. T. Bugenthal & J. F. Pierson (Eds.), *The handbook of humanistic psychology* (pp. 275–289). Sage Publications.

Khosravi, S. (2007). The 'illegal' traveller: An auto-ethnography of borders. *Social Anthropology*, *15*(3), 321–334. https://doi.org/10.1111/j.0964-0282.2007.00019.x

Koskela, H. (2012). You wouldn't wear that body': The problematic of surveillance and gender. In: K. Ball, K. Haggerty, & D. Lyon (Eds.), *Routledge handbook of surveillance studies* (pp. 49–56). Routledge.

Lebuhn, H. (2019). Insurgent citizenship. In: A. M. Orum (Ed.), *The Wiley Blackwell encyclopedia of urban and regional studies* (pp. 1–6). Wiley-Blackwell.

Lincoln, Y. S., Lynham, S. A., & Guba, E. G. (2011). Paradigmatic controversies, contradictions, and emerging confluences, revisited. In: N. Denzin & Y. S. Lincoln (Eds.), *The Sage handbook of qualitative research* (pp. 97–128). Sage Publications.

McCormack, C. (2004). Storying stories: A narrative approach to in-depth interview conversation. *International Journal of Social Research Methodology*, *7*(3), 219–236. https://doi.org/10.1080/13645570210166382

Meethan, K. (2001). *Tourism in global society: Place, culture, consumption*. Palgrave.

Minnaert, L. (2014). Social tourism participation: The role of tourism inexperience and uncertainty. *Tourism Management*, *40*, 282–289. https://doi.org/10.1016/j.tourman.2013.07.002

Minnaert, L., Maitland, R., & Miller, G. (2006). Social tourism and its ethical foundations. *Tourism Culture & Communication*, *7*(1), 7–17. https://doi.org/10.3727/109830406778493533

Minnaert, L., Maitland, R., & Miller, G. (2009). Tourism and social policy. *Annals of Tourism Research*, *36*(2), 316–334. https://doi.org/10.1016/j.annals.2009.01.002

Noy, C. (2004). This trip really changed me: Backpackers' narratives of self-change. *Annals of Tourism Research*, *31*(1), 78–102.

Noy, C., & Cohen, E. (2012). *Israeli backpackers: From tourism to rite of passage*. SUNY Press.

O'Malley, P. (2004). *Risk, Uncertainty and Freedom: Risk, Uncertainty and Government.*, Glasshouse Press.

Ochs, E., & Capps, L. (2001). *Living narrative: Creating lives in everyday storytelling*. Harvard University Press.

Patton, M. Q. (2002). *Qualitative research & evaluation methods* (3rd ed.). Sage publications.

Pennell, J. R., & Maher, T. (2015). Whom will sociology serve? Transforming the discipline by engaging communities. *Humanity & Society*, *39*(1), 47–63. https://doi.org/10.1177/0160597614548421

Pinnegar, S., & Daynes, J. G. (2007). Locating narrative inquiry historically: Thematics in the turn to narrative. In: D. J. Clandinin (Ed.). *The handbook of narrative inquiry: Mapping a methodology* (pp. 3–34). Sage publications.

Polkinghorne, D. E. (1995). Narrative configuration in qualitative analysis. In: J. A. Hatch & R. Wisniewski (Eds.). *Life history and narrative* (pp. 5–23). Falmer.

Pritchard, A., Morgan, N., & Ateljevic, I. (2011). Hopeful tourism: A new transformative perspective. *Annals of Tourism Research*, *38*(3), 941–963. https://doi.org/10.1016/j.annals.2011.01.004

Rahimieh, N. (2007). Border crossing. *Comparative Studies of South Asia, Africa and the Middle East*, *27*(2), 223–232. https://doi.org/10.1215/1089201x-2007-001

Ranciere, J. (2004). Who is the subject of the rights of man? *South Atlantic Quarterly*, *103*(2-3), 297–310. https://doi.org/10.1215/00382876-103-2-3-297

Rojek, C. (1998). Tourism and citizenship. International Journal of Cultural Policy, *4* (2), 291–310. https://doi.org/10.1080/10286639809358077

Rygiel, K. (2010). *Globalizing citizenship*. UBC Press.

Salter, M. B. (2004). Passports, mobility, and security: How smart can the border be? *International Studies Perspectives*, *5*(1), 71–91. https://doi.org/10.1111/j.1528-3577.2004.00158.x

Salter, M. B. (2006). The global visa regime and the political technologies of the international self: borders, bodies, biopolitics. *Alternatives: Global, Local, Political*, *31*(2), 167–189. https://doi.org/10.1177/030437540603100203

Santos, C. A., Belhassen, Y., & Caton, K. (2008). Reimagining Chinatown: An analysis of tourism discourse. *Tourism Management, 29*(5), 1002–1012. https://doi.org/10.1016/j.tourman.2008.01.002

Shamir, R. (2005). Without borders? Notes on globalization as a mobility regime. *Sociological Theory, 23*(2), 197–217. https://doi.org/10.1111/j.0735-2751.2005.00250.x

Sheller, M. (2013). Sociology after the mobilities turn. In: P. Adey, D. Bissell, K. Hannam, P. Merriman, & M. Sheller (Eds.), *The Routledge handbook of mobilities* (pp. 45–54). Routledge.

Sheller, M., & Urry, J. (2006). The new mobilities paradigm. *Environment and Planning A: Economy and Space, 38*(2), 207–226. https://doi.org/10.1068/a37268

Sofield, T. H. B. (2006). Border tourism and border communities: An overview. *Tourism Geographies, 8*(2), 102–121. https://doi.org/10.1080/14616680600585489

Sparke, M. B. (2006). A neoliberal nexus: Economy, security and the biopolitics citizenship on the border. *Political Geography, 25*(2), 151–180. https://doi.org/10.1016/j.polgeo.2005.10.002

Stasiulis, D., & Ross, D. (2006). Security, flexible sovereignty, and the perils of multiple citizenship. *Citizenship Studies, 10*(3), 329–348. https://doi.org/10.1080/13621020600772107

Stephenson, M. L. (2006). Travel and the 'freedom of movement': Racialised encounters and experiences amongst ethnic minority tourists in the EU. *Mobilities, 1*(2), 285–306. https://doi.org/10.1080/17450100600726662

Stephenson, M. L., & Ali, N. (2010). Tourism, travel and islamophobia: Post 9/11 journeys of Muslims in non-Muslim states. In N. Scott & J. Jafari (Eds.). *Tourism in the Muslim world* (pp. 235–251). Emerald Group Publishing Ltd.

Teeple, G. (2005). *The riddle of human rights*. University of Toronto Press.

Tesfahuney, M., & Schough, K. (2016). The nomos of tourism. In. M. Tesfahuney & K. Schough (Eds.), *Privileged mobilites: Tourism as world ordering* (pp. 11–31). Cambridge Scholars Publishing.

Teixeira da Silva, J. A. (2020). Stigmatization, discrimination, racism, injustice, and inequalities in the COVID-19 era. *International Journal of Health Policy and Management. 9, 484-485.* https://doi.org/10.34172/ijhpm.2020.87

Torabian, P., & Mair, H. (2017). (Re)constructing the Canadian border: Anti-mobilities and tourism. *Tourist Studies, Special Issue: Moralities & Mobilities, 17*(1), 17–35.

Turner, B. S. (1993). *Citizenship and social theory*. Sage Publications.

United Nations. (1948). *The universal declaration of human rights.* http://www.un.org/en/documents/udhr/

United Nations Department of Economic and Social Affairs. (2015). *Transforming our world: The 2030 Agenda for Sustainable Development.* Retrieved from https://sdgs.un.org/sites/default/files/publications/21252030%20Agenda%20for%20Sustainable%20Development%20web.pdf

Urry, J. (1995). *Consuming places*. Routledge.

Urry, J. (2000). *Sociology beyond societies: Mobilities for the twenty first century*. Routledge.

Walters, W. (2002). Deportation, expulsion, and the international police of aliens. *Citizenship Studies, 6*(3), 265–292. https://doi.org/10.1080/1362102022000011612

Walters, W. (2006a). Rethinking borders beyond the state. *Comparative European Politics, 4*(2-3), 141–159. https://doi.org/10.1057/palgrave.cep.6110076

Walters, W. (2006b). Border/control. *European Journal of Social Theory, 9*(2), 187–203. https://doi.org/10.1177/1368431006063332

Wilson, D., & Weber, L. (2008). Surveillance, Risk and Preemption on the Australian Border. *Surveillance & Society, 5*(2), 124–141.

Yu, W., Keralis, J. (2020). Controlling COVID-19: The folly of international travel restrictions. Health and Human Rights Journal. https://www.hhrjournal.org/2020/04/controlling-covid-19-the-folly-of-international-travel-restrictions/

OPEN ACCESS

The role of dark commemorative and sport events in peaceful coexistence in the Western Balkans

Metod Šuligoj and James Kennell

ABSTRACT

In order to analyse the ways in which problematic ethno-national behaviour at dark events in the Western Balkans undermines the transition from conflict, to post conflict and then peaceful societies, this research employed a sociological discourse analysis to critique the role dark events play in post-conflict tourism development and peaceful coexistence. Accordingly, by combining previous analyses of dark commemorative events with a new analysis of sport events, this research explains the relationship between event tourism and 16th Sustainable Development Goal (SDG) focused on peace, justice and strong institutions. This media-based analysis shows that dark commemorative and sport events share similar historical contexts, dissonant heritage and the dark leisure practices of attendees. These dark events attract tourism flows across national boundaries in the region, as well as including more widely dispersed diasporas, and international media and politicians. The prevalence of dark events in the region, which feature strongly in tourism flows, require significant attention in order to promote sustainable development. The findings of this research can be used to develop future research into the relationship between dark events and SDG16, using methods that build on this exploratory study and the new model of dark events that it provides.

Introduction

The United Nations have set out 17 Sustainable Development Goals (SDG) intended to "end poverty, protect the planet and improve the lives and prospects of everyone, everywhere" (United Nations, 2015a). This research focuses on the relationship between event tourism and SDG16, "Peace, Justice and Strong Institutions" in the post-conflict, post-Yugoslav states of the Western Balkans (Figure 1; hereinafter Western Balkans). There have been several attempts to analyse the contribution that tourism makes to the SDGs, including in the areas of poverty (Folarin & Adeniyi, 2020), climate change (Dick-Forde et al., 2020), partnerships for development (Nguyen et al., 2019) and gender (Alarcón & Cole, 2019). Tourism is frequently mentioned in the preamble to the SDGs and in the detail of some of them, but it is not explicitly acknowledged in SDG16.

Many international organisations, including the United Nations Educational, Scientific and Cultural Organization (UNESCO), the United Nations World Tourism Organization (UNWTO) and the European Commission have identified tourism as an instrument of peace. However, in the Western Balkans, tourism involving the commemoration of conflict has contributed to increasing regional tensions (Naef & Ploner, 2016), notwithstanding its economic benefits. In the Western Balkans, tourism's contribution to GDP ranges from more than 20% in the very tourism-dependent states of Croatia and Montenegro, to less than 10% in others that are less so (Organisation for Economic Cooperation and Development, 2016).

Tourism offers opportunities to war-torn inland areas that are in the shadow of regionally dominant coastal destinations, and it is important to consider how tourism can contribute to SDG16 in this region. This study responds to Farmaki's (2017) call for research to consider the types of tourism that are appropriate for peace-building, the types of peace that tourism can and does contribute to and, to what extent local post-conflict situations condition these relationships. Previous research has highlighted the potential of special interest tourism to contribute to peace in general (Blanchard & Higgins-Desbiolles, 2013) and to peace specifically in the Western Balkans (Causevic & Lynch, 2013; Naef & Ploner, 2016) but this exploratory research is the first to analyse the role of event tourism. Commemorative events in the Western Balkans with tourist audiences play divisive roles within communities, due to their links to recent conflicts, and this has been explored from a dark events perspective (Kennell et al., 2018). This research additionally argues that sport events can be analysed as a type of dark events, linked to the tragic past through dissonant heritage (Đorđević, 2019; Lalić & Biti, 2008; Mills, 2012; Ramshaw & Gammon, 2015, 2017).

Previous research into the relationship between sport events and peace has frequently focused on their sporting aspects, emphasising the "universally understood language" of sport (Schulenkorf & Edwards, 2010, p. 102), or the role of sport organisations and federations (Peachey & Burton, 2016), and rarely considering their links to tourism. This research employs a sociological discourse analysis of regional media, using insights from dark tourism and dark leisure to analyse sport events alongside commemorative events, showing commonalities between them from a dark events perspective. This extends previous conceptualisations (Frost & Laing, 2013; Laing & Frost, 2016, Kennell et al., 2018) of dark events to further develop this emerging field of enquiry.

Despite the emergence of a substantial literature on tourism and peace, and the well-researched phenomenon of event tourism, the role that events can play in the development of sustainable peace has received little attention (Moufakkir & Kelly, 2013). In the Western Balkans, commemorative and sport events that are promoted for cross-border and international tourist consumption, and for both diasporic and non-diasporic communities, frequently feature behaviours and discourses that exceed the moral boundary-crossing and taboos attributed to dark tourism. These events are, in their extreme, accompanied by dangerous, criminal or violent acts (Bryant, 2011; Kennell et al., 2018; Stone & Sharpley, 2013). The aim of this research is to explore the role of these dark events in post-conflict tourism development in the Western Balkans, in the context of SDG16 and its aspirations for the peaceful coexistence of societies.

Literature review

The Western Balkans as a post-conflict destination

Although the relationship between tourism and peace has been the subject of academic and practitioner debate (Farmaki, 2017; Litvin, 1998), there is a little consensus on the directionality of this relationship. Does an increase in tourism help to promote peace, or is peace a necessary condition for the development of tourism? Advocates of the first position claim that contact between people from different cultures, societies and nations will lead to mutual understanding, helping to reduce the likelihood of conflict breaking out in the future (Durko & Petrick, 2016; Kim et al., 2007). The counter-argument is that this contact is not possible unless there is a peaceful situation in a destination and that, without peace, it is not possible to develop or grow tourism

Figure 1. Map of the post-Yugoslav Western-Balkan states included in the analysis.
Source: d-maps (n.d.) and authors' elaboration

(Pratt & Liu, 2016). From this perspective, tourism is "simply the beneficiary of peace" and not its cause (Litvin, 1998, p. 63).

Research into the sustainable development of tourism in post-conflict areas including Bosnia-Herzegovina (Causevic & Lynch, 2013), Rwanda (Maekawa et al., 2013) and Colombia (Mosquera-Laverde et al., 2019), indicates the importance of understanding local contexts. Previous research has examined the relationship between tourism and peace in the post-conflict Western Balkans (Causevic & Lynch, 2013; Goulding & Domic, 2009; Naef & Ploner, 2016), where processes of the normalization of ethnic relations have not been linear, especially in multi-ethnic areas where national and political elites foster divisions and antagonism. The development of tourism in these contexts, which include sites linked to recent conflicts, requires careful stakeholder engagement (McDowell, 2008), to ensure that these "dark" tourism offers contribute to sustainable development.

The Balkan peninsula has been the focus of many conflicts, and the twentieth century saw a series of wars that had their roots or major battles in the territories of the former state of Yugoslavia (Mujanovic, 2018). The First World War (WWI) began with the assassination of the Archduke Ferdinand, in Sarajevo, the capital of what is now Bosnia-Herzegovina. During the Second World War (WWII), most of the territory of the Balkan peninsula was either occupied by, or aligned with, Nazi Germany (Glenny, 2012). Domestic quisling forces fought against "Partisans" led by communists in an anti-fascist coalition. After intense fighting for national liberation, the violence of a socialist revolution followed.

In the 1990s, the states of the former Yugoslavia were involved in the largest military conflict in Europe since the end of WWII (Šuligoj, 2016). This was accompanied by concentration camps and mass graves, indicating the scale of ethnic cleansing and atrocities that took place, which were brought to an end through the Dayton Peace Accords (U.S. Department of State, 1995). A later conflict took place between Serbia and forces of the independence struggle of Kosovo.[1] The Yugoslav wars ended in 2001 after a civil war between Macedonians and Albanians in the contemporary Republic of North Macedonia. However, the core of all these historical conflicts was in the post-Yugoslav states presented in Figure 1, which are the focus of this research. The relationship between tourism and the legacies of these conflicts is the focus of the following section.

Tourism and sustainable development in the Western Balkans

Despite a growing acceptance that sustainability should be central to the development of tourism, tourism development seems less sustainable than ever (Hall, 2019). Research in this field is dominated by a relatively narrow range of topics which frequently do not engage with the sociopolitical context of sustainability, which is essential for more critical scholarship to develop (Boluk et al., 2019). Research into tourism development in the Western Balkans often considers the "peace dividend" (Butler & Baum, 1999) in post-conflict areas. This mirrors research that has taken place in destinations as diverse as Northern Ireland (Leslie, 1996), Sri Lanka (Fernando et al., 2013) and the Philippines (Schiavo-Campo & Judd, 2005), as well as central and eastern Europe (Silvkova & Bucher, 2017).

In the late 1990s and early 2000s, Western Balkan states with coastal areas were the first to see the peace dividend, using tourism for diversification following their recent conflicts (Vitic & Ringer, 2008). The transition from planned to market economies in these states created new conditions for economic diversification through tourism (Knežević et al., 2020; Mulec & Wise, 2013). Across all of the states of the former Yugoslavia, tourism has seen significant growth in the past two decades, notwithstanding significant problems related to its sustainable development (Bučar, 2017). Croatia has experienced the fastest and most significant growth in international arrivals, due to its coastal tourism offering, although niche tourism elsewhere is also expanding. For example, Serbia has experienced annual arrivals growth rates of more than ten percent since 2017 (SORS, 2020).

Dark tourism in post-conflict areas

Dark tourism involves "encountering spaces of death or calamity that have political or historical significance, and that continue to impact upon the living" (Stone, 2016, p. 23). Post-conflict destinations provide resources on which dark tourism can be built (Zhang, 2017). Contestation over post-conflict tourism resources, however, means that the selection of sites for development can exacerbate or create inequalities and lead to concentrations of wealth (Winter, 2008).

The choices of whether to develop dark tourism in post-conflict destinations are often made by public authorities, due to their politicized and sensitive nature (Light, 2017). Dark tourism sites are often the focus of nation-building projects, due to the role of places of conflict, death and suffering in national narratives (Nora, 1989; Slade, 2003). Where territory is disputed, or conflicting post-conflict narratives exist, tourism has been identified as problematic for peace, as its commodifies and preserves difficult heritage, including in the Western Balkans (Kamber et al., 2016; Pavlaković & Perak, 2017; Šešić Dragićević & Mijatović Rogač, 2014). However, the role that events play in the relationship between tourism and peace has received only limited attention in the literature (Moufakkir & Kelly, 2013; Skoll & Korstanje, 2014). Consequently, we now turn to examining the role of commemorative and sport events in post-conflict areas, to shed light on this topic.

Dark commemorative events in post-conflict areas

Commemoration establishes and fixes in the public sphere the "correct" representations of events deemed significant for particular groups (Jedlowski, 2002). Sites where this take place become "lieu du memoire" (Nora, 1984, 1989); a metaphor for the spatial, material, narrative and non-narrative foundations of memory. Šuligoj (2016) notes, in the case of Croatia, that these are related to national memory and identity. They cannot be unproblematically introduced into dark tourism (Ashworth & Hartmann, 2005; Ashworth & Isaac, 2015).

A number of authors have used the term "dark events" to refer to commemorative events that are associated with death, disaster and suffering (Kennell et al., 2018; Laing & Frost, 2016; Šuligoj, 2019). Within leisure studies, the word dark has been associated with practices that involve deviance from social norms, or to denote associations with illegal or controversial subcultures (Blackshaw, 2018; Spracklen & Spracklen, 2012). Commemorations can also have their own dark sides (e.g. politization, manipulation, nationalism), which place them within a dark leisure framework as previously explained by Stone and Sharpley (2013), Kennell et al. (2018) and Bryant (2011), undermining contributions that they could make to SDG16. Typically, these events are associated with war, conflict or trauma and linked to ideas such as remembrance and education, as well as to political interventions and post-conflict development (Frew & White, 2015; Sather-Wagstaff, 2011). Although they often become tourist attractions, these events should be considered on their own terms as specific types of events, with specific management and representational challenges (Frew, 2012; Frost & Laing, 2013; Kennell et al., 2018). Although events can contribute to peace through bringing together divided groups for dialogue (Moufakkir & Kelly, 2013), this can be problematic in post-conflict contexts, where the coming together of victims and perpetrators, and of communities with competing claims on justice, can lead to volatile and unpredictable situations due to "non-negotiable discrepancies" (Skoll & Korstanje, 2014, p. 100) between the values of participants.

Dark sport events in post conflict areas

Dark events research has not previously considered non-commemorative events. In many states, sport events are politicized occasions, involving associations between teams and political parties, or the use of sport events as platforms for political or nationalist messages (Bairner, 2008; Hodges & Brentin, 2018). Sport events also have a social function and form part of narratives about identity and belonging, mediated through the collective memory of significant matches and occasions (Healey, 1991; King, 2016), as well as the symbolism of specific teams, venues and individuals (Bairner, 2008; Ramshaw & Gammon, 2005). Research into the relationship between sport events and peace has often taken place in the "sport for development and peace" field (Peachy & Burton, 2016). This has emphasised the positive contribution to peace from sport participation, especially by young people (Clarke et al., 2021) and is frequently based on projects that have explicit peace-building roles (Schulenkorf & Sugden, 2011). In the Western Balkan context, many sport events occur outside of the scope of this previous research. Despite its recent history of conflict, the region is well-integrated into international competition networks, in multiple sports. The Western Balkans hosts significant sport tourism events including international matches in football, tennis, handball and basketball, which attract international sports tourist audiences. Notwithstanding this apparent integration, sport events in the region with tourist audiences are frequently marked by deviant behaviour.

Sport event research in transition economies remains sparse (Hodges & Brentin, 2018), but the relationship between sport and the dissolution of Yugoslavia has been the topic of numerous studies, with a focus on the emergence of ethnic nationalisms in the 1990s (Đorđević, 2019; Lalić & Biti, 2008; Mills, 2012). In the Western Balkans, sport associations continue to be heavily politicized (Karadzoski & Siljanoska, 2011) and many sport events are notable for their hooligan violence involving ethno-national displays (Mills, 2012; Misic, 2012; Vejnović, 2014). Violent sport hooliganism is a common feature of football matches internationally, with some notable hooligan groups hailing from this region. However, unlike in most other destinations, hooligan violence is not confined to football matches, and is a feature of many sport events. Although a "sport for development and peace" movement exists in some areas (Giulianotti et al., 2017), and despite the popular perception that sport offers the potential for diffusing or releasing inter-state tension

– that it is "war without the bullets" (Orwell, 1945: 10, cited in Beck, 2013: 73), the dominant regional discourses surrounding sport events, are of discord and nationalism, with repeated references to conflicts, such as the ongoing struggle over the recognition of Kosovo (Brentin & Tregoures, 2016). Wood (2013) and Lalić and Biti (2008) summarize some of the deviant behaviours of fans in this region, which can best be understood from the perspective of dark leisure (Spracklen, 2013), which is manifested in its extreme form as hooligan violence.

Decades of violence at sport events in the region, against a nationalist background and intertwined with corruption and organized crime, are harmful to post-conflict societies (Lalić & Biti, 2008; Mills, 2012). These events become sites for manipulating history, but also for the creation of new national narratives (Mills, 2012), that are almost always related to previous conflicts (Vejnović, 2014). The informal, dark leisure practices connected to recent conflicts position these events outside of previous analyses of sport heritage and sport *as* heritage, which have tended to concentrate on aspects of nostalgia (Ramshaw & Gammon, 2005), cultural heritage (Hinch & Ramshaw, 2014), or the formal ways in which sport events contribute to commemoration (King, 2016). Research into sport heritage is still relatively underdeveloped (Ramshaw & Gammon, 2015) and the emergence of a critical sport heritage field involves an increasing engagement with dissonance and intangible heritage (Pinson, 2017; Ramshaw & Gammon, 2017), but has not focused on its associated tourism.

Sport events are drivers of tourism and this is one of the fastest growing sectors of the industry (Higham, 2018; UNWTO, 2020). Although sport tourism is well researched (Hinch & Higham, 2001; Lesjak, 2018), the dark sides of this type of tourism are frequently overlooked and mostly considered only as aspects of safety and security (Perić, 2018). Links between the dark aspects of these events and sport tourism through media reporting should not be overlooked; journalists inside and outside of the region produce a number of articles on regional sport hooliganism and violence every year.

This research extends previous conceptualisations of dark events to include these sport events in the Western Balkans, due to their specific post-conflict context, in order to consider how dark event tourism contributes to SDG16. In the following section, we set out our methodological approach to analysing dark commemorative and sport events.

Method

The media represent current events to diverse audiences (Abazi & Doja, 2017) and influence popular perceptions of tourist destinations (Volcic et al., 2014). Accordingly, the media has become an important source of information for tourism researchers (Li et al., 2018). The media also play an important role in raising awareness about the SDGs (Nishimura & Shimbun, 2020). Research into the relationship between the media and sustainable tourism has tended to focus on the influence of the media on the behaviours of visitors, and also on the promotion of certain destinations as more or less sustainable to visit (Pasquinelli & Trufino, 2020). However, the relationship between dark tourism, events and sustainable development has not been previously considered. This exploratory research analyses dark events, using a sociological discourse analysis (SDA) of media reports to gain deeper insights into this issue in the Western Balkans.

Discourse defines experiences and constructs shared realities. In turn, discourse is shaped by its social context (Jaworski & Pritchard, 2005) and analysing this enables qualitative research to engage with the complex social structures of tourism development. Discourse analysis is increasingly prevalent in the tourism literature (Qian et al., 2018) as a method for dealing with large amounts of textual data, including written documents and the photographs, brochures and websites that proliferate in tourism (Hannam & Knox, 2005; Li et al., 2018). There is a lack of consensus on what is meant by discourse analysis across different disciplines, with approaches varying from the analysis of individual conversations to the analysis of complex social systems (Jaworski & Pritchard, 2005).

In this research SDA was used, which relies on sociological theories and concepts and "allows us to understand social intersubjectivity because discourses contain it and because social intersubjectivity is produced through discursive practices" (Ruiz, 2009, p. 11). Gee (2014) places this approach to discourse analysis within the umbrella categorization of critical discourse analysis, rather than with the more linguistic forms of discourse analysis that focus more descriptively on the use of language and its meaning.

For SDA, Ruiz (2009) suggests three levels of analysis carried out simultaneously: a textual level (discourse characterization), a contextual level (understanding of discourse) and an interpretive level (explanation of the discourse). Li et al. (2018) employed a similar tripartite model for their analysis of media texts in tourism, spitting their analysis into textual description, process interpretation and social explanation. Gee (2014) used six steps of analysis that help to analyse the significance, practices, identities, relationships, politics and connections inherent in discourse, which has been widely applied in other fields. Gee's method, however, is more suited for the analysis of dynamic discourse with multiple participants, rather than for single authored media texts. This research employed Ruiz's three stages of analysis, which are well aligned with Qian et al. (2018) explanation of the dominant approaches taken to critical discourse in tourism research.

To collect the data for analysis, it was necessary to prepare a corpus of news media texts (Schweinsberg et al., 2017) to be analysed, which served as the sample for this research, following the previous practice of Wise and Mulec (2012) similar study. Searches for news stories were run in Google with a focus on the Western Balkans, using search terms in the main south Slavic languages: Croatian, Serbian, Montenegrin and Bosnian, and covering the period from June 2010–June 2020. This was the first decade without armed conflict and so offered the opportunity to analyse the post-conflict period, which was accompanied by the rise of online news media (Stojarová, 2020). Search terms were developed by generating keywords from the literature review and background research in the region. Table 1 shows a selection of keywords and results, to indicate the volume of news stories related to dark events in the period under study. Subsequently the "Customised date range" and "Sorted by relevance" Google tools (see Segev, 2010) were used for initial screening by headline. News reports produced by this search that did not focus solely on the Western Balkans or which covered non relevant events were excluded.

We identified a heterogeneous list of 150 articles; all of which were then read for deeper analysis. Articles focusing on other geographical areas, duplicate articles, articles with a focus on solely the sporting aspects of an event (technical and match reports), advertisements, paid posts and similar were excluded. Selected articles were then manually segmented using a matrix, where the criteria were "Country of publisher" and 'Conflict." For sport events, the level of competition was also used to select articles for analysis, with only national and international levels of competition being included for analysis as these are most clearly linked to tourism. In matrix quadrants with multiple articles, a reduction was made to ensure representativeness in a purposive sample for further analysis, but to ensure that the matrix preserved its temporal, substantive, media and national dispersion. Our purposeful sample is presented in Tables 2 – 5.

Table 1. Key words and news results.

Search term	News results
Fan violence	20,000
Fight between fans	8,660
Violence at sport event	21,100
Commemorations of the Homeland war	20,600
Anniversary of Operation Storm	62,400
Anniversary of WWI	19,900
Anniversary of WWII	22,100
Commemoration of Anti-fascists	4,620

Following Ruiz's (2009) method, three stages of SDA were performed on the corpus of 30 news stories. The first stage involved textual analysis, aiming to give a descriptive characterisation of the discourse. To this end, headlines of the articles are presented in tables three to six, below. They create a representation of events, although they cannot convey broader narratives. The second stage of contextual analysis required engagement with the deeper meanings of each journalist's narrative (Zandberg, 2010), therefore, entire articles were then brought within the analysis. Counterfactual analysis (Mahoney & Barrenechea, 2017) and sequence elaboration (Mahoney et al., 2009), were employed to deepen the contextual analysis, which also involved a critical comparison of our emerging findings with relevant literature reviewed for this research. The final, interpretive stage of analysis involved drawing connections between the discourse and its more broad social context, which in this case concerned the relationship to SDG16 and the post-conflict environment of the Western Balkans. The following discussion section presents these contextual and interpretive phases of the SDA.

Discussion

Contextual and interpretive analysis of dark commemorative events in the Western Balkans

The first set of stories to be analysed in the SDA were about dark commemorative events, shown in Table 2. Only Serbian media gave prominence to events linked to WWI. The anniversary linked to the Sarajevo assassination that sparked WWI highlighted contemporary divisions in the Western Balkans, as reports showed that Serbs did not want to take part in the main international dark event in Sarjevo (10). An alternative commemorative event organised by Serbs had a notable military presence (3). Politicians and ambassadors participated, peace and cooperation

Table 2. Selected articles on dark commemorative events.

No.	Headlines (translated)	Conflict context	Media/State	Date
1	Serbian sorrow, Croatian celebration	YW	B92/SR	4.08.2019
2	In the monastery of Krušedol, the memory of the victims and exiles in the 'Storm' operation	YW	Radio Televizija Srbije/SR	4.08.2019
3	Stefanović: Serbia is the guarantor of peace and stability	W1	Danas/SR	11.11.2018
4	Here is how the media in Serbia covered the anniversary of the Storm	YW	Večernji list/CR	4.8.2018
5	Anniversary of the breakthrough from Jasenovac – official commemoration	W2	Hrvatska radiotelevizija/CR	22.04.2018
6	President's envoy booed at Jazovka commemoration. When he mentioned that 'the president continues Tudjman's policy of reconciliation' insults started	RV	Jutarnji list/CR	22.06.2018
7	Commemoration in D. Gradina – Dodik: If we had Srpska on the 41st, this would not have happened	W2	Radio Televizija Republike Srpske/BH	23.04.2017
8	Memory column: certainly 100 thousand people in Vukovar, Grabar-Kitarović was walking with a 'girl in a blue coat'	YW	Večernji list/CR	18.11.2016
9	Srebrenica incident: Serbian prime minister hit in the face with a stone	YW	RTV BN/BH	11.7.2015
10	Vujanović tomorrow in Sarajevo on the 100th anniversary of the WWI	W1	Vijesti online/MN	27.06.2014

Key: BH – Bosnia-Herzegovina; CR – Croatia; KO – Kosovo, MN – Montenegro; SR – Serbia; W1 – WWI; W2 – Fascism and WWII; RV – Post-WWII revolutionary violence; YW – The Yugoslav Wars of the 1990s; OT – other disputes and conflicts in the 20th century and beyond.

Table 3. Selected articles on dark commemorative events outside the region.

No.	Headlines (translated)	Conflict context	Media/State	Date
11	Croatia appalled by Tajani's statements – an explanation is sought	RV	Hrvatska radiotelevizija/CR	11.02.2019
12	Commemoration in Bleiburg, one incident	RV	Hrvatska radiotelevizija/CR	18.05.2019
13	Commemoration for thousands of Istrians in the former Rižarna: torture in the death camp in the centre of Trieste,	W2	Glas Istre/CR	27.11.2018
14	Commemoration in Bleiburg: never more police and never fewer participants	RV	RTL/CR	12.05.2018
15	Thousands of people at the ceremony in Maribor. Croatian and Slovenian state leaders at the funeral of the victims from Huda jama	RV	Jutarnji list/CR	27.10.2016

See Table 2 for legend.

were highlighted, the heroism of Serbians was proclaimed, yet other Western Balkan nations were excluded, illustrating the difficulties of integrating these international events into sustainable tourism development for peace.

The achievement of SDG16 in the region is also undermined by events related to WWII and the later socialist revolution. Because of the legacies of both resistance and collaboration, the Western Balkans are internally and externally divided, which is reflected in their memorial practices. This division is seen in the politicization of the anniversary of the liberation of the Jasenovac concentration camp (Pavlaković & Perak, 2017), where three different dark events were organized. There was an official state commemoration organized by the centre-right government, a commemoration of Serbs and WWII veterans, and another for the Jewish community. The Croatian president also visited the memorial park separately (5).

A similar example of division (previously highlighted by Šešić Dragićević & Mijatović Rogač, 2014; Pavlaković & Perak, 2017) is shown by far-right commemorations at the Jazovka pit (6), a site of post-WWII mass executions, which is a "mirror image" of anti-fascist commemorations in Brezovica forest. A journalist reports that the reconciliatory words of the Croatian president's envoy (… *"The President of the Republic under the name of unity continues Tuđman's policy of reconciliation… it is obvious how difficult it is to build unity, …. we must therefore make additional efforts to strengthen unity on what is historically unquestionable, and that is the idea of statehood* ….") were booed. Many criticisms of the communist regime and anti-fascists were shared, such as, *"There, those who call themselves anti-fascists, with all the state honours, orgy over the bones of thousands of killed Croats* … ".

Also based on WWII, is a dark event at the former execution site of Donja Gradina in Bosnia, where the victims of the Croatian Ustasha regime are remembered (7). A journalist explains that it is: *"… the largest execution site from WWII on the territory of the former Yugoslavia, because 700,000 victims of the Ustasha crime died in it"*; it was also reported that, during this event, the leader of the Republika Srpska entity in Bosnia proclaimed about Serbian forces that *"If they went outside Serbia, to some other areas, then they brought freedom to others."* This perspective resonates with memories of conflicts in the 1990s, marked by Serbian expansionism (and factual inaccuracy). Regional politicians, ambassadors, representatives of Roma and Jews, anti-fascists and the armed forces participated in the commemoration organized by governments of Serbia and the Republika Srpska entity. In the programs, which also included a religious ritual, historical facts were presented markedly differently than in (5). The nationalism of these events, and their politicization, mark them as dark events in a dark leisure and dark tourism context (Bryant, 2011; Kennell et al., 2018; Stone & Sharpley, 2013).

The wars in the 1990s saw the collapse of the Yugoslav idea of "brotherhood and unity," whilst at the same time democratizing society and introducing new memorial practices

(Pavlaković & Perak, 2017). The tragedies of Vukovar and Srebrenica, have attained the status of national myths for Croats and Bosniaks (Naef & Ploner, 2016), and have become the focus of commemorative events. The 20th anniversary event of the Srebrenica massacre, which attracted local and international participants, world leaders and international organizations, was overshadowed by a violent attack on the Serbian Prime Minister by angry Bosniaks (9). The Prime Minister's entourage told the media that *"It was very uncomfortable, they were throwing stones, shoes, whatever they could get,"* ... *"They threw stones at his face and broke his glasses, shouting 'Allahu akbar'"* and *"Chetnik, go home!"*. indicating the interconnectedness of different 20th century conflicts through mention of WWII belligerents in the protests.

Commemorations in Vukovar, in contrast, have not yet seen similar violent acts. At the 25th anniversary, the Minister of Croatian Veterans said that ... *"with the arrival of almost 100,000 people today in Vukovar to pay their respects and tribute to its victims, a message was once again sent that 25 years after the crime, the courage and resistance of Vukovar remain in the collective consciousness of the Croatian people."* (8). However, expressions of anti-Serbian rhetoric have frequently occurred at such events (Goulding & Domic, 2009; Šuligoj, 2016). Similarly, the commemoration of 1995s Operation Storm in August each year has special meaning for Croatians and Serbs, but their perspectives are contradictory: Croatians celebrate victory, pride and patriotism, while for Serbs it is a day of sorrow, a symbol of exodus, and linked in the media to Ustashaism (1, 2, 4). Media reporting on these events and the participation in them by large numbers of tourists place these events in a dark tourism and dark leisure context (Kennell et al., 2018) that are not compatible with the achievement of SDG16 in the region, due to their role in perpetuating divisions in this post-conflict context.

One of the peculiarities of Western Balkan commemorations is that events are organized outside of the contemporary borders of the region, a sample of which are shown in table 3. Italian Remembrance Day commemorates post-WWII executions and the "exodus" of the Italian community of the Eastern Adriatic. This takes place at the monument at Basovizza/Bazovica near Trieste, Italy (11), where the then Italian President of the European Parliament provocatively exclaimed *"Long live Trieste, long live Italian Istria, long live Italian Dalmatia"*. This provokes Slavs of the Eastern Adriatic through accusations of the post-WWII persecution and massacre of Italians, while 20th century Italian imperialism and Fascism are ignored (Hrobat Virloget, 2015). Less controversial, and generally calmer, according to regional media, are events at the Risiera di San Sabba/Rižarna, a former Nazi camp in Trieste, where anti-fascists were imprisoned and murdered (13). These events are less subject to co-option by Italian extremists, as Italians were also amongst the victims.

Events 12, 14 and 15 relate to the post-WWII massacres of fleeing enemy soldiers, enemies of the new regime, and of civilians in Austria and Slovenia (Geiger, 2010). A symbolic funeral event was held in Maribor (15), where Slovenian and Croatian political leaders and visitors from both stateses participated, and which took place peacefully. However, a related event in Pliberk/Bleiburg (12, 14) in Austria appears much more problematic. According to the Croatian organizers, it is a peaceful mass event, but it is under strict police control and banned Ustasha iconography and symbols have been prominent at the event (Pavlaković et al., 2018). In 2018, *"about 11,000 people arrived on the Bleiberg field to pay their respects to the fallen civilians and soldiers of the pro-Nazi Independent State of Croatia"* (14). The contemporary cross-border dark tourism dimension of these events is related to the dynamic history of the Upper Adriatic and Central Europe, which remains characterised by conflicts, disputed borders, ideological differences, dissonant heritage and dark leisure. Media discourses, which are also relevant from the SDG perspective (Nishimura & Shimbun, 2020), on these events clearly demonstrate that commemorative events mostly exacerbate these tensions, presenting barriers to the implementation of SDG16 (United Nations, 2015a, 2015b). This consequently indicates that weak national institutions (Goals SDG 16.6 and 16.A) do not effectively enforce the rule of law (Goals SDG16.3 and 16.B) sufficiently and thus do not effectively mitigate the above-mentioned violence (Goal SDG16.1) which is endemic to these events.

Contextual and interpretive analysis of dark sport events in the Western Balkans

Table 4 firstly shows dark sport events within the sample that feature football matches, the most frequently reported sport events in the region, and the most problematic in terms of SDG16, due to the high incidence of ethno-nationalist fan violence and controversies involving domestic and international fans (Blackshaw & Crabbe, 2004; Bodin et al., 2007; Lalić & Biti, 2008; Mills, 2012; Misic, 2012; Vejnović, 2014; Vrcan, 2003; Wood, 2013). Aside from verbal and physical violence, the most frequently reported deviant behaviour at these events are expressions of nationalism, such as those of visiting fans at the Rijeka stadium, Croatia (19). Shouts were heard of *"Miserable Rijeka, you are full of Serbs, don't worry, Rijeka has more willows"* (alluding to a slogan from the Ustasha period of *"Serbs on willows"*); *"We Croats do not drink wine, only the blood of Chetniks from Knin"*; *"Kill, kill the Serb"*. Nationalism is also evident in other reports, i.e. 16 – 21 and 25. They in some cases involve coalitions between different international groups of fans expressing intolerance against another state in the region. For example, in report 16, Kosovan and English fans act against the Serbs. In Pilsen " … *Czech supporters sang 'Kosovo is Serbia' and 'Serbia, Serbia' in front of the hotel where Kosovo football players are accommodated. They also carried slogans with the flag of Serbia and the famous inscription 'No surrender'.*" Some were arrested for rioting (17). In report 22, Bosnian Serb and Russian fans act against Montenegrins, which escalated at the match into violence, chaos and destruction; a reporter perceived this as *"the end of football as a civilized sport in Montenegro, at least for some time."*

During these events, many problematic symbols are reported as displayed by fans to provoke their opponents, glorify historical events or proclaim their own identity: the double-headed eagle of Albania (16 and 25), symbols of Ustasha violence against Serbs (18), the flag of the wartime proto-state of Herzeg-Bosnia (20) and swastikas (23, 24). These symbolic manifestations of nationalism show how contemporary violence at sport events takes place in specific post-conflict contexts (Vrcan, 2003), involving narratives of former domestic belligerent groups such as Partisans, Chetniks and Ustashe, whose activities during WWII continue to be the source of contemporary tensions and conflicts. These violent and provocative fan behaviours are part of the "dark leisure family" (Spracklen, 2013).

Although football-related violence is common in many states, it also features frequently in other sport events in the Western Balkans. Table 5 shows reports on water polo (26, 28), shooting (29)

Table 4. Selected articles on dark sport event that feature football events.

No.	Headlines (translated)	Conflict context	Media/State	Date
16	Journalist from England: Fans in Pristina insulted Serbs, will UEFA react?	OT	Kosovo online/KO	18.11.2019
17	Chaos in the Czech Republic due to the match with Kosovo: Fights, drones, cavalry and 'Kosovo is Serbia'	OT	B92/SR	15.11.2019
18	Shame on Poljud, fans extol the Ustashas from the song 'Jasenovac and Gradiška Stara'	W2	Večernji list/CR	25.2.2019
19	Insults on Rujevica derby, Armada silent in protest, Torcida was much more active, especially in sending vulgar messages	W2	Jutarnji list/CR	18.2.2019
20	Scandal Pod Bijelim Brijegom: The flag of 'Herceg-Bosna' flies by drone at a match in Mostar	YW	Dnevni avaz/BH	24.10.2017
21	The Albanian Football Association: The battle is lost, but the fight continues	OT	Radio Televizija Srbije/SR	9.10.2015
22	Scandal and shame: Montenegrin fans interrupted the match with Russia	OT	Vijesti online/MN	27.03.2015
23	Swastika on the grass, the culmination of Split's spite	W2	Al Jazeera Balkans/BH	13.6.2015
24	Croatian shame has dispersed around the world: Everyone transfer the swastika from Poljud!	W2	Gol.hr (NOVA TV)/CR	12.6.2015
25	Serbia – Albania match suspended	OT	Radio Televizija Srbije/SR	14.10.2014

See Table 2 for legend.

Table 5. Selected articles on non-football dark events.

No.	Headlines (translated)	Conflict context	Media/State	Date
26	25 Partizan fans arrested	OT	Radio Televizija Crne Gore/MN	07.03.2020
27	Budućnost is asking the ABA League to cancel the final series	OT	Al Jazeera Balkans/BH	24.4.2019
28	Red Star water polo players attacked in Split, match cancelled	W2	Radio Televizija Srbije/SR	9.02.2019
29	A group of hooligans attacked Serbian athletes in Split? 'So how they recognized that we were from Serbia, we didn't have Belgrade plates!'	W2	Jutarnji list/CR	07.11.2016
30	Unusual provocation of Serbs at the European Women's Basketball Championship in Vukovar!	YW	Večernji list/CR	15.08.2013

See Table 2 for legend.

and basketball (27, 30). In report 30, for example, groups of Serbian fans cheered with national flags, among which was one with the inscription "*Vukovar*" in Cyrillic, accompanied with shouts "*Kosovo is the heart of Serbia*", "*Vukovar is the heart of Serbia*" and "*Here is Serbia,*" provoking locals and the Croatian police. Other reports show international fans damaging buildings, equipment and vehicles (26), and attacking participants verbally and physically (30). A Serbian sport marksman stated (29): "*When we passed seven young men in the parking area, they followed us. When we got in the car, they ran and caught up with us, they started to yell and threaten. They tried to open the doors and get us out on the road, and then started pounding on the car. Fortunately, I managed to start the car quickly, and after that they ran after us for another fifty meters.*" A similar attack happened to Serbian water polo players in Split (Croatia), forcing one to jump into the sea; some locals wanted to help him, while others commented: "*After all, he is a Chetnik*" (28).

The commemorative and sport events analysed here involve tourist activities and interactions with host communities, some of which are marred by violence outside the events (28, 29) and across the wider destination (17, 21), indicating a darker side of event tourism. Media reports reflect the dark leisure practices associated with these events, which undermine their potential to contribute to SDG16, despite their stimulation of tourism development in the region.

Further discussion and implications

Media, as an important transmitter of information on deviant behaviour (Blackshaw & Crabbe, 2004; Bodin et al. (2007), show that both the content of these events, and their representations, are existentially dependent on narratives of past conflicts. Unlike other kinds of dissonant heritage which rely on locational authenticity, this research shows that the darkness of events does not always depend on the location where historic events occurred. Previous research into dark commemorative events has highlighted the location-specific nature of dark events (Frost & Laing, 2013; Kennell et al., 2018), and dark tourism research typically considers experiences at the locations of tragedy to be at the darker end of the spectrum (Stone, 2006, 2016). In the Western Balkans, these dark events attract international tourism flows. By considering dark sport events as a type of dark event, sustained through the mobile dark leisure practices of travelling participants, sport fans and other bad-faith actors, this research provides a new understanding of dark events, and of dark tourism that moves beyond locational authenticity.

This new perspective is synthesized in Figure 2. The model connects dark commemorative events with this analysis of dark sport events. Both types of events are influenced by politics, manipulation and nationalisms (Bryant, 2011; Kennell et al., 2018; Stone & Sharpley, 2013), significantly related to identity and belonging, (Healey, 1991; King, 2016; Šuligoj, 2016) and the symbolism of participant behaviours (Bairner, 2008; Pavlaković et al., 2018; Ramshaw & Gammon, 2005; Vrcan, 2003). Moreover, visitors' deviant behaviour at dark events are in line with the "dark leisure

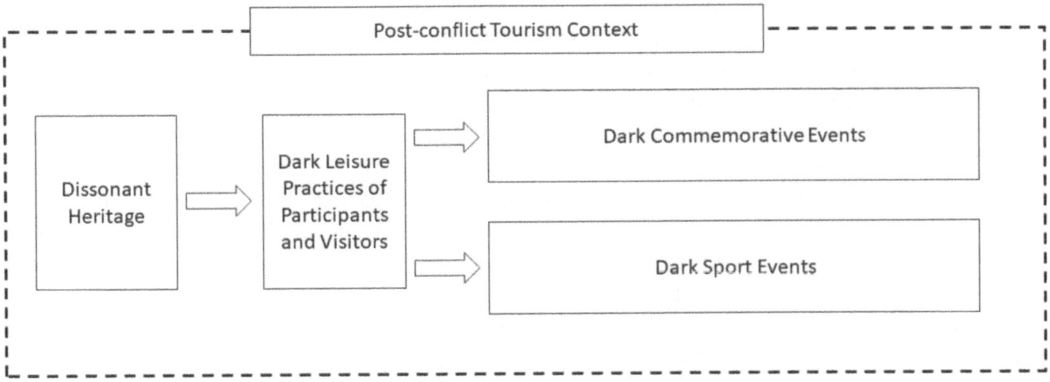

Figure 2. Dark Commemorative and Dark Sport Events in Post-conflict Tourism Contexts, based on media analysis. *Source*. authors' elaboration.

theory" (Spracklen, 2013). Commemorative and sports events in post-conflict tourism contexts differ in their form, yet share a socio-political environment and are both frequently marred by violence and other controversies.

Consequently, the two types of dark events are shown in Figure 2 as products of the post-conflict environment of their destinations, where the dark past has left a legacy of dissonant heritage that that is transmitted to the events through the dark leisure practices of participants and visitors. Our analysis shows that earlier conceptualisations of dark events, which focus on commemorative content, can be extended to other types of events that share the same context, but which do not intentionally make links to the dark past. By considering events in terms of their links to dissonant heritage and the dark leisure practices of attendees, in this case in a post-conflict tourism context where SDG16 is particularly pertinent, we propose the new model in Figure 2 as a contribution to the dark tourism and dark events literature. This can be further explored in future research and extended to include other forms of events in post-conflict tourism contexts where dark leisure practices are prevalent.

In practical terms, event organizers, destination managers and marketers should be aware that dark events transcend event management and sporting issues and can be related to many historical divisions and deviations. To integrate dark events into aspirations for SDG16, Western Balkan governments should ensure the rule of law, strong and effective institutions and, through cross-border campaigns at all levels of societies, promote peaceful coexistence and prevention of violence to achieve the goals of SDG16. Events can be catalysts for bridging community divides and can form part of sustainable tourism development in this context (Moufakkir & Kelly, 2013; Peachey & Burton, 2016), but only when event organisers can design and deliver inclusive events. Currently, in the Western Balkans, divisive state politics manifested at commemorations, and deviant fan behaviours at sport events mean that aspirations for peaceful coexistence in the region are hampered by dark events.

Conclusion

Controversial dark events can be understood from the perspective of memory studies (Nora,1984, 1989), management challenges (Kennell et al., 2018) and dark leisure (Stone & Sharpley, 2013) as well as dark tourism (Stone, 2006; Stone & Sharpley, 2013; Šuligoj, 2016, 2017). Moreover, they can be discussed in domestic as well as international tourism contexts. Although the main non-domestic target audiences for dark commemorative events come from widespread diasporas in the region (cross-border tourism) and elsewhere (see Drnovšek Zorko, 2020), some of the major dark commemorative events, e.g. milestone anniversaries of the Srebrenica massacre or the Jasenovac camp, attract international elites and other visitors. More attractive to international tourists, however, are international sport events, especially various championships, attracting the international sport tourists.

In order to analyse the contribution made by dark events to SDG16 in the Western Balkans, this exploratory research provided an SDA of media reports of these events, sampled from across the region. A corpus of media reports on thirty sport and commemorative events was developed in order to develop a novel analysis, extending the scope of dark events research beyond commemorations (Kennell et al., 2018; Laing & Frost, 2016; Šuligoj, 2019) to include sport events which, in particular in this region, share dark characteristics linked to memory, politics and violence (Bairner, 2008; Brentin & Tregoures, 2016; Healy, 1991; Vejnović, 2014). This research shows that dark events of different types in the Western Balkans are part of a complex and seemingly permanent social system of cause-and-effect relationships, where excesses of verbal/physical violence often occur, including attacks on politicians, fans and property. The events are part of social systems characterised by nationalism and intolerance towards "others" in the region, supported and maintained by senior politicians and media. Future research into the relationship between tourism and sustainable development in post-conflict areas should also consider the influence of dark events, and this research provides a new model that can be used for this purpose.

Dark events can contribute to increasing polarization (Farmaki & Antoniou, 2017), and memorial tensions (Naef & Ploner, 2016) or open up moral concerns (Stone, 2009; Stone & Sharpley, 2013). This article demonstrates the very limited peaceful coexistence of post-conflict communities at these events, in line with claims of Adamović et al. (2017) and Banovac et al. (2014). Ambitions of international actors to develop "Peace, Justice and Strong Institutions" (United Nations, 2015a, 2015b) as well as hopes for sustainable tourism associated with these events are "prisoners of the past." Advocates of peace through tourism suggest that increased contact between communities will lead to peace (Durko & Petrick, 2016; Kim et al., 2007), yet this research contributes new evidence that the peaceful coexistence of post-conflict societies is a pre-condition for sustainable tourism development, in line with the previous findings of Pratt and Liu (2016). In the Western Balkans sustainable tourism development linked to events is more likely to be the "beneficiary of peace" and not its cause (Litvin, 1998, p. 63).

This study followed Farmaki's (2017) call for research into the relationships between types of tourism and peace, in post-conflict contexts. In the Western Balkans, coastal tourism dominates discussions of tourism development, and despite the emergence of special interest and urban tourism in the region, this form of tourism will continue to grow. However, a desire to spread the benefits of tourism to inland communities, and to support cross-border international tourism in the Western Balkans, means that is important to consider how other forms of tourism can contribute to sustainable development/SDG16. Tourist arrivals to the region have been increasing dramatically, and many of the dark events analysed in this research take place in destinations that have a general tourist appeal. At present, the dark nature of many commemorative and sports events in the Western Balkans will make them difficult to integrate into itineraries and products that are offered to general tourists, and destination managers and marketers should be aware of this. For these events to support sustainable tourism development in the Western Balkans, it will first be necessary for political and civil society communities to address the underlying tensions from which many of these events draw their dark character.

One of the main limitations of this research is the reliance on media reporting, which depends on journalists' interpretations and editorial policies; in many destinations these are also heavily influenced by the authorities (Stojarová, 2020). Through developing a heterogeneous purposive sample of media reports for this analysis, we tried to minimise these biases. However, only Western Balkan media were considered, although dark events are also reported internationally in a way that may influence tourism to the region. We are also aware that non-controversial events are less interesting to the media and thus have often been overlooked. Although this research is limited by its focus on media reports of these events, future research into dark events could take a more ethnographic approach to analyse the experiences of participants and managers, to develop a more bottom-up understanding of the issues that have been identified through this exploratory study. In addition, further studies could use big data analytics to take a more quantitative approach to media reporting of dark events in order to generate alternative data from a more broad sample, now that some key issues for the analysis of these events have been identified.

Note

1. This designation is without prejudice to positions on status, and Is In line with UNSCR 1244/1999 and the ICJ Opinion on the Kosovo declaration of independence.

Disclosure statement

No potential competing interest was reported by the authors.

The authors received no financial support for the research, authorship, and/or publication of this article.

ORCID

Metod Šuligoj http://orcid.org/0000-0001-8877-6388
James Kennell http://orcid.org/0000-0002-7877-7843

References

Abazi, E., & Doja, A. (2017). The past in the present: time and narrative of Balkan wars in media industry and international politics. *Third World Quarterly*, *38*(4), 1012–1042. https://doi.org/10.1080/01436597.2016.1191345

Adamović, M., Gvozdanović, A. & Kovačić, (2017). *Process of reconciliation in the Western Balkans and Turkey: A qualitative study*. Compex d.o.o. & Institute for Social Research in Zagreb.

Alarcón, D. M., & Cole, S. (2019). No sustainability for tourism without gender equality. *Journal of Sustainable Tourism*, *27*(7), 903–919. https://doi.org/10.1080/09669582.2019.1588283

Ashworth, G., & Hartmann, R. (2005). Introduction: Managing atrocity for tourism. In G. Ashworth & R. Hartmann (Eds.), *Horror and human tragedy revisited: The management of sites of atrocities for tourism* (pp. 1–14). Cognizant Communication Corporation.

Ashworth, G., & Isaac, R. K. (2015). Have we illuminated the dark? Shifting perspectives on 'dark' tourism. *Tourism Recreation Research*, *40*(3), 316–325. https://doi.org/10.1080/02508281.2015.1075726

Bairner, A. (2008). The cultural politics of remembrance: sport, place and memory in Belfast and Berlin. *International Journal of Cultural Policy*, *14*(4), 417–430. https://doi.org/10.1080/10286630802445880

Banovac, B., Katunarić, V. i., & Mrakovčić, M. (2014). From war to tolerance? Bottom-up and top-down approaches to (re)building interethnic ties in the areas of the former Yugoslavia. *Zbornik Pravnog Fakulteta Sveučilišta u Rijeci, 35* (2), 455–483.

Beck, P. J. (2013). 'War minus the shooting': George Orwell on International Sport and the Olympics. *Sport in History, 33*(1), 72–94. https://doi.org/10.1080/17460263.2012.761150

Blackshaw, T. (2018). Some notes on the language game of dark leisure. *Annals of Leisure Research, 21*(4), 395–406. https://doi.org/10.1080/11745398.2017.1376220

Blackshaw, T., & Crabbe, T. (2004). *New perspectives on sport and 'deviance': Consumption, performativity, and social control*. Routledge.

Blanchard, L. A. & Higgins-Desbiolles, F. (eds.) (2013). *Peace through tourism: promoting human security through international citizenship*. Routledge.

Bodin, D., Robén, L., & Héas, S. (2007). *Sport i nasilje u Europi*. Knjiga trgovina.

Boluk, K. A., Cavaliere, C. T., & Higgins-Desbiolles, F. (2019). A critical framework for interrogating the United Nations Sustainable Development Goals 2030 Agenda in tourism. *Journal of Sustainable Tourism, 27*(7), 847–864. https://doi.org/10.1080/09669582.2019.1619748

Brentin, D., & Tregoures, L. (2016). Entering through the sport's door? Kosovo's sport diplomatic endeavours towards international recognition. *Diplomacy & Statecraft, 27*(2), 360–378. https://doi.org/10.1080/09592296.2016.1169799

Bryant, C. D. (Ed.) (2011). *The Routledge handbook of deviant behaviour*. Routledge.

Bučar, K. (2017). Green orientation in tourism of Western Balkan. In: Renko, S. & Pestek, A. (Eds.), *Green economy in the Western Balkans. Towards a sustainable future* (pp. 175–209). Emerald Publishing Limited.

Butler, R. W., & Baum, T. (1999). The tourism potential of the peace dividend. *Journal of Travel Research, 38*(1), 24–29. https://doi.org/10.1177/004728759903800106

Causevic, S., & Lynch, P. (2013). Political (in) stability and its influence on tourism development. *Tourism Management, 34*, 145–157. https://doi.org/10.1016/j.tourman.2012.04.006

Clarke, F., Jones, A., & Smith, L. (2021). Building peace through sports projects: A scoping review. *Sustainability, 13*(4), 2129. https://doi.org/10.3390/su13042129

Dick-Forde, E. G., Oftedal, E. M., & Bertella, G. M. (2020). Fiction or reality? Hotel leaders' perception on climate action and sustainable business models. *Worldwide Hospitality and Tourism Themes, 12*(3), 245–260. https://doi.org/10.1108/WHATT-02-2020-0012

D-maps (n.d.). Balkans. Retrieved from https://d-maps.com/pays.php?num_pay=181&lang=en.

Đorđević, I. (2019). Politics on the football field. An overview of the relationship between ideology and sport in Serbia. In: M. Martynova & I. Bašić (Eds.), *Prospects for Anthropological Research in South-East Europe* (pp. 153–177). N. N. Miklouho-Maklay Institute of Ethnology and Anthropology Russian Academy of Sciences & Institute of Ethnography Serbian Academy of Sciences and Arts.

Drnovšek Zorko, Š. (2020). Cultures of risk: On generative uncertainty and intergenerational memory in post-Yugoslav migrant narratives. *The Sociological Review, 68*(6), 1322–1337. https://doi.org/10.1177/0038026120928881

Durko, A., & Petrick, J. (2016). The Nutella project: an education initiative to suggest tourism as a means to peace between the United States and Afghanistan. *Journal of Travel Research, 55*(8), 1081–1093. https://doi.org/10.1177/0047287515617300

Farmaki, A. (2017). The tourism and peace nexus. *Tourism Management, 59*, 528–540. https://doi.org/10.1016/j.tourman.2016.09.012

Farmaki, A., & Antoniou, K. (2017). Politicising dark tourism sites: Evidence from Cyprus. *Worldwide Hospitality and Tourism Themes, 9*(2), 175–186. https://doi.org/10.1108/WHATT-08-2016-0041

Fernando, S., Bandara, J. S., & Smith, C. (2013). Regaining missed opportunities: The role of tourism in post-war development in Sri Lanka. *Asia Pacific Journal of Tourism Research, 18*(7), 685–711. https://doi.org/10.1080/10941665.2012.695284

Folarin, O., & Adeniyi, O. (2020). Does tourism reduce poverty in Sub-Saharan African countries? *Journal of Travel Research, 59*(1), 140–155. https://doi.org/10.1177/0047287518821736

Frew, E. A. (2012). Interpretation of a sensitive heritage site: The Port Arthur Memorial Garden. International Journal of Heritage Studies, *18*(1), 33–48. https://doi.org/10.1080/13527258.2011.603908

Frew, E., & White, L. (2015). Commemorative events and national identity: Commemorating death and disaster in Australia. *Event Management, 19*(4), 509–524. https://doi.org/10.3727/152599515X14465748512722

Frost, W., & Laing, J. (2013). *Commemorative events: Memory, identities, conflict*. Routledge.

Gee, J. P. (2014). *An introduction to discourse analysis: Theory and method* (4th ed.). Routledge.

Geiger, V. (2010). Ljudski gubici Hrvatske u Drugome svjetskom ratu i u poraću koje su prouzročili Narodnooslobodilačka vojska i Partizanski odredi Jugoslavije/Jugoslavenska armija i komunistička vlast Brojidbeni pokazatelji (procjene, izračuni, popisi) Case study: Bleiburg i folksdojčeri. *Časopis za Suvremenu Povijest, 42*(3), 693–722.

Glenny, M. (2012). *The Balkans 1804-2012: Nationalism, War and The Great Powers*. Granta.

Goulding, C., & Domic, D. (2009). Heritage, identity and ideological manipulation: The case of Croatia. *Annals of Tourism Research*, *36*(1), 85–102. https://doi.org/10.1016/j.annals.2008.10.004

Giulianotti, R., Collison, H., Darnell, S., & Howe, D. (2017). Contested states and the politics of sport: the case of Kosovo–division, development, and recognition. *International journal of sport policy and politics*, *9*(1), 121–136.

Hall, C. M. (2019). Constructing sustainable tourism development: The 2030 agenda and the managerial ecology of sustainable tourism. *Journal of Sustainable Tourism*, *27*(7), 1044–1060. https://doi.org/10.1080/09669582.2018.1560456

Hannam, K., & Knox, D. (2005). Discourse analysis in tourism research a critical perspective. *Tourism Recreation Research*, *30*(2), 23–30. https://doi.org/10.1080/02508281.2005.11081470

Healey, J. F. (1991). An exploration of the relationships between memory and sport. *Sociology of Sport Journal*, 8(3), 213–227.

Higham, J. (2018). *Sport tourism development*. Channel view publications.

Hinch, T. D., & Higham, J. E. S. (2001). Sport tourism: A framework for research. *International Journal of Tourism Research*, *3*(1), 45–58. https://doi.org/10.1002/1522-1970(200101/02)3:1<45::AID-JTR243>3.0.CO;2-A

Hinch, T., & Ramshaw, G. (2014). Heritage sport tourism in Canada. *Tourism Geographies*, *16*(2), 237–251. https://doi.org/10.1080/14616688.2013.823234

Hodges, A., & Brentin, D. (2018). Fan protest and activism: football from below in South-Eastern. Soccer & Society, *19*(3), 329–336. https://doi.org/10.1080/14660970.2017.1333674

Hrobat Virloget, K. (2015). Breme preteklosti: Spomini na sobivanje in migracije v slovenski Istri po drugi svetovni vojni. *Acta Histriae*, *23*(3), 531–554.

Jaworski, A., & Pritchard, A. (Eds.). (2005). *Discourse, Communication and Tourism*. Channel View Publications.

Jedlowski, P. (2002). *Memoria, esperienza e modernità. Memorie e società nel XX secolo*. Franco Angeli.

Kamber, M., Karafotias, T., & Tsitoura, T. (2016). Dark heritage tourism and the Sarajevo siege. *Journal of Tourism and Cultural Change*, *14*(3), 255–269. https://doi.org/10.1080/14766825.2016.1169346

Karadzoski, M., & Siljanoska, J. (2011, May). Political interference in the sport associations in the countries of the Western Balkans. *Ovidius University Annals, Physical Education and Sport/Science, Movement and Health Series*, *11*(Suppl 2), 631–636.

Kennell, J., Šuligoj, M., & Lesjak, M. (2018). Dark events: Commemoration and collective memory in the former Yugoslavia. *Event Management*, *22*(6), 945–963. https://doi.org/10.3727/152599518X15346132863247

Kim, S. S., Prideaux, B., & Prideaux, J. (2007). Using tourism to promote peace on the Korean Peninsula. *Annals of Tourism Research*, *34*(2), 291–309. https://doi.org/10.1016/j.annals.2006.09.002

King, A. (2016). Sport, war and commemoration: Football and remembrance in the twentieth and twenty-first centuries. *European Journal for Sport and Society*, *13*(3), 208–229. https://doi.org/10.1080/16138171.2016.1226027

Knežević, M. N., Hadžić, O., Nedeljković, S., & Kennell, J. (2020). Tourism entrepreneurship and industrial restructuring: GLOBE national and organisational cultural dimensions. *Journal of the Geographical Institute Jovan Cvijic, SASA*, *70*(1), 15–30. https://doi.org/10.2298/IJGI2001015N

Laing, J., & Frost, W. (2016). Dark tourism and dark events: A journey to positive resolution and well-being. In S. Filep, J. Laing, & M. Csikszentmihalyi (Eds.), *Positive Tourism* (pp. 82–99). Routledge.

Lalić, D., & Biti, O. (2008). Četverokut sporta, nasilja, politike i društva: Znanstveni uvid u Europi i u Hrvatskoj. *Politička Misao: časopis za Politologiju*, *45*(3–4), 247–272.

Lesjak, M. (2018). *Velike športne prireditve in turizem: Teoretični in raziskovalni vidik merjenja vplivov velike športne prireditve*. Založba Univerze na Primorskem.

Leslie, D. (1996). Northern Ireland, tourism and peace. *Tourism Management*, *17*(1), 51–55. https://doi.org/10.1016/0261-5177(96)83161-X

Li, J., Pearce, P. L., & Low, D. (2018). Media representation of digital-free tourism: A critical discourse analysis. *Tourism Management*, 69, 317–329.

Light, D. (2017). Progress in dark tourism and thanatourism research: An uneasy relationship with heritage tourism. *Tourism Management*, *61*, 275–301. https://doi.org/10.1016/j.tourman.2017.01.011

Litvin, S. W. (1998). Tourism: The world's peace industry? *Journal of Travel Research*, *37*(1), 63–66. https://doi.org/10.1177/004728759803700108

Maekawa, M., Lanjouw, A., Rutagarama, E., & Sharp, D. (2013). May). Mountain gorilla tourism generating wealth and peace in post-conflict Rwanda. *Natural Resources Forum*, *37*(2), 127–137. https://doi.org/10.1111/1477-8947.12020

Mahoney, J., & Barrenechea, R. (2017). The logic of counterfactual analysis in case-study explanation. *The British Journal of Sociology*, *70*(1), 306–338. https://doi.org/10.1111/1468-4446.12340

Mahoney, J., Kimball, E., & Koivu, K. L. (2009). The logic of historical explanation in the social sciences. *Comparative Political Studies*, *42*(1), 114–146. https://doi.org/10.1177/0010414008325433

McDowell, S. (2008). Selling conflict heritage through tourism in peacetime Northern Ireland: transforming conflict or exacerbating difference? *International Journal of Heritage Studies*, *14*(5), 405–421. https://doi.org/10.1080/13527250802284859

Mills, R. (2012). It all ended in an unsporting way': Serbian football and the disintegration of Yugoslavia, 1989–2006. *The International Journal of the History of Sport*, *26*(9), 1187–1217. https://doi.org/10.1080/09523360902941829

Mosquera-Laverde, W. E., Vásquez-Bernal, O. A., Gomez-E, C. P. (2019). Eco-touristic foresight in the Colombian post-conflict for the sustainability of the tourist service with emphasis on ecological marketing. Buenaventura Case. *Proceedings of the International Conference on Industrial Engineering and Operations Management Pilsen, Czech Republic, July 23-26, 2019*

Moufakkir, O., & Kelly, I. (2013). Peace through tourism A sustainable development role for Events. In T. Pernecky & M. Lück (Eds.), *Events, society and sustainability: Critical and contemporary approaches* (pp. 130–150). Routledge.

Mujanovic, J. (2018). *Hunger and Fury: The Crisis of Democracy in the Balkans*. Hurst.

Mulec, I., & Wise, N. (2013). Indicating the competitiveness of Serbia's Vojvodina Region as an emerging tourism destination. *Tourism Management Perspectives, 8*, 68–79. https://doi.org/10.1016/j.tmp.2013.07.001

Naef, P., & Ploner, J. (2016). Tourism, conflict and contested heritage in former Yugoslavia. *Journal of Tourism and Cultural Change, 14*(3), 181–188. https://doi.org/10.1080/14766825.2016.1180802

Nguyen, T. Q. T., Young, T., Johnson, P., & Wearing, S. (2019). Conceptualising networks in sustainable tourism development. *Tourism Management Perspectives, 32*, 100575. https://doi.org/10.1016/j.tmp.2019.100575

Nishimura, Y., Shimbun, A. (2020). How the media can be a meaningful stakeholder in the quest to meet the SDGs. Retrieved from World Economic Forum website: https://www.weforum.org/agenda/2020/01/sdgs-sustainable-development-news-media-coverage/

Nora, P. (1989). Between Memory and History: Les Lieux de Mémoire. *Representations, 26*, 7–24. https://doi.org/10.2307/2928520

Nora, P. (1984). *Entre. Mémoire et Histoire. La problématique des lieux*. Université Lumière Lyon 2. https://perso.univ-lyon2.fr/~jkempf/LDM_intro.pdf

Pavlaković, V., Brentin, D., & Pauković, D. (2018). The controversial commemoration: Transnational approaches to remembering Bleiburg. *Politička Misao, 55*(2), 7–32. https://doi.org/10.20901/pm.55.2.01

Pavlaković, V., & Perak, B. (2017). How does this monument make you feel?: Measuring emotional responses to war memorials in Croatia. In T. S. Andersen & B. Törnquist-Plewa (Eds.), *The Twentieth Century in European Memory* (pp. 268–304). Brill.

Peachey, J. W. & Burton, L. (2016) Servant Leadership in Sport for Development and Peace: A Way Forward. *Quest, 69*(1), 125–139, https://doi.org/10.1080/00336297.2016.1165123

Perić, M. (2018). Estimating the perceived socio-economic impacts of hosting large-scale sport tourism events. *Social Sciences, 7*(10), 2–18. https://doi.org/10.3390/socsci7100176

Pinson, J. (2017). Heritage sporting events: Theoretical development and configurations. *Journal of Sport & Tourism, 21*(2), 133–152. https://doi.org/10.1080/14775085.2016.1263578

Pratt, S., & Liu, A. (2016). Does tourism really lead to peace? A global view. *International Journal of Tourism Research, 18*(1), 82–90. https://doi.org/10.1002/jtr.2035

Qian, J., Wei, J., & Law, R. (2018). Review of critical discourse analysis in tourism studies. *International Journal of Tourism Research, 20*(4), 526–537. https://doi.org/10.1002/jtr.2202

Ramshaw, G., & Gammon, S. (2005). More than just nostalgia? Exploring the heritage/sport tourism nexus. *Journal of Sport & Tourism, 10*(4), 229–241. https://doi.org/10.1080/14775080600805416

Ramshaw, G., & Gammon, S. (2015). Heritage and sport. In E. Waterton & S. Watson (Eds.), *The Palgrave handbook of contemporary heritage research* (pp. 248–260). Palgrave Macmillan.

Ramshaw, G., & Gammon, S. J. (2017). Towards a critical sport heritage: Implications for sport tourism. *Journal of Sport & Tourism, 21*(2), 115–131. https://doi.org/10.1080/14775085.2016.1262275

Ruiz, J. R. (2009). Sociological discourse analysis: Methods and logic. *Forum Qualitative Sozialforschung/Forum: Qualitative Social Research, 10*(2). https://doi.org/10.17169/fqs-10.2.1298

Sather-Wagstaff, J. (2011). *Heritage that hurts: Tourists in the memoryscapes of September 11*. Left Coast Press.

Schiavo-Campo, S., & Judd, M. P. (2005). *The Mindanao conflict in the Philippines: Roots, costs, and potential peace dividend* (Vol. 24). Conflict Prevention & Reconstruction, Environmentally and Socially Sustainable Development Network, World Bank.

Schulenkorf, N., & Edwards, D. (2010). The role of sport events in peace tourism. In O. Mouffakir & I. Kelly (Eds.), *Tourism, progress and peace* (pp. 99–117). CABI.

Schulenkorf, N., & Sugden, J. (2011). Sport for development and peace in divided societies: Cooperating for inter-community empowerment in Israel. *European Journal for Sport and Society, 8*(4), 235–256. https://doi.org/10.1080/16138171.2011.11687881

Schweinsberg, S., Darcy, S., & Cheng, M. (2017). The agenda setting power of news media in framing the future role of tourism in protected areas. *Tourism Management, 62*, 241–252. https://doi.org/10.1016/j.tourman.2017.04.011

Segev, E. (2010). *Google and the Digital Divide. The Bias of Online Knowledge*. Chandos Publishing.

Šešić Dragićević, M., & Mijatović Rogač, L. (2014). Balkan dissonant heritage narratives (and their attractiveness) for tourism. *American Journal of Tourism Management, 3*(1B), 10–19.

Slivková, S., & Bucher, S. (2017). DARK TOURISM AND ITS REFLECTION IN POST-CONFLICT DESTINATIONS OF SLOVAKIA AND CROATIA. *19*(1), 22–34.

Skoll, G. R., & Korstanje, M. E. (2014). Terrorism, homeland safety and event management. *International Journal of Hospitality and Event Management, 1*(1), 95–110. https://doi.org/10.1504/IJHEM.2014.062869

Slade, P. (2003). Gallipoli Thanatourism: The meaning of ANZAC. *Annals of Tourism Research, 30*(4), 779–794. https://doi.org/10.1016/S0160-7383(03)00025-2

SORS. (2020). Tourism Statistics. Online. Retrieved from https://www.stat.gov.rs/en-us/oblasti/ugostiteljstvo-i-turizam/turizam/16.01.2021

Spracklen, K. & Spracklen, B. (2012). "Pagans and Satan and Goths, Oh My: Dark Leisure as Communicative Agency and Communal Identity on the Fringes of the Modern Goth Scene." *World Leisure Journal, 54* (4), 350–362. https://doi.org/10.1080/04419057.2012.720585

Spracklen, K. (2013). *Leisure, sports & society*. Palgrave Macmillan.

Stojarová, V. (2020). Media in the Western Balkans: Who controls the past controls the future. *Southeast European and Black Sea Studies, 20*(1), 161–181. https://doi.org/10.1080/14683857.2020.1702620

Stone, P. R. (2006). A dark tourism spectrum: Towards a typology of death and macabre related tourist sites, attractions and exhibitions. *Turizam: Međunarodni Znanstveno-Stručni Časopis, 54*(2), 145–160.

Stone, P. R. (2009). Dark tourism: Morality and new moral spaces. In R. Sharpley & P. R. Stone (Eds.), *The Darker Side of Travel: The Theory and Practice of Dark Tourism* (pp. 56–72). Channel View Publications.

Stone, P. R. (2016). Enlightening the 'dark' in dark tourism. *Interpretation Journal, 21*(2), 22–24.

Stone, P. R., & Sharpley, R. (2013). Deviance, Dark Tourism and 'Dark Leisure': Towards a (re)configuration of morality and the taboo in secular society. In S. Elkington & S. Gammon (Eds.), *Contemporary Perspectives in Leisure: Meanings, Motives and Lifelong Learning* (pp. 54–64). Routledge.

Šuligoj, M. (2016). Memories of war and warfare tourism in Croatia. *Annales. Series Historia et Sociologia, 26*(2), 259–270.

Šuligoj, M. (2017). Warfare tourism: an opportunity for Croatia? *Economic research-Ekonomska Istraživanja, 30*(1), 439–452. https://doi.org/10.1080/1331677X.2017.1305800

Šuligoj, M. (2019). Dark events of the Istrian countryside: An electronic media perspective. *Academica Turistica, 12*(2), 121–132. https://doi.org/10.26493/2335-4194.12.121-132

U.S. Department of State. (1995). Dayton Accords. Retrieved from https://www.state.gov/p/eur/rls/or/dayton/.

United Nations. (2015a). Transforming Our World: The 2030 Agenda for Sustainable Development: Preamble and Declaration. Available from: https://sustainabledevelopment.un.org/post2015/transformingourworld

United Nations. (2015b). *Transforming Our World: The 2030 Agenda for Sustainable Development*. United Nations.

UNWTO. (2020). *Sport Tourism*. Retrieved from https://www.unwto.org/sport-tourism.

Vejnović, D. (2014). *Nasilje i sport. Uzroci, posljedice i strategije prevazilaženja*. Evropski defendologija centar za naučna, politička, ekonomska, socijalna, bezbjednosna, sociološka i kriminološka istraživanja.

Vitic, A., & Ringer, G. (2008). Branding post-conflict destinations: Recreating Montenegro after the disintegration of Yugoslavia. *Journal of Travel & Tourism Marketing, 23*(2–4), 127–137. https://doi.org/10.1300/J073v23n02_10

Volcic, Z., Erjavec, K., & Peak, M. (2014). Branding post-war Sarajevo: Journalism, memories, and dark tourism. *Journalism Studies, 15*(6), 726–742. https://doi.org/10.1080/1461670X.2013.837255

Vrcan, S. (2003). *Nogomet – politika – nasilje*. Jesenski i Turk.

Winter, T. (2008). Post-conflict heritage and tourism in Cambodia: The burden of Angkor. *International Journal of Heritage Studies, 14*(6), 524–539. https://doi.org/10.1080/13527250802503274

Wise, N. A., & Mulec, I. (2012). Headlining Dubrovnik's tourism image: Transitioning representations/narratives of war and heritage preservation 1991–2010. *Tourism Recreation Research, 37*(1), 57–69. https://doi.org/10.1080/02508281.2012.11081688

Wood, S. (2013). Football after Yugoslavia: Conflict, reconciliation and the regional football league debate. *Sport in Society, 16*(8), 1077–1090. https://doi.org/10.1080/17430437.2013.801225

Zandberg, E. (2010). The right to tell the (right) story: Journalism, authority and memory. *Media, Culture & Society, 32*(1), 5–24. https://doi.org/10.1177/0163443709350095

Zhang, J. J. (2017). Rethinking 'heritage' in post-conflict tourism. *Annals of Tourism Research, 66*(C), 194–196. https://doi.org/10.1016/j.annals.2017.06.005

Peacebuilding and post-conflict tourism: addressing structural violence in Colombia

Mónica Guasca, Dominique Vanneste and Anne Marie Van Broeck

ABSTRACT

The contribution of tourism to peacebuilding in areas emerging from destructive conflicts has been primarily examined from the perspectives of economic recovery and reconciliation between antagonist groups. Yet we argue that the study of the tourism-peace nexus in post-conflict spaces should engage further in analysing issues of power, social justice, and development. The present article aims to enrich this debate by discussing the case of Montes de María, a war-stricken Colombian region where empowered communities are harnessing tourism as a means to defend local visions of peace and sustainable development, while neoliberal approaches threaten to create new forms of dispossession in the post-conflict transition. Drawing on qualitative research, we explore the opportunities that tourism presents for the rural communities in the region to resist structural violence and support their emancipatory struggle for self-determination and the defence of their livelihoods.

Introduction

Tourism has been increasingly proposed as a contributor to peace and stability in post-conflict areas. Due to its capacity to generate employment and revenues across different sectors, governments and international institutions have given greater importance to tourism during reconstruction periods, presenting it as a means to diversify the economy and tackle poverty (Novelli et al., 2012; World Tourism Organization [UNWTO], 2018).

Whilst the contribution of post-conflict tourism to peace has been understood primarily through its role in revitalising war-torn economies, its potential to rebuild the social fabric in divided societies has gained prominence in recent years. Following the call to recognise the sector not only as an economic enhancer but also as a social force (Higgins-Desbiolles, 2006), several scholars and practitioners have sought to transcend the income-oriented discourse, focused on the tallying of jobs and profits, and have engaged in examining the significance of tourism for individual and collective healing, inter-group dialogue, and a culture of non-violence (Anson, 1999; Causevic & Lynch, 2011; Haessly, 2010).

Nevertheless, the study of peace through tourism in societies emerging from destructive conflicts remains surprisingly thin and is guided mostly by a conceptualisation of peace based on harmonious relationships. Indeed, most research to date has focused almost exclusively on the repaired social ties between previous enemies, while critical reflections about tourism's role in promoting inclusive development in post-conflict transitions are often missing. Therefore, the opportunities that the sector presents to support peacebuilding beyond economic recovery and reconciliation are less clear, and its relevance in contexts where there is still no or limited contact between opposing sides is overlooked.

We want to address this imbalance by presenting the case of Montes de María, a Colombian region where the armed conflict has left its deepest wounds. In this region, empowered communities are harnessing tourism as a tool to defend local visions of peace and sustainable development, while neo-liberal approaches threaten to create new forms of dispossession in the post-conflict transition. Drawing on qualitative research, this article sets out to critically discuss the role of tourism in supporting peacebuilding in local communities, showing that just as it could work for the benefit of powerful elites, it can also assist the struggles of the marginalised (Blanchard & Higgins-Desbiolles, 2013).

Our aim is to enrich the debate on post-conflict tourism and peacebuilding by engaging further on issues of power, social justice and development, acknowledging the deeply political nature of tourism (Matthews & Richter, 1991). Moreover, responding to the call for more interdisciplinary dialogue between tourism studies addressing peace and political science, our analysis also relies on the critical scholarship of the peace and conflict discipline, which seeks to unravel the significance of local agency in shaping peace from the bottom-up.

The article starts by presenting insights from peace and conflict studies to reach a better understanding of the risks facing local communities in post-conflict transitions. We then briefly present the research settings and describe our methodological approach. Next, the case of Montes de María is introduced, outlining the regional context before, during and after the armed conflict. The subsequent sections discuss how rural communities are using tourism in their struggle to promote inclusive development and peace in their territories. We conclude with reflections on the opportunities and threats for our case study and, more in general, on the study of post-conflict tourism and peacebuilding at a micro-level.

Structural violence in post-conflict contexts: a challenge for positive peace

Although peace involves a kaleidoscope of meanings, there is a growing tendency to understand it not only as a security matter but as a developmental endeavour. Since the influential work of Galtung (1969), the concept of peace has been conceived of as two-fold: it implies both the absence of direct violence (e.g. physical aggression) and of structural violence (e.g. harm caused through systems of oppression), the latter referring mainly to social justice. Hence, removing direct violence from the equation is not enough, and a lasting peace – or 'positive peace' in Galtung's words – is better understood not solely in terms of *absence*, but in terms of *presence* of certain factors that can help address the root causes of the conflict. For instance, the presence of policies that promote equality, of participatory governance, of conflict resolution mechanisms, and so forth.

This conceptualisation of peace resonates with the international commitment conveyed in the 2030 Agenda for Sustainable Development to promote just, peaceful and inclusive societies, taking into consideration that 'sustainable development cannot be realized without peace and security; and peace and security will be at risk without sustainable development' (United Nations [UN], 2015, p.11). From this statement, it becomes evident that the understanding of peace has moved beyond its traditional view 'as a carrier of feelings of universal love and brotherhood' (Galtung, 1969, p.185), to a more holistic view aligned with principles of sustainability; like self-reliance of local communities, dignified livelihoods and environmental protection (Sharpley, 2000).

The interdependent relationship between sustainable development and peace is especially conspicuous in post-conflict territories, where the end of violent confrontation does not necessarily translate into positive peace. Certainly, the term 'post-conflict' is misleading. Even though these spaces are emerging from large-scale organised violence, it does not imply that they are necessarily undergoing war-to-peace transitions, nor that the conflicts that gave rise to the confrontation are resolved (Heathershaw & Lambach, 2008). In fact, nowhere is it more burdensome to promote peace than in post-conflict environments, which are better described as 'no peace, no war' zones (Richards, 2005).

Peacebuilding in post-conflict contexts is a daunting challenge because, in addition to the need to recover from the long-lasting physical and psychological effects of war, affected communities are highly vulnerable to new forms of structural violence under the guise of economic reconstruction. Undoubtedly, economic development is a crucial element for peace in societies that normally face alarming poverty rates in the aftermath of violence. Moreover, it has been demonstrated that low per capita income is correlated with the outbreak of wars (Collier et al., 2008). However, peace is not a natural consequence of economic growth, as this growth can be uneven, consolidating trends of social and spatial inequality and being counterproductive to peace.

Critical peace researchers have provided several examples of war-torn areas where the benefits of ending overt violence did not trickle down to the everyday lives of local communities. Instead, the 'pacification' of those spaces was seen as a suitable opportunity to make investments in favour of privileged groups who imposed economic and social reforms based on neoliberal postulates. This practice constitutes an example of what Klein (2007) termed 'disaster capitalism', pointing to the opportunism of powerful groups to use crises and times of transition to their advantage. For instance, in Liberia, Namibia and Mozambique the benefits of ending the war went to large companies, while unemployment rates rose up to 60 percent (Richmond & Mitchell, 2011). In Bosnia and Herzegovina, East Timor and Kosovo, the post-conflict stage was marked by rising poverty rates and precarious livelihoods in spite of national economic growth (Pugh, 2005). Countries receiving major economic assistance for peacebuilding purposes like Burundi, Somalia, or Rwanda continue to rank low in development indicators (Mac Ginty & Richmond, 2013). These findings have prompted a vigorous debate about issues of power in peacebuilding, and have fuelled the interest in identifying who promotes a specific vision of peace and, ultimately, who benefits from it (Lederach, 1997; Richmond, 2009).

Hence, post-conflict spaces are better understood as 'fields of power' (Heathershaw & Lambach, 2008), where there exists a latent peril that political and economic elites, guided purely by capitalist interests, perpetuate systems of oppression. The risk is significant considering the prevalence of top-down approaches in peacebuilding interventions that place unfettered markets as a basic pillar of peace (Paris, 1997). A clear example comes from the ubiquitous land issues present in most post-conflict contexts, as internally displaced people more often than not struggle to get access to land – which is in many cases at the basis of their livelihoods – because it goes to the hands of domestic or international investors under the belief that land is a resource best regulated by the market, not the state (Grajales, 2020).

The mechanisms that create these new forms of dispossession are facilitated by what Said (1978) called 'imaginative geographies', referring to the skewed representation of places as barren and peoples as too primitive to exercise control over them. Indeed, peace scholars have identified a tendency from international, national and local elites to conceive war-shattered territories as *terra nullius*, as empty containers devoid of ownership, detached from historical uses, and as 'wasted spaces simply waiting to be transformed' (Mitchell & Kelly, 2011, p. 308); while local communities are seen, at best, as traumatised, dysfunctional, and too damaged to take responsibility for their future (Donais, 2009) and, at worst, as 'unenlightened, dangerous [and] uncivilised', as the ones to blame for the violence (Mac Ginty, 2015, p. 840). Therefore, they are not recognised as having any decision-making ability, and their needs and wishes are usually ignored.

Bringing the discussion back to the tourism realm, we must acknowledge that these imaginaries and dynamics of dispossession have also been reinforced by the tourism industry in various post-conflict spaces. Evidence from Sri Lanka (Buultjens et al., 2016), Cambodia (Winter, 2008) Bosnia and Herzegovina (Volcic et al., 2014), Cyprus (Scott, 2012), and Colombia (Ojeda, 2012), demonstrated that tourism can be a source of structural violence, particularly when large operators advanced their interests to the detriment of several local communities. In these contexts, tourism is used as a gateway for the privatisation of common goods, deregulation, displacement, and the commodification of people, places, cultures, and pasts (Büscher & Fletcher, 2017; Devine & Ojeda, 2017).

Nonetheless, this is a rather bleak and incomplete picture of the impacts of tourism in conflict-torn areas. Tourism does not always oppose peace and is not just promoted by greedy actors. As mentioned before, there is an increasing body of research on reconciliation pointing at the success of locally-driven initiatives in bridging the divide between antagonist groups. Such have been the cases of Turkish-Cypriot and Greek-Cypriot communities in Cyprus (Sonmez & Apostolopoulos, 2000), Unionist and Republican parties in Northern Ireland (Simone-Charteris & Boyd, 2010), opposing ethnic groups in Bosnia & Herzegovina (Causevic & Lynch, 2013), and contending communities in Rwanda, Uganda and the Democratic Republic of Congo (Alluri, 2009). Furthermore, tourism has also facilitated contact between civilian communities and former combatants, supporting the social reintegration of demobilised fighters in Namibia and Colombia (Guasca et al., 2020; Metsola, 2006; Van Broeck et al., 2019).

Still, when considering the insights from critical peace studies, it becomes evident that the potential of tourism to contribute to positive peace remains underexplored. Peace scholars warn about the risk of perpetuating structural violence in post-conflict transitions, so the benefits of halting direct violence do not translate into better livelihoods for the people on the ground, who struggle to make their voices heard and are burdened with a set of stigmas. Therefore, there is an increasing need to understand how vulnerable populations can gain recognition as active agents in the reconstruction of their territories. It is then worth delving deeper into the significance of bottom-up tourism initiatives in serving such purpose.

To explore further this potential, we will discuss the case of Montes de María, a region where new forms of structural violence are present in what is now considered a 'post-conflict Colombia', and where rural communities are starting to open their doors to tourists.

Research settings and methods

Montes de María is an inland region in the Colombian Caribbean, including municipalities from the departments of Bolivar and Sucre, and within driving distance of Cartagena, the leading Colombian tourism destination (Figure 1). Two main roads connecting the coast with the interior of the country pass through the region, which makes it a strategic area for the transport of goods and people. It covers 6,466 km^2 – comprising 15 municipalities – and has a population of about 372,504 inhabitants (Departamento Administrativo Nacional de Estadística [DANE], 2018). The region is highly rural, and is home to indigenous (5.1%), afro-descendant (29%) and peasant (65.9%) communities (Duarte, 2016). It is also one of the regions most affected by the internal armed conflict, where development indicators remain below the national average (Programa de las Naciones Unidas para el Desarrollo [PNUD], 2010).

As this study takes on an exploratory stance, we opted for qualitative research methods to better grasp the complexity of the context. Data were collected during fieldwork in August 2018 and October 2019. During the first data collection, we visited various municipalities, which allowed us to identify main actors and projects, and led us to select San Jacinto and San Juan as the most promising research areas for the second data collection. Therefore, our research was mainly focused on these two municipalities, and particularly on peasant communities. Given the impossibility to come back to the area due to the Covid-19 pandemic, in June 2020 we

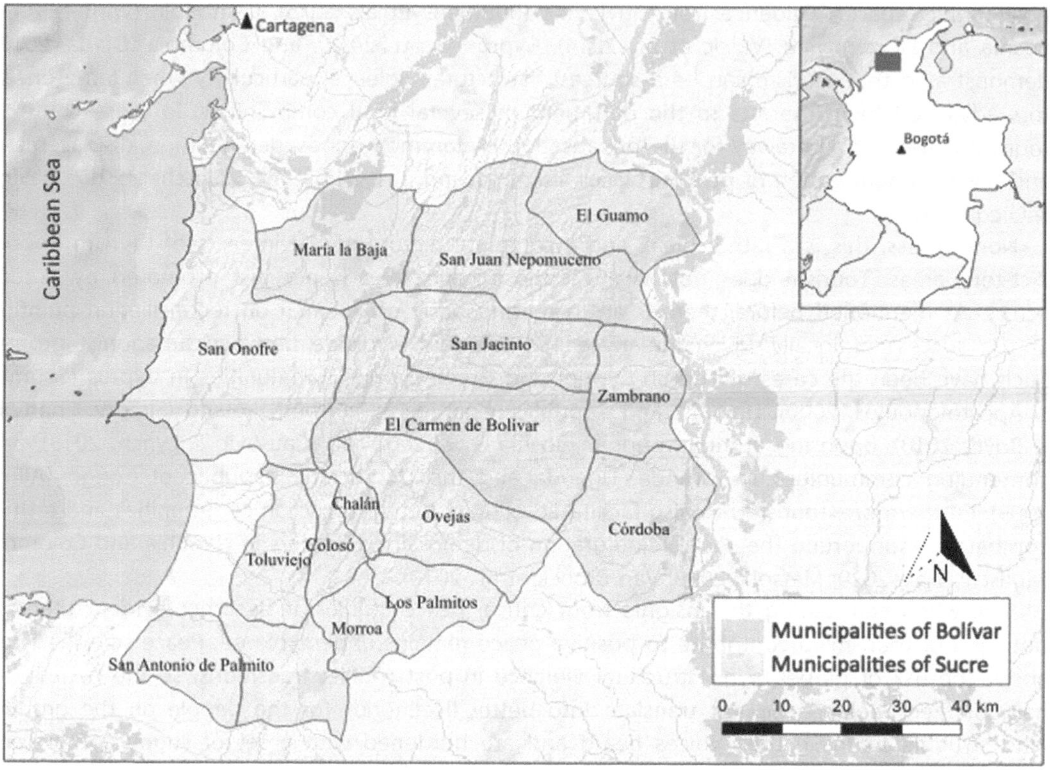

Figure 1. Geographical location of the municipalities belonging to the Montes de María region. Own mapping.

conducted telephone interviews with key informants identified during the second round of fieldwork.

We favoured semi-structured interviews and informal conversations with local tourism entrepreneurs, local inhabitants engaged in tourism and various stakeholders who are supporting tourism development and peace initiatives in the region. We deliberately set the tone of the interviews in a conversational style, and we allowed interviewees to ramble, which proved to be the best way to build a connection with them (many of whom are direct victims of the conflict) and get significant information from what seemed to be trivial accounts. The interview protocols also included questions that allowed us to understand local visions of peace, the micro and macro dynamics in which tourism was developed and the transformations of spaces that were taking place. We did not question their experience during armed conflict for privacy reasons, but sometimes the interviewees mentioned the subject spontaneously.

We used purposive sampling looking for representation of different stakeholders, followed by snowball sampling. In total, we conducted 31 semi-structured interviews in the two municipalities that were prioritised with local entrepreneurs promoting eco-agritourism (18) (from which six are leaders of peasant associations), regional and national government officials (6), local tourism and peace researchers (4), and representatives of conservations projects in the area (3). The average interview time was about one hour. We did not target tourists as research participants, as we were primarily interested in identifying who was promoting what type of tourism, along with their motivations to engage and expected outcomes. Through the interviews we also sought to understand to what extent our respondents linked tourism development in the region and peacebuilding. Did respondents perceive tourism solely as an additional source of income? What narratives were shared in tourism spaces and for what purposes? What concepts did the respondents associate with their interpretation of peace? And, ultimately, which were their hopes and wishes for the future.

The interviews were conducted by the main author who, being a Colombian but not a native of Montes de María, was both an insider and an outsider. The insider role helped to build rapport with the participants, since the interviewer shares the experience of living in a country at war, albeit not exposed to the same level of violence as the majority of those interviewed. The outsider role came in handy when approaching different stakeholders, as neither the interviewer nor the co-authors have political or institutional affiliations in the region or the country. This reassured participants that our objectives were purely research-related and that we were not advancing someone else's agenda. In fact, the main author arrived in the region as a newcomer, with a list of potential contacts but not being acquainted with them. It was by spending time in the region, walking the streets and engaging in conversations with locals that trust was gained. Indeed, the first stage of non-participatory observation was essential for the researcher to understand and adjust to local dynamics, and proved to be valuable for participants who manifested being more comfortable talking to someone 'who is not a complete stranger'.

The data analysis followed an inductive and deductive approach. The interviews were transcribed and coded using qualitative content analysis with the help of NVivo™ software. At this stage, it was essential to reflect on how the main author's personal values and beliefs may have influenced the research, recognising the possible effects of biases and trying to address them. The process of self-reflexivity was aided by a continuous discussion with the co-authors who, not being Colombian, were able to distance themselves from the data and ask questions to challenge the main author's assumptions. In addition, interview data and notes from participant and non-participant observation were triangulated with desk research to validate the information with other sources.

Finally, it should be noted that in the presentation of the results we decided to highlight the voices of some research participants, particularly small farmers engaged in tourism initiatives, as they reflect the opinions and feelings shared by most of our interviewees, who also work closely with these rural communities. The selected quotes have been translated from Spanish by the main author. All participants were assigned pseudonyms to ensure anonymity.

Montes de María: a history of peasant resistance and armed conflict

Montes de María has a long history of territorial disputes and armed confrontation. In the second half of the twentieth century, it hosted several actors who competed for space and resources, including large landowners, peasants, state military forces, drug traffickers and illegal armed groups. What began as a demand for the democratisation of the agrarian property, turned into a bloody war affecting thousands of civilians (Centro Nacional de Memoria Histórica [CNMH], 2013).

The land disputes started in the 1960s, when peasant movements embarked in a decades-long struggle against the monopoly of land ownership. Traditionally, small farmers leased plots from large landowners to cultivate crops such as cassava, yam, maize and avocado, making the region a bountiful source of agricultural produce for the Caribbean (PNUD et al., 2003). However, landowners began to prioritise extensive ranching instead of agricultural activities, leaving many landless farmers without the means to earn a living (Fundación Ideas para la Paz [FIP], 2011; PNUD, 2010). The peasants decided to fight for access to the land, and in 1967 created the National Association of Peasant Users (*Asociación Nacional de Usuarios Campesinos* - ANUC) to jointly demand for the redistribution of rural land. The ANUC became the most influential peasant organisation in Colombia, gaining the support of the government of Lleras Restrepo (1966–1970) that pushed forward an agrarian reform. However, the reform failed to deliver a full land redistribution as landowners pulled strings to halt the process, and the vast majority of farmers were left empty-handed. Simultaneously, landowners sought to curb the peasant movement by persecuting its leaders and expelling the remaining tenants from their properties (around 30.000 families) (FIP, 2011), but the repression had the opposite effect.

Between 1970 and 1975, the peasants decided to invade the land under the motto *'la tierra es pa'l que la trabaja'* (the land belongs to the one who works it), which unfortunately opened the path for the arrival of armed groups in the region (Comisión Nacional de Reparación y Reconciliación [CNRR] & Semana, 2010). The existing guerrilla groups at the time, such as the Workers Revolutionary Party (*Partido Revolucionario de los Trabajadores* – PRT) and the Socialist Renewal Stream (*Corriente de Renovación Socialista* – CRS), saw a political opportunity to enter Montes de María by offering assistance to the repressed peasant movement, while the landowners created private protection groups to safeguard their properties. These private groups would later become paramilitary forces organised under the United Self-Defence Forces of Colombia (*Autodefensas Unidas de Colombia* – AUC) (FIP, 2011). Even though the peasants opposed the guerrilla intervention, their movement was perceived as a breeding ground for subversion, and they were also persecuted by the paramilitaries for their alleged collaboration with the insurgency.

Tensions kept rising during the 1980s, but it was in 1991, with the arrival of the guerrilla group Revolutionary Armed Forces of Colombia (*Fuerzas Armadas Revolucionarias de Colombia* - FARC), that the conflict escalated to full-fledged warfare (PNUD et al., 2003). The FARC filled the void left by the guerrilla groups like PRT and CSR that demobilised in the late 1980s though, unlike those groups, they did not enter the region as allies of the peasant resistance, but as new contenders for political and economic dominance (CNMH, 2013). The decade of the 1990s was marked by a terror regime imposed by the FARC and the AUC, as both groups engaged in a bloody battle for territorial control. The strategic location of Montes de María close to the Atlantic Ocean, coupled with its mountainous geography and the lack of state presence, provided the perfect conditions for fighters to shelter and to manage the distribution routes for drugs and contraband weapons, which provided funding and resources to enlarge their military structures (FIP, 2011).

The armed groups controlled Montes de María until the early 2000s, when the government of Uribe Velez (2002–2010) started a fierce military campaign to recover the territory. The FARC, which by then had 650 combatants in the area, was completely expelled from the region in 2007. The AUC, which had around 600 fighters in the area, handed over weapons and demobilized in 2005 (Aguilera-Díaz, 2013; Plataforma de Organizaciones de Desarrollo Europeas en Colombia [PODEC], 2011). Nonetheless, despite the effectiveness in defeating the armed groups, the national army also committed serious human right violations against civilian peasants. For instance, several small farmers were registered as 'guerrillas killed in combat', without any evidence of their involvement in an illegal group (PNUD, 2010).

Ultimately, rural communities were the most affected by the armed conflict. The fear of dying at the hands of illegal armed groups or state forces led the peasants to abandon their plots and flee to urban areas. It is estimated that during the period 1998–2008, 160,000 people were forcibly displaced in Montes de María, most of them of rural origin (PNUD, 2010). The forced displacement also reversed the progress made to redistribute the rural land, with figures revealing that the peasants lost 63 percent of the land acquired through the agrarian reform (PODEC, 2011).

Although the exodus momentarily broke peasants' ties with their land, it did not affect their sense of belonging to the places they inhabited. Since the departure of the FARC and the AUC, small farmers are coming back to the land they still own, and claiming their right to be active agents in the reconstruction of the region, even though vast swathes of the territory are already in the hands of actors who do not support peasant agriculture, as will be explained later on. Against this backdrop of historical land disputes and mixed interest over land use, the reconstruction of the countryside in Montes de María remains one of the main contentious issues in the post-conflict period.

The countryside is open to 'progress'

As Montes de María grew more secure, rural development policies favoured landowners' investments in extensive livestock farming and business groups that entered the region to advance large-scale agro-industrial and commercial reforestation projects (PODEC, 2011; PNUD, 2010; Borda, 2002). With the incursion of palm oil crops for biofuel and commercial forestry plantations of teak and eucalyptus, new conflicts were added to the traditional dispute over the use of fertile land for cattle ranching and not for agriculture.

Peasant communities argue that this development model only benefits the 'rural-rich', excluding them and creating new forms of dispossession in the name of sustainability (Fajardo, 2014). The approach is perceived as being purely market-driven, seeking to satisfy the global demand for biofuels and to comply with the need for companies to offset their carbon emissions, while the expansion of these monocultures is far from promoting local development (Instituto de Estudios para el Desarrollo y la Paz [INDEPAZ], 2013; Mesa de Interlocución y Concertación de los Montes de María [MIC Montes de María], 2017). For instance, while communities do not have aqueducts and suffer from water shortages, the state has financed thirty irrigation districts for the palm crops, which is now the second most prolific crop, after corn, in terms of planted area in the region (Daniels Puello, 2016; Ojeda et al., 2015). Concerns are also raised about the predominance of energy crops over agricultural smallholdings since more than 60 percent of the population currently suffers from food insecurity in a region with extremely fertile soils for food production (Maza et al., 2017).

Moreover, the expansion of agro-industrial crops and forest plantations coincide with the areas of greatest forced displacement of peasant communities (Fajardo, 2014, MIC Montes de María, 2017). After 2005, massive land purchases were undertaken in the region, as displaced peasants 'voluntarily' sold their plots at bargain prices in what they saw as an opportunity to alleviate their poverty (CNRR & Semana, 2010). Although there is no exact number on the total amount of land purchased, estimates vary between 80.000 and 120.000 hectares, which were bought at about 80 euros a hectare and are now offered for more than 800 euros a hectare (Daniels Puello, 2016; PODEC, 2011). The phenomenon of massive purchases concentrated further on the land, resulting in a rural land concentration index (Gini Index) of 0.70[1] in 2011 (Aguilera-Díaz, 2013).

In the light of the rapid expansion of monocultures, which already occupy 10 percent of the region's area (70.000 hectares), and with 67 percent of the soils suitable for agroforestry being used for livestock activities (Daniels Puello, 2016; MIC Montes de María, 2017), rural communities in Montes de María are betting on a different development strategy based on the conservation of native forests and multi-cropping agriculture. Thanks to this work, it is now possible to find several farms, owned and run by the peasants, that are devoted to agroforestry– named by the locals as '*fincas montemarianas*'.

Resistance from the fincas montemarianas

The peasants who did not migrate during the times of violence and those who returned afterwards continue to work for the defence of the peasant agriculture, and started to engage in initiatives for the protection of the Tropical Dry Forest (TDF). The TDF is one of the most threatened ecosystems in Colombia, as indiscriminate deforestation had led to 90 percent of its loss (Pizano et al., 2017). Environmental organisations have launched a call to protect what remains and have given special attention to Montes de María, because the region hosts the Sanctuary of Flora and Fauna 'Los Colorados' that has 1.000 hectares of TDF, and there are still fragments of forest on private properties (about 12.000 hectares). Several farmers have heeded the call and have committed themselves to leave parts of their terrains for in-situ conservation.

Their conservation initiatives operate under the umbrella of a larger project that seeks to create green corridors by connecting the remaining fragments of forests, allowing animals to move freely. One of the project managers told us that there are currently more than 150 peasants involved, who agreed to maintain the existing patches of TDF and who are also helping to reforest additional hectares. According to him, in the six years that the project has been running, about 90.000 new native trees have been planted, animals who have for a long time not been seen in the area are returning, and waterholes are reappearing.

The revival of family agriculture and the thriving natural reserves present on the *fincas* made it possible to promote Montes de María as a destination for eco-agritourism. Aware of the tourism potential and hoping to take advantage of the region's proximity to Cartagena, local entrepreneurs (including peasant leaders) started to encourage tourism in rural areas and built alliances with travel agencies focused on sustainable tourism. It is important to clarify that, so far, there are no foreign tourism entrepreneurs investing directly in the region.

Tourism is still in its early stages. Although there have been local initiatives since the early 2010s, the sector started to take off in 2017 when Montes de María became visible as a destination to the national government and was included in the programme 'Tourism, Peace and Coexistence'. The programme is a legacy of the government of Juan Manuel Santos (2010–2018), which gave tourism a leading role in developing areas that have historically suffered the most from the armed conflict. In Montes de María, the programme is currently focused on supporting local entrepreneurs in the municipalities of San Jacinto, San Juan Nepomuceno, Carmen de Bolívar and María la Baja (located in the department of Bolívar), among which San Jacinto and San Juan have the most advanced eco-agritourism projects.

Tourism, sustainable development and peacebuilding: the reality on the ground

Tourists who visit *fincas montemarianas* come across a type of tourism that does not overemphasize the violent past and gives more attention to narratives of peasant resilience after the armed conflict. It is a tourism managed by local entrepreneurs that seek to provide additional income to rural communities but which, at the same time, is showing its potential to pursue broader aims, such as making visible the contribution of peasants to the region's renewal, breaking down stigmas of them as linked to underdevelopment while supporting their struggle to defend their livelihoods and the land they consider as their home. Tourism then plays on the side of the rural communities, whose voices have long been repressed, and is inextricably ingrained in their backlash against top-down discriminatory development models that are stripping them of their territories, as we will discuss in the following sections.

Bringing peasant's agency and local visions of development to the spotlight

Throughout our conversations with small farmers and local tourism promoters, it became clear that one of the main goals of tourism is to highlight peasants' resistance, together with their continuous defence of the peasant agriculture and their engagement in the conservation of the Tropical Dry Forest. By way of example, Carlos and Manuel, two peasant leaders and tour guides in the *Vereda Brasilar*, in San Jacinto, proudly describe to tourists how, because of their new agroforestry approach, their farms are now producing food in more sustainable ways and are, at the same time, host to an array of birds and even of animals that were at risk of extinction, like the cotton-top tamarin. 'Everyone who comes here is amazed by the diversity of animals, plants and crops they encounter', they told us.

Apart from getting to learn about the agroforestry techniques and the TDF flora and fauna, tourists are invited to 'live a full day as a farmer', as Sara, an enthusiastic young entrepreneur, explained to us. By way of illustration, in her father's farm, 'tourists not only learn how cocoa is grown, but can also plant a cocoa tree, harvest the pods, toast the grains and prepare their own chocolate'. This approach of encouraging visitors to harvest the food they are going to consume in the tours is shared by other farms that offer agritourism activities, with the purpose of helping them re-establish their connection with food and reminding them that it is the fertile land that provides it. 'They can see they are eating healthy food, directly from the ground', one farmer explained.

The tourism narratives also feature the peasants' efforts to recover the environmental memory of the territory. Farmers comment on how they are recovering native trees, such as the *Brasil* or the *Caracolí*, that have attracted new sources of water and have helped them better withstand periods of drought. Their efforts are also intended to recover native seeds, which are more resistant to adverse weather conditions. Their explanation on how this process is taking place is accompanied by reflections on the importance of food security in peacebuilding because as a social leader expressed, 'peace with hunger does not last' (Bungard, n.d). For example, Manuel shares with visitors his work to recover seeds that were lost during the time of violence, when people abandoned their plots. He explained:

> *We keep the seeds in calabashes so that they do not get damaged. We do not put any chemicals in it, only ash [...] Now, the big laboratories cannot force us to buy the seed at the price they want, because we have our own creole seeds.*

Tourism is then being used as a tool for environmental and social pedagogy, where small farmers remind visitors of the importance of restoring the natural balance of the environment and our connection as humans with nature. Further, these reflections get constantly mixed with references to their pride of being peasants and their mettle to remain in the territory, as Manuel expressed to us:

> *I am extremely proud. We acquired this land 31 years ago and here we are. They have come to buy them. The war came, we had to flee, lost animals, lost harvest. We have not recovered yet. Perhaps we may never recover one hundred percent. But we are in the struggle. We are proud to continue in our land. We are still proud to get some bananas and bring them to town, a bundle of yams, some melons, some watermelons, some beans [...] That is why we are here. We are proud to have a piece of land and do things to improve the quality of life for ourselves, our families, our communities, but we want that to transcend.*

It is precisely the desire for their message of resistance to go beyond the borders of their farms what has become one of the main motivations to engage in tourism. Of course, having an additional source of income is also a strong incentive to engage in tourism activities, but our findings suggest that the possibility to have a forum to make visible their contribution to the transformation of the region is equally attractive to them. Moreover, describing their work and lives in the countryside also supports their self-knowledge and reaffirmation of their individual and collective identities.

Disrupting imaginaries

Montes de María suffers from the 'imaginative geographies' of post-conflict spaces, which leads to ignoring local agency and the immaterial significance of the territory. Albeit tourism is primarily intended to highlight rural communities' resilience while focusing on their current projects, small farmers also expect that it helps counter decades of stigmatization in which they have been portrayed first as dangerous members of insurgent groups, and later as helpless victims associated with backwardness and underdevelopment (MIC Montes de María, 2017; Ojeda et al., 2015).

As it is difficult to ignore that these farms are steeped in a history of conflict, the visits also include some accounts of the times of overt violence. For example, peasants share with visitors how they became the scapegoats of the different armed groups and were persecuted by all sides, who considered them collaborators. Several farmers explained:

> *We did not belong to any group, but if the military came you could not tell them to leave. If the guerrillas came, you could not tell them that either. If the paramilitaries came, what could you do? How do you tell an armed man to go away? And then they said that we were helping the one or the other.*

Yet peasants clarify that their intention is not to awaken feelings of sadness by sharing stories of the conflict, but to send a message of hope. Indeed, a phrase that we encountered in most of our conversations is that they do not want to be seen solely as victims, but as 'survivors of the conflict'. Victor, a peasant leader who offers tours in his cocoa farm, told us for instance how the history of cocoa in Montes de María cannot be told without referring to the armed conflict, as this crop survived thanks to peasant defiance. With a note of pride in his voice, he explained how they confronted the FARC commanders who wanted to destroy their cocoa trees because they were financed by United States agencies: 'If we had been afraid, now there would be nothing sown here', he says. Tourism entrepreneurs who are not peasants themselves are also betting on a tourism that does not reinforce the image of victims in a war-shattered region, as expressed by the director of a local travel agency: 'we want tourists to fall in love with our territory, not to feel sorry for it'.

After the departure of the armed actors, the stigma turned into that of victims without the ability to take advantage of the region's potential. According to our interviewees, this was an underlying motive for the government's support of large business groups and landowners over peasant agriculture. The top-down development model, put in place to benefit large investors, is based on the assumption that local communities are incapable of taking the reins of their lives and see these lands as a resource ready to be inserted into global market dynamics (Borda, 2002). Victor asserted that small farmers in Montes de María, and all over the country, are still regarded with contempt, 'as illiterate, as *montañeros* (rustic and unrefined), as if we were the worst'. But he sees in tourism a way to shed that image by offering tours to researchers and students interested in agricultural practices, clarifying that for rural communities it is not about a confrontation between the modern and the traditional, it is about complementarity:

> *We have a tour called 'Cocoa and Peace', where we have a practical exchange with scientists. We can have a debate on the spiritual and artisanal practices that we have and on the scientific knowledge that they have. It is an exchange of ideas and knowledge. After that, we go hiking, we have lunch next to some natural pools and we end [the tour] with a chocotherapy and with the sounds of cumbia [folkloric music from Colombia].*

The idea of a peasant as unable to bring progress to the region goes hand in hand with the perception of her or him as an 'employee of large industries and not as the owner of the land' (PNUD, 2010, p.52). One peace researcher with years of experience working with rural communities in Montes de María referred to the agro-industrial development model being implemented in the region by political and economic elites as 'a way to proletarianize peasants, while they want to be the entrepreneurs of their territory, not employees with a minimum wage and job insecurity'. According to him, if the industrial megaprojects continue to expand, 'soon we will have barren fields, which can neither be cultivated nor does anyone want to visit. So, I think that tourism, if well-managed, can help to protect the territory'.

In fact, all of our research participants living in Montes de María agreed that the recently adopted top-down model based on monocultures is hampering local development and producing environmental damage. Taking the example of teak, Manuel elaborated the argument as follows: 'we realised that teak sterilises the soil. You plant an avocado tree and when you cut it, you can plant other crops, and they grow. But if you plant a teak tree and then cut it, no other crops grow there'. And Carlos added: 'we have large carbon gas capture woods, such as *Caracolí, Sabajón, Ceiba*. Teak is being planted just for business. It is about sowing and destroying. That is not our goal. For us, it is about sowing and preserving'.

Therefore, our findings indicate that eco-agritourism initiatives in the *fincas montemarianas* are opening spaces where rural communities – which account for 66 percent of the region's population (Duarte, 2016) – can disrupt the imaginaries of them as unskilled passive agents.

These communities have found in tourism a platform to vouch for a more holistic view of development that 'generates economic growth and distributes its benefits equitably; the one that protects the environment instead of destroying it; the one that empowers people instead of discriminating them', as described in a route of action proposed by more than 150 grassroots organisations in the region (PNUD et al., 2003: p.142, own translation). At the same time, they are demonstrating that victims in post-conflict areas 'despite the adversities are people who work their fingers to the bone', as a national officer put it.

Defending livelihoods and the territory

In Montes de María, the role of tourism in peacebuilding cannot yet be understood from the perspective of reconciliation between antagonist groups. Certainly, tourism has promoted collaboration among members of the community, helping to re-establish social ties that were broken by mass displacement or simply by fear to interact with others in times of war. However, to our knowledge, there are still no eco-agritourism projects involving ex-combatants from any armed group. When asked about the willingness to set up collaboration with demobilised combatants, tourism entrepreneurs invariably answered positively. So, at least from their responses, we infer that there is room for cooperation, although it might be an option for the medium to long-term, since the wounds that armed actors left in the region are still very recent.

Instead, tourism's contribution to peace needs to be addressed from the lenses of social justice. In the case of rural communities in Montes de María, it starts by guaranteeing their right to stay and have a decent living on land that they conceive for habitation and not merely for production (MIC Montes de María, 2017). Access to land is a key element for peacebuilding. First, because it provides the means of subsistence for more than half of the population, and second because the land has been a historical aspiration and is strongly anchored with peasant identity (Lederach, 2017).

Indeed, the significance of the territory for small farmers, in terms of appropriation as much as property, was an element echoed throughout our conversations. For instance, Carlos continuously stressed his attachment to the countryside: 'This is our home. If it is not here, where are we going to go? I like the countryside. It is calm. We have cleaner air and accompanied by the chirps of the birds'. And in the words of Victor:

> *[This territory] is one's life. I cannot imagine myself anywhere else. I am 45 years old and I think 'well, my pension is going to be this farm, with a good house'. I never aspired to live in a city. I would not change this for anything else. Not even for a building.*

In such scenario, tourism is inextricably embedded in a broader emancipatory struggle to reaffirm ownership over the territory. By becoming hosts and sharing their stories and knowledge with visitors, peasants are undergoing a process of 're-appropriation of their bio-cultural territories' (Porto-Gonçalves & Leff, 2015, p. 65), which is allowing them to reify the power they still hold on those spaces; a power that emanates from land tenure but, above all, from rooting, ancestral presence and affection for that territory. Defending it is, for these small farmers, a matter of protecting life itself (Escobar, 2014).

The struggle to preserve their livelihoods anchored to the territory implies thinking about the future of these spaces as well. Farmers fear that, as they age, new generations will not continue with their efforts to defend their right to land because the violence uprooted them from the countryside. A local tourism entrepreneur mentioned that 'there are young people who, being in their twenties, knew nothing of the countryside. They could not go because their parents, out of fear, would not let them [...] [With tourism] we have taken away that fear. People now go without fear to the countryside'. Another entrepreneur told us that the aim of encouraging young people residing in the municipal towns to visit the countryside is to 'awaken their attachment to the territory', and to show them that their options are not limited to agriculture; tourism can also be a possibility to make a living.

The idea to involve young generations in the development of eco-agritourism is partly driven by the farmers' desire to guarantee the continuity of their work, but there is also the pressing need to generate sources of income for young people and prevent them from swelling the ranks of armed groups. The lack of education and employment opportunities for youth is evident, and the consequences are reflected, among other issues, in the high teenage pregnancy rates and the increasing drug consumption among adolescents. Many unemployed youngsters are now offering services as motorcycles taxis (*mototaxismo*), earning less than a minimum wage and without any further career prospects. This situation makes them extremely vulnerable to recruitment by gangs, traffickers, dissident groups or active guerrilla groups, like the National Liberation Army (*Ejército de Liberación Nacional* - ELN). This fear, shared by most of our interviewees, is reflected in the words of Victor:

> *I cannot stand another war like that. Now we are older and the health is not the same. It is scary. It is very scary that it could happen again. That is why we need to talk to young people. They [illegal groups] always look for young people for their things, to be included in those groups. [We need to] talk to young people so that they have it clear and do not support those bad habits.*

Final remarks

For our case study, discussing the role of tourism in peacebuilding in terms of its potential is both hopeful and distressing. On the one hand, although still in its infancy, we see how tourism can embed the symbolic, filling spaces marked by violence with new meanings, helping to consolidate the ties of rural communities with their lands and supporting their recognition as decisive actors in the region's recovery. It is a take on tourism that, we believe, could contribute to sustainable development from an economic, social and environmental perspective. On the other hand, there is still no regulatory framework that protects grassroots approaches to tourism, increasing the risk of large companies entering the territory and advancing projects that are not in line with the sustainability principles that local communities want to promote.

Hence, our aim is not to give a romanticised view of the sector because, as mentioned before, some types of tourism can also neglect the well-being of host communities, becoming a source of inequality. In addition, it is also possible that many bottom-up initiatives are not successful in consolidating tourism, as communities might not be aware of maybe unable to address the inherent difficulties in advancing community-based approaches to tourism (Salazar, 2012), and the myriad of challenges to develop tourism in post-conflict settings (Novelli et al., 2012; Richter, 1999).

Furthermore, the national and regional outlook for the tourism potential in Montes de María to be fulfilled is not the most favourable. Throughout our interviews, we heard repeated concerns about the setbacks in the implementation of the peace agreement with the FARC, which could lead to the rearmament of ex-guerrillas. We also heard about the emergence of new paramilitary groups in nearby regions. In particular, several of our research participants expressed their worry about the systematic killings of social leaders all over the country, and worry that promoting community-based tourism could make them a target for extreme-right neo-paramilitary groups.

It is undeniable that Colombia is currently facing significant threats to peace and security, jeopardising the progress made so far. New waves of violence are targeting social and environmental leaders, who are the backbone of peace processes at the community level. Since the signing of the peace agreement with the FARC in 2016, and as at September 2020, the United Nations Verification Mission in Colombia has confirmed the assassination of 351 social leaders and human rights defenders, urging the government to not turn a blind eye to this tragedy (UNVMC, 2019, 2020). Due to the inaction of the government and the power vacuum left by the FARC (which the State has not been able to fill), regions such as Montes de María are vulnerable to new criminal groups that seek to take control of the territory, possibly to reactivate drug trafficking routes in alliance with Mexican cartels. This is a real threat for tourism development, but above all for communities that fear being caught in the crossfire once again.

One additional matter to consider is the impact of the Covid-19 pandemic for these grassroots initiatives. In this respect, our respondents remain optimistic about the future, hoping that this situation becomes a catalyst for domestic tourism and an occasion to change the mental maps of Colombian people, who still see Montes de María as a 'no-go area'. We also believe that it might open the opportunity to focus much more on the rights of local communities, and on alternatives forms of tourism away from masses and more aligned with the tenants of sustainability (Higgins-Desbiolles, 2020).

Conclusion

The case of Montes de María is not exceptional. It is but one of many examples of post-conflict spaces where new forms of victimisation emerge in the aftermath of direct violence, and where empowered communities (be they peasant, indigenous, Afro-descendant, etc.), use tourism not just to make a living, but also to try to restore their relationships with their territories. They are defending their cosmovision, defending their right to decide over their lives and the spaces they inhabit. These are communities that, as Escobar (2014) puts it, are 'advancing ontological struggles [...] for the defence of the many worlds that inhabit the planet' (p. 77).

Unfortunately, there is limited discussion on how bottom-up tourism initiatives in places recovering from war contribute to positive peace – framed in terms of presence of elements that support decent livelihoods and social justice. In this article, we sought to address that gap and argued that peace through tourism in post-conflict contexts is not limited to spurring economic growth and assisting reconciliation between opposing groups. Both elements are undoubtedly significant to peace, but the implications of tourism are greater than this. We wanted to highlight its potential to help tackle structural violence when tourism serves as a tool for the reparation of affected communities and not for their re-victimisation. Specifically, we illustrated how it could support expressions of resistance from local communities against new forms of dispossession, putting a spotlight on their active agency in rebuilding their territories.

Furthermore, we want to stress that placing local communities as key actors in peacebuilding also involves defining peace through a bottom-up lens (Koopman, 2011). Rather than trying to compartmentalise peace in terms of indicators or study these areas with a pre-conceived notion of what peace should entail, the aim should be to amplify the voices of people on the ground and to understand what peace means to them in their everyday lives. It also serves as a reminder to pay more attention to how spaces are reconfigured in post-conflict transitions, particularly in conflicts that had a strong territorial dimension. Nonetheless, a bottom-up approach to peacebuilding does not mean putting everything in the hands of locals and freeing external actors from their responsibility to help. For sure, external and top-down assistance is needed to make a transition towards a more peaceful tomorrow.

Finally, we have to acknowledge that tourism is just a piece in the peace puzzle. Societies that have suffered the heart-breaking reality of conflict carry with them the loss of their human, cultural, natural, and physical capital, and are left with legacies of the violence that can take decades to overcome. We must be realistic about the sector's limitations and should not reinforce overambitious discourses that promote it as a universal driver of peace. Therefore, we call for more attention to conflict-affected areas when studying the tourism-peace nexus, as they offer a fertile ground to identify the complex interrelations between positive peace, tourism and sustainable development. As we engage further with critical inquiry, being able to produce knowledge that can drive social change in war-torn societies is a call difficult to ignore; especially in present times when violent conflicts seem to escalate worldwide.

Note

1. The Rural Land Gini Index is a measure of the degree of concentration of rural land at a defined time in a specific spatial reference unit. It takes into account the number of people who exercise the legal right of ownership over the land. The indicator takes the value of 0 when rural land is not concentrated, meaning that the land is equally distributed. The increase in value of the indicator represents a greater degree of concentration of the rural property. The indicator takes the value of 1 when the land is fully concentrated (Instituto Colombiano Agustín Codazzi [IGAC], n,d; own translation).

Acknowledgments

The authors would like to thank the three anonymous reviewers for their helpful comments.

Disclosure statement

No potential conflict of interest was reported by the authors.

ORCID

Mónica Guasca http://orcid.org/0000-0002-2679-262X

References

Aguilera-Díaz, M. M. (2013). *Montes de María: Una subregión de economía campesina y empresarial*. Banco de la República.

Alluri, R. M. (2009). *The role of tourism in post-conflict peacebuilding in Rwanda*. Swisspeace.

Anson, C. (1999). Planning for peace: The role of tourism in the aftermath of violence. *Journal of Travel Research*, *38*(1), 57–61. https://doi.org/10.1177/004728759903800112

Blanchard, L. A., & Higgins-Desbiolles, F. (Eds.). (2013). *Peace through tourism: promoting human security through international citizenship*. Routledge.

Borda, O. F. (2002). *Historia doble de la costa, Volumen IV, retorno a la tierra*. Universidad Nacional de Colombia, Banco de la Republica, El Ancora Editores.

Bungard, K. (n.d.). *"Soy un optimista por naturaleza. Mi ideología es la esperanza": Ricardo Esquivia Ballestas*. https://diariodepaz.com/2019/09/04/soy-un-optimista-por-naturaleza-ricardo-esquivia/

Büscher, B., & Fletcher, R. (2017). Destructive creation: Capital accumulation and the structural violence of tourism. *Journal of Sustainable Tourism*, *25*(5), 651–667. https://doi.org/10.1080/09669582.2016.1159214

Buultjens, J. W., Ratnayake, I., & Gnanapala, W. K. A. C. (2016). Post-Conflict tourism development in Sri Lanka: Implications for building resilience. *Current Issues in Tourism*, *19*(4), 355–372. https://doi.org/10.1080/13683500.2014.1002760

Causevic, S., & Lynch, P. (2011). Phoenix tourism. *Annals of Tourism Research, 38*(3), 780–800. https://doi.org/10.1016/j.annals.2010.12.004
Causevic, S., & Lynch, P. (2013). Political (in)stability and its influence on tourism development. *Tourism Management, 34*, 145–157. https://doi.org/10.1016/j.tourman.2012.04.006
Centro Nacional de Memoria Histórica (Ed.). (2013). *¡Basta ya! Colombia, memorias de guerra y dignidad* (Segunda edición corregida). Imprenta Nacional.
Collier, P., Hoeffler, A., & Söderbom, M. (2008). Post-conflict risks. *Journal of Peace Research, 45*(4), 461–478. https://doi.org/10.1177/0022343308091356
Comisión Nacional de Reparación y Reconciliación & Ediciones Semana. (2010). *La tierra en disputa. Memorias del despojo y resistencias campesinas en la costa Caribe 1960–2010*. Semana.
Daniels Puello, A. D. J. (2016). La transformación de la estructura productiva de los Montes de María: De despensa agrícola a distrito minero-energético. *Memorias. Revista Digital de Historia y Arqueología Desde el Caribe colombia, 29*, 52–83.
Departamento Administrativo Nacional de Estadística. (2018). *Censo Nacional de Población y Vivienda 2018*. https://www.dane.gov.co/index.php/estadisticas-por-tema/demografia-y-poblacion/censo-nacional-de-poblacion-y-vivenda-2018
Devine, J., & Ojeda, D. (2017). Violence and dispossession in tourism development: A critical geographical approach. *Journal of Sustainable Tourism, 25*(5), 605–617. https://doi.org/10.1080/09669582.2017.1293401
Donais, T. (2009). Empowerment or imposition? Dilemmas of local ownership in post-conflict peacebuilding processes. *Peace & Change, 34*(1), 3–26.
Duarte, C. (Ed.). (2016). *Desencuentros territoriales: Caracterización de los conflictos en las regiones de la Altillanura, Putumayo y Montes de María*. Instituto Colombiano de Antropología e Historia.
Escobar, A. (2014). *Sentipensar con la tierra: Nuevas lecturas sobre desarrollo, territorio y diferencia* (Primera edición). Ediciones Unaula.
Fajardo, D. (2014). *Las guerras de la agricultura colombiana, 1980–2010*. ILSA Instituto Latinoamericano para una Sociedad y un Derecho Alternativos.
Fundación Ideas para la Paz .(2011). *Análisis regional de los Montes de María*. http://ideaspaz.org/media/website/MontesdeMariaweb.pdf
Galtung, J. (1969). Violence, peace, and peace research. *Journal of Peace Research, 6*(3), 167–191. https://doi.org/10.1177/002234336900600301
Grajales, J. (2020). Losing land in times of peace: Post-war agrarian capitalism in Colombia and Côte d'Ivoire. *The Journal of Peasant Studies*, 1–21.doi:10.1080/03066150.2019.1691535.
Guasca, M., Van Broeck, A. M., & Vanneste, D. (2020). Tourism and the social reintegration of Colombian ex-combatants. In Tavares da Silva, J. Breda Z. & Carbone F. (Eds.). *Role and impact of tourism in peacebuilding and conflict transformation* (pp. 66–86). IGI Global.
Haessly, J. (2010). Tourism and a culture of peace. In Moufakkir, O. & Kelly, I. (Eds.). *Tourism, progress and peace* (pp. 1–16). CABI.
Heathershaw, J., & Lambach, D. (2008). Introduction: Post-conflict spaces and approaches to statebuilding. *Journal of Intervention and Statebuilding, 2*(3), 269–289. https://doi.org/10.1080/17502970802436296
Higgins-Desbiolles, F. (2006). More than an "industry": The forgotten power of tourism as a social force. *Tourism Management, 27*(6), 1192–1208. https://doi.org/10.1016/j.tourman.2005.05.020
Higgins-Desbiolles, F. (2020). Socialising tourism for social and ecological justice after COVID-19. *Tourism Geographies, 22*(3), 610–623.
Instituto Colombiano Agustín Codazzi. (n.d.). *Índice de GINI de la tierra rural*. https://sinchi.org.co/files/Base%20de%20Datos%20Inirida/PDF/Índice%20Gini%20de%20la%20tierra%20rural.pdf
Instituto de Estudios para el Desarrollo y la Paz. (2013). *Documento de línea base agroindustria de la palma aceitera. Caso María la Baja (Bolívar)*.
Klein, N. (2007). *The shock doctrine: The rise of disaster capitalism*. Macmillan.
Koopman, S. (2011). Let's take peace to pieces. *Political Geography, 30*(4), 193–194. https://doi.org/10.1016/j.polgeo.2011.04.013
Lederach, A. J. (2017). "The Campesino Was Born for the Campo": A multispecies approach to territorial peace in Colombia. *American Anthropologist, 119*(4), 589–602. https://doi.org/10.1111/aman.12925
Lederach, J. P. (1997). *Sustainable reconciliation in divided societies*. USIP.
Mac Ginty, R. (2015). Where is the local? Critical localism and peacebuilding. *Third World Quarterly, 36*(5), 840–856. https://doi.org/10.1080/01436597.2015.1045482
Mac Ginty, R., & Richmond, O. P. (2013). The local turn in peace building: A critical agenda for peace. *Third World Quarterly, 34*(5), 763–783. https://doi.org/10.1080/01436597.2013.800750
Matthews, H. G., & Richter, L. K. (1991). Political science and tourism. *Annals of Tourism Research, 18*(1), 120–135. https://doi.org/10.1016/0160-7383(91)90043-B
Maza, F., Herrera, G., & Jiménez, T. (2017). Palma de aceite y seguridad alimentaria en el caribe colombiano: El caso del municipio de María la Baja. *Palobra*. 17:122–143.

Mesa de Interlocución y Concertación de los Montes de María. (2017). En busca de la permanencia digna en el territorio. *Economía y Región, 11*(2), 339–359.
Metsola, L. (2006). Reintegration' of ex-combatants and former fighters: A lens into state formation and citizenship in Namibia. *Third World Quarterly, 27*(6), 1119–1135. https://doi.org/10.1080/01436590600842407
Mitchell, A., & Kelly, L. (2011). Peaceful spaces? "Walking" through the new liminal spaces of peacebuilding and development in North Belfast. *Alternatives: Global, Local, Political, 36*(4), 307–325. https://doi.org/10.1177/0304375411431761
Novelli, M., Morgan, N., & Nibigira, C. (2012). Tourism in a post-conflict situation of fragility. *Annals of Tourism Research, 39*(3), 1446–1469. https://doi.org/10.1016/j.annals.2012.03.003
Ojeda, D. (2012). Green pretexts: Ecotourism, neoliberal conservation and land grabbing in Tayrona National Natural Park, Colombia. *Journal of Peasant Studies, 39*(2), 357–375. https://doi.org/10.1080/03066150.2012.658777
Ojeda, D., Petzl, J., Quiroga, C., Rodríguez, A. C., & Rojas, J. G. (2015). Paisajes del despojo cotidiano: Acaparamiento de tierra y agua en Montes de María, Colombia. *Revista de Estudios Sociales No.35, No.35*, 107–119. https://doi.org/10.7440/res54.2015.08
Paris, R. (1997). Peacebuilding and the Limits of Liberal Internationalism. *International Security, 22*(2), 54–89. https://doi.org/10.1162/isec.22.2.54
Pizano, C., González-M, R., Hernández-Jaramillo, A., & García, H. (2017). *Agenda de investigación y monitoreo en bosques secos de Colombia (2013–2015): Fortaleciendo redes de colaboración para su gestión integral en el territorio.* Instituto Humboldt.
Plataforma de Organizaciones de Desarrollo Europeas en Colombia. (2011). *Análisis del plan de consolidación de Montes de María. Una mirada desde el desarrollo, la democracia, los derechos humanos y la cooperación internacional.*
Porto-Gonçalves, C. W., & Leff, E. (2015). Political ecology in Latin America: The social re-appropriation of nature, the reinvention of territories and the construction of an environmental rationality. *Desenvolvimento e Meio Ambiente, 35:65–88.* https://doi.org/10.5380/dma.v35i0.43543
Programa de las Naciones Unidas para el Desarrollo, Corporación territorios, & U. Cartagena. (2003). *PROMONTES. Programa de Desarrollo y Paz de los Montes de María.*
Programa de las Naciones Unidas para el Desarrollo. (2010). *Los Montes de María: Análisis de la conflictividad.*
Pugh, M. (2005). The political economy of peacebuilding: A critical theory perspective. *Journal of Peace Studies, 10*(2), 23–42.
Richards, P. (2005). *No peace, no war: the anthropology of contemporary armed conflicts.* James Currey.
Richmond, O. P. (2009). A post-liberal peace: Eirenism and the everyday. *Review of International Studies, 35*(3), 557–580. https://doi.org/10.1017/S0260210509008651
Richmond, O. P., & Mitchell, A. (2011). Peacebuilding and critical forms of agency: From resistance to subsistence. *Alternatives: Global, Local, Political, 36*(4), 326–344. https://doi.org/10.1177/0304375411432099
Richter, L. K. (1999). After political turmoil: The lessons of rebuilding tourism in three Asian Countries. *Journal of Travel Research, 38*(1), 41–45. https://doi.org/10.1177/004728759903800109
Said, E. (1978). *Orientalism.* Vintage, 25.
Salazar, N. B. (2012). Community-based cultural tourism: issues, threats and opportunities. *Journal of Sustainable Tourism, 20*(1), 9–22. https://doi.org/10.1080/09669582.2011.596279
Scott, J. (2012). Tourism, civil society and peace in Cyprus. *Annals of Tourism Research, 39*(4), 2114–2132. https://doi.org/10.1016/j.annals.2012.07.007
Sharpley, R. (2000). Tourism and sustainable development: Exploring the theoretical divide. *Journal of Sustainable Tourism, 8*(1), 1–19. https://doi.org/10.1080/09669580008667346
Simone-Charteris, M. T., & Boyd, S. W. (2010). Northern Ireland re-emerges from the ashes: The contribution of political tourism towards a more visited and peaceful environment. In Moufakkir, O. & Kelly, I. (Eds.). *Tourism, progress and peace* (pp. 179–198). CABI.
Sonmez, S. F., & Apostolopoulos, Y. (2000). Conflict resolution through tourism cooperation? The case of the partitioned island-state of Cyprus. *Journal of Travel & Tourism Marketing, 9*(3), 35–48.
United Nations Verification Mission in Colombia. (2019, December 26). *Report of the secretary-general.* https://colombia.unmissions.org/sites/default/files/n1942147.pdf
United Nations Verification Mission in Colombia. (2020, September 25). *Report of the secretary-general.* https://colombia.unmissions.org/sites/default/files/res-2545-e-.pdf
United Nations. (2015). *Transforming our world: The 2030 agenda for sustainable development.* Division for Sustainable Development Goals.
Van Broeck, A. M., Guasca, M., & Vanneste, D. (2019). Tourism Initiatives as a Lever for Reintegration of Former FARC Combatants in Colombia. *Via. Tourism Review* (15). https://doi.org/10.4000/viatourism.3551.
Volcic, Z., Erjavec, K., & Peak, M. (2014). Branding Post-War Sarajevo: Journalism, memories, and dark tourism. *Journalism Studies, 15*(6), 726–742. https://doi.org/10.1080/1461670X.2013.837255
Winter, T. (2008). Post-conflict heritage and tourism in Cambodia: The burden of Angkor. *International Journal of Heritage Studies, 14*(6), 524–539. https://doi.org/10.1080/13527250802503274
World Tourism Organization. (2018). *Tourism for development – Volume I: Key areas for action* (p. 144).

Disrupting structural violence in South Africa through township tourism

Meghan L. Muldoon and Heather L. Mair

ABSTRACT

This paper describes the potential that tourism encounters have in disrupting structural violence (conceptualized as silencing and invisibilization) in South Africa. Based on PhotoVoice research undertaken with a small number of residents of three Cape Town-area townships, we consider how residents' descriptions of their encounters with the tourists can be seen as helping to "polish the wounds of the past" as they shared a sense of being seen and heard. We apply the African philosophical lens of Ubuntu, described by one participant as "I am because we are," to consider the role of tourism in promoting peace simultaneously at the individual as well as the collective level. While far from unproblematic, this research finds hope in the ways township tourism is disrupting structural violence, thereby supporting the emancipatory aims of post-apartheid South Africa.

Introduction

Tourism for peace is a notion enjoying a welcome resurgence (Becken & Carmignani, 2016; Blanchard & Higgins-Desbiolles, 2013; Farmaki, 2017; Haessly, 2010; Moufakkir & Kelly, 2010). Welcome because too often tourism is relegated to little more than an economic exchange or the hedonistic privilege of the world's wealthy elites (Blanchard & Higgins-Desbiolles, 2013). Indeed, as Higgins-Desbiolles (2006) reminds us, tourism is a social force. As such, more critical scholarly attention needs to be paid to the complex social, historical, political, environmental, and economic interplays that occur in touristic practices. This paper describes research that sought to understand tourism in the townships of South Africa, and the role(s) it may play in disrupting certain forms of structural violence in that country. Our objective here is in highlighting how relational encounters in tourism disrupt and challenge structural violence in the form of silencing and invisibilization of the Black majority as a barrier to peace.

Here we examine encounters from a poststructural and relational approach as described by Jóhannesson et al. (2015) as "complex and emerging" where "the roles and characteristics of related actors are co-constituted" (p. 5). Rather than considering encounters as clashes between two or more competing actors, relational encounters describe the worldmaking forces of tourism (Hollinshead, 2004), the "*doing* of tourism and its entanglements to other things" (Jóhannesson et al., 2015, p. 5). In other words, the heterogeneous nature of encounters in tourism lead to the creation and re-creation of social phenomena and to shaping tourism encounters in the future (Jóhannesson et al., 2015).

The concept of Ubuntu is employed here in order to help us conceptualize notions of peace and the role of tourism encounters in contributing to a more just world. A Nguni Bantu word and a philosophical principle throughout many parts of Africa, Ubuntu is defined as humanity towards others or as the belief in a universal bond of sharing that connects all humanity (Dreyer, 2015; Murove, 2012). Through the lens of Ubuntu, we understand peace through township tourism as operating both to restore the dignity and humanity of the people victimized by the atrocities of apartheid as well as the larger conceptualization of peace as a unified humanity and a breaking down of the barriers that divide us. As such, tourism encounters in the townships operate to promote peace at both the individual and the collective levels, because Ubuntu tells us that there is no individual without the collective.

Issues of lasting peace, human dignity, and social justice are complex, and it would be naïve to suggest that tourism presents a solution to these deeply embedded challenges. However, we argue that tourism, as a social force, can be a powerful disruptor of hidden norms and structural inequalities. This research offers opportunities for understanding tourism's role, while far from unproblematic, in contributing to a lasting peace and sustainable development in a country still scarred by centuries of colonial oppressions and racialized violence.

Literature review

Structural violence

Galtung (1969) was the first peace theorist to propose the notion of structural violence, which he defined as "when human beings are being influenced so that their actual somatic and mental realizations are below their potential realizations" (p. 168). In other words, structural - or indirect - violence is conceived as any constraint on human potential due to economic, political, educational, or social structures. Unequal access to political power, resources, economic opportunity, legal standing, higher education, and health care are all forms of structural violence, and all operate in present day South Africa (Nyamnjoh, 2016).

Haessley (2010, p. 3) defines structural violence as "social, political, and economic systems and policies that result in an inequitable distribution of resources or cause damage to the environment." Many forms of structural violence, such as racism and sexism, are so deeply embedded within our societies that they become almost invisible. Farmer (2001, p. 79) explains that in instances of structural violence:

> ... neither culture nor pure individual will is at fault; rather, historically given (and often economically driven) processes and forces conspire to constrain individual agency. Structural violence is visited upon all those whose social status denies them access to the fruits of scientific and social progress.

The processes are structural because they are embedded within the institutions and political and economic orders that govern our societies (Farmer, 2001). They are violent because they prevent people from enjoying their full dignity as human beings and lead directly to physical, mental, and spiritual illnesses, physical violence, and premature death (Rylko-Bauer & Farmer, 2016).

In South Africa, systemic racism was made manifest through the segregationist policies of apartheid, which abnegated Black people's education, health care, safe housing and land ownership, mobility, and political influence; the effects of these policies continue to reverberate and inform daily life to this day. Further, the mental and emotional effects of state-sanctioned exclusion and poverty include "a loss of dignity and autonomy, of purpose and coherent structure to life, of a sense of safety and the onset of feelings of hopelessness" (Swartz et al., 2012, p. 29-30). These negative impacts are experienced over generations, as people who were deprived of education and employment struggle to educate their children and provide them with pathways out of poverty (Swartz et al., 2012).

While the causes and implications of racialized poverty in South Africa are complex and multifaceted, in this paper we draw attention to two interrelated forms of structural violence – silencing and invisibilization – which may be disrupted and challenged by the presence of, and engagement with, tourists in the townships.

Silencing

Herzog (2018) describes two forms of structural violence, which operate simultaneously to deny marginalized people's access to democratic freedoms and, therefore, fundamental social justice: silencing and invisibilization. Silencing occurs when a certain portion of the population is excluded from public discourse, i.e., in the spaces where decisions related to "knowledge, definitions, practices, and the distribution of power" take place (Herzog, 2018, p. 16). Silencing also occurs through processes of victimhood, as witnessed during the Truth and Reconciliation Commission (TRC, 1996–2003). South Africa's TRC has been characterized in myriad ways, ranging from a necessary compromise to prevent the country from spiralling into economic collapse and civil war, to the ultimate betrayal of Black South Africans (Goodman, 2017; Marsh & Knaus, 2015 Villa-Vicenio & Verwoerd, 2000). The TRC allowed citizens to testify to what they suffered under apartheid, but the Commission made the decision to ultimately focus only on individual, as opposed to structural, culpability of violence (Villa-Vicenio & Verwoerd, 2000). For many, this meant re-opening old wounds for public consumption without ultimately holding those accountable to task. It also meant that Black South Africans became coded purely as 'victims,' as the TRC created no public space for accounts of resistance, bravery, or self-sacrifice (Reynolds, 2013; Wright, 2017). This silencing occurred, not through "a destructive narrative experience, but rather the reduction of experience to a passive state of victimhood" (Wright, 2017, p. 164). Because of the TRC's emphasis on individual and isolated incidences of violence, narratives of personal and collective struggle and resistance were excluded from the official narrative, as were the impacts of long-term structural violence (Reynolds, 2013; Wright, 2017).

Invisibilization

Invisibilization can be both physical and social (Herzog, 2018). Physical invisibilization occurs through the prevention of one group from entering the public sphere. Under apartheid, this was effected through the imposition of the Group Areas Act (1955) and other Acts, which forced Black and Coloured South Africans to live in townships on the urban periphery and restricted their movements beyond the townships. Today, Blacks are no longer required to carry a *dompas*[1], however most formal employment is located in the city centre, far from the townships where most Black South Africans continue to live, meaning that substantial portions of people's incomes must be spent on transportation, limiting their capacity to actively participate in the public sphere (Goodman, 2017). Herzog (2018) draws on Foucault's notion of 'the order of the discourse' to demonstrate that certain clusters of people in society – including politicians, journalists, and educators – enjoy dominant voices in society while others are seldom heard. While Black politicians have been in the ruling majority since 1994, whites continue to dominate the economic and educational spheres in South Africa (Goodman, 2017). Processes of social invisibilization also occur through the dominant group's decision to 'look through' the Other (Herzog, 2018; Honneth, 2001). For Honneth (2001), this is an intentional act, which requires "gestures or ways of behaving that make clear that the other is not seen, not merely accidentally, but rather intentionally" (Honneth, 2001, p. 112). This is an act of structural violence, which is intended to be readily apparent to anyone present in the social situation (Honneth, 2001).

The deeply embedded nature of silencing and invisibilization make it difficult for anti-racist educators to attempt to challenge and dismantle these structures. Even long after the legal systems of colonialism are dismantled, white and European ways of knowing and doing continue to be privileged and serve to undermine Indigenous knowledges and cultures. An example of this can be seen in a speech delivered by Julius Malema – the then leader of the Africa National Congress (ANC) Youth League, now the founding leader of the Economic Freedom Fighters (EFF) political party – at Limpopo University. According to Malema, "[t]he ANC changes lives. It can change you from a hobo into someone very important. This ANC has taught those who are insulting it today to use a fork and knife, to taste red wine, to wear expensive suits" (As cited in Posel, 2010, p. 159). This not-so-subtle privileging of European culture, mannerisms, and dress silences African values and behaviours, and reinforces the race-based power imbalances that order life in South Africa. This type of privileging of whiteness can be witnessed even in the prepared speaking points of a politician whose political party operates on a platform of staunch pan-Africanism, and Malema himself has been quoted as saying that the EFF is "cutting the throat of whiteness" (Areff, 2018).

Finally, processes of silencing and invisibilization also operate through discursive (re)constructions of racial divides through the primarily white-owned mainstream media. The media have been accused of perpetuating the 'crime epidemic' and 'culture of violence' in post-apartheid South Africa (Van Dongen, 2005, p. 531). This violence is invariably associated with the Black and Coloured populations, and while violent crime is a serious problem, the discourse around this violence is rarely contextualized within entrenched poverty and disenfranchisement (Van Dongen, 2005). The voices of township residents are rarely critically portrayed, nor are images of the townships shown as anything other than sites/sights of violence. Rather, and as we shall see in conversation with one of the participants in this research, these media representations are instead used to further justify discrimination and social distancing between the apartheid-era defined racial divides (Van Dongen, 2005).

As is discussed below, the tourism encounter in the highly racialized space of the township can help to disrupt these processes of silencing and invisibilization in the day-to-day lives of many South Africans. Nonetheless, the question remains as to whether this is an appropriate space within which to begin challenging deeply embedded processes of silencing and racism. In the following sub-sections, we consider notions of tourism and peace and the power of the tourism encounter in spaces of poverty.

Tourism and peace

Peace, as a desired state of being, encompasses much more than the mere absence of war or physical violence, otherwise known as 'negative' peace (Galtung, 1969). 'Positive' or active peace is comprised of a conscious and defined social contract (Rummell, 1982), one that is absent of structural violence, and constituted by social care, justice, and ecological sustainability (Haessly, 2010; Pratt & Liu, 2016). Current evaluations of the concept of peace argue it is an active process, not a passive state of being, and one in which all actors in society are implicated in order to attain understanding, equity, human dignity, and security (Pratt & Liu, 2016). According to Rees,

> Peace with justice is characterized by an absence of violence, whether it is direct and observable or indirect and invisible. A just peace would be marked not only by an end to hostilities between peoples, but also by fairness in social, economic, and political arrangements. It would be a fairness characterized by equality and opportunity and by associated influences in the building of civil societies... An easier way to describe a just peace is to identify its opposite – the violence which hinders individual freedom, and which diminishes the quality of people's lives. Such violence may be direct, as in acts of terrorism, or indirect as in the persistence of poverty which curtails individuals' life chances and life expectancies. Relationships of domination and subordination are often a cue for violent conflict and transformation of such unequal power relationships is key to peace with justice (2003, pp. 20–21).

The concept of tourism for peace has a long history and has traditionally been contingent upon the belief that proximity is anathema to bigotry. As far back as 1867, the American satirist Mark Twain wrote, "[t]ravel is fatal to prejudice, bigotry, and narrow-mindedness" (as cited in D'Amore, 2009). American presidents Eisenhower and Kennedy both called for an expansion of tourism in the aftermath of war as a pathway towards understanding and respect amongst nations and as "one of the great forces for peace and understanding of our time" (Kennedy, as cited in D'Amore, 2009). Their comments support the popular belief that tourism "actively contributes to post-conflict stability" (Becken & Carmignani, 2016, p. 65; Novelli et al., 2012). Scheyvens (2002) has argued for "justice tourism" that is embedded within notions of equality and ethicality and based on the notion that "travelers can be part of the liberation process" (p. 104; Higgins-Desbiolles, 2006). A growing number of tourists are seeking ethical and impactful forms of tourism, and the emergent market of 'peace tourism' is increasingly providing opportunities for tourists to participate in activities and locations that celebrate peace with one's self, one's society, and with nature (Becken & Carmignani, 2016, p. 65; D'Amore, 2009). This brings us to a central factor in conceptualizing tourism and peace: the tourism encounter.

Tourism encounters

The central tenet underlying tourism's potential as a social force for peace is the "cultivation of understanding among people" (Farmaki, 2017, p. 528); that increased contact and familiarity among peoples will lead to increased empathy and decreased conflict and misunderstanding globally (Amir, 1969; Pettigrew, 1998; Pratt & Liu, 2016). Travel is also encouraged in its role as a peacemaker for global cooperation because it reduces "psychological and cultural gaps" between people (Nyaupane et al., 2008, p. 650), and because it leads people to develop a sense of themselves as agents of change for good in the world (Askjellerud, 2003; Causevic, 2010; Farmaki, 2017). Much of the belief in the transformative potential of the tourism encounter is based in contact theory (Askjellerud, 2003; Causevic, 2010; Farmaki, 2017). The theory posits that increased proximity between peoples and greater opportunities for social contact will lead to fewer negative stereotypes of the Other, greater understanding of differences between cultures and behaviours, and a greater likelihood for collaboration and peace (Becken & Carmignani, 2016; Caton, 2012). Gibson and Claassen (2010, p. 525) described immobility, the alternative to tourism, as "an invitation to xenophobia." Other scholars have described contact theory's role in leading to a greater peace as 'naïve' and have instead advocated for a more nuanced 'peace-sensitive tourism' (Wintersteiner & Wolmuther, 2014). The danger in reifying the tourism encounter, as a form of 'direct experience,' is that it becomes imbued with power in and of itself, uninformed or mediated by the structural inequalities within which the encounter is embedded (Pedwell, 2012; Tucker, 2016, p. 35).

Indeed, in their work researching 'everyday encounters' of race, Valentine (2008) found that "proximity does not equate with meaningful contact" (p. 334), and that in fact encounters with raced Others may lead to biases hardening and becoming more entrenched. Valentine further found that even in instances of intentional meaningful contact at the micro-scale, effectively scaling these encounters up to influence the prejudices of the dominant race seems unlikely (2008, p. 334). Clearly, an encounter in and of itself is insufficient to disrupt race-based structural violence. In their work, both Gibson (2010) and Tucker (2009, 2016) advocate a "reflexive interrogation" (Tucker, 2009, p. 444) of potentially productive, but often embodied as 'negative,' emotions of shame, guilt, fear, or pity that often arise over the course of the tourism encounter. This perspective is further drawn upon below, through the voice of two participants in this research who advocate for greater interrogation of power and subjective positionalities in the tourism encounter.

In applying the notion of Ubuntu to encounters in tourism, which shall be discussed in greater detail below, we see that such encounters are imbued with transformative potential because my humanity and well-being are directly connected to yours (Murove, 2012). This transformative potential, as seen through the lens of Ubuntu, is also embedded within colonization, because through the tourism encounter, people who lived through apartheid (and beyond) are able to share their voices and their experiences and to celebrate each other's humanity within the confines of a physical place literally designed to strip residents of theirs. Before moving on to this deeper discussion of Ubuntu, we first turn our attention to those racially segregated spaces – townships – and the ways in which tourism is impacting them post-apartheid.

Slum tourism and the townships

Modern slum tourism has long been understood as having begun under the auspices of social justice activists in the early 1990s who wished to shed an international spotlight on deplorable living conditions in the townships of South Africa and the *favelas* of Brazil (Freire-Medeiros, 2013; Frenzel et al., 2015). Increasingly listed on 'must do' visitors' lists of cities in the global South (Booyens & Rogerson, 2019), pre-Covid-19 estimates suggested that as many as one million international tourists visited the slums of South Africa, Brazil, and India every year (Frenzel, 2016, 2017).

Slum tourism has become a significant profit generator for tourism operators in the major cities of the countries listed above, as well as a source of income for many informal businesses located within the slums, which would not exist absent the tourists (Booyens & Rogerson, 2019; Koens & Thomas, 2016). Increasingly of interest are the socio-political implications of an ongoing presence of international tourists in urban slums (Holst, 2018; Meschkank, 2011; Muldoon, 2018; Nisbett, 2017). Broadly speaking, slum tourism is a "mass tourism phenomenon occurring only in a few destinations and a niche form in a growing number of other destinations" (Frenzel et al., 2015, pp. 237–238), including Kenya, the Philippines, the Caribbean, Indonesia, Vietnam, and elsewhere.

Frenzel (2012) suggests that politically motivated tourists are an important factor contributing to the development of slum tourism. The three countries where slum tourism first flourished (Brazil, South Africa, and India) all are post-colonial states and exemplars of high degrees of inequality amongst their populations. Academic research has found that while many residents are often wary of the motivations of international visitors to the slums, over time they come to feel that the tourists have friendly and caring intentions (Muldoon, 2018; Rolfes, 2010; Rolfes et al., 2009).The contention that slum tours focus only on the negative and exploit the poverty of the residents is not borne out in the research, which has found rather that the tours primarily feature historical and cultural assets, new developments, and personal success stories (Freire-Medeiros, 2013; Frenzel, 2012, 2016; Muldoon, 2018; Rolfes, 2010; Rolfes et al., 2009). Research by Rolfes et al. (2009) sought to examine this aspect of township tourism, where tourists were asked to rank their impressions of the townships according to a number of binaries, including clean/dirty, happy/sad, safe/violent, etc. The findings of the study demonstrated that the tourists' impressions of the townships were significantly more positive following their tour, suggesting that the encounters they experienced gave them a more nuanced understanding of spaces of poverty and the people who live there.

As a form of slum tourism[2], township tourism has seen a massive upswing in the years following South Africa's independence. Koens and Thomas (2016) estimate that as many as 25 percent of tourists to South Africa book a township tour. While in the immediate aftermath of apartheid, township tours were constructed as primarily political (McEachern, 2002), they have evolved to take on more of a cultural tourism aspect. The tours generally take place over the course of an afternoon or a full day and are traversed in vans, on bicycles or by foot. They often feature a trip to a pre-school or an orphanage, a visit to a 'typical' home, arts and crafts vendors, and lunch or a locally brewed beer (called *umqombothi*) in a neighbourhood *shebeen*. Many now also offer overnight homestays, cultural excursions, or even bungee jumping from the famed Soweto water tower.

Through tourism, townships are constructed as spaces of difference *par excellence* (Hollinshead, 1992, p. 49), where international tourists are invited to come and explore bounded, impoverished communities alien to their home environments. While concerns about exploitation continue to be expressed (Frenzel, 2012, 2016; Muldoon & Mair, 2016; Nisbett, 2017), township tours are viewed by many as one pathway out of poverty and obscurity for Black communities post-independence (Frenzel, 2012; Rogerson, 2004; Rolfes et al., 2009).

Ubuntu

As white, foreign scholars our desire here is not to westernize or (re)colonize the African notion of Ubuntu, We do not pretend to be experts in this area, and Ubuntu was not part of the worldviews with which we were raised. However, we do find it to be a useful framework to help us in conceptualizing notions of peace and justice in the tourism encounter, and to consider why these encounters have transformative potential. Ubuntu originated in Southern Africa and is founded on the notion that all humans are relational beings, and our very humanity is embedded within our relationships with others (Dreyer, 2015; Murove, 2012). As relational beings, we rely on our relationships with one another for our well-being (Murove, 2012). This stands in contrast to the individualistic understanding of humanity, which has dominated western theology for many centuries. Ubuntu means humanness – it means behaving towards others with care and compassion, placing it in direct contrast to the dehumanizing objectives of colonialism (Murove, 2012). Ubuntu is perhaps best understood through the Zula adage, *Umuntu ngomuntu ngabantu*, which translates as a person is a person because of other persons (Murove, 2012). Therefore, the colonizers could not be people because they did not treat all people with dignity and respect. In fact, Stanlake and Samkange (1980) made this observation during the colonial period:

> We know also, it [Ubuntu] means more than just a person, human being or humanness because when one sees two people, one white and the other black, coming along, we say "*Hona munhu uyo ari kufamba nomurung*" ... (There is a munhu [person] walking with a white man) (As cited in Morove, 2014, p. 38).

Ubuntu is also helpful in shining light on why people often appear to be very welcoming to tourists in the townships. According to Gelfand (1973, p. 57), "Unhu (humanness) is derived from parents, from tribal practice, and inheritance from the distant past. The parents teach their children Unhu. The good man [sic] has Unhu. He welcomes visitors to his home where he receives them in the correct way due to the particular visitor" (as cited in Murove, 2012, p. 39). Finally, Ubuntu has played an important role in the political realm of many African nations post-independence, often operationalized in an effort to restore the dignity of Africans, which was severally marred over centuries of colonialism (Dreyer, 2015). Zondi (2014) refers to the "doctrine of *ubuntu*" which "has shown capacity to translate itself into a strategic impetus for replacing colonial humanism on which the fragmenting notion of separate development of races was founded with a decolonial one premised on restoring authentic sociality through the dictum: 'I am because you are'" (p. 104).

Economic encounters

While not the explicit focus of this research, we would be remiss in not speaking to the overwhelming role that intergenerational poverty and economic injustice play in prohibiting a true and lasting peace in many nations, including South Africa. 'The Great Compromise' of Nelson

Mandela and the ANC in 1994 was at the time perceived as necessary in order to prevent economic collapse and civil war (Marsh & Knaus, 2015). The government adopted policies such that "the world would use the word democracy to describe [South Africa] while the whites would keep our wealth and property" (Marsh & Knaus, 2015, p. 202). The continuing legacies of economic apartheid are readily apparent in today's South Africa. Mortality rates among Black infants are 6 times that of white babies. Illiteracy rates are 1% among white South Africans and 46% among Blacks. Black workers earn an average of 1/8 the monthly incomes of whites (Foster, n.d.).

Much has been made of the economic potential of slum tourism to allow tourists to spend their money in the communities that need it the most, ultimately helping to lift them out of poverty (Freire-Medeiros, 2013; Frenzel, 2016). Unfortunately, this hope has not been borne out in reality. Many studies have found that the economic reach of tourism to the townships has benefitted a small number of micro-enterprises, local artisans, and tour guides, but has had little impact on displacing real poverty at the community level (Booyens & Rogerson, 2019; Frenzel, 2016; Koens & Thomas, 2016; Muldoon, 2018). While this research is explicit in our emphasis on the transformational potential of the encounter in tourism in the townships, it is important to acknowledge that true peace is ultimately not possible without economic justice.

Context of the research

Our focus here on structural violence and peace in the context of township tourism stems from community-based research undertaken over a period of five months in 2016 with residents of three townships surrounding Cape Town, South Africa. The first author conducted a PhotoVoice methodology with 22 participants, eight women and 14 men aged between 23 and 60, all of whom identified as Black and Xhosa, with one exception, a White Afrikaans man. The first township, T1, is on the outskirts of Cape Town and is home to approximately 400,000 residents. Six female participants from T1 were employed at a women's empowerment center and two males were tour guides in the community. The second township, T2, is often framed as the oldest township in Cape Town. Here, participants included one female tour guide, one male freelance photographer, and two small business owners, one female and one male, who profit from the tours. The third township, T3, was the smallest and located the furthest from the city center and was also temporarily home to the first author. T3 was home to three licensed tourism guides, all males, with varying degrees of success in making a living through guiding, as well as the Afrikaans male resident of the township. In addition, a focus group was undertaken with six male residents of T3, who wished to support this research, but had had little substantive experience with tourism.

PhotoVoice is a feminist research methodology, which puts cameras in the hands of participants and invites them to share photographs in order to represent their own particular worldviews (Wang & Burris, 1997; Wang & Redwood-Jones, 2001). This methodology has been found to be effective in situations where the researcher is from a different cultural background than that of the participants, therefore creating a starting point from the participants' situated positions as experts in their own lives (Muldoon, 2018; Wang & Burris, 1997; Wang & Redwood-Jones, 2001). It has also been used to demonstrate that all people, regardless of their marginality or relative powerlessness in society, have a voice and are able to express it when given the tools to do so (Wang & Burris, 1997).

All participants in this study had some experiences of tourism to the townships, some as direct recipients of tourism employment and some as observers of the impacts of tourism. In accordance with PhotoVoice methods, participants were given digital cameras and asked to take photos in their communities that represented *what tourism looks like* and *what tourism should look like*. After having been given time to collect their photographs (between two to five weeks), participants were asked to meet with the first author for participant-guided interviews, in which the photographs were shared and contextualized for the first author. These conversations about the photographs then led to further discussions regarding economic power, storytelling, mobilities, race, gender, and the legacies of apartheid in the townships. Most interviews were conducted in English, although two participants chose to speak in isiXhosa and a member of the community translated *in situ*. The audio recordings of those interviews were then translated and transcribed by two language specialists in the Department of Linguistics at the University of Stellenbosch. Interview durations ranged from 45 minutes to five hours. All interviews were audio recorded, and the English-language recordings were transcribed by the first author, who also conducted the interviews with participants in their homes, places of employment, or public meeting places.

Analysis was undertaken using a Foucauldian approach to critical discourse analysis. Photographs, interview texts, and observational data were analyzed iteratively throughout the data collection process as well as upon its completion (Hesse-Beiber, 2007). Line-by-line and focused coding (Charmaz, 2006) were employed as the transcripts were read and re-read and the data examined through lenses of power, silencing, and resistance. While there is no proscribed method in analysing discourse (Rose, 2001), this research followed a series of steps described by Rose which help provide guidance to the analysis process:

1. Looking at your sources with fresh eyes
2. Immersing yourself in your sources
3. Identifying key themes in your sources
4. Examining the effects of truth
5. Paying attention to their complexity and contradictions
6. Looking for the invisible as well as the visible
7. Paying attention to details (2001, p. 158).

The discussion that follows is focused upon two themes from the literature which we have deemed to be of significance to the present study: silencing and invisibilization. The goal is to offer insights into how some forms of tourism can be considered as supporting peace. In addition to data collected through PhotoVoice, the first author also employed participant observation and reflexive writing while collecting and analyzing the data. Both authors prefaced this study with previous research into the representation of townships in social media (citation details withheld for review process).

Findings

In this section we present the findings of the research related to two forms of structural violence: silencing and invisibilization. We consider the ways in which Black South African voices were silenced during apartheid and their presence erased from the nation's history books. Through relational tourism encounters, township residents described opportunities to share their stories and have them carried beyond the boundaries of the township. Through the valuing of their experiences by the tourists, township residents shared that they felt their stories taking flight and beginning to help "polish the wounds of the past" (T1, male, mid-30s). This was described as the spirit of Ubuntu, finding well-being and joy through our shared humanity with one another.

Structural silencing and invisibilization

Through their use of photographs, participants in this study shared the ways in which they perceived guests and hosts to be working together through tourism encounters in the townships to break down discursive racial barriers between whites and Blacks in South Africa. This was expressed through notions of disrupting dominant media narratives of the townships, bringing whites and Blacks together, sharing one's stories beyond the township, and self-conscientization, all of which are described in greater detail below.

As mentioned above, the mass media in South Africa remains in primarily white hands (Mail & Guardian, 2019). The prevailing narrative of the townships in the media is that they are dangerous 'no-go' zones. One participant shared that the local university informs incoming students each year that it is dangerous to visit the township (T3). From his perspective, each isolated incidence of violence is reported in the media and no other stories are told about the township. According to this participant,

> The information that they print about [the township] they only get from the police station ... people that have power at that moment, you know, decide which stories form part of the history of that place. They decide what to call you and how the whole world will perceive you. If they decide to call you a hell, then everyone that respects them because they are already respected people because of the power they have, then everyone will know that place as hell. So, who wants to go to hell? (T3, male, mid-30s)

Over the course of this study, the first author observed that many of the non-township residents she encountered questioned her interest in visiting the townships, and many times warned her of the dangers of traveling to the townships. Even the residents of one township would warn the first author about the dangers of traveling to a different township. Therefore, the media construction of the dangers of the spaces of the Other prevail, even amongst those who are harmed and misrepresented by that narrative. Tourists posting photographs and publishing blog posts about their visits to the townships disrupt and challenge these narratives, showing tourists and residents dancing together, sharing a meal, kicking a ball, and laughing. These published narratives of time spent in the townships also cast an international spotlight on the living conditions in the townships, potentially leading to pressure on governments to enact changes (Scheyvens, 2010), as is discussed further below.

Another way in which township tourism may disrupt structural silencing and invisibilization is by providing opportunities for tourists and residents to come together and interact with one another in a Black space, something that many participants reported as a positive outcome of the tours. Many participants shared that without the tours, people living in townships would have little to no interaction with white people. As one participant stated,

> It's helping our country a lot. Because if there were not any international people coming even younger kids, like from that generation would still not know how to relate to white people. Because they don't see them, they only know that my mom works for the white person who is the boss. So that understanding or the mentality that you were given in the past that white people are superior, you need to respect them, so it's like kind of fading out, now the society is seeing them like people, you know they're like people like us, they can laugh at our jokes, they can sit down, they can eat what we eat (T1, male, mid-20s).

Given that barely a generation ago, it was illegal for Black South Africans to address whites in the street, tourism in the townships creates opportunities to interact for children who, for all intents and purposes, are growing up in a segregated society. Another participant who works as a tour guide reflected,

> ... even though you think this is about you, but you don't even know about that kid, but you made that kid feel special at that time, you know, taking your time, playing with your expensive shoes. Letting go and just, you know, being there. So it means a lot to them ... Just give them love, give them attention (T1, male, mid-20s).

One participant shared that many in the community felt fear when the tourists began coming to the townships, and that "they thought it was the same, same people and same, same history, and same, same things coming back again" (T2, male, mid-40s), meaning the same white South African people who had supported the apartheid regime. Once they began to understand that these were different people, they began to welcome them and "they always say hello, they tell them come in, this is where I live, they even share drinks" (T1, female, mid-50s).

For many of the participants, having the tourists enter their homes, walk in the streets with them, and share in their stories, has meaning in and of itself. Having been forced to the outskirts of the cities and obscured behind physical barriers, such as fences and highways, many Black

Figure 1. Tourists and residents interacting at the Women's Empowerment Center, creating opportunities to learn about one another's lives (source: T1 resident, female early 50 s).

South Africans felt erased from their country's narrative under apartheid. For many, just the act of speaking with the tourists brings joy in being made visible, as with this participant (Figure 1):

> Yes, you see this [photograph], why I am taking. I am taking because we are lucky. The people who come from another places [sic] or another world they come to us and we are the friends of these people. They are not, no you are Blacks, I am white. No, no, no, no they say hi how are you and then we make friends and that's why I like it here. Yeah. Really, really, really, their hearts is white, white, not Black like me, saying ah those people, I like you cause you are here, I'm talking with you (T1, female, early-50s).

For some older residents, these interactions take on further meaning and have a role to play in helping them to "polish the wounds of the past." One tour guide explained that as the elders have come from a divided society (Figure 2);

> Now healing those wounds is about saying we are all human beings.... .let's help you enjoy life, let's help you engage with people that you were not, that you were separated from, meaning different colours. Let's speak, let's talk about who we are and who you are and let's learn from each other. And in the meantime, let's enjoy and celebrate humanity, which is known as Ubuntu. Ubuntu in South Africa is I am because you are, you are because I am (T1, male, mid-30s).

Additionally, township tourism may be understood as disrupting structural silencing as tourists place value on the township spaces and the people who live within them. Many participants spoke of the importance of being able to share their stories, to see their stories being appreciated, and knowing that they will be carried on beyond the township. This is in part perceived as having one's story live on as part of the narrative of South Africa, and also speaks to the self-worth that is created in having someone, to whom you were once forbidden to speak, seek out your home and listen to your life experiences. As one participant shared:

Figure 2. Tourists and residents dancing together at the seniors' center, creating moments of joyful encounter between tourists and residents who remember being prohibited from speaking to white people under apartheid (source: T1 tour guide, male, mid 30 s).

> I think that the idea that a person that I see as a superior person or a person that is better than me, the idea that a person like that can come and walk in the same street as I live makes me, even if there's no money it does something for my self-esteem. You know, it means a lot. It means, it means I'm a person too. It means people, people don't look at me the way I look at myself, it means some people realize that I exist in this world (T3, male, mid-30s).

While it is deeply problematic that anyone would define their self-worth according to being seen by a 'superior person,' as is discussed in greater detail below, it would be disingenuous to simply dismiss his observations because we do not agree or because they make us uncomfortable. He goes on to tell us (Figure 3).

> Now I have to tell [first author] that we are from here, we come from this place, we grew up like this. By doing that I am re-learning. I am learning to love myself again, I am learning to appreciate myself more. So, it becomes a re-conscientization of the other as well - now you become the ambassador of your own township without even realizing it (T3, male, mid-30s).

In this section we have presented some of the ways in which township tourism can work to disrupt the silencing and invisibilizing forms of structural violence, which present as barriers to a lasting post-apartheid peace in South Africa. The following discussion contextualizes these findings into the larger conversation regarding the potential of tourism as a driver for peace.

Discussion & conclusion

Structural violence poses a barrier to peace everywhere. It prevents people, particularly those who are marginalized according to their race, age, gender, sexual orientation, or ability from achieving their true health, economic potential, and well-being (Galtung, 1969; Rees, 2003; Rylko-Bauer & Farmer, 2016). Silencing and invisibilization were powerful tools used to marginalize and oppress Black South Africans under apartheid. Today, Black South Africans have the power to vote, and the ANC has been the ruling political party since 1994. However, through a combined lack of land ownership, an expansion of housing construction in the townships, as well as

Figure 3. A resident shows a tourist around the township – by showing the tourists around the township one guide felt that he was rediscovering the beauty and value of his own community (source: T3 resident, male, early 30 s).

ongoing barriers to movement beyond the townships, Black presences in the dialogical and the physical public spaces of South Africa continue to be silenced and/or made invisible. As one participant shared, "Black people in town in [the city] should be out by 6 or 7, or else, if anyone is Black from [the township] that is there should be behind a bar or waiting some tables" (T3, male, mid-30s). Public space continues to be white space.

Tourism encounters have the power to disrupt this silencing in a number of ways. Firstly, they place township spaces and their residents in the international spotlight. Slum tourism is increasingly becoming a topic of conversation, of study, and of planned travel. Townships have always been invisibilized spaces – encounters in tourism force them out into the open, which places pressure on governments to acknowledge these spaces and do something to improve their conditions (Scheyvens, 2010). While the print and television media may continue to emphasize stories of violence and disruption emerging from the townships, the increasingly influential social mediascape is presenting townships in an alternate light. Through blog postings (Muldoon & Mair, 2016), Tweets, Instastories, and TikTok videos (search #townshiptourism in Twitter, Instagram, and TikTok) alternatives to the mainstream media's discourse of townships feature tourists and residents singing, dancing, sharing food, and enjoying their time together in the townships. The relational aspect of the encounter is central – narratives are co-created and re-created through our encounters with one another, demonstrating the worldmaking potential of tourism (Hollinshead, 2004). The opportunities for interactions and the sharing of stories created via tourism encounters disrupt the discourse of violence and present a counter-narrative of townships as welcoming and joyful spaces.

The philosophy of Ubuntu is embedded within and throughout the findings described above. It is for this reason that we have conceptualized the transformative potential of tourism for peace as both individual and collective, wherein peace is characterized as a valuing of individuals' stories and a healing of the wounds of the past, as well as a breaking down the barriers that divided us and sharing in a common humanity. Within Ubuntu these are one and the same - a person is a person because we are persons. In fact, in a very real way, Ubuntu *means* peace – my humanity, my well-being, my future are bound to yours, therefore peace is the only option that allows us to preserve our very humanness. Through the lens of Ubuntu, colonization dehumanized both the colonized and the colonizer: the colonized through the intentionally inhuman operationalization of people as commodities of capitalism, and the colonizer through their own dehumanization in treating other people with cruelty and avarice. Encounters in tourism restore humanness to both. The resident/formerly colonized person rediscovers their humanity, their dignity, and their voice in welcoming visitors to their home, showing them the fruits of their labours, and breaking bread with them. The tourists/colonizing foreigners affirm their humanity in the valuing of Others' stories, in honouring their struggles, and in celebrating our coming together with joy and dancing.

As foreign researchers in the townships, it was at first difficult to apprehend the full import ascribed to the sharing of one's stories. Through the lens of Ubuntu, we can see how in sharing our stories we live on beyond our immediate spheres – we become a part of our shared humanity. Preventing Black South Africans from sharing their stories, from having their experiences form any part of the national narrative was a form of cultural violence inflicted over many generations. Tourism in the townships has transformative potential because it allows people to take back the narrative, to define their experiences and their identities on their own terms, to decide how their lives are told for themselves. One participant confided that it does not matter what the tourist *does* with a person's story once it has been told to them – it is the act of sharing the story, of framing the narrative itself, that has power. Nevertheless, through the influx of tourism in the townships the narrative is being changed on a larger scale, as tourists' narratives and photographs grow in prominence and visibility and compete with mainstream media narratives. Through ongoing interactions and the sharing of one's humanity, sentiments like the one expressed above of the white person being 'superior' will invariably become less prominent. Tourism to the townships is a form of Ubuntu, the inviting of another person into your life, of affirming your own humanity by welcoming and celebrating the humanity of others.

We return to the question of tourism for peace and ask whether tourism to the townships has the potential to be a force for peace in post-apartheid South Africa. Issues of lasting peace, human dignity, and social justice are complex, and it would be naïve to suggest that tourism presents a standalone solution to these deeply embedded challenges. However, with this paper we posit that tourism as a social force is a powerful disruptor of hidden norms and structural inequalities. Through tourism, we reaffirm our shared humanity and disrupt and challenge the structural barriers that divide us. Goudie et al. (1999) recognize township tourism as a "catalyst for social change and healing in South Africa" (p. 29). This research finds that township tourism presents many opportunities for undermining or overthrowing structural silencing and invisibilization and contributing to a lasting peace in a country still scarred by centuries of colonial oppressions and racial violence.

Notes

1. Literally 'dumb pass,' this was an identification document that Black and Coloured South Africans were required to carry with them at all times, and which delimited where travel was permitted. Failure to carry your *dompas* could result in arrest and imprisonment.
2. Townships are marginalized communities segregated by race in South Africa's urban areas. While they are often characterized by impoverished living conditions they are rarely labeled as 'slums.' However, much of the tourism literature does not distinguish between slums, *favelas*, and townships, and the term 'slum' is often applied to encompass a range of marginalized urban neighbourhoods.

Disclosure statement

No potential conflict of interest was reported by the authors.

ORCID

Meghan L. Muldoon http://orcid.org/0000-0001-5905-6218

References

Amir, Y. (1969). Contact hypothesis in ethnic relations. *Psychological Bulletin*, *71*(5), 319–342. https://doi.org/10.1037/h0027352

Areff, A. (2018, March 4). We are cutting the throat of whiteness: Malema on plans to remove Trollip. *News24*. Retrieved September 17, 2020, from https://www.news24.com/news24/southafrica/news/we-are-cutting-the-throat-of-whiteness-malema-on-plans-to-remove-trollip-20180304

Askjellerud, S. (2003). The tourist: A messenger of peace? *Annals of Tourism Research*, *30*(3), 741–744. https://doi.org/10.1016/S0160-7383(03)00049-5

Becken, S., & Carmignani, F. (2016). Does tourism lead to peace? *Annals of Tourism Research*, *61*, 63–79. https://doi.org/10.1016/j.annals.2016.09.002

Blanchard, L., & Higgins-Desbiolles, F. (2013). *Peace through tourism: Promoting human security through international citizenship*. Routledge.

Booyens, I., & Rogerson, C. (2019). Re-creating slum tourism: Perspectives from South Africa. *Urbani Izziv*, *30*, 52–63.

Caton, K. (2012). Humanism and tourism: A moral encounter. In M. Mostafanezhad & K. Hannam (Eds.), *Moral encounters in tourism* (pp. 185–198).Ashgate Publishing Ltd.

Causevic, S. (2010). Tourism which erases borders: An introspection into Bosnia and Herzegovina. In O. Moufakkir, & I. Kelly (Eds.), *Tourism, progress, and peace* (pp. 48–64). CABI.

Charmaz, K. (2006). *Constructing grounded theory: A practical guide through qualitative analysis*. Sage Publications.

D'Amore, L. (2009). Peace through tourism: The birthing of a new socio-economic order. *Journal of Business Ethics*, *89*(S4), 559–569. https://doi.org/10.1007/s10551-010-0407-3

Dreyer, J. S. (2015). Ubuntu: A practical theoretical perspective. *International Journal of Practical Theology*, *19*(1), 189–209. https://doi.org/10.1515/ijpt-2015-0022

Farmaki, A. (2017). The tourism and peace nexus. *Tourism Management*, *59*, 528–540. https://doi.org/10.1016/j.tourman.2016.09.012

Farmer, P. (2001). *Infections and inequalities: The modern plagues*. University of California Press.

Freire-Medeiros, B. (2013). *Touring poverty*. Routledge.

Frenzel, F. (2012). Beyond 'Othering': The political roots of slum tourism. In F. Frenzel, K. Koens, & M. Steinbrink (Eds.), *Slum tourism: Power, poverty and ethics* (pp. 49–65).Routledge.

Frenzel, F. (2016). *Slumming it*. Zed Books.

Frenzel, F. (2017). Tourist agency as valorisation: Making Dharavi into a tourist attraction. *Annals of Tourism Research*, *66*, 159–169. https://doi.org/10.1016/j.annals.2017.07.017.

Frenzel, F., Koens, K., Steinbrink, M., & Rogerson, C. M. (2015). Slum tourism: State of the art. *Tourism Review International*, *18*, 237–252. https://doi.org/10.3727/154427215X14230549904017.

Galtung, J. (1969). Violence, peace, and peace research. *Journal of Peace Research*, *6*(3), 167–191. https://doi.org/10.1177/002234336900600301

Gelfand, M. (1973). *The Genuine Shona: Survival Values of an African Culture*. Gweru: Mambo Press.

Gibson, J. L., & Claassen, C. (2010). Racial reconciliation in South Africa: Interracial contact and change over time. *Journal of Social Issues*, *66*(2), 255–272.

Goodman, P. (2017, October 24). End of apartheid in South Africa? Not in economic terms. *The New York Times*. Retrieved June 24, 2020, from https://www.nytimes.com/2017/10/24/business/south-africa-economy-apartheid.html
Goudie, S., Khan, F., & Kilian, D. (1999). Transforming tourism: Black empowerment, heritage and identity beyond apartheid. *South African Geographical Journal*, *81*(1), 22–31. https://doi.org/10.1080/03736245.1999.9713658
Guardian, M. (2019, January 4). *Who runs SA's media is a black-and-white issue*. Retrieved September 20, 2020, from https://mg.co.za/article/2019-01-04-00-who-runs-sas-media-is-a-black-and-white-issue/
Haessley, J. (2010). Tourism and a culture of peace. In O. Moufakkir, I. Kelly (Eds.). *Tourism, progress, and peace* (pp. 1–16).CABI: Wallingford, Oxfordshire.
Herzog, B. (2018). Invisibilization and silencing as an ethical and sociological challenge. *Social Epistemology*, *32*(1), 13–23. https://doi.org/10.1080/02691728.2017.1383529
Hesse-Beiber, S. (Ed.) (2007). *Handbook of feminist research: Theory and praxis*. Sage Publications.
Higgins-Desbiolles, F. (2006). More than an "industry": The forgotten power of tourism as a social force. *Tourism Management*, *27*(6), 1192–1208. https://doi.org/10.1016/j.tourman.2005.05.020
Hollinshead, K. (1992). White' gaze, 'red' people – shadow visions: The disindentification of 'Indians' in cultural tourism. *Leisure Studies*, *11*(1), 43–64. https://doi.org/10.1080/02614369100390301
Hollinshead, K. (2004). Tourism and new sense: Worldmaking and the enunciative value of tourism. In M. C. Hall & H. Tucker (Eds.), *Tourism and postcolonialism* (pp. 39–56). Bew York: Routledge.
Holst, T. (2018). *The affective negotiation of slum tourism: City walks in Delhi*. Routledge.
Honneth, A. (2001). Invisibility: On the epistemology of recognition. *Aristotelian Society Supplementary Volume*, *75*(1), 111–126. https://doi.org/10.1111/1467-8349.00081
Jóhannesson, G., Ren, M. C., & van der Duim, R. (Eds.). (2015). *Tourism encounters and controversies: Ontological politics of tourism development*. Ashgate.
Koens, K., & Thomas, R. (2016). "You know that's a rip-off": Policies and practices surrounding microenterprises and poverty alleviation in South African township tourism. *Journal of Sustainable Tourism*, *24*(12), 1641–1654. https://doi.org/10.1080/09669582.2016.1145230
Marsh, T., & Knaus, C. (2015). Fostering movements or silencing voices: School principals in Egypt and South Africa. *International Journal of Multicultural Education*, *17*(1), 188–210. https://doi.org/10.18251/ijme.v17i1.969
McEachern, C. (2002). *Narratives of a nation: Media, memory and representation in the making of the New South Africa*. Nova Science Publishers.
Meschkank, J. (2011). Investigation into slum tourism inn Mumbai: Poverty tourism and the tensions between different constructions of reality. *GeoJournal*, *76*(1), 47–62. https://doi.org/10.1007/s10708-010-9401-7
Moufakkir, O., & Kelly, I. (2010). *Tourism, progress and peace*. CABI.
Muldoon, M. (2018). *Gazing back: A feminist postcolonial lens on tourism in the townships of South Africa* [PhD thesis]. Department of Recreation and Leisure Studies, University of Waterloo.
Muldoon, M., & Mair, H. (2016). Blogging slum tourism. *Tourism Analysis*, *21*(5), 465–479. https://doi.org/10.3727/108354216X14653218477525
Murove, M. F. (2014). Ubuntu. *Diogenes*, *59*(3–4), 36–47. https://doi.org/10.1177/0392192113493737
Nisbett, M. (2017). Empowering the empowered? Slum tourism and the depoliticization of poverty. *Geoforum*, *85*, 37–45. https://doi.org/10.1016/j.geoforum.2017.07.007
Novelli, M., Morgan, N., & Nibigira, C. (2012). Tourism in a post-conflict situation of fragility. *Annals of Tourism Research*, *39*(3), 1446–1469. https://doi.org/10.1016/j.annals.2012.03.003
Nyamnjoh, F. B. (2016). *#RhodesMustFall: Nibbling at resilient colonialism in South Africa*. Langaa Research and Publishing Common Initiative Group.
Nyaupane, G. P., Teye, V., & Paris, C. (2008). Innocents abroad: Attitude change toward hosts. *Annals of Tourism Research*, *35*(3), 650–667. https://doi.org/10.1016/j.annals.2008.03.002
Pedwell, C. (2012). Economies of empathy: Obama, neoliberalism, and social justice. *Environment and Planning D: Society and Space*, *30*(2), 280–297. https://doi.org/10.1068/d22710
Pettigrew, T. F. (1998). Intergroup contact theory. *Annual Review of Psychology*, *49*(1), 65–85. https://doi.org/10.1146/annurev.psych.49.1.65
Posel, D. (2010). Races to consume: Revisiting South Africa's history of race, consumption, and the struggle for freedom. *Ethnic and Racial Studies*, *33*(2), 157–175. https://doi.org/10.1080/01419870903428505
Pratt, S., & Liu, A. (2016). Does tourism really lead to peace? A global view. *International Journal of Tourism Research*, *18*(1), 82–90. https://doi.org/10.1002/jtr.2035
Rees, S. (2003). *Passion for peace: Exercising power creatively*. University of New South Wales Press.
Reynolds, P. (2013). *War in worcester: Youth and the apartheid state*. Fordham University Press.
Rogerson, C. M. (2004). Urban tourism and small tourism enterprise development in Johannesburg: The case of township tourism. *GeoJournal*, *60*(3), 249–257. https://doi.org/10.1023/B:GEJO.0000034732.58327.b6
Rolfes, M. (2010). Poverty tourism: Theoretical reflections and empirical findings regarding an extraordinary form of tourism. *GeoJournal*, *75*(5), 421–442. https://doi.org/10.1007/s10708-009-9311-8

Rolfes, M., Steinbrink, M., & Uhl, C. (2009). *Township as attraction: An empirical study of township tourism in Cape Town. Praxis Kultur -und Sozialgeographie, 46*. Universitätsverlag Potsdam.
Rose, G. (2001). *Visual methodologies*. Sage Publications.
Rummell, R. L. (1982). *Understanding conflict and war, Vol. 5: The just peace*. Sage Publications.
Rylko-Bauer, B., & Farmer, P. (2016). Structural violence, poverty, & social suffering. In D. Brady & L. Burton (Eds.). *The Oxford handbook of the social science of poverty* (pp. 47–74). Oxford University Press.
Scheyvens, R. (2002). *Tourism for development: Empowering communities*. Prentice-Hall.
Scheyvens, R. (2010). *Tourism and poverty*. Routledge.
Stanlake, J., & Samkange, T. (1980). Hunnuism of Ubuntuism: A Zimbabwea indigenous political philosophy. Salisbury, Rhodesia: Graham Publishing.
Swartz, S., Harding, J. H., & De Lannoy, A. (2012). Ikasi style and the quiet violence of dreams: A critique of youth belonging in post-apartheid South Africa. *Comparative Education, 48*(1), 27–40. https://doi.org/10.1080/03050068.2011.637761
Tucker, H. (2009). Recognizing emotion and its postcolonial potentialities: Discomfort and shame in a tourist encounter in Turkey. *Tourism Geographies, 11*(4), 444–461. https://doi.org/10.1080/14616680903262612
Tucker, H. (2016). Empathy and tourism: Limits and possibilities. *Annals of Tourism Research, 57*, 31–43. https://doi.org/10.1016/j.annals.2015.12.001
Valentine, G. (2008). Living with difference: Reflections on geographies of encounter. *Progress in Human Geography, 32*(3), 323–337. https://doi.org/10.1177/0309133308089372
Van Dongen, E. (2005). Remembering in times of misery: Can older people in South Africa 'get through'? *Ageing and Society, 25*(4), 525–541. https://doi.org/10.1017/S0144686X05003806
Villa-Vicenio, C., & Verwoerd, W. (Eds.). (2000). *Looking back, reaching forward: Reflections on the Truth and Reconciliation Commission of South Africa*. Zed Books.
Wang, C., & Burris, M. A. (1997). Photovoice: Concept, methodology, and use for participatory needs assessment. *Health Education & Behavior, 24*(3), 369–387. https://doi.org/10.1177/109019819702400309
Wang, C., & Redwood-Jones, Y. (2001). Photovoice ethics: Perspectives from Flint photovoice. *Health Education & Behavior, 28*(5), 560–572. https://doi.org/10.1177/109019810102800504
Wintersteiner, W., & Wohlmuther, C. (2014). Peace sensitive tourism: How tourism can contribute to peace. In W. Wintersteiner & C. Wohlmuther (Eds). *International handbook on tourism and peace* (pp. 31–61).Centre for Peace Research and Peace Education.
Wright, T. (2017). Justice, silence, complexity: Recent forays into the reconstitution of apartheid experience. *African Studies, 76*(1), 163–176. https://doi.org/10.1080/00020184.2017.1285673
Zondi, S. (2014). The rhetoric of ubuntu diplomacy and implications for making the world safe for diversity. *African Journal of Rhetoric, 6*(1), 106–142.

Part II

Peace Tourism Pedagogies

OPEN ACCESS

"We are reconciliators": when Indigenous tourism begins with agency

Nicole Curtin and Steven Bird

ABSTRACT

The notion that Indigenous tourism can advance reconciliation contrasts with prevailing 'tourism as industry' discourses. Commodification processes treat tourists as consumers, rather than as visitors to a place, or visitors to the *people* of a place. How can Indigenous tourism deliver sustainable benefits to the hosts and communities that receive visitors? This study adopts critical Indigenous methodology with constructivist grounded theory, as we source and validate theoretical constructs of sustainability in Indigenous tourism with Aboriginal tourism operators themselves. Three practices emerge, namely hosting, connecting, and sharing. Through hosting, operators set the scene for culturally safe interactions. Through connecting, hosts and tourists recognise their shared humanity. Through sharing, local identities, cultures, and histories are brought to the surface. These three practices of hosting, connecting, and sharing arise from the agency, and thereby reinforce the agency, of Aboriginal tourism operators. In order for Indigenous tourism operators and communities to derive sustainable benefits from receiving visitors, such engagements must be founded on recognition and respect for Indigenous agency. These practices imply reciprocity and point to local understandings of reconciliation, not as an endpoint, but as a practice in the here and now. We argue that this represents a strengths-based model of Indigenous tourism.

Introduction

Indigenous tourism operators bring their own perspectives to the practice and scholarship of tourism. In the words of the Australian Aboriginal tourism operators who participated in this research[1], as the "first humans here" and "through doing what the old fellas did", they are "going back to [their] old practices of welcoming people to [their] Country". Through "sharing vital information" with their visitors about "the authentic truthful stories of the Country, of the people, and the history", they prioritise cultural continuity and caring for Country. They aim to "teach others about it so that future generations can enjoy it too".

This position sits in stark contrast with the prevailing "tourism as industry" discourse (Higgins-Desbiolles, 2006, p. 1206). Here, tourists are primarily understood as consumers, and only secondarily as visitors to a place, and seldom as visitors to the *people* of a place. Commercialisation presents risks for Indigenous hosts and communities who may sacrifice their authenticity in order to meet market demands (Coronado, 2014). This leads to the question: how can the rights and wellbeing of local people and places be prioritised above those of tourists? Higgins-Desbiolles et al. (2019) criticise excessive commodification, and recommend that tourism be "returned to ideas of hospitality and connection" (p. 14).

Indigenous tourism operators report using their platform to advance their agenda: "we are reconciliators". Through practicing reconciliation in the present, and through sharing and sustaining their cultures, they support peaceful and inclusive societal relations in alignment with the United Nations Sustainable Development Goal 16, which is concerned with peace, justice, and strong institutions. At the same time they promote the importance of 'caring for Country'. These practices align with the 'peace through tourism' agenda where peace is not so much the absence of war but rather a visioning for a desired future (Haessly, 2010). Although there are conflicting narratives about if and how true reconciliation can be achieved in colonised nations (Elder, 2017), there are opportunities to realise more just and equitable societies, and for tourism to fulfil its potential as a social force (Higgins-Desbiolles, 2006).

To understand how Indigenous tourism can benefit and support the wellbeing of Indigenous communities, we adopted critical Indigenous methodology with constructivist grounded theory to privilege Indigenous knowledges and perspectives in an emergent, responsive, and flexible research process (Charmaz, 2014; Denzin & Lincoln, 2008). As non-Indigenous researchers, we endeavoured to undertake respectful and ethical research through a collaborative, participatory, and reciprocal approach (Whitford & Ruhanen, 2016). Accordingly, this work is founded upon the relationships that we have fostered with the people who have shared their stories with us. The first author led these engagements, and gathered and coded the data.

Throughout this paper where possible, our terminology reflects that used by participants. 'Indigenous' extends to Aboriginal and Torres Strait Islander Australians, whereas 'Aboriginal' refers to Indigenous peoples of mainland Australia. Capitalisation conventions in Australia extend to cultural terms such as Country and Elder. The term Country means "a place of origin in spiritual, cultural and literal terms. It refers to a specific clan, tribal group or nation of Aboriginal people and encompasses all the knowledge, cultural norms, values, stories and resources within that particular area — that particular Indigenous place" (Fredericks, 2013, p. 3).

This paper is organised as follows. First, we position Indigenous tourism within its larger context, define the processes and goals of reconciliation in Australia, and then consider sovereignty within Indigenous tourism. We then detail the research design and present the findings, foregrounding Indigenous voices in the process. In the ensuing discussion, we describe our analysis and present a strengths-based model of Indigenous tourism.

Background

Indigenous tourism and sustainability

We adopt the definition of Indigenous tourism proposed by Butler and Hinch (2007) as "tourism activity in which Indigenous people are directly involved either through control and/or by having their culture serve as the essence of the attraction" (p. 5). Indigenous tourism is "commonly viewed as a means of facilitating socio-economic benefits to Indigenous individuals, communities and host regions" (Carr et al., 2016, p. 1068). The Territory Economic Reconstruction Commission (TERC, 2020) identified Aboriginal tourism as presenting "a significant economic opportunity for the [Northern] Territory, coupled with invaluable social and economic benefits that will flow directly to Aboriginal Territorians" (p. 70). In line with these perceived benefits, Robert Taylor, CEO of the Western Australian Indigenous Tourism Operators Council (WAITOC), states that:

> Aboriginal tourism is a unique growth industry in Western Australia. As the world's oldest living culture, Aboriginal culture offers tourism experiences unique to Australia not available anywhere else in the world. A strong, diverse and self-supporting Aboriginal business sector is key to empowering Aboriginal and Torres Strait Islander people, and placing Aboriginal business owners, their families, and communities in the

driver's seat of their economic future (Western Australian Indigenous Tourism Operators Council [WAITOC] & Clean State, 2020, p. 2)

Whitford and Ruhanen (2016) argue that it was Altman's (1989) warning for tourism to not be considered as a panacea for improving the economic status of Indigenous peoples that served as a "precursor to calls for underpinning Indigenous tourism with the principles of sustainability" (p. 1085). Specifically, the practice of Indigenous tourism carries risks for local communities to have their rights eroded and to suffer oppression and exploitation (Cole & Morgan, 2010; Ruhanen & Whitford, 2019). Commercialisation and commodification processes can subvert local control over representations of culture and limit access to financial benefits of operations conducted on sovereign lands (Coronado, 2014). Indeed, the Larrakia Declaration on the Development of Indigenous Tourism asserts that: "for Indigenous tourism to be successful and sustainable, Indigenous tourism needs to be based on traditional knowledge, cultures and practices and it must contribute to the well being of Indigenous communities and the environment" (Pacific Asia Travel Association [PATA] & World Indigenous Tourism Alliance [WINTA], 2014, p. 12). This raises the question: how can Indigenous tourism deliver sustainable benefits to the hosts and communities that receive visitors?

One approach is to think in terms of balancing the rights of locals with the rights of visitors (Williams & Gonzalez, 2017). Another approach addresses the imperative of sustainable development to ensure that tourism "does not degrade or alter the environment (human and physical) ... to such a degree that it prohibits the successful development and wellbeing of other activities and processes" (Butler, 1999, p. 35). Yet another approach, which we take up here, is to consider the "larger context within which [tourism] operates in order to arrive at real sustainability" (Higgins-Desbiolles, 2020a, p. 19). We are mindful that, in order to "gain a more nuanced understanding of what constitutes sustainability in the Indigenous tourism context, care should be taken to avoid tendencies to oversimplify sustainability and its capacity to enhance Indigenous tourism" (Whitford & Ruhanen, 2016, pp. 1089–1090). One way to take up this challenge is to consult the operators themselves. When we do this, we find that two other concepts often come up, namely reconciliation and sovereignty. We consider these next.

The processes and goals of reconciliation in Australia

Reconciliation Australia (2016) identifies the aim of the social and political movement of reconciliation as achieving a just and equitable nation. The process, although contentious, seeks to address the disparities between Indigenous and non-Indigenous Australians in health status and outcomes, and determinants of health[2], including inequities in imprisonment and children in out-of-home care (Australian Institute of Health and Welfare [AIHW], 2020; Dudgeon et al., 2014), to facilitate respectful, trusting, and positive societal relationships (Reconciliation Australia, 2016), and for key constitutional changes to occur (Reconciliation Australia, 2020). As explained by Indigenous Australian politician Linda Burney, "The Australian constitution is the only constitution of a first world nation with a colonial history that does not recognise its first people" (RMIT ABC Fact Check, 2019).

To develop a framework to assess progress towards reconciliation, a working group consisting of Indigenous and non-Indigenous members from Reconciliation Australia (2016) conducted a meta-analysis of approaches to defining and measuring reconciliation both in Australia and internationally. Reconciliation Australia state that reconciliation depends on progress in the dimensions of race relations, equality and equity, institutional integrity, and historical acceptance. Together these establish unity: "In a reconciled Australia, national unity means Aboriginal and Torres Strait Islander rights, histories and cultures are valued and recognised as part of a shared national identity" (Reconciliation Australia, 2020, p. 5).

However, "the matter of a national identity can be a problematic one for colonized countries when coming to terms with their history of colonization and the usually brutal truths upon which that settlement history was founded" (Galliford, 2010, p. 228). The concept of reconciliation occupies a contested space in Australia, and "a mix of narratives, often contradictory, are continuing to shape the ongoing reconciliation discourse and are demonstrating the plural nature of the process" (Elder, 2017, p. 91). Although on one hand reconciliation can be described as political packaging, there is evidence of vitality in reconciliation processes across the Australian community (Brennan, 2004). Garvey (2015) explains that Indigenous and non-Indigenous Australians alike "may both invoke their adherence to social justice, or their commitment to reconciliation, however, the expression and experience of each may differ as one considers the life experience that each player brings to their conceptualization" (p. 341f).

Although there are differing ideas on if and how reconciliation may be achieved, Indigenous tourism may help such societies to address their colonial legacy (Higgins-Desbiolles, 2003) through connecting with an understanding of reconciliation as "an act of public consciousness and will, a socio-political atmosphere that everyone 'breathes from', which opens up the possibility not only for social justice but positive intercultural relations" (Galliford, 2010, p. 233). Tourism encounters offer a space where dialogue can "continue with all sectors of the Australian community to discuss the unresolved issues of sovereignty, treaty and agreement-making" (Reconciliation Australia, 2016, p. 21).

Considering sovereignty within Indigenous tourism

In line with the goal of recognising the sovereignty of Indigenous Australians through reconciliation processes, there is hope that shifting tourists' mindsets from 'consumers' to 'guests' will allow the rights and wellbeing of local people and places to be prioritised (Higgins-Desbiolles et al., 2019). This leads us to understand tourism as "the process of local communities inviting, receiving and hosting visitors in their local community, for limited time durations, with the intention of receiving benefits" (Higgins-Desbiolles et al., 2019, p. 11). From this position, a respectful starting point would be to explore tourism and its workings with Indigenous tourism operators themselves. To date, much of the inquiry in Indigenous tourism has been conceptual and not based on ground-up engagement with Indigenous operators (cf. Whitford & Ruhanen, 2016). We can hope that Aboriginal voices will help to "put tourism in its place, at the service of local communities and societies" (Higgins-Desbiolles, 2020b, p. 19).

Indigenous people are key stakeholders in Indigenous tourism and in Indigenous tourism research (Nielsen & Wilson, 2012). As such, they are able to theorise about sustainability in Indigenous tourism and to articulate their own concerns and interests in ways that can inform our investigation. This is in keeping with the tenet that Indigenous research must take seriously the worldviews and knowledge practices of Indigenous participants (Christie, 2013). It also aligns with practices of decolonising the scholarship of Indigenous tourism (Pritchard & Morgan, 2007).

A consequence of taking local authority seriously is to *ask the operators*: "it is imperative that academic researchers gain a more comprehensive understanding of Indigenous tourism from the perspective of Indigenous stakeholders" (Whitford & Ruhanen, 2016, p. 1091). The most common approach to engaging Indigenous operators is through case studies (Whitford & Ruhanen, 2016), yet individual studies may be dismissed "as particularistic, as localised, or as having no wider legitimacy" (Chambers & Buzinde, 2015, p. 3). Therefore, there is a need to look across cases to allow for theorising that may have broader applicability. Accordingly, this research explores how Aboriginal tourism operators understand sustainability in Indigenous tourism, and what we show to be closely linked concepts of reconciliation and authority. We expect this to serve as a check on externally-driven theorising, and to point the way forward for further collaborative investigation.

Research design

Methodology

In establishing the methodology for the research presented here, we begin with Whitford and Ruhanen's (2016) suggested directions for Indigenous tourism research being based upon collaborative and reciprocal relationships. These relationships facilitate iterative, adaptable, and flexible knowledge creation and "increased engagement and empowerment of Indigenous peoples [to provide] more opportunities to focus on the potential scope of Indigenous tourism" (p. 1091).

Accordingly, we look to critical Indigenous methodology as an overarching framework because it "begins with the concerns of Indigenous people" (Denzin & Lincoln, 2008, p. 2) and privileges Indigenous knowledges and perspectives. Critical Indigenous methodology provides guidance on how to conduct ethical research in Indigenous contexts, as mandated in Australia (Australian Institute of Aboriginal & Torres Strait Islander Studies [AIATSIS], 2020; National Health & Medical Research Council, 2018). Reciprocal relationships are the foundation of respectful research in Indigenous contexts (Smith, 2012), and through adopting critical Indigenous methodology, we prioritise relationship building and careful checking of our representations of the ideas and experiences of the participants. Reflexive journaling was also an important strategy in attempting to reveal, minimise, and challenge the subjectivity and positionality of the first author.

As tourism is a socially constructed and determined phenomenon, we have adopted the methodology of constructivist grounded theory (Charmaz, 2014). Jennings and Junek advocate for the use of grounded theory "to achieve more in-depth knowledge and understanding of the complexities of tourism" (Jennings & Junek, 2007, p. 198). Further, Bainbridge et al. (2019) advocate for the application of constructivist grounded theory as an exploratory research approach in Indigenous topics where a solid evidence-base is lacking, such as sustainability in Indigenous tourism.

Constructivist grounded theory offers inductive processes for developing a substantive local theory that is 'grounded' in the data and that faithfully represents and amplifies the voices of participants (Charmaz, 2014). Bainbridge et al. (2019) note that there "are strong correlations between a constructivist grounded theory approach as a social justice tool in critical inquiry and its utility as a tool for decolonizing methodologies" (p. 622). The generation of a grounded theory is iterative and emergent, with data gathering and analysis occurring simultaneously, each informing the other, guided by the insights and concerns of the participants (Charmaz, 2014).

Data gathering

Data gathering was guided by the grounded theory technique of 'theoretical sampling' (Charmaz, 2014). As this research considered reconciliation, we started by speaking with Aboriginal operators, before considering the voices of non-Indigenous visitors. These participants were regarded as the domain experts while we positioned ourselves as naive enquirers.

It was necessary to apply a number of data gathering methods in order to yield "rich data to get beneath the surface of social and subjective life … to reveal participants views, feelings, intentions, and actions, as well as the contexts and structures of their lives" (Charmaz, 2006, pp. 13–14). This 'all is data' approach – with analysis of the totality of gathered data – allowed for triangulation, giving robustness to the researchers' understandings of the emergent conceptual categories (Charmaz, 2006). Indeed, constructivist grounded theory licenses investigators to draw on whatever methods are suitable for collecting rich data (Charmaz, 2014).

During fieldwork, researchers spent time with operators in visits of a few hours up to a few days while they conducted their tourism activities, both individually and with other visitors on Country in urban, regional, and remote settings. Field notes described observations from the first author's participation in 26 Indigenous tourism activities throughout this research, including cultural tours, workshops, festivals, cultural centres, performances, and presentations, and also drew on ad hoc conversations with other Aboriginal tourism operators and visitors.

During fieldwork with Aboriginal tourism operators, yarning was undertaken to prioritise Indigenous ways of communicating (Walker et al., 2014). Yarning often includes storytelling, a pervasive feature of oral cultures, and allowed the researchers to spend time with, and listen carefully to, operators while engaging in free flowing conversations and two-way sharing (Garvey, 2015). Additionally, 14 semi-structured interviews with open-ended questions were conducted with Indigenous tourism stakeholders (6 Aboriginal tourism operators, 3 non-Indigenous tourism guides, 4 non-Indigenous domestic tourists, and an industry representative). The average duration of semi-structured interviews was 60 minutes, plus other incidental conversations which also informed the researchers' interpretations of the data. Other relevant data was also sourced, such as promotional, marketing, and website materials; media and social media data; and transcribed presentations by operators.

Data analysis

Data were stored, organised, coded, annotated, and sorted with Atlas.ti qualitative data analysis software. Constructivist grounded theory analysis and narrative analysis were used in conjunction, the former deconstructing data through a systematic coding process while the latter ensuring that coding does not decontextualise participants' stories (Lal et al., 2012). The grounded theory technique of 'constant comparison' enabled the identification of similarities and differences across the data (Charmaz, 2014). It drew attention to relationships between codes and categories, leading to a theory that closely represents the data.

Memoing was crucial for recording reflections, ideas, and further questions arising from the analysis, and moved the analysis beyond descriptive to conceptual formulations. Memoing also assisted with reflexivity and critical thinking. Literature was referred to during the analysis to increase 'theoretical sensitivity' to construct a substantive ground-up theory through inductive reasoning, not only within and across the data, but also across relevant bodies of literature (Charmaz, 2014). During the analysis phase, findings were communicated to participants to assess their accuracy, relevance, and suitability (Charmaz, 2014). This 'member checking' assisted with the co-creation of knowledge and collaboration with participants. We offered participants the choice to give input to the analysis to the extent that they desired. Data gathering and analysis proceeded until "theoretical sufficiency" (Dey, 1999, p. 257), the point at which sufficient depth of understanding enabled the researchers to represent the data in a substantive theory, was achieved (Charmaz, 2006).

Presentation of findings

The paper's composition reflects the choices of participants concerning how they would be involved and how they presented information. The findings represent the sum of all analysed data and are focussed on the aspirations, aims, and actions of Aboriginal tourism operators. In presenting the findings, we take seriously Whitford and Ruhanen's (2016) call to understand Indigenous tourism from the perspective of Indigenous stakeholders, and so each heading in the findings section opens with the voices of Aboriginal tourism operators, inviting readers to draw their own conclusions before a summary is provided, and an interpretation is offered in the subsequent discussion section. Although we have selected, structured, and sequenced the participants' quotes, we have retained extended context, and have been guided by member checking, giving us confidence that we have represented the operators' lived experiences faithfully.

Quotes from six Aboriginal tourism operators from urban, regional, and remote Australia, shown in Figure 1, are included to illustrate the findings. Aboriginal tourism operators in Western Australia (WA) are members of the Western Australian Indigenous Tourism Operator's Council (WAITOC). In the Northern Territory (NT), Injalak Arts is a non-profit Aboriginal

Figure 1. The locations of Aboriginal tourism operations that form the basis of this study.

Corporation which employs Indigenous staff, including Indigenous tour guides. At their request, and in line with the project's ethical clearance, the operators are identified in this paper and have their quotes attributed to their names. Further, spellings adopted reflect those used by the participants. This explains the variations of spellings of the Aboriginal group of the South West of Western Australia: Nyungar/Noongar/Nyoongar.

- Bart Pigram: Yawuru owner/operator of Narlijia Experiences Broome (Broome, WA)
- Clinton Walker: Ngarluma and Yindjibarndi owner/operator of Ngurrangga Tours (Karratha, WA)
- Kerry-Ann Winmar: Whadjuk Nyungar owner/operator of Nyungar Tours (Perth, WA)
- Marissa Verma: Noongar owner/operator of Bindi Bindi Dreaming (Perth, WA)
- Mick Hayden: Njaki Njaki Nyoongar owner/operator of Njaki Njaki Aboriginal Cultural Tours (Merredin, WA)
- Roland Burrunali: Bininj tour guide with Injalak Arts and Craft (Gunbalanya, NT)

Findings

The voices of Aboriginal tourism operators are foregrounded by introducing in vivo codes[3] as headings to emphasise the lived experiences of participants, followed by direct quotes and then by a brief summary, plus further observations that will inform our later analysis. Quotes are incorporated not as italicised indented paragraphs but as mainline paragraphs, consistent with our commitment to foreground Indigenous voices.

"Doing what the old fellas did"

Mick: It's just the knowledge from the old fellas, what they shown us and from talking. And to me, that's the same. Like I said, while I'm still here fit and able I'm doing what the old fellas did.

Bart: The welcoming of people, or the sometimes not welcoming, but the communication, the trades, the old trades, the whole thing … we've got a social structure, kinship and skin groups that actually link all of us in that community together. Everybody is somebody. And every tribe was linked to another tribe. There was no secluded tribe ever in Australia. They had relationships, they had trade, they had interactions with certain people on certain boundaries at certain times of the year. And you could really put that down to our tourist season in Broome. From May to September is when these guys migrate to our area and we share, we give them bush tucker, I teach them the language, I teach them how to move around the place with respect. You could say in a sense that's been happening for thousands of years, because that's what we were doing synchronised in Australia, we were all linked to somebody. … We're just going back to our old practices of welcoming people to our Country.

Through the above quotes, Indigenous tourism can be seen to be sitting within an existing frame of a continuing cultural practice of receiving visitors. As the foundation of this practice, visitors seek and are granted permission before entering the custodial lands of another group. This sets the scene for a different starting point for Indigenous tourism, one that is grounded in the sovereignty of hosts.

"They have to abide by our practices"

Bart: When people come here they have to abide by our social, environmental, and cultural practices.

Roland: I take tourists to the hills, showing for them rock art and we showing for like, people living in the hills. And, we take tourists showing for some loose rocks. Or, sometimes, tourists like for slippery rocks and we explain sometimes there's rocks too low.

In being welcomed to Country, there are expectations set for visitor behaviour. Visitors receive information to conform with "social, environmental, and cultural" protocols to respect local authority. Bart explains that there are cultural restrictions that limit the knowledges and places, such as sacred areas, that he can share with visitors. Sometimes, these restrictions declare a place to be off-limits (as in the case of closing the Uluru climb; Parks Australia, 2019). A part of welcoming visitors to Country is also sharing safety information with them.

"I've branched out"

Bart: I've branched out to do Welcome to Countrys [ceremonies] because of the language and the understanding of what a Welcome to Country is. Weddings and other things. But sort of doing cultural awareness stuff … I call it "cultural interpretation", *ngan-ga jayida*, which means our story, or our word.

Marissa: Business is fun and can take you anywhere and everywhere.

Through being cultural authorities and having diverse business offerings, many interactions with visitors beyond guided tours are occurring. Other services offered by operators include: officiating Welcome to Country, and other, ceremonies; conducting cultural awareness training and presentations in school, corporate, and local government settings; running workshops for activities such as cooking, gardening, and arts and craft; providing language translation services; photography; and environmental and cultural consultation. This variety means that visitors have diverse opportunities to connect with Aboriginal tourism operators without necessarily being defined, or considering themselves, as tourists. An example of this is a sewing and cooking workshop that Marissa co-facilitates in collaboration with Ballardong and Whadjuk Nyungar woman Rebecca Barlow of Deadly Denim. For the operators, being involved in multiple areas of community engagement creates networks and partnerships across industries and increases visibility.

Through offering multiple services, tourism operators strengthen the viability of their businesses. For some, the seasonal and sporadic tourism element of their business does not provide a full time income. Transcending reliance on tourists supports financial sustainability. Several operators also explained that having diverse interactions with visitors sustains their personal interest and energy. They talk about enjoying their lifestyle, with their day to day business operations presenting opportunities for doing a variety of activities with different people while being on Country and connected with their environment. Clinton contrasts this with the potential for getting bored or burnt out if tours are conducted as rote performances.

"A safe environment for both parties"

Clinton: Most people don't want to just rock up into an Aboriginal community and say, "tell us all your history". Whereas if you go on a tour, you're providing a service for someone, you've got a customer who's willing to pay for it, and they get to learn about it in a safe environment for both parties. And then it breaks down those barriers and then people are more willing to interact more, you know with Indigenous, and vice versa.

Bart: Well, a lot of them [visitors] actually haven't had a chance to sit down and talk honestly and openly with Aboriginal people. They haven't had a safe environment to discuss these things. ... We get asked from the most obvious question to the most challenging question. And this is where I respect a lot of our tour guides, because we have to challenge ourselves and to be able to answer the easy questions and answer the most difficult questions that politicians struggle to answer. And we do it with a cool head because we've got to operate a tour.

Interactions between hosts and visitors in tourism settings have defined boundaries. Western-style transactions dictate the provision of a pre-agreed, time-limited service. These boundaries mean that tourism interactions are markedly different to social interactions, and that visitors can choose to be either passive or active participants. This seems especially important for visitors who "actually haven't had a chance to sit down and talk honestly and openly with Aboriginal people". These boundaries assist operators in creating "a safe environment for both parties".

Hosts talk about aiming to engage with their visitors in a positive way so that "people are more willing to interact". The operators appear skilled in opening spaces for dialogue and discussion about often uncomfortable truths concerning the history, legacy, and ongoing practices of colonisation. It is important for operators to meet people where they are at, particularly when delivering information that visitors find confronting: "we do it with a cool head because we've got to operate a tour". The notion of safe spaces implies opportunities for grappling with ideas and emotions, and not always having answers to "the most challenging question".

"I make it really personable"

Mick: I just want them to take one thing about from when I start my tours to the end of it, just learn one thing and that's all I ask. If they learn heaps more, that's well and good, and to me, I done my job. So, I'm imparting some of my knowledge that I know onto others and they'll go away with a different point of view and say, oh there's something I learnt that I didn't know before.

Bart: A lot of the time, the tour basically turns into a discussion, and a sharing of values and ideas and concerns really. So, I hope I make it really personable, so I try to get them to open up and I want them to go away thinking maybe a little bit of hope, maybe a little bit of faith in the Aboriginal peoples' future in Australia. But out of anything, I hope that they take away more understanding than what they came to Broome with.

Marissa: No job is too hard that I can't do to teach somebody something.

Within the context of safe spaces, operators speak of prioritising personal connection with their visitors, aiming to "make it really personable". The above quotes capture a hope to facilitate movement along a continuum of understanding. This approach contrasts with one where there is a prescribed body of knowledge that operators aim to impart in order to achieve a desired goal. Through meeting visitors where they are at, the operators are adaptable, noting that often a tour "basically turns into a discussion". By being responsive to the range of interests, questions, and prior knowledge and experience of their visitors, the operators aim to advance their visitors' knowledge in whichever way leads to greater understanding: "I just want them to take one thing about from when I start my tours to the end of it". Several operators expressed feeling empathetic towards their visitors who lack understanding of Aboriginal peoples and cultures, and attribute this to inadequate Indigenous education across the country.

"We can all relate to one another"

Mick: For everybody and anybody, we're all not the same, we're all different. And that's the same case with Aboriginal people. We all may have different coloured skin, but we all bleed the same. We all got red blood. And we're no different.

Clinton: I don't see people as different. I see them as the same, equal. We just come from different backgrounds. We come from different cultures. That's the only thing that's different. We might look different, but at the end of the day, we're human beings. That's what I want people to understand. … We're still people, and that everybody has struggles and that we can all relate to one another even though we come from different backgrounds.

Bart: I think it just comes down to the whole human element of it all. And then obviously the first humans here is us mob, so we've got the oldest culture here, and so we're just sharing it. … I think that the whole reconciliation thing needs to go back to the actual human element and the understanding of cultures, and that's what we do.

The operators express transcending differences and finding common ground with their visitors through the universal experience of being human: "it just comes down to the whole human element of it all". This common ground may be experienced through shared activities or through discussions in safe spaces: "we can all relate to one another even though we come from different backgrounds". Speaking to the commonalities that exist across cultures and acknowledging that "at the end of the day, we're human beings", there is hope for empathy and understanding.

"I think of them as friends"

Bart: I thought that being a people person I could actually try to do these tours with a bit more personal connection.

Clinton: There's actually a family who come on my tours nearly every year. And they always come back. The guy who's the Dad of that family, he used to work up here [Karratha] and he heard about my stuff, and I worked with him I think in the past. … I'd taken him out on a tour and shown him around and they said "oh, we loved it, we want to come back next year. What can you recommend?" And I said "well, if you come back next year, we can do this and do that". And they come back nearly every year and I do a different tour for them every year. And they follow me on all my social media platforms and they always message me and stuff like that. And I've had others who have done the same thing, they come back all the time. What do you call it, repeat customers? But I don't even think of them as repeat customers, I think of it [them] as friends.

Through prioritising "personal connection", the potential exists for building meaningful relationships with visitors. The notion of meaningful relationships in the Indigenous tourism context seems to refer to both parties feeling seen, respected, and valued by the other, even if only fleetingly. There is, however, the potential for creating ongoing relationships with visitors that can transcend the service provider/customer dynamic, as illustrated in Clinton's example, with social media facilitating continuing dialogue.

"We're showing our way"

Roland: They [tourists] want to know everything, like the rock art. Because any *balanda* [non-Indigenous person], we're showing our way, culture, *yoh*.

Bart: … introducing our guests to Broome; our language and our significance and our values of this place.

Kerry-Ann: I want to tell my family history, and even before colonisation, some of my family history on how they did things.

The operators display their identity by introducing themselves, their culture, and their Country: "we're showing our way". They often incorporate their ancestral language, sharing place names and keywords. They also regularly include words from their ancestral language in their business names. Operators express a preference for engaging with visitors on Country and through sharing their intimate knowledge of their land they are able to highlight "our language and our significance and our values of this place". For example, Mick spoke about the significance of Merredin Peak and the surrounding reserve to his family; Kerry-Ann spoke about the history of the Swan River; and Clinton spoke about the abundance of rock art and the World Heritage nomination of the Murujuga National Park.

Embedded within their connection to their Country are the cultural practices and stories which the operators share with their visitors. For Kerry-Ann, this means telling "some of my family history on how they did things". This also relates to activities that they share such as hunting and foraging for bush tucker and seafood, making damper or grinding ochre. The operators express an intention to promote the sustainability values that these practices are imbued with.

"I give them the same experience that I've had growing up"

Clinton: What I'm good at is being myself. I don't like to pretend to be anyone else other than myself. So, I work to my own strengths. ... When I take people out on tours with me, I give them the same experience that I've had growing up. The tours that they do is the same thing that I would do without tourists. So, it's genuine, authentic things that I normally would do. And I make it really personal, so I tell them all my stories from when I was growing up and I take them to places where I used to go. And I think because it's a much more personal story coming from myself they can relate to it a bit more I suppose.

For many operators, giving a tour is about being who they are. Thus Clinton talks about doing "the same thing that I would do without tourists". He recounted being asked to play the didgeridoo and replying that the didgeridoo is not a part of his culture, and offering instead to show visitors how to throw a boomerang.

"The true story needs to be told"

Bart: [explaining the meaning behind his business name Narlijia Experiences Broome] I like telling everything straight down the line. ... And *narli* means proper, true or real in our language [Yawuru]. And *-jia* is a possessive suffix, meaning for you. So, actually, the truth is not just a concept, it's actually possession; for you [true for you]. Giving you the authentic truthful stories of the Country, of the people and the history. The true story needs to be told here. ... we're battling against the misconception of the actual place.

Clinton: In Aboriginal culture, where I grew up, not even just in colonial times, but even ancient times, but we tell the whole story, rather than just tell the bits that we want to hear. And then that way, when you know the full story you can make better decisions about life. And that's where the Western society within mainstream Australia can learn a lot from Aboriginal people. The more we talk about it, the less it becomes stigmatic. We don't feel ashamed. It's like, it's open now and everyone is aware of it. No longer need to be ashamed of things. And that's all we want as a people. We know all our history, but do you know it? We want everyone to know it so that we're on the same page and then we can move forward.

Mick: Well, it's the truth isn't it? It's gotta be told.

These quotes underscore the need for truth telling, and not conveniently omitting anything: "we tell the whole story". The operators share "authentic truthful stories of the Country, of the people, and the history". As a consequence, the content of their tours may be confronting, which highlights the importance of safe spaces. Not only do the operators challenge misconceptions of themselves, their cultures, and their Country, but truth telling also brings attention to the social and environmental issues facing a community. We can thus understand truth as a starting place: "then we can move forward".

"Sharing vital information"

Kerry-Ann: People get the wrong interpretation of Aboriginal people.

Bart: This is a "sharing of vital information" industry. Personally, as a tour guide, and socially for me, it's fantastic. It's very very beneficial. But for the rest of the community it's beneficial, because they start to understand, they start to respect, and then everybody is a bit more aware of why that person is doing that, or who that person is, what that place is. So, they start to get a first level respecting our cultural values, and that's not just socially, that's environmentally, that's culturally, all that sort of stuff. And that's what our culture is all about. It's not a culture based on industry, it's not a culture based on capitalism or anything like that. It's a big big circle of environment, people, practice, the whole thing. It's a big big culture.

Clinton: Sometimes when people talk about the negative aspects of our history, they tend to be very negative. But making people feel bad is not going to make the situation better. Making people understand is how you do it, and that the way I do it. … I think that if you massage it into people then I think that it's ok. They can deal with it better. Because we're talking about a nation where things have been swept under the rug and when they hear those things they do get on the defensive. I tell people "none of you were here, none of you did this, but we're all here and we're all living here today, and we are the product of this business". We have to understand what's happened and we learn from it all.

Education is implicated as a part of the problem, of the lack of understanding of Aboriginal culture: "people get the wrong interpretation of Aboriginal people". Education is also implicated as a part of the solution, of respect and understanding: "we have to understand what's happened and we learn from it all". The operators identify as teachers of Aboriginal cultural knowledges, with Bart considering his role as an educator as doing the other half of his cultural responsibilities and acting as a conduit, carrying on the teachings of his Elders. He attributes his Western education as assisting him to articulate stories and cultural knowledges to diverse audiences so that "everybody is a bit more aware". This relationship between hosts as teachers and visitors as learners creates a space where truth telling leads to understanding: "making people understand is how you do it". Their content is thus seen as "vital information" and equally, the operators are a vital source of this information.

"I like to tell my Aboriginal story my way"

Kerry-Ann: I saw a lot of businesses being set up with tourism and they would be giving a spiel on Aboriginal culture, but they weren't Aboriginal people. So, I thought, well, I like to tell my Aboriginal story my way. And we have our history and I didn't think that was fair for non-Indigenous people to share our culture in that sense when we're here and we can share.

Clinton: I think to have a proper authentic experience you need an Aboriginal person delivering it. Because it's just more relevant. The person has the lived experience of those things and you're getting it out of the horse's mouth. It's not second-hand information.

Aboriginal ownership and control is important to the operators. They decide how to establish and operate their businesses, develop their offerings, and promote themselves and develop their brand. This business position appears to grow from the operators identifying as cultural authorities: "you're getting it out of the horse's mouth". They have a continuing connection to, and the authority to showcase, their Country. In some instances, they are Elders within their communities, or else, they enjoy the support of their Elders.

Discussion

The in vivo codes in the findings section can be categorised as three broad practices: hosting, connecting, and sharing. Notions of hospitality and connection were already identified by Higgins-Desbiolles et al. (2019) and are confirmed by the findings. Sharing is an additional practice identified in this research. We categorise each heading from the findings into the relevant identified practice and consider each in turn.

Hosting

Hosting is the means by which operators set the scene for culturally safe interactions with visitors. It is understood as a continuation of a pre-existing cultural practice in receiving visitors: "doing what the old fellas did". Hosting comes with expectations on visitor behaviour: "they have to abide by our practices", and so hosting is seen as an expression of authority. Operators meet Indigenous and non-Indigenous, domestic and international visitors in a variety of settings thanks to the diversity of their offerings: "I've branched out". Hosting is how operators set up "a safe environment for both parties" creating opportunities for discussions about truths that are often confronting and uncomfortable.

Tourism has long been proposed to facilitate cross-cultural contact (Tomljenovic, 2010). In the Australian case, Grant (2019) explains that "most Australians have never met an Indigenous person and may be utterly unaware if they had. People like me – urban, mixed race, middle class – appear like any other Australian; another face on our multi-ethnic streets" (p. 204). This notion is confirmed by the biennial Reconciliation Australia (2020) reconciliation barometer survey in which 65% of respondents stated that they never or rarely socialise with Aboriginal or Torres Strait Islander Australians (p. 27).

The findings speak not only to the existence of cross-cultural contact but the nature of this contact when it does occur. The importance of Indigenous tourism interactions occurring in a safe space for hosts and visitors is demonstrated. This *tourism safe space* may parallel what we see in the mental health literature, where safe spaces are places of non-judgement, of feeling heard, and of expressing oneself freely (Bulman & Hayes, 2011). These safe spaces can allow hosts and visitors to sit with the tensions of the unfinished business of the nation's colonial history (Grant, 2019). Indigenous tourism encounters appear pertinent in this respect as Australian society does not provide many safe spaces where Aboriginal people can be free from racism and marginalisation, or for non-Indigneous Australians to openly ask questions about Aboriginal knowledges or cultures (Reconciliation Australia, 2018).

Connecting

Connecting is how hosts meet visitors where they are at: "I make it really personable". Hosts find common ground with their visitors: "we can all relate to one another". Aside from cross-cultural contact, these encounters may establish meaningful relationships: "I think of them as friends".

Operators express a social agenda in promoting stronger relationships between Indigenous and non-Indigenous Australians. The operators participate in the 'race relations' goal of reconciliation for "positive two-way relationships built on trust and respect [to] exist between Aboriginal and Torres Strait Islander and non-Indigenous Australians throughout society" (Reconciliation Australia, 2016, p. 7). Operators connect with their visitors by finding ways of relating to one another, which can be as simple as identifying shared interests or experiences. This is a case of recognising shared humanity (cf. 'universal recognition', McBride, 2013) which sets up the conditions for empathy, of placing oneself in the other's position. Connecting in this way may create a sense of intimacy between visitors and hosts, enhancing the meaning of the toured objects and associated content (Galliford, 2010).

Sharing

Together, hosting and connecting create the conditions for *sharing*. Sharing relates to the content presented by the operators through "showing our way". Through sharing, the hosts manifest their authenticity: "I give them the same experience that I've had growing up". Visitor perceptions of Aboriginal peoples, places, and cultures may be informed or challenged: "the true story needs to be told". The operators identify as teachers through "sharing vital information".

Sharing is an important addition to the notion of Indigenous tourism being built on hospitality and connection (Higgins-Desbiolles et al., 2019). It is through sharing – of personal and cultural stories – that operators are defining and displaying their authenticity. This is also an expression of agency; operators control the presentation of their indigeneity (Ruhanen & Whitford, 2019).

By distinguishing sharing from connecting, our participants highlight the importance of grassroots truth-telling (cf. Indigenous agency as expressed through the proposed Makarrata process in the 'Uluru Statement from the Heart'). This supports the 'historical acceptance' dimension of reconciliation whereby "widespread acceptance of our nation's history" (Reconciliation Australia, 2016, p. 7) is achieved by acknowledging the past and facilitating understanding, in which operators act as educators. Here we can also see operators carrying forward their cultural and environmental agendas. The Reconciliation Australia (2018) survey further confirms the link between education and attitudes towards reconciliation: "people among both the general community and Indigenous peoples who cite personal experience or education sources (such as school or other research) as their main source of information about Aboriginal and Torres Strait Islander people are most likely to view the relationship [between Indigenous and non-Indigenous Australians] as very important" (p. 12). When built upon the practices of hosting and connecting, sharing can extend beyond education to promote "dialogue in which Aboriginal people are positioned as equal partners who are safe to speak and whose perspectives are valued" (Taylor & Habibis, 2020, p. 14).

Connecting and sharing are complementary aspects of recognition. As previously mentioned, when hosts and visitors connect, *universal recognition* is at work, as Clinton explains, "we might look different, but at the end of the day, we're human beings". Subsequently, when hosts are sharing and visitors are respectfully listening and receiving, *particular recognition* is at work (McBride, 2013), as Mick describes, "we're all not the same, we're all different". During these interactions, hosts move between exhibiting their universal and their particular identities, a feature of the multicultural space as elaborated by McBride (2013): "the desire to be different and the desire to be the same are simply two sides of the same coin. What matters then [is] whether these are the terms under which one wants to be recognized, and whether others will, or more accurately, *can* recognize you in this way" (p. 41).

Beginning from a place of agency

We can interpret the practices of hosting, connecting, and sharing as arising from the operators' authority: "I like to tell my Aboriginal story my way". The operators perform their self-determination, enacting their agency through their business operations. Their authority enables them to take ownership of how they represent themselves and their cultures, as Kerry-Ann explains: "We want to share our own culture. We want to tell our own story our own way". As seen in the case of sharing, this often means challenging existing views or offering alternative views on dominant Australian narratives through truth telling. Agency thereby serves as the basis for how the hosts are operating their businesses and interacting with their visitors.

Based on the findings, we propose that tourism engagements be founded upon recognition and respect for Indigenous agency to ensure that Indigenous operators and communities can derive sustainable benefit from receiving visitors. This contrasts with cases where tourism is imposed upon Indigenous communities or where other organisations appropriate and profiteer from Indigenous cultures, knowledges, and imagery (Ruhanen & Whitford, 2019). It can thus be predicted that hosts and communities can be vulnerable to exploitation in circumstances where Indigenous tourism operations do not arise from agency.

We have presented examples of hosts enacting their agency through setting the terms and boundaries for interacting with their visitors through hosting, connecting, and sharing. Viewed in this way, recognising and respecting agency implies reciprocity and sets the scene for how reconciliation processes can be animated through Indigenous tourism.

Practicing reconciliation

The findings present examples of hosts practicing reconciliation at a grassroots level and in the present, rather than waiting for the political processes of reconciliation to run their course. As Bart explains: "I believe that reconciliation – because it's about people coming together – I believe that we need to do it. Politicians don't need to do it and sign a paper, we need to do it; each and every one of us need to actually do it. Because it's actually the concept of people coming together. And we do it; we do it every day. This is our lifestyle. That's why I said, we are reconciliators, because we are actually, physically, socially, we are doing it. That's how we get paid, by practicing reconciliation, not by talking about it."

Operators describe, through interacting with visitors, experiencing Indigenous tourism and reconciliation simultaneously. *Hosting* creates safe spaces that support *connecting* with visitors to promote the reconciliation dimension of 'race relations' and *sharing* with visitors to promote the reconciliation dimension of 'historical acceptance'.

When it comes to the question of how operators conceptualise reconciliation, we found that they are not thinking in terms of an endpoint. Bart describes operators as agents of reconciliation: "we are reconciliators". In this way, reconciliation is atelic: operators model what reconciled interactions can look like. As key to this, Mick describes "having a sit down and listening", i.e. coming together with visitors for the *doing* of *being* reconciled (cf. animating the two-way processes of reconciliation; Reconciliation Australia, 2016). By practicing reconciliation, operators manifest their agenda with its cultural, environmental, and social values, thereby bringing a distant goal into the present. This practice echoes Garvey's (2015) application of an arena metaphor to describe movements within, towards, and away from contested cross-cultural social settings, and Grant's (2019) sentiment that national reconciliation is not so much an endpoint but a commitment, with ongoing issues, discomforts, and tensions to be worked out.

Sustaining culture

Keeping culture strong appears to be a key expression of agency and its primary means is through intergenerational transmission. As Marissa explains: "We've gotta start with our mob first and teach our kids to be strong. I was one of them kids too. I was real shame. Because I didn't know my culture. But now I do, and I have to step up and teach other kids to do that."

Several operators explain how motivating it is for their children and other relatives to witness the respect that is accorded to them on account of their cultural knowledge. They describe seeking opportunities to involve their family in tourism activities to build awareness of the importance of this knowledge. We posit that agency is what drives intergenerational transmission. Indigenous hosts leverage tourism – and the recognition it brings – to motivate the next generation about the relevance, importance, and feasibility of maintaining their culture.

Beyond this, sustaining culture relates to the transmission of ecological knowledges for caring for Country. Clinton explains: "The main reason that I got into tourism is so that I could preserve this knowledge and pass it on and teach others about it so that future generations can enjoy it too." As such, cultural practices are tied to places and both must be maintained. The operators are harnessing business as an opportunity to educate visitors about how to care for Country.

Implications for sustainable Indigenous tourism development

The operators aspire to establish sustainable business models, both using and supporting family involvement. As Kerry-Ann explains: "I want to hand this [business] down to my family. So, I want my children to maintain it as much as possible … because if I'm not here and not doing it then someone's got to do it. And I want it to be a business where they can thrive and be successful and be comfortable in their own traditions and their own ways of doing things, but still live in a multicultural country." The agency of operators in choosing to sustain their cultures lies at the root of authenticity. Cultural sustainability is supported by the operators sharing the practices that they are independently motivated to continue. The availability of such meaningful experiences for visitors is an indicator of sustainable tourism (Zeng, 2017).

Having diverse offerings moves businesses away from tourist dependency, another indicator of sustainable tourism development (Higgins-Desbiolles et al., 2019). The operators move towards financial viability through transcending reliance on seasonal clientele. Such diversity also supports community engagement, with visitors afforded more points of contact with operators. Business operations become personally sustainable for operators who express enjoying the varied lifestyles that their businesses provide, as opposed to getting burnt out or bored.

Practicing reconciliation through Indigenous tourism can support more sustainable societal relationships and peaceful communities. In addressing peace, justice, and strong institutions, the United Nations Sustainable Development Goal 16 aspires to support inclusive societies. We interpret this as favouring a reconciliation agenda and sustaining cultural diversity through keeping Indigenous cultures strong. Moreover, in aspiring to support strong institutions, we interpret SDG 16 as favouring the agency of operators in developing their businesses in ways that achieve viability and withstand exploitation. Ruhanen and Whitford (2019) make a similar point: "From a sustainability perspective, as long as questions concerning ownership, control, authenticity and the commercialisation of Indigenous cultural heritage remain an ongoing issue, the ability for Indigenous communities involved in tourism to develop in sustainable ways also remains questionable" (p. 184). In other words, when Indigenous operators begin from a place of strength, sustainability may be one of the outcomes.

A strengths-based model of Indigenous tourism

Analysis of the findings reveals four levels: agency, hosting, connecting, and sharing. From a position of enacting their *agency*, *hosting* sets up the space where operators and visitors are *connecting* over the things they hold in common, thus establishing the trust and rapport for hosts *sharing* the unique stories of their place, their culture, and of who they are. We depict this as shown in Figure 2.

We consider this to be a strengths-based model because operators build on what they have in the present. This model contains no "absence, lack or failure" (Fogarty et al., 2018, p. 6). During conversations, the operators did not refer to their actions according to a deficit framing but according to the perceived benefits of their actions flowing to themselves, their visitors, and their wider communities as they enact their social, cultural, and environmental agendas.

The model brings us full circle, from agency to agency. The operators are bringing to life their desired future through practicing reconciliation and sustaining culture. This represents a shift from telic to atelic, from idealising a future state to acting in the present. The future goal of reconciliation is seized on as a present action: "we are reconciliators". This accords with Grant's (2019) position that *being reconciled* is the practice of working out our reconciliation. It parallels the situation in international development where agency has been analysed as both a means and an end (Sen, 2001). Indeed, sustainability itself is both a process and a goal. Indigenous peoples sustain their cultures into the future by practicing them in the present.

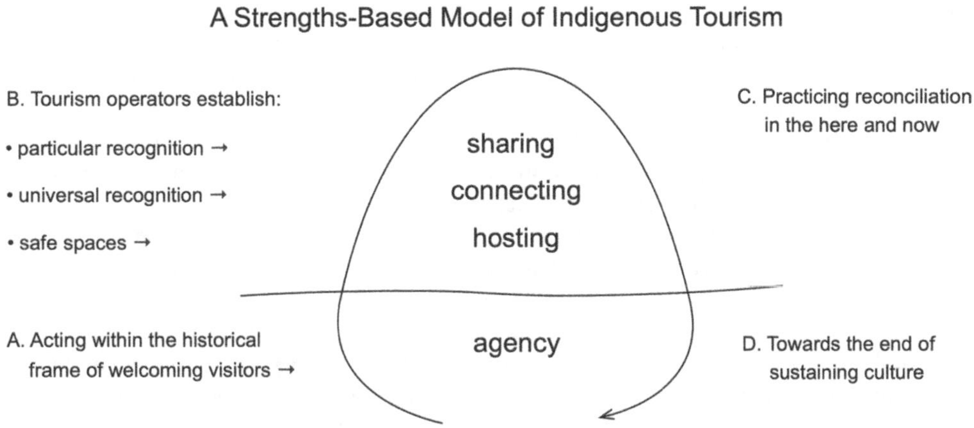

Figure 2. A strengths-based model of Indigenous tourism: each level sets up the conditions required for the higher levels, agency > hosting > connecting > sharing. Agency has deeper roots in authority and sovereignty. Operators express their agency in continuing the historical practice of welcoming visitors (A), which they implement as three types of practice (B). This is how they enact reconciliation (C) and sustain their culture (D), reinforcing agency and completing the cycle.

Further considerations

This substantive grounded theory is a theoretical interpretation of processes that are specific to, but not necessarily limited to, the area of investigation (Charmaz, 2014). There are nuances that this grounded theory does not account for. The model expresses the main ideas and experiences of the operators, without asserting completeness or wider generalisability. Its utility comes from helping to generate an understanding of some of the dynamics of Indigenous tourism according to Aboriginal tourism operators located in Western Australia and the Northern Territory. As Indigenous groups are not homogenous, careful interpretation of the findings is necessary. There is, however, the potential for the model to be tested, validated, and elaborated in diverse Indigenous tourism contexts (Charmaz, 2014).

Positioning this model within the larger context requires critical reflection on the challenges, difficulties, tensions, and complexities of the forces of commercialisation, the politics of reconciliation, the dynamics of social cohesion, the political and policy environment, the available support structures, and the evolving marketplace: "For an Indigenous enterprise to be sustainable, the Indigenous groups involved need not only be given control of the tourism enterprises in which their culture is the central element, they also need to be given the capacity to build their own future, as well as the ability to exert political control over their cultural and natural resources" (Ruhanen & Whitford, 2019, p. 184). There are many other forces at work that are not accounted for by the model. Sustainability depends on engagement: how do visitors experience the two-way processes of reconciliation? Sustainability also depends on government and industry: how can they recognise and support the agency of Aboriginal tourism operators? Sustainability in this strengths-based model depends on maintaining the cycle from agency back around to agency, which may be interrupted at any point. All of these dynamics vary from place to place, and operators may articulate them differently.

Conclusions

This research has foregrounded the experiences and perspectives of sustainable Indigenous tourism practices in the words of Aboriginal tourism operators. The focus has been broadened from the needs and wants of visitors to include the vantage point of Aboriginal hosts. We sought to understand how Indigenous tourism can benefit the Indigenous hosts and communities who receive visitors. The flexible and iterative methodologies, critical Indigenous methodology and constructivist grounded theory, provided a framework for a collaborative research practice. The findings reveal the actions of Aboriginal tourism operators in manifesting a strengths-based approach that is faithful to their understandings of the nature and potentials of tourism. The findings also highlight the operators' environmental, social, and cultural agendas and the subsequent benefits they perceived to flow to themselves, their visitors, and their communities.

This strengths-based model of Indigenous tourism, a substantive grounded theory, demonstrates how Aboriginal tourism operators enact their agency. Operators express their *agency*, setting the terms and boundaries for interacting with visitors through three kinds of practices: hosting, connecting, and sharing. Indigenous tourism can be seen to be sitting within an existing frame of *hosting*, the continuation of an Indigenous cultural practice of receiving and welcoming visitors to Country. *Hosting* creates safe spaces for *connecting*: through finding common ground and establishing meaningful relationships, hosts and their visitors can experience their shared humanity through universal recognition. Through *sharing* their culture with visitors, operators are promoting truthful understandings of Aboriginal histories and cultures, and expressing their particular identities. This is how operators are practicing reconciliation and sustaining their culture, thereby reinforcing their agency, and bringing us full circle from agency back to agency. These reconciliation processes of race relations and historical acceptance have implications for the advancement of peace, justice, and inclusive societies as articulated in the United Nations Sustainable Development Goal 16.

This strengths-based model of Indigenous tourism articulates the agendas and actions of Aboriginal tourism operators, showing the key role that Indigenous tourism plays in education and reconciliation. In sum, this research supports the idea that reconciliation is an ongoing practice rather than a future endpoint arising from the agency of Indigenous tourism hosts and the participation of visitors. We suggest that for Indigenous tourism to be beneficial for hosts they must have the freedom and support to enact their agency, and that it is this agency that lies at the heart of sustainability.

* * *

Clinton: I grew up around amazing people, amazing Elders, and they were such great storytellers that they inspired me to want to be like them. And I did it in a different way and that's through my tour business and I can now tell those stories that I was taught out on Country. My job means I get to live my culture. I get to live the lifestyle that I was taught growing up. It's taking me back to my roots of who I am.

Kerry-Ann: This is a part of reconciliation. Or reconciling. Knowing our First Nations people. This is their Country, this is their land. We should respect them and respect their lands as well.

Bart: So in a sense, we are vital to reconciliation because we're actually doing it from the grassroots; from the ground level as Aboriginal cultural tour guides. We're doing it; we're doing it with enjoyment and business enterprise. We're having fun with it.

Notes

1. Quotations in this paragraph denote excerpts from findings, reported in full below.
2. "Health is influenced by social determinants – the circumstances in which people grow, live, work, and age ... and individual health risk factors" (AIHW, 2020, p. 43). Social determinants of health include factors such as education, employment, income, housing, child protection, justice, weight, and substance use, and are also influenced by access to health care (AIHW, 2020).
3. A coding practice of using direct quotes.

Acknowledgements

We acknowledge the Bininj, Larrakia, Noongar, Ngarluma, Yindjibarndi, and Yawuru peoples as the Traditional Owners of Country where this research was conducted and written, and we pay our respects to Elders past and present. We are grateful to Bart Pigram, Clinton Walker, Kerry-Ann Winmar, Marissa Verma, Mick Hayden, and Roland Burrunali for hosting, connecting, and sharing with us, and for their time and generosity in helping us to understand more about their Country and culture. This research was supported by a commonwealth Research Training Program scholarship to the first author. This project has approval from the Charles Darwin University Human Research Ethics Committee (clearance number H19019). We are grateful to Gavin Morris, Rachel Tumminello, Ruth Wallace, Valerie Leishman, and several anonymous reviewers for their feedback on earlier versions of this article.

Disclosure statement

No potential conflict of interest was reported by the authors.

ORCID

Nicole Curtin http://orcid.org/0000-0003-1996-5403
Steven Bird http://orcid.org/0000-0003-3782-7733

References

Altman, J. (1989). Tourism dilemmas for aboriginal Australians. *Annals of Tourism Research, 16*(4), 456–476. https://doi.org/10.1016/0160-7383(89)90002-9

Australian Institute of Aboriginal and Torres Strait Islander Studies [AIATSIS] (2020). *AIATSIS code of ethics for Aboriginal and Torres Strait Islander research*. AIATSIS.

Australian Institute of Health and Welfare [AIHW] (2020). *Aboriginal and Torres Strait Islander health performance framework 2020 summary report* (Cat. no. IHPF 2). https://www.indigenoushpf.gov.au/getattachment/65fbaaf3-100c-4df5-941c-a8455922693c/attachment.aspx

Bainbridge, R., McCalman, J., Redman-MacLaren, M., & Whiteside, M. (2019). Grounded theory as systems science: Working with indigenous nations for social justice. In *The SAGE handbook of current developments in grounded theory*. (pp. 611–629). SAGE Publications Ltd. https://www.doi.org/10.4135/9781526485656

Brennan, S. (2004). Reconciliation in Australia: The relationship between Indigenous peoples and the wider community. *The Brown Journal of World Affairs, 11*(1), 149–161.

Bulman, J., & Hayes, R. (2011). Mibbinbah and Spirit Healing: Fostering safe, friendly spaces for Indigenous males in Australia. *International Journal of Men's Health, 10*(1), 6–25. https://doi.org/10.3149/jmh.1001.6

Butler, R. W. (1999). Tourism: An evolutionary perspective. In J. G. Nelson, R. Butler, and G. Wall (Eds.), *Tourism and sustainable development: A civic approach* (2nd ed., pp. 33–63). University of Waterloo.

Butler, R., & Hinch, T. (2007). *Tourism and Indigenous peoples: Issues and implications*. Butterworth-Heinemann.

Carr, A., Ruhanen, L., & Whitford, M. (2016). Indigenous peoples and tourism: The challenges and opportunities for sustainable tourism. *Journal of Sustainable Tourism, 24*(8–9), 1067–1079. https://doi.org/10.1080/09669582.2016.1206112

Chambers, D., & Buzinde, C. (2015). Tourism and decolonisation: Locating research and self. *Annals of Tourism Research, 51*, 1–16. https://doi.org/10.1016/j.annals.2014.12.002

Charmaz, K. (2014). *Constructing grounded theory* (2nd ed.). Sage.

Charmaz, K. (2006). *Constructing grounded theory: A practical guide through qualitative analysis*. Sage Publications.

Christie, M. J. (2013). Generative and 'ground-up' research in Aboriginal Australia. *Learning Communities: International Journal of Learning in Social Contexts, 13*, 3–12. https://espace.cdu.edu.au/eserv/cdu:40561/Christie_40561.pdf

Cole, S., & Morgan, N. (2010). *Tourism and inequality: Problems and prospects*. CABI.

Coronado, G. (2014). Selling culture? Between commoditisation and cultural control in Indigenous alternative tourism. *PASOS Revista de Turismo y Patrimonio Cultural, 12*(1), 11–28. https://doi.org/10.25145/j.pasos.2014.12.002

Denzin, N. K., & Lincoln, Y. S. (2008). Critical methodologies and indigenous inquiry. In Norman K. Denzin, Y. S. Lincoln, and L. T. Smith (Eds.), *Handbook of critical and Indigenous methodologies* (pp. 1–20). Sage. https://doi.org/10.4135/9781483385686

Dey, I. (1999). *Grounding grounded theory*. Academic Press.

Dudgeon, P., Milroy, H., & Walker, R. (2014). *Working together: Aboriginal and Torres Strait Islander mental health and wellbeing principles and practices* (2nd ed.). Australian Government Department of the Prime Minister and Cabinet.

Elder, C. (2017). Unfinished business in (post) reconciliation Australia. *Australian Humanities Review, 61*, 74–93. http://australianhumanitiesreview.org/wp-content/uploads/2017/06/AHR61_Elder.pdf

Fogarty, W., Lovell, M., Langenberg, J., & Heron, M.-J. (2018). *Deficit discourse and strengths-based approaches: Changing the narrative of Aboriginal and Torres Strait Islander health and wellbeing*. Lowitja Institute.

Fredericks, B. (2013). We don't leave our identities at the city limits': Aboriginal and Torres Strait Islander people living in urban localities. *Australian Aboriginal Studies, 2013*(1), 4–16. https://eprints.qut.edu.au/61823/

Galliford, M. (2010). Touring 'country', sharing 'home': Aboriginal tourism, Australian tourists and the possibilities for cultural transversality. *Tourist Studies, 10*(3), 227–244. https://doi.org/10.1177/1468797611407759

Garvey, D. C. (2015). *A causal layered analysis of movement* [paralysis and liminality in the contested arena of Indigenous mental health [Doctoral dissertation]. Curtin University. https://espace.curtin.edu.au/handle/20.500.11937/249

Grant, S. (2019). *Australia day*. HarperCollins.

Haessly, J. (2010). Tourism and a culture of peace. In O. Moufakkir & I. Kelly (Eds.), *Tourism, progress and peace* (pp. 1–16). CABI.

Higgins-Desbiolles, F. (2003). Reconciliation tourism: Tourism healing divided societies!. *Tourism Recreation Research, 28*(3), 35–44. https://doi.org/10.1080/02508281.2003.11081415

Higgins-Desbiolles, F. (2006). More than an "industry": The forgotten power of tourism as a social force. *Tourism Management, 27*(6), 1192–1208. https://doi.org/10.1016/j.tourman.2005.05.020

Higgins-Desbiolles, F. (2020a). Socialising tourism for social and ecological justice after COVID-19. *Tourism Geographies, 22*(3), 610–623. https://doi.org/10.1080/14616688.2020.1757748

Higgins-Desbiolles, F. (2020b). The "war over tourism": challenges to sustainable tourism in the tourism academy after COVID-19. *Journal of Sustainable Tourism, 29*(4), 551–569. https://doi.org/10.1080/09669582.2020.1803334

Higgins-Desbiolles, F., Carnicelli, S., Krolikowski, C., Wijesinghe, G., & Boluk, K. (2019). Degrowing tourism: Rethinking tourism. *Journal of Sustainable Tourism, 27*(12), 1926–1944. https://doi.org/10.1080/09669582.2019.1601732

Jennings, G., & Junek, O. (2007). Grounded theory: Innovative methodology or a critical turning from hegemonic methodological praxis in tourism studies. In I. Ateljevic, A. Pritchard, & N. Morgan (Eds.), *The critical turn in tourism studies* (pp. 197–210). Taylor and Francis Group. https://doi.org/10.1016/B978-0-08-045098-8.50017-9

Lal, S., Suto, M., & Ungar, M. (2012). Examining the potential of combining the methods of grounded theory and narrative inquiry: A comparative analysis. *The Qualitative Report, 17*(21), 1–22.

McBride, C. (2013). *Recognition*. Polity Press.

National Health and Medical Research Council [NHMRC]. (2018). *Ethical conduct in research with Aboriginal and Torres Strait Islander Peoples and communities: Guidelines for researchers and stakeholders*. Commonwealth of Australia.

Nielsen, N., & Wilson, E. (2012). From invisible to Indigenous-driven: A critical typology of research in Indigenous tourism. *Journal of Hospitality and Tourism Management, 19*(1), 67–69. https://doi.org/10.1017/jht.2012.6

Pacific Asia Travel Association [PATA] and World Indigenous Tourism Alliance [WINTA] (2014). *Indigenous tourism and human rights in Asia and Pacific Region: Review, analysis & checklists*. PATA.

Parks Australia (2019). Uluru climb closure. https://parksaustralia.gov.au/uluru/discover/culture/uluru-climb/

Pritchard, A., & Morgan, N. (2007). De-centring tourism's intellectual universe, or traversing the dialogue between change and tradition. In I. Ateljevic, A. Pritchard, & N. Morgan (Eds.), *The critical turn in tourism studies* (pp. 11–28). Taylor and Francis Group. https://doi.org/10.1016/B978-0-08-045098-8.50006-4

Reconciliation Australia (2016). The state of reconciliation in Australia in 2016. https://www.reconciliation.org.au/wp-content/uploads/2017/11/State-of-Reconciliation-Report_SUMMARY.pdf.

Reconciliation Australia (2018). 2018 Australian reconciliation barometer. https://www.reconciliation.org.au/wp-content/uploads/2019/02/ra_2019-barometer-brochure_web.single.page_.pdf.

Reconciliation Australia (2020). 2020 Australian reconciliation barometer. https://www.reconciliation.org.au/wp-content/uploads/2020/11/australian_reconciliation_barometer_2020_-full-report_web.pdf.

Ruhanen, L., & Whitford, M. (2019). Cultural heritage and Indigenous tourism. *Journal of Heritage Tourism, 14*(3), 179–191. https://doi.org/10.1080/1743873X.2019.1581788

RMIT ABC Fact Check (2019). *Fact check: Linda Burney says Australia is the only first world nation with a colonial history that doesn't recognise its first people in its constitution. Is she correct?* https://www.abc.net.au/news/2019-10-10/fact-check3a-is-australia-the-only-first-world-nation-with-a-c/11583706

Sen, A. (2001). *Development as freedom*. Oxford University Press.

Smith, L. T. (2012). *Decolonizing methodologies: Research and indigenous peoples* (2nd ed.). Zed Books.
Taylor, P. S., & Habibis, D. (2020). Widening the gap: White ignorance, race relations and the consequences for Aboriginal people in Australia. *Australian Journal of Social Issues*, *55*(3), 354–371. https://doi.org/10.1002/ajs4.106
Territory Economic Reconstruction Commission [TERC] (2020). *Final report: A step change to win investment and create jobs*. https://ntrebound.nt.gov.au/__data/assets/pdf_file/0020/952301/terc-final-report.pdf.
Tomljenovic, R. (2010). Tourism and intercultural understanding or contact hypothesis revisited. In O. Moufakkir and Kelly, I. (Eds.), *Tourism, progress and peace* (pp 17–34). CABI.
Walker, M., Fredericks, B., Mills, K., & Anderson, D. (2014). "Yarning" as a method for community-based health research with Indigenous women: the Indigenous Women's Wellness Research Program . *Health Care for Women International*, *35*(10), 1216–1226. https://doi.org/10.1080/07399332.2013.815754
Western Australian Indigenous Tourism Operators Council [WAITOC], & Clean State (2020). Aboriginal tourism and business development: A Clean State & WAITOC stimulus and recovery proposal. https://d3n8a8pro7vhmx.cloudfront.net/ccwa/pages/11996/attachments/original/1599531759/WAITOC-Clean_State_Proposal_V2.pdf?1599531759.
Whitford, M., & Ruhanen, L. (2016). Indigenous tourism research, past and present: Where to from here? *Journal of Sustainable Tourism*, *24*(8-9), 1080–1099. https://doi.org/10.1080/09669582.2016.1189925
Williams, L. K., & Gonzalez, V. V. (2017). Indigeneity, sovereignty, sustainability and cultural tourism: Hosts and hostages at 'Iolani Palace, Hawai'i. *Journal of Sustainable Tourism*, *25*(5), 668–683. https://doi.org/10.1080/09669582.2016.1226850
Zeng, B. (2017). Cultural centre, destination cultural offer and visitor satisfaction. *Sustainability*, *9*(11), 1912–1984. https://doi.org/10.3390/su9111984

OPEN ACCESS

Exploring a unifying approach to peacebuilding through tourism: Abraham and Israel/Palestine

Jack Shepherd

ABSTRACT

This work emerges from the encouragement of peace studies scholars to seek out commonalities that can unite rival sides in a conflict. Based on this call, I propose the *unifying approach to peacebuilding through tourism* as one where tourism initiatives use unifying points (such as figures, sites, stories and symbols) that help conflicting sides see commonalities and thus facilitate cross-cultural understanding. In particular, I look at how the story of Abraham (communal father of Jews, Christians and Muslims) appears to be used as a *guiding fiction* for the work of two tourism initiatives in Israel and Palestine. Based on mixed qualitative data, I use Creative Analytic Practice to present two semi-fictional stories that explore this use of Abraham, exposing the potential benefits and risks associated with striving for a unifying peace in an arena as charged as the Israeli-Palestinian context. These stories reveal that the unifying approach provides a valuable vocabulary of progress that challenges the dominant sectarian narratives of tourism in the region, yet simultaneously rubs awkwardly against the sharp divisions and imbalances of the Israeli-Palestinian context.

Introduction

The peace through tourism (PTT) debate is growing in both volume and sophistication. Despite the trauma of 2020, the year did produce news of another book celebrating tourism's role in peacebuilding (Da Silva et al., 2021). It also saw the continuing support of a number of world organisations, such as the United Nations (UN), for the notion that tourism promotes peace. On World Tourism Day 2020, for example, the UN lauded tourism's ability to bring people together and promote "solidarity and trust" (United Nations, 2020). Such a public conversation on tourism's role in peacebuilding has arguably become even timelier as the future of tourism is reconsidered in the light of the COVID-19 pandemic. Critical tourism scholars have been at the front line of this reconsideration, arguing for a rethink of the status quo, growth-first agenda of tourism and a reaffirmation of tourism's social power (Higgins-Desbiolles, 2021).

However, the promise of tourism in peacebuilding often fails to match its plaudits. Tourism is repeatedly shown to aggravate conflict in some of the world's most contested spaces, most notably in its ability to amplify sectarian narratives (Clarke, 2000; McDowell, 2008). It is therefore essential that tourism scholars provide what this special issue calls for – critical reflections on the peace through tourism debate.

Consequently, I offer one such critical reflection on the efforts of two tourism initiatives in Israel/Palestine that attempt to use tourism for peacebuilding. In particular, I explore how their attempts at a *unifying approach to peacebuilding*, represented by their employment of the unifying figure of Abraham, biblical ancestor of both Israelis and Palestinians, as a *guiding fiction* for their work, stands up in the face of an on-going conflict between Israelis and Palestinians.

Although some may struggle to see how PTT is possible in such a contested space, peace scholars shine a hopeful way forward for such an investigation. Galtung (1969) affirmed that peace is more than the absence of physical violence, which he termed a negative peace; it is also about building social justice - or positive peace. Galtung does not suggest a negative peace presupposes a positive one, and therefore in a place like Israel/Palestine, his conceptualisation shows how we can work for positive peace even in settings of violence. Moreover, Koopman (2011) argues that peace is a process; that it can exist even in the midst of war. In this way, they justify discussing peacebuilding in environments where evidence of peace is hard to find.

A unifying approach to peacebuilding (through tourism)

According to Lederach (1997), peacebuilding is an inclusive concept that "encompasses, generates, and sustains the full array of processes, approaches and stages needed to transform conflict toward more sustainable, peaceful relationships" (p.20). Although the term peacebuilding is often used in post-conflict situations, I appreciate the processual and performative nature of the term "building", which speaks to the belief that peace is not a static point to be attained or lost, but rather an ongoing process of building, doing and redoing peaceful relations (Koopman, 2011). Based on Lederach's definition, I introduce the notion of a *unifying approach to peacebuilding* as an approach to transforming conflict towards more sustainable and peaceful relationships through the use of unifying points (such as figures, sites, stories and symbols) that help conflicting sides see commonalities and thus facilitates cross-cultural understanding.

Finding unifying points around which conflicting sides can rally is one of the cornerstones of peacebuilding. Abu-Nimer (2003, p.18–19) writes, "one of the guiding principles in the practice of mediation … is to identify commonalities and build on the small agreements achieved by disputants". He sees such a focus on commonalities as an effective way of encouraging conflicting sides to reflect on how they can relate to one another and recognise a relational responsibility for resolving the conflict (Ibid).

The recent peace agreements between Israel and the United Arab Emirates and Bahrain in August 2020 reflect this approach. Named the Abraham Accords, these agreements describe how the unifying figure of Abraham, common ancestor of both Arabs and Jews, encourages these previously antagonistic nations to commit to "a spirit of coexistence, mutual understanding and mutual respect" (United States Department of State, 2020). Another religious figure that has served as a unifying point in conflict is the figure of Saint Patrick in Northern Ireland. Nagle (2014) explains how even in the midst of 'The Troubles', Saint Patrick served as a unifying symbol of peace by activists on both Republican and Unionist sides. Moreover, in Syria, scholars argued that Syrian heritage sites, such as Palmyra, hold significant unifying potential for Syrian students. Their findings forward that such unifying points can "reclaim a common national identity among diverse groups of Syrians" split by years of civil war (Abdullah et al., 2020, p.64).

As a point of reference, the unifying approach stands in stark contrast to another approach to peacebuilding called agonistic peacebuilding. Agonists see conflict as an essential part of social life. For them, peacebuilding is not an exercise in finding commonalities. Instead, peacebuilding is about ensuring political struggle occurs democratically rather than antagonistically (Shinko, 2008). Agonists are suspicious about peacebuilding based around unifying points. Nagle (2014) vocalises these concerns in saying, "While reasonable, the assumption that common symbols and rituals in divided societies generate peaceful intercommunal relations is deeply problematic" (p.480). He explains that despite being seen by some as a unifying point in Northern Ireland, Saint Patrick was also a point of conflict, with both Protestants and Catholics claiming ownership of the saint.

This paper posits that the unifying approach to peacebuilding can be applied within a tourism context. Based on the discussion above, I define a unifying approach to PTT as an approach to PTT where tourism initiatives use unifying points (such as figures, sites, stories and symbols) that help conflicting sides see commonalities and thus facilitate cross-cultural understanding. Such an approach is seldom reflected in either tourism scholarship or in the public institutions tasked with promoting PTT. Tourism scholars have largely focused on tourism's role in promoting first-time encounters and dialogue between conflicting groups, an idea rooted in Allport's (1954) contact theory. The results of these inquiries are mixed, yet often do not show significant shifts in attitudes (e.g. Anastasopoulos, 1992). The institutions, such as the UNWTO, have largely toed the same line, often coupled with a liberal peace approach (Megoran, 2011) that stresses tourism's potential to promote economic development.

Although I am proposing a new concept in this paper, the unifying approach to PTT is reflected in a number of existing cases. In South Korea, for example, tourists to the Demilitarized Zone between North and South Korea are encouraged to buy into the narrative of a unified Korea instead of focusing on the current conflict. As such, tourists take the "Peace Train" from Seoul to Dorasan station with its "Unification Platform", where they are told trains will one day continue over the border to Pyongyang (Korea Tourism Organization, 2020). Jung (2019, p.317), however, remains sceptical of the potential of such tourism to promote peace, suggesting it risks remaining "tourism longing for peace". In Israel/Palestine, Gelbman (2019) describes the Bethlehem to Jerusalem Peace Run as an event where Israelis, Palestinians and Italians put aside their differences and unite through their love of sport. Gelbman appears convinced in the run's potential to highlight a mutual hope for peace in this troubled region. Sadly, however, the run ceased its operations as of 2011. In Europe, an interesting project is developing along the former trench lines of World War One. Tourists from all sides of The Great War are being invited to walk the new 1000 km long Western Front Way hiking trail and reflect on the unifying points of both a lasting European peace and a shared history of relatives lost in Flanders' fields (Western Front Way, 2020).

Guiding fictions

In this paper, I seek to demonstrate how the use of Abraham at two tourism initiatives in Israel/Palestine reflects a unifying approach to PTT. A key element of this analysis is the way in which these two tourism initiatives use the story of Abraham as a *guiding fiction* in service of their peacebuilding efforts. By *guiding fiction*, I mean a fiction that is used for its potential to guide one towards a beneficial goal.

Although I explain later how the story of Abraham has the potential to do this, it is important to note that referring to the story of Abraham as a guiding fiction does not deny the very real attachments to Abraham felt by billions of his followers. Referring to Abraham and his story as a fiction is based on the fact that there still exists no verifiable historical evidence for his existence outside the religious texts that mention him (Gopin, 2002). Suggesting the story of Abraham can serve as a guiding fiction does not deny its utility, power or meaning; in fact, I argue it does quite the opposite, in opening up the possibility for multilayered meaning and therefore value to both believers and non-believers. After all, as is made clear in literary theory, a fiction is not an act of duping, falsehood or fallacy, but "something made for its own sake" (Smythe, 2005, p.300).

My understanding of guiding fictions is grounded in the ideas of German philosopher Hans Vaihinger in *The Philosophy of As If* (Vaihinger, 1924). The main thrust of this magnum opus was that fictional ideas are invaluable tools "for finding our way about more easily in the world" (p.15). For example, although we know that negative square numbers, the ego or the tourist gaze are not independent objective realities, mathematicians, psychologists and tourism scholars each use these concepts to make important scientific observations. As Vaihinger writes, ideas "whose theoretical untruth or incorrectness, and therewith … falsity is admitted is not for that reason practically valueless or useless, for such an idea, in spite of its theoretical nullity, may have great practical importance" (Ibid, p.8). Therefore, in terms of science, fictions are vital. Kant proffered the moral value of fictions in his categorical imperative, arguing that people should behave *as if* their actions "were to constitute a universal law" (Smythe, 2005, p.299). Recently, Harari's (2014) popular history of humanity, *Sapiens*, rebooted such ideas, stressing the fictional nature of many moral concepts such as human rights. Harari's main thesis was, however, that fictions unify. They help us cooperate in large numbers through our shared belief in fictional constructs like money, nations and empires. They are, for Harari, the main reason for our success as a species.

Of course, not all guiding fictions are beneficial. Although they help us cooperate, as Harari makes clear, such cooperation does not always end well. One only needs to think of the darkest chapters of human history to see they were the products of a guiding fiction of some sort, from fictions of eugenics theory to national glory. It is therefore important that whilst accepting the potentially positive power of guiding fictions, we remain open to the risk that the fiction may lead us further away from our intended positive goal.

The benefits and risks of guiding fictions are clearly seen in the subject matter of this journal. Sustainable tourism is without doubt the most successful guiding fiction in our discipline, with over 5000 papers published on the topic in the last twenty years (Sharpley, 2020). I call it a guiding fiction for its fictive premise that we can continue to develop tourism so that "it remains viable over an indefinite period and does not degrade or alter the environment (human and physical)" (Butler, 1993, p.29). This premise falls flat when we realise that sustained growth cannot occur sustainably on a planet with finite resources. Despite criticism, it remains unimaginably successful, given that its core message is undeniably attractive - that tourism can continue developing in a way that does not degrade the resources it depends upon (human or physical) – and inexplicit – an ideological Play-Doh that can be moulded to suit the individual. The quaffable and malleable nature of this guiding fiction is what makes it powerful in engaging people in sustainability issues (Berno & Bricker, 2001). In true guiding fiction fashion, Hardy et al. (2002, p.438) argue that defining sustainable tourism "is less important than the journey towards it".

Yet such ambiguity can also be problematic. The lack of substance in the concept can lead it to serve as a "smoke screen" (McCool et al., 2013, p.222), covering up a failure to tackle the environmental, social and economic problems created by tourism, with an ecolabel here and a carbon offset there. In essence, that it helps maintain the status quo, when "the status quo clearly cannot hold" (Higgins-Desbiolles, 2018, p.158). Recently, scholars have said that sustainable tourism fails to face up to the bare reality that tourism is "addicted to growth" (Ibid, p.157) and in need of de-growth, not continual, sustain-ed/-able development (Higgins-Desbiolles, 2021; Sharpley, 2020).

The peacebuilding potential of Abraham in Israel/Palestine

Several commentators have noted how the Israeli-Palestinian conflict is rooted in a divergence of Zionist and Palestinian interpretations of the land that makes up the former British mandate of Palestine (1920-1948). The Zionist narrative is that this land is the historic home of the Jewish people who have been in diasporic exile for two thousand years. The State of Israel, established in 1948, is therefore more than just a life raft for the Jewish people after centuries of persecution, it is the rightful home of the descendants of the Israelites (Oz, 1995; Shavit, 2013). The Palestinian narrative is that the land is theirs by virtue of having resided there for millennia; that Zionism represents a settler-colonial movement by largely European Jews; and that the removal of the indigenous people of Palestine in 1948 to create the State of Israel (the Nakba) was an act of ethnic cleansing that is still ongoing today (Khalidi, 2020; Said, 1992). Clearly, such narratives leave little room for common ground and contribute to the sense that the Israeli-Palestinian conflict is intractable. Indeed, the conflict has raged for nearly a century, and the Israeli military occupation of the West Bank is now over 50 years old. Yet despite the strength of these two narratives, the crude reality is that "Palestinians and Israeli Jews are now fully implicated in each other's' lives" (Said, 1992, p.49). Neither Israeli Jews nor Palestinians are going anywhere, despite what the other might wish.

Within this context, Abraham represents a rare potential unifying point (Gopin, 2002). Abraham is a central figure in the Jewish, Christian and Muslim holy texts. Although the exact details of his life vary in the three religious traditions, there is a common narrative of a man and his family moving through the Middle East at the behest of God. Abraham is essential for each religion. For Jews, it is through Abraham that their covenant is made with God as God's 'chosen people' (Genesis 17). For Muslims, Abraham is considered the first Muslim through his complete submission to God (The Qur'an 1:3). For Christians, Abraham represents the power of and reward for complete faith in God (Romans 4).

When used in the context of peacebuilding, Abraham's story offers the possibility of serving as a guiding fiction, helping conflicting sides work towards the beneficial goal of peace. Three aspects of Abraham and his story can be leveraged for this purpose.

Firstly, Abraham is recognised by followers of the three most prominent religions in Israel/Palestine (Judaism, Islam and Christianity) to be the communal father of their faith. Although each religion sees their connection to Abraham differently, i.e. Muslims believe they are descended from Abraham's first son, Ishmael, whereas Jews stress their progeny from Isaac, the potentially unifying power of the figure of Abraham is undeniable (Gopin, 2002). We see this aspect of Abraham leveraged in the aforementioned Abraham Accords.

Secondly, the story of Abraham places great importance on hospitality. Abraham spent much of his life on the road, travelling across the Middle East with his family and animals. Although always a stranger in other people's lands (Gopin, 2002), the sacred texts show Abraham as deeply hospitable. His tent was always open to strangers, where he would welcome passers-by, wash their feet and feed them (Genesis 18:1-8; The Qur'an 51:24-30). Abraham's hospitality appears to go well beyond Kantian expectations of the cosmopolitan right of strangers not to be treated as enemies, instead demonstrating an unconditional welcoming and ethics of care (Ramadan, 2011). As Gopin (2002, p.23) asserts, Abraham represents the "eager embrace of the unknown other". This central message of the Abrahamic story, in particular the symbol of his open tent, has already inspired peacebuilding activities in Israel/Palestine. For example, outside of Bethlehem stands the Tent of Nations project. Using the symbol of Abraham's open-sided tent, the project seeks to be a place of dialogue and peacebuilding on the threatened land of Daoud Nassar, a Palestinian farmer who invites people of all backgrounds to participate in tree planting and harvesting (Megoran, 2011). Similarly, in 2013, Jewish anti-occupation activists tried to set up a mock-up of Abraham's tent in the heart of the deeply contested city of Hebron, Palestine. They hoped to embody Abraham's message of openness and tolerance in a city marked by closure and violence (Hammami et al., In Press).

Thirdly, the Bible describes Abraham as a mediator of conflict. In Genesis 13, a conflict arises between the herders of Abraham and his nephew Lot's livestock, as the land they are on does not support both herds. Seeking to avoid conflict, Abraham asks Lot to part from him saying, "Let's not have any quarreling between you and me, or between your herders and mine, for we are close relatives". Israeli novelist and peace activist Amos Oz has suggested this gesture could serve as a metaphor for "a pragmatic peace in an imperfect world" (Oz, 1995, p.65).

Using the figure and story of Abraham in this way is not, however, risk free. As stated, there are differences in the understanding and interpretation of Abraham amongst the three religions. Moreover, those whose work reflects the use of Abraham as a guiding fiction have faced significant opposition. The Tent of Nations project is routinely targeted by violent Israeli settlers (Megoran, 2011), and the Israeli army not only prevented the Hebron activists setting up their Abrahamic tent, but also arrested them (Hammami et al., In Press).

Inviting tourists into Abraham's tent

Tourism has largely reinforced the two separate Israeli and Palestinian narratives (Clarke, 2000; Strömbom, 2019). Tourists visiting the region are usually presented with "polished, national narratives, telling single-faceted, partisan stories about the past" (Strömbom, 2019, p.77), in an effort to turn tourists into political agents for one particular side in the conflict. It is therefore unusual when tourism promotes ideals of unity, such as the story of Abraham.

This study looks at two tourism initiatives that reflect the unifying approach to PTT in that they appear to use Abraham as a guiding fiction to promote cross-cultural understanding in the Israeli-Palestinian context. These initiatives are the Abraham Hostels in Israel and the Abraham's Path in Palestine.

Abraham Hostels is the largest hostel chain in Israel with a capacity of over a thousand beds. The company operates four hostels in Eilat, Jerusalem, Nazareth and Tel Aviv. The first Abraham Hostel opened in Jerusalem in 2010, although the Nazareth hostel opened in 2005 under a different name (Gelbman & Laven, 2016). In their online presence, the hostel group champions Abraham as a "unifying symbol among the area's most prominent faiths" and one who embodies a "fitting representation of what we hope to bring to the world" (Abraham Hostels, 2020). Guided by the story of Abraham, the hostel has advanced its vision to provide unbridled hospitality to guests, to be a welcoming space for people of all backgrounds, and to stand for "pluralism, acceptance of the other, equality and democracy, hope and unity" (Abraham Hostels, n.d.). As part of this mission to engage the Other, the company also runs a popular tour offering called Abraham Tours that provides opportunities for tourists to meet Palestinians living in both Israel and Palestine. Although Yang (2020) argued these tours are too apolitical and thus normalise the Israeli occupation, Gelbman and Laven (2016), as well as Shepherd and Laven (2021), have demonstrated the hostels' ability to encourage cross-cultural dialogue between Israelis of different ethnic groups, as well as between tourists and Israelis at large.

The Abraham's Path is a long-distance hiking path and the brainchild of Harvard professor and negotiator William Urry. Seeing the conflicts of the Middle East as in need of a helpful "third side", Urry seized upon the figure of Abraham as someone whose "basic message is unity" (Urry, 2010) and through whom the people of the Middle East can redefine one another as kin. Consequently, Urry established the Abraham's Path Initiative (API) in 2004, which set out to create a hiking path that would trace Abraham's journey from his supposed birthplace near the modern Turkish city of Şanlıurfa, through what is now Syria, Jordan, Israel and Palestine, to his tomb in Hebron (Isaac, 2017). The idea was that the path would serve as "a step towards a reunion of the often-warring members of the patriarch's three-branched family (Leary et al., 2009, p.5) and change people's perceptions of the Middle East through the Abrahamic act of giving and receiving hospitality.

In this paper, I focus only on the Palestinian sections of the trail – 330 km of waymarked paths running the length of the West Bank (Isaac, 2017). These sections are, at the time of writing, managed by a Palestinian NGO called Masar Ibrahim al-Khalil (MIAK), which translates as the Path of Abraham the Friend – a nod to the Qur'an, which describes Abraham as a friend of God

(The Qur'an 4:125). The MIAK's online presence draws heavily on Abraham's reputation for unconditional hospitality and the survival of such cultural heritage in 21st century Palestine (Masar Ibrahim Al Khalil, 2020). The path has been subject to previous academic investigations, both for its significance for tourism development in Palestine (Isaac, 2017) and as a site of social non-movement (Sellick, 2019).

Together these sites represent attempts to invite tourists into the Abrahamic guiding fiction; to encourage visitors to see through the existing conflict in the region and focus on the message of unconditional hospitality and inter-cultural unity that Abraham embodies. This approach makes sense in an industry where stories are central to both the practice, promotion, experience and recollection of tourism (Moscardo, 2020).

Method

To explore what happens when stakeholders and tourists are, metaphorically speaking, invited into Abraham's tent, I crafted two semi-fictional stories from data collected in Israel/Palestine, using an approach to research called Creative Analytic Practice (CAP). These two stories, entitled *One day at the Abraham Hostels* and *One day on the Abraham's Path* follow a character called Tristan as he visits both these initiatives and engages with tourists and stakeholders.

CAP is an approach to analysis and representation that generates contextualized creative representations of synthesized data (Berbary, 2019). It seeks to blur the boundaries of science and art, being scientifically "true to a world known through empirical work and study" (Richardson, 2004, p.212), whilst embracing the creative arts to produce work that is evocative and open to interpretation.

Although data collection for CAP research can resemble conventional qualitative approaches, the analysis, and in particular the representation of the research, is very different (Berbary, 2019). The reasoning for this difference centres on the narrative turn and the crisis of representation. For Bochner (2001, p.134), the narrative turn represents a shift "away from a singular, monolithic conception of social science towards a pluralism that promotes multiple forms of representation and research". It champions narrative as a form of human meaning making (Glover, 2003), and research representations that provide thick descriptions that evoke a sense of being there for readers, so they too can connect with those whose stories are being told (Berbary, 2019). The narrative turn also champions the role of the researcher as an impassioned storyteller, not as a detached observer (Bochner, 2001). Recognising the influence the academic storyteller has over their interpretation of the stories of others, CAP rests heavily on the crisis of representation, which challenges the assumption in traditional qualitative research that we can represent the authentic voice of others (Grant, 2018). Given our inability to truly represent this voice, CAP stresses that as we subjectively engage with our research and (re)represent others, we construct but one possible depiction of our data. In this way, CAP is open about "our abilities and inabilities to capture and re-present reality" (Berbary, 2019, p.164). Conversely, CAP appears to reflect what Grant (2018) terms the "hermeneutics of suspicion", that the "meaning of a text can't, and indeed shouldn't, simply be read off in a straightforward way, and that myriad interpretations are possible" (p.108).

CAP takes many different creative forms. Although rarely used in our field, CAP in tourism studies has taken forms such as autoethnography (Shepherd et al., 2020), ethno-drama (Wright, 2020), and storying (Grimwood et al., 2019). I seek to forward this research agenda and contribute to a growing demand for creative and disruptive methodologies in tourism studies (Ivanova et al., 2020).

Although the stories below are a reflection of three years extensive fieldwork in Israel and Palestine, they most closely reflect data collected in the region in May and November 2019. During this time, I gathered three forms of data: participant observation data, autoethnographic data, and data from a questionnaire. These three sets of data represent three key research perspectives - tourism stakeholder, tourist, and tourism academic – that should illuminate our understanding of the use of Abraham as a guiding fiction for a unifying peace.

Participant observation was conducted at both initiatives. In Israel, I spent two weeks at three of the Abraham Hostels – sleeping at the hostel, speaking with guests and staff, and taking tours that the company offered. I focused my participant observation and my dialogue with staff and guests on how the story of Abraham is leveraged as a guiding fiction for the work of the hostel, and how tourists and staff respond to this appeal. The tours I took, such as a tour of the Gaza border and a food tour celebrating Tel Aviv's often-ostracised African communities, reflected the hostel's mission to engage tourists with the Israeli 'Other'. To compare and contrast my own and the other tourists' experiences, I conducted in-depth, unstructured interviews with two of the hostel's senior management; interviews that were recorded and then transcribed.

In Palestine, I spent four days walking the Abraham's Path with a group of 15 hikers as part of a thru-hike - a group hike along the entire length of the Palestinian section of the trail. As in Israel, I focused my conversations with the hikers and our guides (who changed frequently) on their perspectives towards the aims of the API and their use of Abraham. Together with the hikers, I slept at Palestinian guesthouses and with host-families, who I also engaged with when language permitted. One guesthouse owner, whose English was fluent, agreed to an in-depth, unstructured interview, which was recorded and transcribed. Throughout observations at both initiatives, I kept detailed notes of relevant comments made by tourists and guides, and took over 200 photographs.

In addition, I kept an autoethnographic field diary in which I wrote my reflections on and emotions of these experiences. I call it an autoethnographic field diary because, in line with Ellis et al. (2011, §1) definition of autoethnography, I explicitly tried to describe and analyse how my personal experience (auto) at these sites helped me understand the cultural experience (ethno) in question. This field diary amounted to 30 pages of data.

The third source of data was a questionnaire I sent out to the hikers I joined on the Abraham's Path. This short questionnaire asked specific open-ended questions related to their motivation for walking the trail, their experiences on it, and their thoughts on the use of Abraham as it relates to peacebuilding in the region. The questionnaire, completed by eleven hikers after they had returned from Palestine, provided me with reflective prose data that was difficult to obtain on the trail.

I analysed the data according to the recommend steps laid out by Berbary (2019) for CAP analysis. I began by reading the data repeatedly until I was deeply familiar with all its stories. Then I identified the stories that needed to be told. This selection process was based on what spoke most clearly to the aims of my research, on where there were clear moments of similitude or divergence, and on when the data spoke to me for its aesthetic appeal or emotional gravitas. Following this, I decided on the genre of semi-fictional storying as the best way to synthesise these moments. As Glover (2003, p.159) argues, all narrative work includes fictive elements in the quest to be aesthetically pleasing. Relatedly, my choice responded to a need to blend space and time in order to bring together disparate key moments in the data. It also stemmed from an inspiration – that of the ability of fiction writers, such as Colum McCann and his sensational novel *Apeirogon* (McCann, 2020), to tap into the emotive core of the overlapping storylines of the Israeli-Palestinian context.

This semi-fictional story was then given a plot – a way of connecting a series of events with meaning, with a start and an end (Polkinghorne, 1995). In both stories, this plot revolves around Tristan, a character who is largely the product of my autoethnographic data, as he spends a day in both initiatives. In basing Tristan on my experience, I hope to make clear my own involvement in the events of the story as well as my own positionality in this context (Bochner, 2001). Although Tristan remains the central figure in the plot, he is continually brought into dialogue with other characters who represent informants in the data. Their words largely come verbatim

Figure 1. Abraham – The First Backpacker. Photo taken by the author. November 2019.

from the data and reflect identifiable individuals in the data, yet at times, composite characters are used (Berbary, 2019). Naturally, the plot of both stories focus on the engagement of Tristan, tourists and stakeholders with the invitation to participate in the Abrahamic guiding fiction; to "enter the tent" for peacebuilding in a context of conflict. In this way, the stories are shaped by this invitation, in much the same way that an interviewer shapes the voices of respondents through their line of questioning (Jackson & Mazzei, 2009).

Once complete, I shared my stories with informants and asked for their views on how I represented their words and our shared experiences. I also asked them to choose their own pseudonyms – both to give them ownership of their characters and to protect their identities. The feedback was positive and I inserted the informants' suggestions for improvement into the stories.

Results

One day at the Abraham Hostels

When Abraham set out for Canaan, he did not face a full body scan, a comprehensive bag search or a series of penetrating questions from border security. Tristan, however, gets such a privilege each time he travels to the Holy Land. He tries to think positively about it; at least his laptop gets a swab dusting twice a year.

Hailing a taxi from the airport, he sits in the back and relishes the warm breeze slicing through the open window as the driver hurtles towards the city in undecided lanes. It's 4am, so it's no surprise the driver is silent. Yet the landscape is speaking; palm tree-lined roads only half-covering radar stations that seem a little too far from the airport.

Reaching the Abraham hostel, Tristan only manages a short, sweaty sleep. He has an early meeting with Moran, one of the founders of the Abraham Hostels. Racing out for a desperately needed iced Americano from the aptly named 'Way Cup Coffee', Tristan is reminded of why he is here (Figure 1).

"Our vision is to bring Abrahamic hospitality to the 21st century," beams Moran. "Abraham was the first backpacker in the Middle East, and we want to bring this image, this hospitality, back to life in the Middle East. Hospitality is the core of our business and Abraham was known for it!"

Put at ease by Moran's Hawaiian shirt and approachable disposition, Tristan relaxes into his coffee in Moran's surprisingly pokey office. He takes it all in. Hospitality, Middle East, Abraham. Wait, Middle East? Israel is hardly famous as a cross-border destination.

"We are operating in Israel, Palestine, Jordan and now in the Sinai, Egypt. So we see the Middle East as more of our arena, even though we started in Israel. We are not just an Israeli, or a Jewish, or a Zionist company."

"So do you have plans to set up Abraham hostels in other countries in the region?"

"We are really trying to touch the ground in Jordan but it's not acceptable now," he says, sighing. "Inshallah, someone will open one!"

"But at least if you do, Abraham is the perfect figure right?"

"Exactly!" he says. "Abraham is the communal father for Jews, Christians and Muslims. So no matter where you're from in this region, this is your story. Tourism, branding and marketing is all about making a story that will connect people. So why not bring that story to life? We are all from the same origin; we all love to host other people. This is the story of our life so why not emphasise it."

"But do you see any risks in using a religious figure in a region so marked by religious division?"

"If you want I can change it," says Moran with a wry smile. "Well, to be honest, I don't see it as a religious figure; I see it as a historical figure. You can't please everyone though. Some people don't like that we are liberal either, that we put Pride flags over the hostel, but that's their problem, I don't care."

A staff member comes in and interrupts the conversation. She is wearing a black t-shirt with coloured letters that say 'Come As You Are'. Her name badge has two flags on it: Israel and Palestine. It seems it is okay to be both here.

Giving Moran the space to talk privately with his staff, Tristan goes upstairs to stroll the lobby.

There, the help desk is engulfed in tourists planning weekends in Petra, sunrise in Masada, even trips to the Pyramids! Abraham Tours might not stretch from Haran to Hebron but that's certainly not for a lack of trying. Yet Tristan is drawn to a series of posters selling tours with names such as 'Meet the Ultra-Orthodox Jews', 'The Other Tel Aviv', 'The Best of the West Bank', 'The Gaza Border Reality Tour'.

The Gaza Border Reality Tour, where the reality of the Gaza Border got in the way. 'The first hour led by the Israeli guide, and the second hour led by an academic or activist Palestinian resident of Gaza, by video or phone call'. The phone call lasted 2 minutes. Electricity is limited in Gaza – as is water, as is food, as is patience. Two months after taking the tour, Tristan saw a news article in the Israeli newspaper *Haaretz*, "Hamas Arrests Gazans for Holding Zoom Video Chat With Israelis".

Lost in memories not yet made, Tristan hears a voice from behind.

"There you are! Let's go get some *rugelach*."

As they walk down bustling Allenby Street, Tristan asks Moran, "You said you want to bring Abraham's story to life. Does that mean you see the hostel as working for peace between Israelis and Palestinians?"

Moran pauses for a minute. "Hmm, I'm not so naïve. 15 years ago, maybe I would have said yes. I am convinced we can become great neighbours. I don't say friends but neighbours. What we are doing, it's not about creating world peace. Our agenda is that tourists will meet new people and that we can share some of our values with our guests, with our staff, with the communities we are operating in. Those Abrahamic values I talked about, right, that hospitality, that openness. You know, the truth is that people tend to come with their own agenda and leave with their own agenda. We just want tourism to become more sustainable so that everyone has a better chance of surviving in this crazy capitalist world. And in that respect, I think we bring great hope. At least that is what people tell me."

People like Mahida. A Muslim and of Palestinian origin, she manages the Fauzi Azar Inn - Abraham's hostel in Nazareth. Alongside her Jewish colleague. Tristan catches her in her office before she leaves for the last bus of the day to Nazareth, before Shabbat sets in and all of Israel draws to a sudden halt.

"I respect this place a lot," Mahida says, "because they make it feel like it doesn't matter if you are Jewish, Muslim or Christian."

Despite this, Tristan wonders what the people of Nazareth make of the Fauzi Azar Inn now being an Abraham hostel. Sensing this might be sensitive material, Tristan asks his question delicately.

Mahida's eyes flicker around the room. "Well, it not that they don't like the name Abraham in Nazareth, but you know, it's a Jewish company and Nazareth is all Arab. So there are many questions asked, like how can that Jewish company be working here? Many people can't accept this, a Jewish company in the heart of the Old City of the Arab's town. But for me, it doesn't matter. We are all human, and if someone wants to support the community, it doesn't matter if you're Jewish or Arab."

She pauses and smiles nervously. "I'm sorry; I don't like to speak much about politics."

Mahida leaves for her bus that will take her up the heavily developed coastline towards Nazareth. The roads will be congested as everyone rushes home for Friday's Shabbat dinner. Some will cook for one extra person in case an unexpected guest arrives, honouring the memory of Abraham who sat at the front of his tent awaiting unknown guests. Eventually, the bus will reach Nazareth, but only once it has passed by Jewish, Circassian, Bedouin, Arab and Druze communities. The many nations of Israel.

> God told Abraham, "You will be a father of many nations."
>
> They say you cannot choose your family.
>
> Not all families are functional

One day on the Abraham's Path

Despite being early November, the Christmas decorations are up in the Shaheen household. They are last year's decorations but the family never took them down. As members of Palestine's diminishing Christian minority, their decorations represent a stubborn affirmation of their faith. Their youngest son, Omar, watches their guest Tristan brush his teeth. Tristan smiles at him; he enjoyed playing football with Omar last night, even if the British tourist's goal attempts broke his mum's plastic garden chair, which Omar had used as an impromptu goal post.

Omar is too young to know what else the British broke here, but he will learn. The streets of the village may be short, but their memories are long.

For Omar, Christmas will be different this year. One month ago, his older brother was ripped out of bed and arrested by Israeli soldiers in the middle of the night. The family was never told what crime their son had committed, but they knew well enough the crime of being Palestinian; of their existence violating the sacred mythology of 'a land without a people'.

This thought weighs on Tristan's mind as he straps on his backpack for another day on the Abraham's Path. It also jars with the bucolic atmosphere of the village that morning – the crowing of roosters, the hum of old tractors moving into the olive groves, and the thick smell of the earth recoiling from the sun.

Nowhere is that smell more powerful than the group's destination for the day. A town 250 m below sea level, Jericho.

At the head of the group this morning is Ricardo, a good-humoured French Canadian whose Coquille Saint Jacques from pilgrimages past bounces on his backpack straps as he walks. He likes to walk in front. Seeing the open road clears his head. "It's my own personal therapy," he says, placing one hiking stick in front of the other, as he walks in the footsteps of Abraham, of David, and most importantly for him and his friend Luigi, of Jesus Christ.

Figure 2. Wadi Diba. Photo taken by the author November 2019.

Other members of the group are less interested in the biblical dimensions of the trail. They are here to learn about the Palestine of today.

"So, guys, up here on your right you can see the settlement of Na'ama," points today's guide Zahed. "They took over 5000 dunams of our land to build it."

The group has become used to these observations by now. The landscape narrated as undulating hills of dunams stolen and olive trees uprooted.

Nathalie, who is new to the group, plucks up the courage to ask a question others have learnt not to ask: "Would you ever consider having the path go through these settlements? Or connect up with other sections in Israel?"

"I don't want that to happen. No," Zahed says, "even if it did, I am not allowed to go there."

"But wasn't that the original idea? The Abraham's path … "

"There is no Abraham's Path. This is Masar Ibrahim."

"Doesn't that mean the same thing?" wonders Tristan.

"Ok, *yani*, Abraham's Path, it's a nice idea and all. It is true that Abraham is the father of Muslims, Christians, even Jews … but for all I care this could be called the Olive Trail or the Palestine Trail. People don't come here for Abraham, they come for Palestine."

Many nod enthusiastically. It is true; most did not come here for Abraham. Yet Jewish-Canadian Jacob is quietly disappointed. Like Nathalie, he is walking the path, in part, because of William Urry's call for a helpful "third side" to the conflict. Jacob expected more conversations about peace. At the same time, he is clear-eyed about what he has seen since he set out from Rummanah nine days ago. He is still trying to square Urry's vision of inter-faith unity with what he saw yesterday: a charred house in Duma, where Israeli settlers incinerated a Palestinian family in July 2015. Standing beside that burnt out skeleton of a home, the silence was deafening.

The day's walk is tough going under an oppressive sun. The little shade of the Wadi Diba, or 'Valley of Hyenas' (Figure 2), providing welcome relief during a simple lunch of *hummus* and *labneh*.

Unlike the other days on the path, the group are yet to receive a thimble-sized cup of cardamom-spiked coffee; a practice that has become almost routine to the hikers as they pass each small, threatened village. Therefore, when they reach the village of Al-Mughayyir that afternoon, they are glad to be welcomed by a coffee-bearing village elder.

He passes around the coffee. "Welcome to Al-Mughayyir. We welcome you to Palestine in the name of Ibrahim, the father, the friend," embarrassing the hikers in their sweat stained clothes with his bright white *qamis*.

Tristan, however, is distracted by graffiti in Hebrew on the wall behind him (Figure 3).

Figure 3. Graffiti in Al-Mughayyir. Photo taken by the author. November 2019.

Seeing Tristan's puzzled face, Clayton, who mastered Hebrew at one of America's top divinity schools, whispers, "It says 'Death to Arabs'."

The coffee bitters in Tristan's mouth. The words, the graffiti, and its echoes through history, haunt the hikers as they thank the village and move on.

Having been a heavy day on both the hikers' backs and hearts, they are more than ready for a beer by the time they reach Jericho. Frankly, they are grateful to have made it, given Israeli settlers are constantly erasing the red stripes and green oak tree motifs that mark the path (Figure 4).

Luckily for them, beer can be found at Saqer's guesthouse, their accommodation for the night. Unassuming from the outside, the guesthouse is a miniature Eden inside, with a small swimming pool and a garden filled with palms and pomegranate trees. While some of the group retire to their rooms to rest, the others sit with Saqer on the balcony to read and savour well-deserved bottles of Taybeh beer.

Although the group had spent all day discussing the 'issue', Tristan seems unable to relax.

"I'm really struggling to understand how we fit together the Ted Talk idea of the Abraham's Path and all the shit we saw today. Demolition orders, burnt houses, settlements, arrests. I don't know, I just … ." he trails out, seeking reassurance.

Emma replies, "Well, yeah, it certainly is odd to think about unity when there is an ongoing war, if that's what you mean." Her eyes widen. "But I would say that in this case, I don't think that idea of Abraham risks depoliticizing or unity-washing the conflict. I've seen that in other projects and I don't feel that's what's happening here." She swigs her beer before continuing. "And to be honest, I'm not sure how you could do this walk and feel more connected with the settlers."

Quinn snickers, putting down his iPad, "Quite! I think what this walk allows tourists to see is the reality of the occupation and the expansion of apartheid in the West Bank. I don't think it's going to help peacebuilding but it definitely helps understand the hopelessness of the current situation."

The group sits in a moment of muted agreement.

"It brings me hope actually," Luigi counters. "That story, it's a perfect unification for so many people and you can see how important it is for locals. I guess I just want to believe that it might bring some understanding of shared values."

Jacob was listening keenly, unsure how to respond. Staring into his already empty bottle of Taybeh Gold, he adds, "It might not bring peace no, but if it's a story that brings Muslims, Christians and Jews together then maybe it's worth the risk of 'unity washing'." He pauses. "That

Figure 4. Spoiled markings. Photo taken by the author. November 2019.

said, if we are all rallying around Abraham, what about the folks that aren't Abrahamic, who aren't Christian, Muslim or Jew? We aren't all 'children of Abraham'."

Saqer emphatically puts down his coffee. "Well, it's a terminology," he says, "so the trail could be called something totally different. It could be called the Abraham's Path and still be the shittiest trail in the world so it's what we make out of the trail itself."

"Secondly," he continues, lighting a cigarette, "it's a story, a historical one, an imaginary one. From a marketing perspective, it's very attractive because it applies to the Holy Land perfectly. But I have my doubts. Maybe it gives a good idea about co-existence for people who do the trail as clients, but for people who work on the trail, a story like this is not sufficient to change their perspective of the reality that they see every day and no marketing campaign is going to convince them that things are going to be better unless something actually happens on the ground."

"You know," he says, leaning in, "there is always this conflict, this schizophrenia, between tourism and daily life, because in tourism you have to have a happy narrative, one that makes people feel good about themselves, which is not necessarily the case when it comes to the daily reality of Palestinians. So I feel there is this political agenda, one that promotes peace and co-existence, which is nice and makes a very good pitch for reaching foreigners, but it's quite far from the reality people live with."

"It's a nice story but it has its limits," Tristan says.

"Yeh, man, I'm sorry, I am very to the point, but this place is so old that even religions start to smell bad."

Unlike the *maqluba,* whose spiced scent floats up from Saqer's kitchen. Sadly for Tristan, however, dinner is back in Jerusalem tonight.

When his taxi arrives, Zahed helps him put his bag into the boot of the car. With a pat on Tristan's back, he says, "When you get to the checkpoint, tell them we are all sons of Abraham. Talk about peace blah blah blah. They like that."

The taxi speeds out of Jericho and onto Route 1. Tristan ignores the flashing warning lights on the car's dashboard, too engrossed by a landscape that suddenly seems so alien to the past days' walking. Gone are the patchwork of olive-pregnant valleys. Now, the occupation rakes the roadside; signs only in Hebrew and Israeli wineries squatting on hillsides. Passing the Alon settlement, Tristan notices a billboard advertising an Israeli tour company: "Genesis Land - The best place to start any tour of Israel is with the story of Abraham in the Judean Desert". Some call this spaciocide.

When Tristan finally reaches the checkpoint, he is grateful that tonight is a quiet night. No traffic, no tear gas, no shouting. The checkpoint's turnstiles turn their slow, claustrophobic contortions, allowing Tristan to give his passport to the soldier. He thinks it best not to mention Abraham, or peace.

Conclusion

Although entitled "conclusion", this section does not intend to conclude this conversation. Rather, as befitting CAP work, I encourage a multitude of interpretations, based on the polysemy of the stories and the hermeneutics of suspicion (Berbary, 2019; Grant, 2018). However, I do feel there is room to provide critical reflections on what I see these stories as telling us about a unifying approach to PTT, attempted here through the guiding fiction of Abraham.

If such an approach begins with a unifying point, then Abraham is unquestionably a valid one. The stories demonstrate that despite the social chasm between Israelis and Palestinians, and the differing perceptions of Abraham's centrality to the work of both initiatives, those on all sides promote the idea of Abraham as a common ancestor. He thus represents a commonality, which Abu-Nimer (2003) stresses is a first step of peacebuilding. Perhaps the greatest benefit of the Abrahamic guiding fiction is that it provides a valuable vocabulary for those who wish to challenge the dominance of sectarian narratives in Israeli and Palestinian tourism (Clarke, 2000; Strömbom, 2019). This vocabulary is enthusiastically employed by Moran who uses it to assist him in carving out an Israeli tourism space that celebrates engagement with the Other. With the narrative buttressing of Abraham, 'Come As You Are' moves from being a woke corporate slogan to a tangible and historic element of a shared local culture that prioritises the welcoming of strangers and an ethics of care through hospitality. The Abrahamic vocabulary allows both initiatives to break the boundaries of the current nationally routed tourism imaginary, and makes accessible the idea of a multi-cultural Middle East, a vision that tourists can help co-create, and one that Said (1992) argues is ultimately inescapable. Hazburn (2012) has forwarded that such a de-territorialisation of tourism in the Middle East will improve tourism's potential to promote peace.

Yet, using the guiding fiction of Abraham to operationalize the unifying approach is also problematic. Just as Higgins-Desbiolles (2018) highlighted how the drive for sustainable tourism rubs up against the unsustainability of a market-driven tourism industry, the stories demonstrate that the unity aimed for with the guiding fiction of Abraham rubs violently against the harsh reality of the Israeli-Palestinian conflict. Asking us to see unity in the midst of conflict is, as Emma points out, extremely difficult. Try, for example, standing next to that burnt out house in Duma or sharing tea with the mother of the Shaheen household and talk about unity. In such circumstances, the call for unity risks seeming tone-deaf, especially when it comes from foreigners not living under a military occupation. The guiding fiction, therefore, struggles in narrative transportation - to get us to a point where we "cease to pay attention to the world around [us] … and become part of the story world itself" (Moscardo, 2020, p.9). There appears to be too great a dissonance between the story world of Abraham invoked by both initiatives and the objectively real world of the Israeli-Palestinian conflict.

The case of Zahed is most instructive here. Despite the API arguing for Jews, Christians and Muslims to walk in the footsteps of Abraham for peace, Zahed explains that, unlike Ricardo and Jacob, he cannot do this. He cannot walk in Israel. Although he does not feel the need to explain what is to him obvious, this is because of Israeli restrictions on Palestinian mobility. Here, the invitation to follow in Abraham's footsteps on the Abraham's Path only underlines Zahed's immobility. It also underlines the simultaneous inability of Palestinians – juxtaposed against the ability of foreign tourists - to move freely in the region, as Tristan so seamlessly does. This imbalance in the guiding fiction's applicability is further highlighted by Saqer's description of a "happy narrative", suitable for tourists, but inappropriate for Palestinians living in the conflict. If the benefits of the guiding fiction of Abraham are uneven, then it struggles to fit within Galtung's (1969) concept of positive peace, which prioritises equality and social justice.

The unifying approach to PTT can therefore be criticised for encouraging tourists and locals to suspend their disbelief and focus on a utopian future of inter-cultural unity that may prove unreachable. In our case, its apparent inability to challenge the power differentials between Israelis, Palestinians and tourists is also problematic, although far from unique in PTT initiatives in the region (Strömbom, 2019). Yet the power of such an approach lies in its ability to promote a future vision that does not rely on the "nullification of one by the other" (Said, 1992, p.244). It challenges the supremacy of existing sectarian narratives; narratives given too great a platform in Israeli-Palestinian tourism (Clarke, 2000; Strömbom, 2019). Although it might struggle to challenge the status quo, it does, to use Hannah Arendt's words, "gives access … to reasons for rejecting the status quo", which she argues, "is sometimes all the revolt one needs or can muster" (Shuster, 2018, p.7). Mustering the stomach for peace in an ongoing conflict is challenging, yet this is precisely where peacebuilding begins. After all, as Koopman (2011) argues, peace and conflict do not occlude one another. Sellick (2019, p.878) confirmed this during her time on the Masar Ibrahim, in saying that, "peaceful and violent relations are entangled within [this] same space."

My hope is that this paper offers insights for stakeholders and researchers into the positive and negative potential of using a unifying approach to PTT, particularly when doing this through a guiding fiction. Supporting Moscardo's (2020) call for more attention towards the use of stories in tourism, I propose that stakeholders in contested spaces look closely at the ability of local stories to promote fruitful visions of cross-cultural unity in the midst of conflict. Abraham, in particular, being well suited in the Middle East given the popularity of religious tourism in the region, catering to tourists who might be receptive co-creators of such a vision in their desire to honour Abraham. For researchers, I hope that the concept of a unifying approach to PTT proves helpful in our on-going debate about PTT, and can serve as a theoretical counter to agonistic PTT put forward by Strömbom (2019). I also anticipate that such concepts encourage other PTT scholars to delve deeper into the wealth of knowledge that can be found in peace studies scholarship.

Acknowledgements

I would like to thank all the informants that feature in this study, both for their company in the field and their engagement with the study. A special thanks to Daniel Laven, Kevin Keystone, Simon Farrar, colleagues at Mid Sweden University, and the anonymous reviewers for their insightful comments on this manuscript.

Disclosure statement

No potential conflict of interest was reported by the author(s).

ORCID

Jack Shepherd http://orcid.org/0000-0002-6082-5085

References

Abdullah, A., Ahmad, A., Al Alabdullah, S., Chikhou, R., & Mamo, A. D. (2020). Supporting peacebuilding in Syria through universities: The role of cultural heritage. *Education and Conflict Review*, *3*, 60–66.

Abraham Hostels. (2020). *About Us*. https://www.abrahamhostels.com/tel-aviv/about-us/

Abraham Hostels. (n.d.). *Abraham Impact*. https://www.abrahamhostels.com/impact-en/

Abu-Nimer, M. (2003). Toward the theory and practice of positive approaches to peacebuilding. In C. Sampson, M. Abu-Nimer, & C. Liebler (Eds.). *Positive Approaches to Peacebuilding* (pp. 13–23). PACT Publications.

Allport, G. W. (1954). *The Nature of Prejudice*. Addison-Wesley.

Anastasopoulos, P. (1992). Tourism and attitude change: Greek tourists visiting Turkey. *Annals of Tourism Research, 19*(4), 629–642. https://doi.org/10.1016/0160-7383(92)90058-W

Berbary, L. (2019). Creative Analytic Practices: Onto-epistemological and Theoretical Attachments, Uses, and Constructions within Humanist Qualitative Leisure Research. *Leisure Sciences, 41*(3), 148–166. https://doi.org/10.1080/01490400.2018.1526999

Berno, T., & Bricker, K. (2001). Sustainable tourism development: The long road from theory to practice. *International Journal of Economic Development, 3*(3), 1–18.

Bochner, A. (2001). Narrative's virtues. *Qualitative Inquiry, 7*(2), 131–157. https://doi.org/10.1177/107780040100700201

Butler, R. W. (1993). Tourism – an evolutionary perspective. In J. G. Nelson, R. W. Butler, & G. Wall (Eds.), *Tourism and sustainable development: Monitoring, planning, managing* (pp. 27–44). University of Waterloo.

Clarke, R. (2000). Self-presentation in a contested city: Palestinian and Israeli political tourism in Hebron. *Anthropology Today, 16*(5), 12–18. https://doi.org/10.1111/1467-8322.t01-1-00024

Da Silva, J. T., Breda, Z., & Carbone, F. (Eds.). (2021). *Role and impact of tourism in peacebuilding and conflict transformation*. IGI Global.

Ellis, C., Adams, T., & Bochner, A. P. (2011). Autoethnography: An overview. *Forum: Qualitative Social Research, 12*(1), 10.

Galtung, J. (1969). Violence, peace, and peace research. *Journal of Peace Research, 6*(3), 167–191. https://doi.org/10.1177/002234336900600301

Gelbman, A. (2019). Sport tourism and peace: Crossing the contested wall. *Tourism Geographies, 21*(1), 163–189. https://doi.org/10.1080/14616688.2018.1519718

Gelbman, A., & Laven, D. (2016). Re-envisioning community-based heritage tourism in the old city of Nazareth. *Journal of Heritage Tourism, 11*(2), 105–125. https://doi.org/10.1080/1743873X.2015.1044993

Glover, T. D. (2003). Taking the narrative turn: The value of stories in leisure research. *Loisir et Société / Society and Leisure, 26*(1), 145–167. https://doi.org/10.1080/07053436.2003.10707610

Gopin, M. (2002). *Holy war, holy peace: How religion can bring peace to the Middle East*. Oxford University Press.

Grant, A. (2018). Introduction: Voice, ethics and the best of autoethnographic intentions (or writers, readers, and the spaces in-between). In L. Turner, N. P. Short, A. Grant, & T. E. Adams (Eds.). *International perspectives on autoethnographic research and practice* (pp. 107–122). Routledge.

Grimwood, B., Stinson, M., & King, L. J. (2019). A decolonizing settler story. *Annals of Tourism Research, 79*, 102763. https://doi.org/10.1016/j.annals.2019.102763

Hammami, F., Harvey, D. C., Laven, D., & Walters, D. (In Press). Heritage and peacebuilding: challenges, possibilities and sustainable practices. In K. Fouseki, M. Cassar, G. Dreyfuss, & K. Eng (Eds.), *Routledge handbook of sustainable heritage*. Routledge.

Harari, Y. (2014). *Sapiens: A brief history of humankind*. Harper.

Hardy, A., Beeton, R. J. S., & Pearson, L. (2002). Sustainable tourism: An overview of the concept and its position in relation to conceptualisations of tourism. *Journal of Sustainable Tourism, 10*(6), 475–496. https://doi.org/10.1080/09669580208667183

Hazburn, W. (2012). Itineraries of peace through tourism: Excavating territorial attachments across the Arab/Israeli frontier. *Peace and Change, 37*(1), 3–36.

Higgins-Desbiolles, F. (2018). Sustainable tourism: Sustaining tourism or something more? *Tourism Management Perspectives, 25*, 157–160. https://doi.org/10.1016/j.tmp.2017.11.017

Higgins-Desbiolles, F. (2021). The "war over tourism": challenges to sustainable tourism in the tourism academy after COVID-19. *Journal of Sustainable Tourism, 29*(4), 551–569. https://doi.org/10.1080/09669582.2020.1803334

Isaac, R. (2017). Taking you home: The Masar Ibrahim Al-Khalil in Palestine. In C. M. Hall, Y. Ram, & N. Shoval (Eds.). *The Routledge international handbook of walking* (pp. 172–183). Routledge.

Ivanova, M., Buda, D.-M., & Burrai, E. (2020). Creative and disruptive methodologies in tourism studies. *Tourism Geographies*. https://doi.org/10.1080/14616688.2020.1784992

Jackson, A. Y., & Mazzei, L. A. (Eds.). (2009). *Voice in qualitative inquiry: Challenging conventional, interpretive, and critical conceptions in qualitative research*. Routledge.

Jung, K.-S. (2019). The construction of the cold war landscape and tourism. *Journal of Asian Sociology, 48*(3), 387–420.

Khalidi, R. (2020). *The hundred years' war on Palestine: A history of settler colonialism and resistance, 1917-2017*. Metropolitan Books.

Koopman, S. (2011). Let's take peace to pieces. *Political Geography, 30*(4), 193–194. https://doi.org/10.1016/j.polgeo.2011.04.013

Korea Tourism Organization. (2020). *DMZ Peace Train*. https://english.visitkorea.or.kr/enu/ATR/SI_ENG_2_2_1.jsp

Leary, K., Sebenius, J. K., Weiss, J. (2009). *Negotiating the Path of Abraham. (Working Paper 10-049)*. Harvard Business School. https://www.hbs.edu/faculty/Publication%20Files/10-049.pdf

Lederach, J. P. (1997). *Building peace: Sustainable reconciliation in divided societies*. United States Institute of Peace.

Masar Ibrahim Al Khalil. (2020). *FAQ*. https://masaribrahim.ps/plan/faq/#1532331395031-569eeb0e-8628
McCann, C. (2020). *Apeirogon*. Random House.
McCool, S., Butler, R., Buckley, R., Weaver, D., & Wheeller, B. (2013). Is concept of sustainability Utopian: Ideally perfect but impracticable? *Tourism Recreation Research*, *38*(2), 213–242. https://doi.org/10.1080/02508281.2013.11081746
McDowell, S. (2008). Selling conflict heritage through tourism in peacetime Northern Ireland: Transforming conflict or exacerbating difference? *International Journal of Heritage Studies*, *14*(5), 405–421. https://doi.org/10.1080/13527250802284859
Megoran, N. (2011). War *and* peace? An agenda for peace research and practice in geography. *Political Geography*, *30*(4), 178–189. https://doi.org/10.1016/j.polgeo.2010.12.003
Moscardo, G. (2020). Stories and design in tourism. *Annals of Tourism Research*, *83*, 102950. https://doi.org/10.1016/j.annals.2020.102950
Nagle, J. (2014). From the politics of antagonistic recognition to agonistic peacebuilding: An exploration of symbols and rituals in divided societies. *Peace & Change*, *39*(4), 468–494. https://doi.org/10.1111/pech.12090
Oz, A. (1995). *Israel, Palestine and peace*. Vintage.
Polkinghorne, D. (1995). Narrative configuration in qualitative analysis. *International Journal of Qualitative Studies in Education*, *8*(1), 5–23. https://doi.org/10.1080/0951839950080103
Ramadan, A. (2011). Hospitality and postnational peace. *Political Geography*, *30*(4), 195–196. https://doi.orq/10.1016/j.polgeo.2011.04.006
Richardson, L. (2004). Creative Analytical Practice (CAP) ethnography. In M. S. Lewis-Beck, A. Bryman, & T. F. Liao (Eds.), *The SAGE encyclopedia of social science research methods* (pp. 213–213). SAGE.
Said, E. (1992). *The question of Palestine* (Reissue ed.). Vintage Books.
Sellick, P. (2019). Boots on the Ground': walking in occupied Palestinian territory. *Mobilities*, *14*(6), 875–889. https://doi.org/10.1080/17450101.2019.1670916
Sharpley, R. (2020). Tourism, sustainable development and the theoretical divide: 20 years on. *Journal of Sustainable Tourism*, *28*(11), 1932–1946. https://doi.org/10.1080/09669582.2020.1779732
Shavit, A. (2013). *My promised land*. Scribe.
Shepherd, J., & Laven, D. (2021). Hostels in hostile territory: Tourist spaces of transformative dialogue in the Israeli-Palestinian context. In J. T. Da Silva, Z. Breda, & F. Carbone (Eds.), *Role and impact of tourism in peacebuilding and conflict transformation* (pp. 195–217). IGI Global.
Shepherd, J., Laven, D., & Shamma, L. (2020). Autoethnographic journeys through contested spaces. *Annals of Tourism Research*, *84*, 103004. https://doi.org/10.1016/j.annals.2020.103004
Shinko, R. (2008). Agonistic peace: A postmodern reading. *Millennium: Journal of International Studies*, *36*(3), 473–491. https://doi.org/10.1177/03058298080360030501
Shuster, M. (2018). Hannah Arendt on the evil of not being a person. *Philosophy Compass*, *13*(7), e12504–13. https://doi.org/10.1111/phc3.12504
Smythe, W. E. (2005). On the psychology of 'As If. *Theory & Psychology*, *15*(3), 283–303. https://doi.org/10.1177/0959354305053215
Strömbom, L. (2019). Exploring prospects for agonistic encounters in conflict zones: Investigating dual narrative tourism in Israel/Palestine. *Alternatives: Global, Local, Political*, *44*(2–4), 75–93. https://doi.org/10.1177/0304375419857421
The Bible. (New International Version). (2011). https://www.biblegateway.com/versions/New-International-Version-NIV-Bible/ (Original work published 1973).
The Qur'an. (2020). *Noble Qur'ran*. https://quran.com/
United Nations. (2020). *World Tourism Day 27 September*. https://www.un.org/en/observances/tourism-day
United States Department of State. (2020). *Abraham accords peace agreement: Treaty of peace, diplomatic relations and full normalization between The United Arab Emirates and The State of Israel*. https://www.state.gov/wp-content/uploads/2020/09/UAE_Israel-treaty-signed-FINAL-15-Sept-2020-508.pdf
Urry, W. (2010). William Urry: The Walk from "No" to "Yes" [Video]. TED Conferences, October). https://www.ted.com/talks/william_ury_the_walk_from_no_to_yes?language=en
Vaihinger, H. (1924). *The philosophy of 'As If': A system of the theoretical, practical and religious fictions of mankind* (C. K. Ogden, Trans.) Kegan Paul and Co. (Original work published. 1911).
Western Front Way. (2020). *Home*. https://www.thewesternfrontway.com/
Wright, R. K. (2020). Que será, será!': creative analytical practice within the critical sports event tourism discourse. *Tourism Geographies. Advance Online Publication*, *23*(1–2), 296–317. https://doi.org/10.1080/14616688.2019.1648542
Yang, C. (2020). Staging Israel/Palestine: The geopolitical imaginaries of international tourism. *Environment and Planning C: Politics and Space*, *38*(6), 1075–1090.

Promoting sustainable tourism futures in Timor-Leste by creating synergies between food, place and people

Tracy Berno, Gobie Rajalingam, Agueda Isolina Miranda and Julia Ximenes

ABSTRACT

Following 400 years of Portuguese colonial administration and 24 years of Indonesian occupation, post-conflict Timor-Leste's government has been underpinned by its commitment to reconciliation and democracy. Sustainable development approaches have catalysed the country's transition from civil unrest to state-stability, but much work remains if lasting economic benefits are to be felt by the country's majority rural population. While rhetoric indicates that Timor-Leste is committed to the United Nation's Sustainable Development Goals and achieving economic diversification, the continued dependence on oil presents an imminent economic and societal challenge. Tourism in Timor-Leste is acknowledged as being one of the nation's most important non-oil sectors, and its growth requires the availability of quality tourism products and experiences. A key activity that can create synergies between culture, history and nature is an exploration of food, place and people towards eliciting the "story" of the food of Timor-Leste. This qualitative research utilised interviews and focus groups to explore how food stories and experiential culinary tourism can allow for the development of a positive narrative of resilience in contemporary Timor-Leste. The empirical findings suggest that food providers and restaurants owners share a responsibility in decisions about what food and food stories become tourism offerings in Timor-Leste.

Introduction

Situated approximately an hour's flying time north of Darwin, Australia, the Democratic Republic of Timor-Leste, is one of the world's youngest nations. Timor-Leste is geographically a small country comprising approximately 15,000 square kilometres of the eastern half of Timor Island, a small number of offshore islands and an exclave situated on the northern coast of West Timor. Timor-Leste is home to just over a million people and has one of the youngest populations in the Asia-Pacific region (median age 17.4 years; Government of Timor-Leste, 2019). After centuries of colonial administration and decades of hostile occupation, Timor-Leste became a sovereign, independent nation in 2002. However, despite its rich natural resources and significant economic, health and education gains since 2002, Timor-Leste remains classified by the United Nations as a 'least developed country' and continues to face the myriad of challenges associated with this (Figure 1).

In 2014, nearly 42 percent of the citizens of Timor-Leste were living in poverty (World Bank, 2020). In 2018, Timor-Leste ranked 131 out of 189 countries and territories on the Human Development Index, which assesses long-term progress on life expectancy, education and per

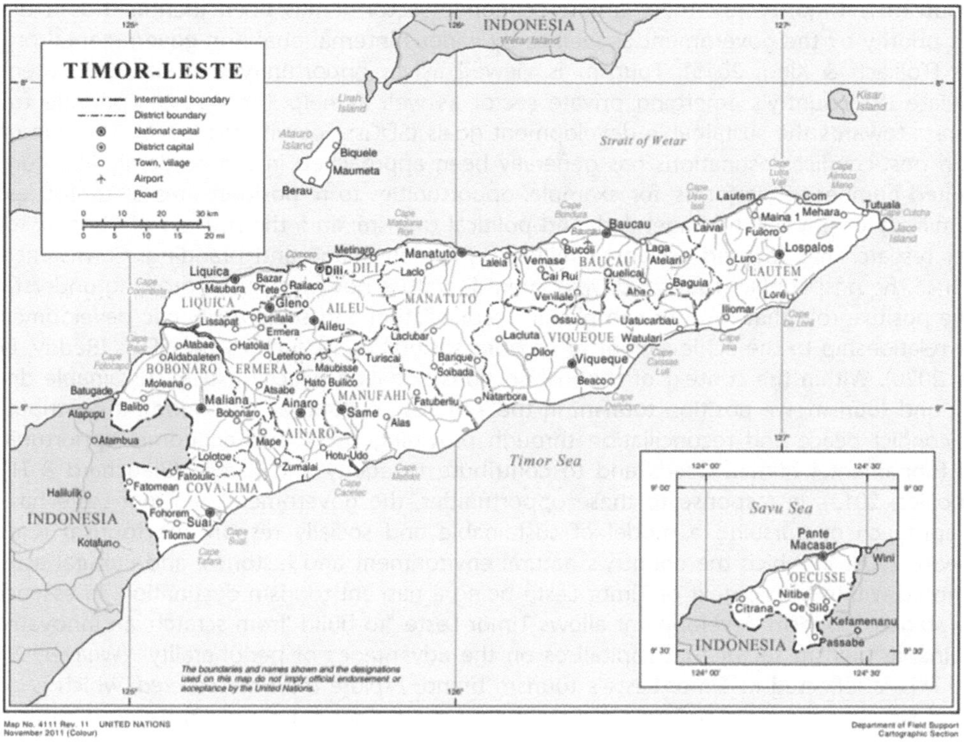

Figure 1. *Map of Timor-Leste* (United Nations, 2011).

capita income. There are also numerous urgent gender concerns, including pervasive gender-based violence, high maternal mortality, and significant gender gaps in labour market and local governance participation (United Nations Development Program, 2018). When employment fell to 89.6 percent in 2010 (from 91.5% in 2004), women and rural areas were most affected, with the employment rate for women falling 5.3 percent compared to 0.2 percent for men (Ministry of Finance, 2010). Gender inequality is a significant obstacle to socio-economic development and needs to be addressed as an imperative for Timor-Leste's sustainable development (Government of Timor-Leste, 2010).

Innes (2017), in his staple thesis, suggests that the reliance on the export of natural resources to more advanced economies has a significant impact not just on the economy of the exporting country, but also on its social and political systems. This dependency is a significant risk for the economy of Timor-Leste, which is heavily oil-dependent, a level of dependency only surpassed by South Sudan. It is estimated that oil revenues provide 90 percent of the government's income, which places Timor-Leste at significant fiscal risk if the economy does not diversify (Domínguez, 2014; Palatino, 2014). Timor-Leste's oil and gas are rapidly dwindling resources and the Petroleum Fund of Timor-Leste, which provides income for the nation, may potentially run out in the next decade due to the Government of Timor-Leste's (GOTL) consistent and unsustainable withdrawals. Economic diversification has therefore become an imperative concern for GOTL, as articulated in its Strategic Development Plan 2011–2030, which aims to leverage the strong 10 percent annual growth in the non-oil economy since 2007 by building a modern and diversified economy.

Coffee and candlenut are the two most important agricultural cash crops in Timor-Leste. In 2019, approximately US $18.3 million in coffee and $450,000 of candlenut were exported from Timor-Leste (National Directorate of Statistics, 2019). Although significant agricultural exports, they pale into insignificance compared to those of oil (coffee comprises approximately 1% of the country's annual budget; Marx, 2016).

Tourism is Timor-Leste's third largest economic sector. It has been identified as a development priority by the government as well as by various international non-governmental organisations (Tolkach & King, 2015). Tourism is viewed as an opportunity that has the potential to stimulate the country's emerging private sector as well as help Timor-Leste continue to make progress towards the sustainable development goals (SDGs; Courvisanos, 2018). The role of tourism in post-conflict destinations has generally been approached in the academic literature from a limited number of platforms, for example: opportunities to re-position sites of conflict as political, military or physical heritage; dark and political tourism; and, the process of Phoenix tourism. Other research has focused on post-conflict destination image and branding. Consistent across much of the post-conflict tourism development literature however, is the growing understanding of the positive role that tourism can play in reconciliation and socio-economic development and their relationship to the achievement of broader societal goals such as the SDGs (Reddy, Boyd & Nica, 2020). Within the context of the intersections between peace, justice, sustainable development and tourism, we position tourism in the Timorese context as an important contribution to post-conflict peace and reconciliation through providing social and economic opportunities to fulfil fundamental human needs and to contribute to equity and justice (Blanchard & Higgins-Desboilles, 2013). In response to these opportunities, the government of Timor-Leste has taken the approach of pursuing a model of sustainable and socially responsible tourism (Cabasset-Semedo, 2009) in which the country's natural environment and historical and cultural assets are promoted within the context of Timor-Leste being a nascent tourism destination. In essence, this early stage of tourism development allows Timor-Leste "to build 'from scratch' an innovative and sustainable tourism sector that capitalises on the advantages of peripherality" (Weaver, 2018, p. 118). This is reflected in Timor-Leste's tourism brand, *Explore the Undiscovered*, which is centred around two key themes: (1) Timor-Leste's unique cultural identity; and (2) its position as a relatively unexplored destination (Government of Timor-Leste, 2017b).

These themes were built on tourism products and experiences linked to Timor-Leste's traditional culture, history, people, and nature. Although Timor-Leste's history is well documented, there is an overwhelming lack of information with regard to what these cultural and natural assets consist of and how they can be consumed and experienced from a tourism perspective. This includes questions such as, "What comprises Timorese food?", "Where can traditional, high-quality Timorese food and coffee be consumed?", "What is Timorese culture?", and "How can travellers access and experience Timorese culture?" A key activity that can potentially create synergies between culture, history and nature is an exploration of place and people through their "food stories" (Berno et al., 2019; Berno & Fountain, 2020). Given Timor-Leste's long history of colonial administration and hostile occupations, food stories and experiential culinary tourism has the potential to become a conduit for the development of a positive narrative of resilience in contemporary Timor-Leste, which, in turn, can help facilitate the creation of shared futures in an otherwise divided post-conflict society. Tourism in Timor-Leste is seen as a possible catalyst for economic growth. By connecting culture, history and identity under an umbrella of "culinary tourism", food and food tourism can unite a population heavily reliant on agriculture creating new economic opportunities for growth and prosperity.

Timor-Leste: the "impossible history and possible futures"[1]

As suggested by Appadurai (1996), place is produced through the intersection of social relations, expressions of identity and the practice of culture. Timor-Leste is a complex place with a complex history through which the nation's culture and identity has evolved via Portuguese colonisation, World War II, the Indonesian invasion and occupation, and the United Nations mission up until the current transitional era of independence (Currie, 2018). As a result, the collective memory (and identity) of the people of Timor-Leste incorporates a history of continuous foreign occupations with devastating consequences that continue to echo in the lived experiences of the people of the nation (Gomes, 2013). "Timor-Leste's story, both sad and triumphant, is of a long, painful struggle for political freedom, independence and unity" (Dayley, 2019, p. 206).

Commencing with the arrival of Portuguese friars in the sixteenth century, Timor-Leste was subjected to various forms of foreign presence until the end of twentieth century. By the late 1700s, Timor-Leste was ostensibly under Portuguese control, a colonial legacy that would last until the second World War when Dutch and Australian troops established East Timor as an Allied military staging ground. This was not to last long as the Allies were defeated in a battle on the island by the Japanese in 1942, thus heralding a brutal occupation by Japanese forces during which tens of thousands of Timorese were killed. At the end of World War II, Timorese leaders led an unsuccessful attempt for independence and the Portuguese returned for what has been described as "three more decades of unwanted colonial rule" (Dayley, 2019, p. 207).

As a result of civil political upheaval in Portugal, East Timor nearly gained independence in 1975. This independence did not take hold however, and in early December of that year, fearing the emergence of a "Cuba of the South Pacific" (Smith, 2004, p. 146), Indonesia launched military attacks from West Timor. This eventuated in a full-scale invasion at the end of 1975 in which Indonesia employed vicious tactics and mass atrocities in their efforts to force annexation (Dayley, 2019). During the invasion, close to 40,000 people fled to West Timor and 4,000 fled to Portugal or Australia (Khamis, 2015). From the original population of 650,000, it is estimated that 100,000 Timorese lost their lives during the invasion, making it one of the worst contemporary mass atrocities in terms of lives lost per capita (Dayley, 2019). The ongoing Indonesian occupation with its ensuing violence and years of starvation and poverty-driven illness resulted in nearly one-quarter of East Timor's pre-1975 population dying during their 25-year rule of the nation (Timor-Leste Commission for Reception, Truth and Reconciliation (2013) cited in Dayley, 2019).

In 1999, the people of East Timor voted in a referendum choosing between independence or limited autonomy within Indonesia. Seventy-eight percent chose independence, resulting in Indonesian forces retaliating by destroying close to 80 percent of buildings in Timor-Leste, effectively disrupting governance, commerce and agriculture for months (Hill, 2001). It was not until late 1999 with the international condemnation of Indonesia's actions and the arrival of an 8,000 strong Australian-led multinational military force that Indonesia finally withdrew (Khamis, 2015). Finally, in 2002, after enduring centuries of colonial rule and decades of violent occupations, Timor-Leste became a sovereign, independent nation. This, however, was not the end of foreign presence in the country. Timor-Leste was still to endure years of 'occupation' by United Nations security forces, a significant Australian-led peacekeeping contingency and dozens of international aid organisations (Dayley, 2019).

The history of Timor-Leste of course pre-dates occupation and conflict, and this history also influences contemporary culture. Traditional knowledge and understanding also has an important influence, for example *Lulik,* which is a concept that is integral to Timorese culture and continues to be widely used and influences daily life for the Timorese. Although *lukik* literally translates as "forbidden", "holy" or "sacred", the concept has much deeper and wider applications. *Lulik* is a set of fundamental, philosophical orientations that determines how Timorese should behave in social and environmental interactions. *Lulik* is a philosophy that ensures peace and tranquillity for Timorese society as a whole and it continues to be a core value in contemporary Timorese society (Bovensiepen, 2014; McWilliam et al., 2014; Trindale, 2011) for intentions and actions across all realms of Timorese life (McWilliam et al., 2014). During the Indonesian occupation, it is said that Timorese protected themselves from the Indonesian military and pro-Indonesia militias by 'opening the land' and releasing *lulik* powers in the form of an army of land spirits (Bovensiepen, 2011) demonstrating that when managed and respected, *lulik* is a powerful resource in Timorese culture (Bovensiepen, 2014). As a result of this complex history, Timor-Leste's culture has evolved to become a "unique personality" (Currie, 2018, p. 451). One characteristic of this is a strong history of resilience, with some researchers suggesting that this resilience is seen as a defining aspect of Timorese identity (Currie, 2018; Pannell, 2011).

The food of Timor-Leste

In a fashion similar to that of the socio-political history of Timor-Leste, the agricultural origins of food in the nation has also evolved over time, reflecting various exogenous influences that have now become embedded in Timorese food culture. Up until the time of European contact, the main sources of food in the Timorese diet were from hunting, fishing and gathering. As a result of European contact in the sixteenth century, watermelon, papaya, chilli, potato, tomato, egg-plant, cabbage, sweet potato and peanuts were introduced (Erskine et al., 2016). However, it was three crops originating in the Americas and brought by the Europeans that completely changed the agriculture and foodscape of Timor. These crops were maize, pumpkin/squash and cassava, which, along with mung beans, now comprise the staples of the Timorese diet (Fox, 2003).

Maize is by far the most important food crop for the Timorese. It is believed that maize was introduced by the Portuguese. Once introduced, it quickly spread throughout the island to become the mainstay of the Timorese diet. Similarly, pumpkin/squash quickly grew to become an invaluable food source both during times of abundance and in times of scarcity. Cassava on the other hand was hardly grown in Timor until the 20th century, and in fact, there was not even a Timorese word for the plant. It was with the Japanese occupation during World War II that cassava production increased, as the Japanese forced the Timorese to plant it. Once introduced, however, cassava quickly became one of the island's most important food crops, particularly during periods of drought (Fox, 2003).

Rice is believed to be one of the earliest food crops cultivated in Timor-Leste. It wasn't until the latter half of the 20th century, however, that rice became the focus for further Timorese agricultural transformation. This came as a result of the Portuguese introducing several varieties of high-yielding rice that they had developed. These new developments in rice agriculture were still in the initial stages when the Indonesians moved in to occupy East Timor in the 1970s. At that time, Indonesia was involved in a massive national programme of rice intensification, and these rice development policies were extended to include East Timor (Fox, 2003). Rice is now a staple of the Timorese diet, though most rice currently consumed in Timor-Leste comprises subsidised imports from Vietnam and Cambodia (Desmond, 2015).

Coffee was another introduced crop, initially brought to East Timor by the Dutch in the mid-eighteenth century. However, it was not until the 1860s that the development of coffee became a priority for the Portuguese (Khamis, 2015). The success of coffee as an export commodity during this time led to a dual economy of "rich crops" (introduced profitable plantation crops like coffee) and "poor crops" (pre-existing subsistence crops; Erskine et al., 2016; Shepherd & Palmer, 2015), which persists to the present day. Although suffering significant setbacks during the Indonesian occupation in particular, coffee now accounts for 80–90 percent of Timor-Leste's non-oil exports. Although comprising a mere 0.2 percent of the global coffee supply, along with the acknowledged quality of its taste, Timor-Leste coffee has one comparative advantage – it is the largest source of single-origin organic coffee in the world (Khamis, 2015).

Subsistence farming continues to be an important source of food for the approximately 70 percent of Timorese who live in rural areas (Liur, 2018), along with foraging for wild foods (Erskine et al., 2015). Foraged foods, in particular, played an important role during the Indonesian occupation. For many of the Timorese resistance fighters who lived for years in the jungle, these foraged "hero foods" contributed to their survival (Szanto, 2017, p. 3). Pannell (2011) suggests that it was this diverse and flexible nature of being able to use traditional food gathering practices that helped facilitate the forest-based resistance to Indonesian forces. Although wild foods are still an important part of the Timorese diet and continue to be consumed as a complement to cultivated foods (Erskine et al., 2015), these once highly regarded "heroic elements" (Pannell, 2011, p. 223) of survival through agricultural diversity are under threat from the nation-building activities of the newly democratic Timor-Leste (Pannell, 2011).

Despite the endurance of local products in Timorese cuisine, contemporary Timor-Leste food reflects its conflict-filled past. Dietler (2007) has suggested that there is an intimate link between foodways and the embodiment of identity, and between commensality and politics. As a result of this, he considers the domain of food an important arena for addressing the struggles of colonisation, appropriation and resistance. In essence, Dietler is suggesting that contemporary foodways and identities are in part the product of colonial encounters and that, concomitantly, food is a medium for the enactment of colonisation. As Desmond (2015, para. 15) has suggested, "[t]he Timorese have been too busy warding off starvation to develop a sophisticated national cuisine" or as Pannell (2011, p. 230) suggests, "[f]or local people, the entire period of the Indonesian occupation is associated with such events and emotions [violence and terror], mediated by their experience of 'running into the forest' and 'hiding in the mountains'".

Zajac (2014, paras. 18–19) further reflects on the link between the food of Timor-Leste and its past:

> What can be said about the food of Timor Leste can be said of cultural identity in Timor Leste. Contemporary cuisine is a blend of European, Asian and modern Western flavours and recipes. Historical and political influencers have left their mark; a fusion of religious beliefs, gender roles and relations, and ways of celebrating and mourning … Every way you look at it, food can be linked to something else about Timor Leste; its climate, maternal health and how to turn a profit at the market; its infrastructure, trade or relationship with Indonesia.

We argue that in addition to the linkages between food to "something else about Timor Leste" suggested by Zajac (2014), above, another potential link (and source of opportunities) is between food and tourism as a tool for sustainable development and the achievement of the SDGs.

Sustainable tourism futures in Timor-Leste

As discussed above, tourism in Timor-Leste is viewed as a key development opportunity to address the heavy reliance on oil and gas production as the country's main economic driver. Although Timor-Leste has many attributes that bode favourably for it as a tourism destination, its history of occupation, civil unrest, limited infrastructure and the ensuing negative reputation of being an unsettled and dangerous destination to visit has inhibited any significant tourism development to date (Tolkach, 2017).

Tourism is a natural source of revenue for Timor-Leste due to its biodiverse marine ecosystems, unique flora and fauna, rich and diverse traditional cultures, and unique history of resistance against Portuguese colonisation and Indonesian occupation. Over the course of six years, tourism has increased its contribution to the country's economy from USD $14.6 million in 2014 (The Asia Foundation, 2014) to USD $23.2 million in 2019 (The Asia Foundation, 2020a). The tourism sector can diversify the economy by growing markets for local agriculture (especially coffee, a main non-oil export), providing employment, reducing regional economic imbalances, and increasing foreign exchange inflows. However, Timor-Leste's tourism sector is at a pivotal moment where, while some progress has been made, new challenges lie on the horizon. Progress to date has included the launch of the official Timor-Leste tourism website and tourism brand *Explore the Undiscovered*; extensive discourse on sustainable and inclusive tourism; creation of industry associations among hoteliers, tour operators and the marine industry; and development and approval of a national tourism policy *Growing Tourism to 2030 – Creating a Sense of National Identity* (Government of Timor-Leste, 2017c).

Since 2014, The Asia Foundation in partnership with the country's Ministry of Tourism has collected survey data annually in the departure lounge of Dili International Airport. Data from the most recent survey in 2019, found that the largest market of leisure travellers to Timor-Leste was Australia (40%), with secondary and tertiary markets being derived from Indonesia (10%) and the USA (8%). Research suggests that a large percentage of visitors entering the country with Class 1 Tourist Visas are in fact visiting Timor-Leste for business purposes. More than half of the tourists to Timor-Leste are visiting for work, meetings, or business-related activities. The survey outcomes suggest that only 30 percent of those issued with Class 1 Tourist Visas visited Timor-Leste on holiday (The Asia Foundation, 2020a).

Beyond the natural geographical source markets of Australia and Indonesia, there are several niche tourism markets within which Timor-Leste has a relative competitive advantage. Niche tourism markets are specific segments, usually developed around a well-defined product, that can be tailored to meet the interests and needs of different tourists. In Timor-Leste these niche tourism markets were proposed as pertaining to dive tourism, cruise tourism, ecotourism, and food and coffee tourism (The Asia Foundation, 2018), and have been reaffirmed by visitor data which indicated that the most popular leisure activities among travellers were diving and snorkelling (69%), beach activities (51%), visiting historical sites (39%), cultural activities (32%) and hiking (27%; The Asia Foundation, 2020a).

While, as in other countries, the aim of tourism development in Timor-Leste is to increase its contribution to economic development and growth, if conducted in a sustainable and inclusive manner, tourism also has the potential to benefit diverse stakeholders and to positively address a broad set of societal issues. This includes tourism's impacts on gender equality, social-cultural issues and the environment, given that sustainable development requires resilient communities and a healthy environment. Fifty-four percent of people employed in the global tourism industry are women, higher than the proportion of women employed in other sectors of the economy (39%; World Bank, 2019). However, women in the global tourism industry earn 15 percent less than men (United Nations World Tourism Organization, 2019). In Timor-Leste women comprise 62 percent of the accommodation and food service workforce; however, the average annual wage of this sector is USD $2,100 less than the average income of other sectors (Government of Timor-Leste, 2017e).Through Sustainable Development Goal number 5, it is possible that "tourism can empower women, particularly through the provision of direct jobs and income generation in tourism and hospitality-related enterprises" (UNWTO, 2019, p. 172). While Timor-Leste's tourism policy aims to achieve widespread economic and social growth, tourism development in the country will have neither sustainable nor inclusive impacts if gender-specific constraints to women's participation in the sector are not addressed. Timor-Leste's Maubisse Declaration (Government of Timor-Leste, 2017d) articulates a whole-of-government approach to increasing rural women's participation in economic and productive sectors in tourism, commerce and industry, through access to markets, financial management and opportunities for developing their own businesses (Government of Timor-Leste, 2017a). Such institutional initiatives are necessary, especially in a country where patriarchal social norms hinder women's participation in society and ability to be empowered by tourism. In Timor-Leste women face more difficulties than men in accessing economic resources, such as land or contracts. Women were estimated to own 40 percent of microenterprises and around 16 percent of formalised businesses in 2010 (State Secretary for the Support & Promotion of Women in the Social Economy, 2017).

In the socio-cultural realm, while tourism can arguably contribute to peace and greater understanding between communities and improved standards of living, it can also potentially be a force for rapid and undesirable social change in host communities. In terms of the environment, while tourism may incentivise the preservation of 'unspoiled' natural attractions, uncontrolled or badly managed tourism development can threaten natural areas, including depleting natural resources such as water, energy, and food (which may already be limited in host communities), and increasing physical impacts such as the degradation of natural ecosystems.

Sustainable and inclusive tourism development has the potential to benefit neglected actors and aspects of the economy, while positively addressing a broad set of societal issues. Niche tourism is not only in demand (The Asia Foundation, 2020a) but can be developed in a focused manner allowing for high value experiences by tourists. One such niche that draws on Timor-Leste's unique history, environmental biodiversity and can contribute to empowerment is that of food tourism.

Food and tourism in Timor-Leste

Currie (2015) suggested that Timorese history comprises a number of "stories" (p. 282) that are important and unique attributes of the country, which could potentially be promoted to tourists. This includes not just the history of events, but also the strength and resilience of the people who survived this history as told through their stories (Currie, 2015). Food can play an important role in this respect as food, beyond being a biological necessity, "is inseparable from powerful material and symbolic realities" (Abarca & Colby, 2016, p. 2). As discussed previously, place is a product of the intersection of social relations, expressions of identity and the practice of culture (Appadurai, 1996). Similarly, the connection between food and people is not static; it is dynamic and continuously changing (Abarca & Colby, 2016). Food and its cultural expression (cuisine) are a function of place and time. In this way, the connections between food, history and memory "produce narratives of personal and collective cultural and social identity ... by which people's emotional, psychological, social, economic, political, historical, and cultural realities are embodied social and cultural realities" (Abarca & Colby, 2016, p. 4). It is not our food that defines social and cultural subjectivities, rather it is the narratives, the food stories that we tell (Abarca & Colby, 2016). A key activity within tourism that can create synergies between culture, history and nature is an exploration of food, place and people, towards eliciting the "story" of the food of Timor-Leste (Berno, 2019).

As Szanto (2017, p. 2) asks in relation to food in Timor-Leste, "when dealing with the incredible layerings of food, place, history, society, development, migration, biodiversity, emotions, war, cuisine, markets, oral narrative, elections, currencies, and dozens of other agencies, what stories are we really entitled to tell, other than our own?". We sought to explore how food stories and experiential culinary tourism can contribute to the development of a positive narrative of resilience in contemporary Timor-Leste, one that not only champions gender equality and economic opportunities, but one that can also facilitate the creation of shared futures in an otherwise divided post-conflict society.

Methods

The research methodology comprised of a qualitative approach, utilising both focus group discussions and semi-structured interviews. The overarching research question was "What is the potential for food as a tourism attribute and attraction in Timor-Leste?" Through the interviews and focus groups, the following sub-questions were explored: what is Timorese cuisine and what are its origins; what are the food stories of Timor-Leste; and, how does food and its stories manifest in tourism in Timor-Leste?

Research was conducted in the sub-districts of Maubisse Villa and Fatubessi in municipality Ainaro, and in the nation's capital, Dili. Maubisse Villa was chosen as a fieldwork site as the Maubisse-Aileu Region has been identified as a regional hub for destination tourism development (including prospective sites for food and coffee tourism).

Fatubessi was chosen due to its historic significance during the 1975 – 1999 Indonesian occupation. Fatubessi is home to an intricate cave system that was used as a place of shelter by FALINTIL (Forças de Defesa de Timor Leste) combatants and resident communities. Given this, a focus group discussion was held with the Fatubessi community with the objective of understanding linkages between food and history during this time. Fifteen people participated in the focus group discussions, comprising of community members and local administrators. Participants were asked to consider the history and origins of Timorese food, if and how it has changed over time and whether they believed that Timorese food could be highlighted as part of the tourist experience.

Seven semi-structured interviews were also held in the sub-district of Maubisse Villa in municipality Ainaro and an additional seven in municipality Dili. These interviews focused on restaurant owners and managers and their decisions about what they offered on their menus and their views on food tourism in Timor-Leste. Interviewees in Maubisse Villa had participated in culinary training conducted in 2019 on how to use local produce in their cooking, with an aim to develop new dishes representative of time and place (Timor-Leste Food Innovators Exchange, 2019). To understand the barriers and facilitators to food tourism in the region, questions about the level of uptake and continued use of this training were also asked. In addition to this, two of the seven interviews in Maubisse Villa were with military veterans with a focus on whether survival foods during the resistance had evolved or been maintained over time.

Given the extensive number of restaurants in Dili, a sample of seven restaurants was selected representing those that offered Timorese, Indonesian or Portuguese cuisines. Interviewees from these restaurants were either owners or managers. Menus were collected from each restaurant and content-analysed alongside information collected from the semi-structured interviews.

Results

What is Timorese cuisine?

The focus groups and semi-structured interviews revealed that "Timorese cuisine" is multi-faceted with several layers of meanings. Although most focus group participants and interviewees talked about "local foods", this did not necessarily refer to Timorese cuisine. Rather, "local foods" referred to the use of fresh local products and/or produce, which are then used in Indonesian, Portuguese, Timorese or other cuisines.

Both traditional Timorese and Portuguese dishes are considered to be "Timorese cuisine". Although Indonesian food is widely available and commonly consumed, the participants did not consider it to be Timorese cuisine. Traditional Timorese cuisine – dishes such as batar da'an (a squash, bean and maize stew), dried cassava, etc. – tended to be viewed by participants as food that should be consumed at home or in the village. Although there were some notable exceptions in Dili, there was an absence of traditional dishes in public spaces like restaurants, hotels and guesthouses. Participants reported that traditional Timorese dishes are common in homes, but they are not eaten on a daily basis. This, they reported, was due to a lack of supply of traditional ingredients (many are region-specific and not readily available in the markets, are seasonal or need to be foraged) and the lengthy preparation times for many of the dishes. These "wild foods" (Castro, 2013, p. 22) are considered to be inconvenient in comparison to foods such as rice and noodles. This is a theme that recurred in interviews with restaurant owners, discussed further below. A similar classification to what was described as traditional dishes are those that are considered "ceremony foods". These are traditional Timorese dishes that are used for traditional ceremonial and ritual purposes; for example, tukir (meat cooked in bamboo tubes), hakmerik rice (rice, cassava leaves and candlenut) and etu sedok (a rice and bean dish). These foods are also ones that tend to be eaten in private spaces such as homes and villages.

Participants explained that knowledge about these traditional dishes, particularly preparation methods, is taught intergenerationally through oral tradition, and very few of them are recorded in the form of recipes or cookbooks. This transmission of knowledge for traditional and ceremonial foods is particularly important as many of these traditional dishes utilise toxic or highly bitter ingredients ("wild foods") that require careful and time-consuming preparation to render them safe and palatable. For example, koto moruk or 'poison beans' must be boiled 12 times, with the water changed in between each boiling.

As part of the fieldwork undertaken in Maubisse, two veterans of the resistance were interviewed about "survival foods" – the foods that they were able to forage seasonally and eat for survival during the Indonesian occupation. Both veterans spoke of eating wild leaves, bitter beans, foraged tubers and fruits, and animal protein (only on occasions when they were able to hunt). These were the foods that they consumed from 1975 to 2000 out of necessity for survival. The veterans said that very few of these "survival foods" were served anymore. Some of the ingredients are currently consumed – for example, koto-tisi leaves and beans, maek (elephant foot yam) and sago – but they are prepared in more contemporary and innovative ways.

Other foods of the occupation era are no longer consumed as they are hard to find (they must be foraged) and there is a lack of knowledge about them. The veterans also said that they chose not to share these survival foods with their families and children for two reasons: (1) no one asks about the survival foods anymore; and (2) they do not want their children to suffer as these foods are viewed as foods of hard times. These findings are similar to those of Castro (2013, p. 22) who found that

> Wild foods are not being used on a daily basis as they are believed to be foods for bad times: famine, extremely bad economic situations or war. Their consumption is undesirable. Nonetheless, we witnessed the selling of ready-to-eat (midar) koto-moruk in Atsabe market during our fieldwork stay in Malabe. Wild foods and other famine foods that require a laborious preparation process which make them inconvenient, may be shifting from a marginal presence in people's diet to a market product.

These data suggest that traditional Timorese cuisine continues to be relevant in contemporary society, but that with a few exceptions (discussed below), it is relatively unseen in tourism and hospitality contexts.

A further classification of Timorese food that participants talked about was "celebration foods". These comprise mainly Portuguese dishes that are made with local ingredients and are present at occasions such as parties and festivities. Portuguese food is also common on restaurant menus and in shops in Timor-Leste. When asked about whether Portuguese food is considered to be Timorese food, participants indicated that yes, it is. Indonesian food, particularly warung-style dishes is also very common on menus in Timor-Leste. Although Indonesian food is not classified as Timorese cuisine, all of the guest houses included in the research offered it to their guests. When asked about the predominance of these Indonesian dishes on their menus, guest-house owners said that is what they believed their guests preferred. They also emphasised the availability of the ingredients locally and that, unlike inconvenient traditional dishes, Indonesian cuisine is convenient.

Telling the story

Interviews with seven restaurant owners or managers in Dili confirmed many of the findings from the fieldwork undertaken in Maubisse and Fatubessi. Several consistent themes emerged from these interviews. Three of the restaurants served traditional Timorese cuisine; the others served Indonesian and/or Portuguese food. Regardless of what style cuisine they offered, all but one of the owner/managers agreed that Timorese food had the potential to be a tourism attraction.

A consistent theme for the restaurants serving Indonesian or Portuguese food was that they served "local food" (meaning dishes using local products and produce), but despite most seeing value in doing so, they were reluctant to serve traditional Timorese food. Echoing the concerns of guest house owners in Maubisse and Fatubessi, these restaurateurs spoke of customer preference for Indonesian or Portuguese foods, the higher cost of and lack of access to traditional ingredients and the significantly longer time that it took to prepare them, particularly bean dishes. A significant theme around knowledge and food stories emerged about serving traditional Timorese cuisine. These four restauranteurs from the Indonesian and Portuguese restaurants all spoke of a lack of knowledge of traditional cooking methods and of the need to be able to tell the "food story" of traditional Timorese dishes if they were to serve them – something they did not believe that they had the skills and knowledge to do. The owner of a Portuguese restaurant indicated that although he did not know the stories of Timorese food himself, he recognised the relationship between food, place and people when he said,

> *most of the food available here in Timor has gone through the evolution from Indonesian occupation and also Portuguese time. I believe that there is an interesting story about Timor-Leste that is a link to food during the resistance, but Timorese people need to know and learn more on their traditional food.*

There is only a small number of restaurants and catering operations in Dili that serve Timorese cuisine. Three of the restaurants included in the research have as their primary focus traditional Timorese cuisine. All of these restaurateurs saw potential for traditional Timorese cuisine to be a tourism attraction, with one of them saying that there was *"a huge potential [for]developing food as a tourism product."* Prior to the COVID-19-related decline in tourist numbers, these restaurants were known as establishments where tourists could sample Timorese cuisine. Owners/managers of these Timorese restaurants emphasised the use of fresh, local products with estimates being that 90 to 99 percent of the products they used were local. All three restaurants drew on traditional knowledge about Timorese ingredients and cuisine in planning their menus, often getting information from extended family and/or their restaurant staff. One restaurant in particular encouraged staff to be creative and come up with innovative ways to present traditional Timorese foods and to tell the stories of these new dishes. Being able to tell the "food story" is something that all three of the restaurants emphasised, particularly in relation to the links to Timorese history:

> *The important thing is to know the story of each food being served ... the evolution also took part in the Timorese history, so when we serve the food, the food team always explain the food by introducing the food and the ingredient and the link to Timor history.* (Owner of a restaurant serving Timorese cuisine)

Discussion

As a least-developed country and a small island nation highly dependent on oil to drive its economy, Timor-Leste faces many challenges. Tourism development has been identified as one way to diversify the economy, while creating benefits for a diverse group of stakeholders and positively addressing a broad set of societal issues. This includes tourism's impacts on gender equality, social-cultural issues, and the environment. It has only been relatively recently that attention has been paid to the relationship between food production systems, culinary traditions, tourist consumption and sustainability. These relationships, however, can be critical to sustainability (Berno, 2020; Berno et al., 2014). This research, therefore, sought to explore how the food of Timor-Leste as experienced through tourism can contribute to the development of a positive narrative of resilience in contemporary Timor-Leste.

Food and foodways reflect the rich and complex histories of the regions and countries in which they are situated. The "food stories" of Timor-Leste tell the tale of colonisation, invasion, struggle, conflict and occupation. They also tell a story of resilience, and with one restaurant actively exploring food innovation within Timorese culinary traditions, there is also evidence of 'new nation identity building' through cuisine. But what do these stories mean for tourism?

The potential for Timorese cuisine in tourism

The majority of participants in the research believed that Timorese food could be part of Timor-Leste's tourism offerings. This use of food in tourism articulates well with the "culture" and "exploration" aspects of the current tourism marketing strategy. It can also potentially contribute to addressing broader development goals, including the SDGs, by creating opportunities by growing markets for local agricultural products, increasing retention of the tourist dollar through reducing food-related economic leakage and providing direct jobs and income generation for women in food-related enterprises, particularly in rural areas such as Maubisse and Fatubessi.

Although there has been the recent emergence of restaurants featuring traditional Timorese cuisine, this has been centralised in Dili where there is a higher concentration of both tourists and expatriates. There have been training programmes offered for restaurants in municipalities on how and why to serve traditional Timorese cuisine to travellers (see, for example, Timor-Leste Food Innovators Exchange, 2019), but uptake has been limited with only 20 percent of restaurants using the training to create new menu dishes in the longer term (The Asia Foundation, 2020b). Interestingly the same proportion (20%) of restaurants that had received training said that they were able to explain the ingredients and describe the food items (The Asia Foundation, 2020b). The data suggest that there are two key issues for operators not currently serving Timorese cuisine: (1) having the knowledge and confidence to prepare the food appropriately; and (2) the ability to tell the "food story". This suggests that technical cookery training alone may not be enough to support operators in offering Timorese cuisine. In addition to teaching culinary skills, there is also a need to intertwine technical skills and product development with experiential components of Timorese cuisine. This goes beyond just being able to describe a dish, and speaks to the articulation of social relations, expressions of identity and the practice of culture as it is manifested through food and place (Appadurai, 1996) and told through "food stories".

The centralisation of tourism and Timorese restaurants in Dili suggests that a partnership approach to food tourism may be required, particularly in relation to food tourism in rural regions such as Maubisse. Demand generation will need to be done in Dili, as it is the international gateway and conduit to Timor-Leste for tourists. Dili-based purveyors of traditional Timorese foods have greater foot traffic and opportunities to refine their product/experience. This is something that can be shared with rural partners as part of an enhanced skills and knowledge training, as discussed above.

Despite the affirmation of the potential role of Timorese cuisine in tourism, not all operators were prepared to feature Timorese dishes on their menus. Several guesthouse owners and restauranteurs were reluctant to put these traditional dishes on their menus, as it was their perception that there was limited demand for them. This demand factor is relatively unknown, however. In Timor-Leste, tourist expenditure on food and beverage is assessed, but specifically what type of food and beverage is purchased and consumed is not. Although there is anecdotal evidence from owners and managers of Timorese restaurants that there is tourist demand for their cuisine, data on what tourists want to eat and why is needed.

Who can tell the "food stories"?

There exists a strong link between history and tourism in Timor-Leste. The *Timor-Leste Strategic Development Plan (2011–2030)* (Government of Timor-Leste, 2010) identifies historical and cultural tourism, inclusive of the history of the resistance against the Indonesian occupation, as having potential for tourism. This potential for a historic tourism process to preserve and promote (from a tourism perspective) the history associated with the Indonesian occupation has also been suggested by the establishment of the Timorese Resistance Archive and Museum and the more recent Centro Nacional Chega! Indeed, stories of the resistance and its heroes are so embedded in the tourism product that the Boneca de Ataúro (a women's cooperative that produces embroidered dolls, bags, education toys and other sewn items on Atauro Island) makes and sells camouflage clad resistance leader dolls as tourist souvenirs. The data, however, suggest that this ability to 'package and sell' stories of the resistance does not necessarily extend to its food. The reluctance of the veterans to share the food and food stories of the occupation suggests that there remains unaddressed trauma associated with these memories – especially when the stories linking food to the resistance are being told by veterans. This raises important considerations about who is entitled to tell the story of war/food/resistance, wherein those that can, perhaps cannot, unless a reconciliatory process is first followed. Szanto (2017, pp. 5–6) questioned what food stories we are entitled to tell other than our own and recognised,

> the important role that storytelling has in preserving and enacting culture, in claiming power back from the 'official' writers of history, in owning one's identity, in teaching and learning, and in celebrating the intensely heterogeneous experience of making, sharing, and eating food.

Only the people of Timor-Leste can tell their food stories, and it is their domain as to what and with whom they share.

Tourism in Timor-Leste can simultaneously facilitate economic growth and create synergies between culture, history and nature. While the country's history has the potential to be developed into a niche tourism product in of itself, the people who bear the experience relating to this time continue to progress through a process of post-conflict healing and reconciliation. Food, as a product, transcends time and place. By exploring Timorese cuisine and the stories that it holds, there exists an opportunity to discover its unique identity. The research in this paper emphasises the need for an inclusive process of developing Timorese food not as a product, but as a form of experiential culinary tourism. It will be a process that, through its co-creation, can facilitate the development of shared inter-generational futures and, in doing so, develop a niche tourism product for Timor-Leste that extends from the fragility of its history through to the resilience of its people.

Note

1. Szanto, 2017, p. 3.

Disclosure statement

No potential conflict of interest was reported by the authors.

Funding

This work was supported by the New Zealand Ministry for Foreign Affairs and Trade under Grant WPG-0101849 to The Asia Foundation's Tourism Development Program in Timor-Leste.

ORCID

Tracy Berno http://orcid.org/0000-0003-1385-8332
Gobie Rajalingam http://orcid.org/0000-0003-0533-8652

References

Abarca, M., & Colby, J. (2016). Food memories seasoning the narratives of our lives. *Food and Foodways, 24*(1–2), 1–8. https://doi.org/10.1080/07409710.2016.1150101

Appadurai, A. (1996). *Modernity al large: Cultural dimensions of globalization*. University of Minnesota Press.

Berno, T., & Fountain, J. (2020). 'From the land': Identifying and communicating Selwyn's food story. In *CAUTHE 2020: 20: 20 vision: New perspectives on the diversity of hospitality, tourism and events*. Auckland University of Technology. https://www.cauthe2020.org/

Berno, T. (2019). *A review of The Asia Foundation's Food and Coffee Tourism Development Plan for Timor-Leste*. Asia Foundation.

Berno, T. (2020). Linking food biodiversity and food traditions to food tourism in Small Island Developing States (SIDS). In D. Hunter, T. Borelli, & E. Gee (Eds.), *Biodiversity, food and nutrition: A new agenda for sustainable food systems* (pp. 236–254). Earthscan.

Berno, T., Devlin, N., Ezaki, A., Wilson, D., & Wolf, E. (2014). Sustainability in food and drink tourism. In E. Wolf & W. Lange-Faria (Eds.), *Have fork will travel: Handbook for food tourism* (pp. 309–322). World Food Travel Association.

Berno, T., Dodd, D., & Wisansing, J. (2019). *The development of ASEAN gastronomy network and region of gastronomy master plan*. Ministry of Sport and Tourism, Thailand.

Blanchard, L. A., & Higgins-Desbiolles, F. (2013). Introduction: Peace matters, tourism matters. In L. A. Blanchard & F. Higgins-Desbiolles (Eds.), *Peace through Tourism: Promoting Human Security through International Citizenship*, (pp. 21–36). London: Routledge.

Bovensiepen, J. (2011). Opening and closing the land: Land and power in the Idaté highlands. In A. McWilliam & E. Traube (Eds.), *Land and life in Timor-Leste: Ethnographic essays* (pp. 47–60). ANU Press.

Bovensiepen, J. (2014). Lulik: Taboo, animism, or aransgressive aacred? An exploration of identity, morality, and power in Timor-Leste. *Oceania*, *84*(2), 121–137. https://doi.org/10.1002/ocea.5049

Cabasset-Semedo, C. (2009). Thinking about tourism in Timor-Leste in the era of sustainable development: A tourism policy emerging from grass-roots levels. In C. Cabasset-Semedo & F. Durand (Eds.), *East Timor: How to build a new nation in Southeast Asia in the 21st century* (pp. 213–232). IRASEC.

Castro, A. (2013). *An approach to the food habits of three communities in Timor-Leste* [Final Report]. CARE International.

Courvisanos, J. (2018). *Implementing the sustainable development goals in Timor-Leste: Recommendation to the New VIII Constitutional Government of the Democratic Republic of Timor-Leste*. Timor-Leste Institute for Development Monitoring and Analysis. https://www.laohamutuk.org/misc/TLSA2017/CourvisanosMar2018en.pdf

Currie, S. (2015). *Considerations for marketing Timor-Leste as a tourism destination*. Ministry of Tourism. https://www.timorleste.tl/documents/considerations-for-marketing-timor-leste-as-a-tourist-destination/

Currie, S. (2018). Beyond a 3s approach to marketing Island Nations? Destination marketing and experiences from Timor-Leste. *The Contemporary Pacific*, *30*(2), 438–459. https://doi.org/10.1353/cp.2018.0034

Dayley, R. (2019). *Southeast Asia in the new international era*. Routledge.

Desmond, P. (2015, March 5). Fish on a stick [Blog post]. *Dili dally*. https://patnpiptimor.wordpress.com/2015/03/05/fish-on-a-stick/

Dietler, M. (2007). 11. Culinary encounters: Food, identity, and colonialism. In *The archaeology of food and identity (Occasional Paper No. 34)*. Southern Illinois University.

Domínguez, G. (2014, August 20). Over-reliance on oil and gas threatens East Timor's economy. *DW*. https://www.dw.com/en/over-reliance-on-oil-and-gas-threatens-east-timors-economy/a-17868268

Erskine, W., Ximenes, A., Glazebrook, D., da Costa, M., Lopes, M., Spyckerelle, L., Williams, R., & Nesbitt, H. (2015). The role of wild foods in food security: The example of Timor-Leste. *Food Security*, *7*(1), 55–65. https://doi.org/10.1007/s12571-014-0406-9

Erskine, W., Ximenes, A., Glazebrook, D., da Costa, M., Lopes, M., Spyckerelle, L., Williams, R., & Nesbitt, H. (2016). Wild plant food in Timor-Leste. In *ACIAR Proceedings Series (No. 146)* (pp. 99–104). Australian Centre for International Agricultural Research.

Fox, J. J. (2003). Drawing from the past to prepare for the future: Responding to the challenges of food security in East Timor. In H. da Costa, C. Piggin, C. J. da Cruz, & J. J. Fox (Eds.), *Agriculture: New directions for a new nation – East Timor (Timor-Leste)* (pp. 105–114). Australian Centre for International Agricultural Research.

Gomes, D. (2013). *Internship report*. Haburas Foundation. https://www.academia.edu/5049147/Internship_report_Community_Based_Tourism_TimorLeste

Government of Timor-Leste. (2010). *Timor-Leste Strategic Development Plan (2011–2030)*. Democratic Republic of Timor-Leste.

Government of Timor-Leste. (2017a). *Business Activity Survey of Timor-Leste*. National Directorate of Economic and Social Statistics.

Government of Timor-Leste. (2017b, July 11). *Developing a brand for tourism to Timor-Leste. Presentation to Council of Ministers by the Ministry of Tourism, Arts and Culture, VI Constitutional Government*. Democratic Republic of Timor-Leste.

Government of Timor-Leste. (2017c). *Growing tourism to 2030 – creating a sense of national identity*. Democratic Republic of Timor-Leste.

Government of Timor-Leste. (2017d). *Maubisse declaration 2018–2023: Improving rural women's lives*. https://asiapacific.unwomen.org/-/media/field%20office%20eseasia/docs/publications/2019/08/tl-maubisse-declaration-2018-23-english-compressed.pdf?la=en&vs=5609

Government of Timor-Leste. (2017e). *Timor-Leste's roadmap for the implementation of the 2030 agenda and the SDGs*. Democratic Republic of Timor-Leste.

Government of Timor-Leste. (2019). *Report on the implementation of the Sustainable Development Goals: From ashes to reconciliation, reconstruction and sustainable development, voluntary national review of Timor-Leste*. Government of Timor-Leste.

Hill, H. (2001). Tiny, poor and war-torn: Development policy challenges for East Timor. *World Development, 29*(7), 1137–1156. https://doi.org/10.1016/S0305-750X(01)00035-3

Innes, H. (2017). *Essays in Canadian economic history*. University of Toronto Press.

Khamis, S. (2015). Timor-Leste coffee. *Food, Culture & Society, 18*(3), 481–500. https://doi.org/10.1080/15528014.2015.1043109

Liur, S. (2018, December 23). *Timor-Leste traditional food* [Blog post]. https://sophieintimor.wordpress.com/2018/12/23/timor-leste-traditional-food/

Marx, S. (2016, June 22). *Timor-Leste's non-oil economy must look to tourism*. Asia Foundation. https://asiafoundation.org/2016/06/22/timor-lestes-non-oil-economy-must-look-tourism/

McWilliam, A., Palmer, L., & Shepherd, C. (2014). Lulik encounters and cultural frictions in East Timor: Past and present. *The Australian Journal of Anthropology, 25*(3), 304–320. https://doi.org/10.1111/taja.12101

Ministry of Finance. (2010). *Highlights of the 2010 census – main results in Timor-Leste*. Ministry of Finance.

National Directorate of Statistics. (2019). *External trade statistics annual report 2019*. Government of Timor-Leste.

Palatino, M. (2014, November 21). East Timor's oil resource: Boon or bane? *The Diplomat*. https://thediplomat.com/2014/11/east-timors-oil-resource-boon-or-bane/

Pannell, S. (2011). Struggling geographies: Rethinking livelihood and locality in Timor-Leste. In A. McWilliam & E. Traube (Eds.), *Land and life in Timor-Leste* (pp. 217–239). ANU E-Press.

Reddy, M. V., Boyd, S. W., & Nica, M. (2020). Towards a post-conflict tourism recovery framework. *Annals of Tourism Research, 84*, 102940. https://doi.org/10.1016/j.annals.2020.102940.

Shepherd, C., & Palmer, L. (2015). The modern origins of traditional agriculture: Colonial policy, swidden development, and environmental degradation in eastern Timor. *Bijdragen Tot de Taal-, Land-en Volkenkunde/Journal of the Humanities and Social Sciences of Southeast Asia, 171*(2–3), 281–311.

Smith, A. (2004). East Timor: Elections in the world's newest nation. *Journal of Democracy, 15*(2), 145–159. https://doi.org/10.1353/jod.2004.0036

State Secretary for the Support and Promotion of Women in the Social Economy. (2017). *Strategy to improve women in economy 2015–2017*. Government of Timor-Leste.

Szanto, D. (2017). Bumping up against things in Dili, Timor-Leste. *The Eco-gastronomy Project*. https://www.unisg.it/assets/DILI_bumpingup.pdf

The Asia Foundation. (2014). *A survey of travelers to Timor-Leste 2014*. The Asia Foundation.

The Asia Foundation. (2018). *Timor-Leste tourism barometer*. The Asia Foundation.

The Asia Foundation. (2020a). *A survey of travelers to Timor-Leste 2019*. The Asia Foundation.

The Asia Foundation. (2020b, June 26). *Culinary uptake survey of Maubisse guesthouses that had received culinary training*. Unpublished manuscript.

The Timor-Leste Commission for Reception, Truth and Reconciliation. (2013). *Chega! The Final Report of the Timor-Leste Commission for Reception, Truth and Reconciliation (CAVR): Full Report*. KPG in Cooperation with STP-CAVR.

Timor-Leste Food Innovators Exchange. (2019). *Timor-Leste Food Innovators Exchange (TLFIX) and The Asia Foundation culinary training pilot with Maubisse guesthouses* [Report]. Unpublished manuscript.

Tolkach, D. (2017). History and politics of tourism in Timor-Leste. In R. Butler & S. Wantanee (Eds.), *Tourism and political change* (pp. 103–120). Goodfellow Publishers.

Tolkach, D., & King, B. (2015). Strengthening community-based tourism in a new resource-based island nation: Why and how? *Tourism Management, 48*, 386–398. https://doi.org/10.1016/j.tourman.2014.12.013

Trindade, J. (2011). *Lulik: The core of Timorese values*. Paper presented at Communicating New Research on Timor-Leste. The 3rd Timor-Leste Study Association (TLSA) Conference, Dili, Timo-Leste. https://www.academia.edu/7450617/Lulik_The_Core_of_Timorese_Values

United Nations Development Program. (2018). *Human Development Report 2019 Inequalities in Human Development in the 21st Century Briefing note for countries on the 2019 Human Development Report*. UNDP.

United Nations World Tourism Organization. (2019). *Global report on women and tourism* (2nd ed.). United Nations World Tourism Organization.

United Nations. (2011). *Timor-Leste (colour)*. http://www.un.org/Depts/Cartographic/map/profile/timor.pdf

Weaver, D. (2018). Creative periphery syndrome? Opportunities for sustainable tourism innovation in Timor-Leste, an early stage destination. *Tourism Recreation Research, 43*(1), 118–128. https://doi.org/10.1080/02508281.2017.1397838

World Bank. (2019). *World Bank open data*. World Bank.

World Bank. (2020). *Timor-Leste overview*. https://www.worldbank.org/en/country/timor-leste/overview

Zajac, H. (2014). The tales of food in Timor-Leste. *Right Now: Human Rights in Australia*. http://rightnow.org.au/opinion-3/the-tales-of-food-in-timor-leste/

OPEN ACCESS

Living in the *Wake* of Rural Irish Troubles: building an institution for sustainable peace through emotive out-of-place tourism

John Erwin and Tristan Sturm

ABSTRACT

During the thirty-year period (1968–1998) known as the Troubles in Northern Ireland, 3500 people died and thousands more suffered physical disabilities and psychological trauma. Belfast, among other conflict cities, helped inspire the term 'dark tourism' in 1989. The country continues to be in conflict but is officially in a period of peace. Northern Ireland has been the theme for much peace and reconciliation research, but literature to date concentrates on polarised discourses rather than rural representation. Toward meeting and expanding the UN SDG16 peace goals, this research explores how what we term emotive peace tourism can be used as a methodology to affect emotional reconciliation registers in a unique rural Northern Ireland visitor experience. Bringing domestic tourists from Catholic and Protestant communities into face-to-face contact through a liminal participative 'out-of-place' visitor experience, we choreographed and performed a series of "Troubles" events: a guided night walk through a checkpoint, an IRA Wake, a UDA Funeral, and a Mixed-Marriage. Contributing to the debate as to whether tourism is more of a 'peacekeeper' than a 'peacemaker', our research demonstrates that an *in situ* liminal, emotive peace tourism experience, can generate sustainable tolerance, respect, trust, sympathy, and empathy towards others in post-conflict Northern Ireland and likely elsewhere.

Introduction

After decades of felt structural and cultural discrimination, Catholics in Northern Ireland by the late 1960s began organizing 'civil rights protests [which] soon escalated into serious unrest with the aim of nationalist (mainly Catholic) paramilitaries to see the border removed and a united Ireland' (Royle, 2012, p. 96). A 1972 civil rights march in Derry/Londonderry resulted in the British Army killing 13 marchers. This incident is known as Bloody Sunday and advanced support for the Irish Republican Army (IRA) which led to an intense 30-year war known as the Troubles. The IRA campaign involved bombings of commercial, civilian, and security forces. Loyalist paramilitaries responded with their own attacks on the IRA and civilians. Multi-party peace talks assisted by the USA, Britain and the Republic of Ireland took place in 1996 and culminated in the Good Friday Agreement of 1998. Although officially at peace, Northern Ireland still experiences elements of conflict between Protestants/Unionists/Loyalists who see themselves as British and look to Westminster while Catholics/Nationalists/Republicans identify themselves as Irish and direct their political gaze to Dublin.

Towards UN Sustainable Development Goal 16 (SDG16) to 'promote peaceful and inclusive societies for sustainable development, provide access to justice for all and build effective, accountable and inclusive institutions at all levels,' in this paper we put forth emotive peace tourism as an alternative to 'dark tourism' and 'affective tourism', as a lens to promote the specificity of sustainable peace and reconciliation within the socio-political troubles of rural Northern Ireland. Nuancing the important umbrella concept of 'affective tourism' which Buda (2015a, p. 3) defines as 'the ways in which affects, emotions, and feelings are accessed, felt, experienced and performed in encounters between touring bodies and places', emotive peace tourism captures the under-applied functional potential of using *emotions* to foster peace, as opposed to tourists chasing affects or emotions (cf. Buda, 2015b). We define emotive peace tourism as a methodology for producing sustainable peace by stimulating emotionally driven empathy for others in a relational experience between domestic tourist groups in a conflict or post-conflict place. Post-conflict tourism scholars of Northern Ireland have referred to dark tourism, political tourism, thanatourism, and politico-religious tourism to describe the emergent form of tourist interest in the Troubles. Poignant in so far as it may illicit emotions for the mediation of peace and yet problematic, the term dark tourism may also limit positive representations for sustainable peace and the inclusion of many of the SDG16 toolkit concerns including conflict resolution, means of implementation, and participation (cf. Boluk et al., 2019).

Using a performative, multi-sensory, participant-based ethnography methodology, we designed a sensual, emotive, participatory visitor experience known as 'Living in the *Wake* of Rural Irish Troubles' (LWRIT). The authors developed the experience to investigate the role of emotive peace tourism as a means of spreading sustainable peace, cross-community understanding and reconciliation in rural Northern Ireland. The experience involved cross-community 'tourists'—taking local residents 'out-of-place' and into a liminal tourist frame-of-mind and place—in a guided night walk that led to three emotive performances: (1) an IRA Wake, (2) UDA Funeral, and (3) Mixed-Marriage. The aim of this research is concerned with how this performative visitor experience can contribute to self-reflective and sustainable reconciliation towards the SDG16 targets. We ask, can a participative experience contribute to reconciliation and peace? And does taking cross-cultural locals as tourists to a liminal face-to-face contact zone provide an opportunity for cultural exchange, thereby tempering the fear of others?

Conceptual background

Emotions

Emotion has its origin in the Latin word *emovere* which means 'to move out or move away' and had an early association with 'physical movement and migration' (Buda, 2015a, p. 24). Ahmed (2004, p. 11) argues that emotions are not just about movements but also involve attachments, connections, and divisions which 'connects bodies to other bodies… attachments take place through movement.' These characteristics of emotions which have been called 'the adventure of the self' are concomitant with the tourist experience (Chard, 2002, p. 188). They are therefore a powerful position from which to examine the emotive implications of tourism for peace. Building on the work of geographers and anthropologists, Buda (2015a) applied an emotional/affectual approach to tourism in areas of conflict. Robinson (2016, p. 22) argues that 'the emotions are necessary, and in many ways inevitable, for engaging with the world. The tourist is part of the world, not detached from it, and is engaged in processes of ordering knowledge of the world'.

Tourism offers a vehicle for physical motion and emotional subjectivities to be experienced by a tourist. Recent research by Picard (2016, p. 2) notes that tourism 'offers itself as a field *par excellence* for the study of the articulation between personal subjective experience of the world and collective emotional and cognitive cultures'. Psychologists recognise that emotions are generated by external and internal stimuli. External stimulus in the tourism industry is produced by interactions with physical and cultural attractions as well as local people and Picard (2016, p. 5) suggests that tourists recognise 'hidden facets of their lives in the visited other'. Inner stimulus can procure deep emotions through personal memories of previous experiences, perhaps from childhood or youth. As we explore below, this is poignant in relation to older tourists who have experienced the Troubles.

Although research on tourists' emotions is developing, there are many gaps in the literature. Scholars acknowledge that research on emotions in tourism have focused on consumer behaviour (Bigne & Andreu, 2004) or affective neuroscience (Pearce, 2012). In contrast to this focus Cohen and Cohen (2012, p. 2194) propose that 'further sociological accounts of affect and emotions in tourism is needed.' This is consistent with Light's (2017, p. 294) comments that there is a need to 'focus on what happens to visitors and the cultural work they undertake when visiting a place of death and suffering'. Toward filling these gaps, we examine the importance of emotions in the creation of a sustainably peaceful and just society in a staged contact zone.

We focus upon how emotions create cross-cultural comprehension and peace through a visitor experience consistent with calls to incorporate SDG16 into tourism research (UNWTO., 2018; Boluk et al., 2019). Tucker (2016, p. 41) has identified several areas which justify additional research: 'What effects might this empathy through tourism have' and does it 'catalyse some sort of action'? Empathy is associated with other emotions such as sympathy, respect and compassion, but empathy has a much stronger identification with others. Sympathy and empathy are linked with generational peace capacities but are often subject to confusion. The OED defines sympathy as 'feelings of pity and sorrow for someone else's misfortune'. The word empathy, however, originates from the Greek *em* ('in') and *pathos* ('feeling') meaning 'the ability to understand and share the feelings of another'. Tucker (2016, p. 38) concludes that 'it is important to emphasise empathy's altogether much stronger notion of *identification* with the other' and therefore deserves separate consideration.

Scholars accept that empathy is both an experiential and emotional apprehension of another's frame of reference (Halpern, 2001). Hollan (2011, p. 2) believe that empathy incorporates 'an emotional context for what the empathiser imagines about the other's experience'. Empathy has also been recognised as 'a bridge between differences and an emotional pre-requisite for engaging in positive dialogue' (Tucker, 2016, p 34). Empathy through tourism could reduce ethnocentrism and Higgins-Desbiolles (2003, p. 35) state that the United Nations and World Tourism Organisation believe world peace could be promoted through 'knowledge of other places' and 'empathy with other peoples'. In other words, tourists can be a powerful force for cross cultural understanding, empathy and peace' (Caton, 2012, p. 122). It is from this definitional point that we emphasize that emotive tourism, explored through our experimental LWRIT contact zone institution, is an empathetic endeavour for opening-up liminal space for understanding, reconciliation, and the SDG 16 goals of 'peace, justice and strong institutions.'

Contact zones and liminality

Our emotionally driven tourist experience is situated in the literatures of liminality and contact zone theories and extends them to consider the functional potential of empathy within post-conflict communities. Where groups encountered conflict, Allport (1954) through 'contact theory' suggests that relations could be ameliorated through increased contact. Pratt (2008, p. 7) advanced contact theory into colonial encounters, what she called 'contact zones' as 'social spaces where desperate cultures meet, clash and grapple with each other'. A contact zone for Pratt (1991, p. 39) is characterised by emotions of 'rage, incomprehension, and pain' and 'exhilarating moments of wonder and revelation, mutual understanding, and new wisdom'.

Building off the concept of liminality within the pilgrimage experience by Turner and Turner (1978), Pratt (2008) identifies the significance of disparate cultures meeting in shared social spaces which she calls 'transculturation'. Contact zones offer travellers and 'travelees' an opportunity to develop relations in 'co-presence, interaction, interlocking understandings and practices' (Pratt, 2008 p. 8) and may be a means to create third spaces for common empathetic understanding in places of conflict. Geographers have also recognised the potential that encounters pose to investigate multiculturalism (Nagel & Hopkins, 2010), colonisation, and indigenous peoples (Stouraiti, 2012) as sites 'of destabilization … co-constitution, improvisation and interaction' (Wilson, 2017, p. 457).

Pratt's conceptualization of contact zones has had limited application to tourism. Clifford (1997) applied 'contact zones' in his argument that museums are places of contentious and collaborative interactions (cf. Boast, 2011). In contrast to this literature, Schorch (2013) investigates the contact zone from the perspective of a visitor experience, humanising the contact zone 'through interpretative actions, movements and performances made by museum visitors' and thereby 'open a hermeneutic terrain of cultural negotiation and contestation' (Schorch, 2013, p. 68). He emphasises that 'cultural world making' is the result of interpretation and is critical to the tourist experience.

Liminality has its origins in van Gennep (1960 [1909]) *Rites of Passage*, which identifies three phases: the separation from social norms, the liminal threshold which is an immersive occurrence and which then takes the participant to an enriched experience during the liminal process. Turner (1974) and Shields (2013) have extended the concept of liminality to include temporal and spatial transformations. Liminality has become a popular subject with tourism scholars in recent years and Andrews and Roberts (2012) comment on the significance of the spatio-temporal association with liminality and its significance to the tourist experience (Bristow & Jenkins, 2020). Toward creating a sustainable institution for peace, the confluence of contact zones and liminality are the methodological justification for cross-cultural tourists visiting LWRIT. Through LWRIT we provided a place to experience and transform, not just social norms, but emotions and empathy between communities. This research asked local residents to step into the shoes of a tourist—*in situ* 'out-of-place'—and face the alterity of others (Republican/Nationalist or Loyalist/Unionist) in order to scrutinise and confront their own presuppositions, memories, experiences, and fears.

Tourism and peace

Moufakkir and Kelly (2010, p. xxv) identify a deficit of research on the association between tourism and peace, acknowledging that as 'new forms of tourism emerge, there are new challenges and opportunities' for empirical studies to investigate tourism and peace. The need for further research has also been identified by Blanchard and Higgins-Desbiolles (2013, p. 261) as they call for 'more diverse, contentious and critical dialogue about peace tourism'. They suggest that peace tourism should be studied through a multidisciplinary lens using 'methodologies beyond the limited capacities of positivistic research'. Below, we push the peace tourism literature further toward methodological 'geographies of hope' in rural Northern Ireland (Joronen & Griffiths, 2019; Sturm, 2021).

For Northern Ireland, Simone-Charteris and Boyd (2011, p. 467) advance that the cultural differences of Catholics and Protestants 'should not be concealed but promoted to the tourist' for the out-of-state overseas market. Two types of post-conflict tourism have been identified by Boyd (2019), the 'Phoenix' era which according to Causevic and Lynch (2011) signals tourism's role in 'normalising social relations after political conflict' (p. 796), and 'normalcy' which Boyd (2019) characterises as a period with similar characteristics to a destination without conflict. Boyd suggests that Northern Ireland has developed normal products, but dark tourism continues to be utilised by communities that have been involved in conflict. Simone-Charteris et al. (2018) examine the generational support for dark tourism in Northern Ireland, however the significance of domestic visitors to dark tourism sites has received little attention in recent years. The Tourism NI report *Northern Ireland: Domestic Tourism Strategy* (2020, p. 1) concludes that 'the domestic market is a critical part of Northern Ireland's tourism economy'. The Report goes on to state that 'undoubtedly "day trippers" are a significant part of domestic tourism' (p. 11). In part, it is to this gap in the research that we investigate the role of *in situ* visitors as 'out-of-place' tourists. Moufakkir and Kelly (2010, p. xviii) identify participatory peace at the top of their Peace Hierarchy and suggest 'participatory peace is what makes peace work in a sustainable way', and it starts with ordinary citizens.

Dark tourism and political tourism have been largely investigated in the urban context (Causevic & Lynch, 2011; Simone-Charteris et al., 2018). For example, Skinner's (2016) work on embodied (and to some extent liminal) guided tours of the Falls Road in Belfast. A review of the literature within the spatial social sciences revealed a focus on the urban Troubles and a disregard for the rural conflict, especially in remote rural border areas (Mesev et al., 2009). Murtagh (1998, p. 221) recognized this urban focus, observing that 'the geography of urban ethnic conflict has been extensively researched and described in Northern Ireland. By comparison, the literature on rural areas is weak'. We have therefore utilized a rural contact zone to address this gap in the literature.

Methodology

Existing visitor experiences in Northern Ireland do not offer sufficient emotionally-driven context towards instrumental and sustainable peace. Therefore, utilizing the lead author's consultancy and tour guiding experience, we designed a living history visitor experience (LWRIT) that aimed to stimulate affects and senses with the intent to generate emotions in a rural performative contact zone (Rodaway, 1994; Anderson, 2014). We understand affect for this research to be the rhythms and flows of feelings in a group dynamic situation and therefore the participatory performances offer actors and visitors a co-creative and embodied engagement (McCormack, 2003). It was paramount that the experience would reflect the views of rural Catholic and Protestant families over the duration of the 1968-1998 Troubles. The design of the experience was inspired by Kelly and Nkabahona (2010, p. 237) dictum: 'interpretation is central to the objective of reconciliation through tourism'.

Before writing the performance scripts, eight semi-structured interviews were conducted with four Protestant and four Catholic rural residents across a range of ages and balanced genders. Audio recordings (when permitted) or field notes were made of the interviews. Also drawn upon was the lead author's personal observations and emotional experiences from attending rural Loyalist/Unionist and Republican/Nationalist wakes and funerals throughout the Troubles. Based on these interviews and visits to other visitor attractions (including a survivor's group guided tour of rural atrocities near the border) and archival research, we wrote scripts for the guided night-time walk and the three performances: (1) IRA Wake, (2) UDA Funeral, and (3) Mixed-Marriage (see Figure 1). The scripts were not necessarily an accurate narrative of history but had to reflect the views that the rural communities had of Irish history. They were reviewed with the rural interviewees who had initially prompted the ideas for the scripts. They were also reviewed with senior politicians from Protestant (DUP MLA) and Catholic (Sinn Fein MP) communities. The amateur dramatic society that agreed to help stage the performances—their playwright and actors—also reviewed and amended the scripts. We designed the performances to utilise bodily senses, participation, and objects in order to produce an emotionally diverse visitor experience which would cultivate an emotive understanding of the visitors' differences. We utilized 'object agency' through props where 'people can influence others through objects and images' to

Figure 1. From top: (1) guided night walk, (2) an IRA Wake, (3) UDA funeral, and (4) mixed-marriage. 234 x 276mm (300 x 300 DPI).

stimulate emotions (Svasek 2014, p. 253). All three performances were designed to elicit an extensive range of senses to pique the human sensorium.

The venue for the experience would play a pivotal role in the experience's delivery. It had to be in a rural setting and one that would present ample opportunities for emotional development. A small cottage on a farm, south west of Belfast was selected. Fifteen actors from a local amateur dramatic society and the authors of this paper were engaged in the enactment and represented mixed religion, race, age and gender. Ninety-two participant visitors attended which included visitors from mixed religious and political affiliations including political representatives, clerics, unemployed, farmers, community workers, lecturers, teachers and students. By design we wanted domestic visitors to experience their own history through the gaze of another by liminally distancing them from their history and encouraging them to have an empathetic emotive experience. The research projected them into the shoes of a tourist, taking them out of their mundane everyday lives. The setting was designed to be familiar, yet disorienting, in the function of creating a liminal contact zone, that is, taking them 'out-of-place' in three ways: (1) to take them out of their mundane everyday lives; (2) taking them to a rural 'foreign' place of the Troubles; and (3) taking them 'out-of-place' as tourists in their own historical subjectivity. As 'out-of-place' tourists, the locals could experience their history with some psycho-spatial distance to elicit a liminal emotional experience.

This idiosyncratic, innovative, multidisciplinary and intersectional praxis research between emotions theory and sustainable peace demanded a mixed method research plan. The study focused on qualitative methods to explore emotions and was anchored firmly in a performative, multi sensory, participant-based ethnography. The analysis of this research was based upon a night-time guided walk that led the groups to three performances as summarised in Table 1 below across 6 evenings (October 14, 17, 18, 21, 24, 25, 2019) with an average of 15 participants per evening.

Visitors were treated to light refreshments during Performance 3 each evening and were invited to complete the emotions questionnaire (89 visitors and 15 actors completed the questionnaires). They identified the five most important emotions for them and then briefly explained their choice for each of the performances. Visitors were then invited to participate in focus group discussions where they could interact without distraction (6 visitor focus group discussions, 1 actor focus group discussion). Focus group discussions targeted two questions: 1. Did the performances influence your understanding of why the other community was involved in violence/ defence? 2. How did the performance influence your tolerance/intolerance, respect/disrespect, trust/distrust, sympathy, empathy of the other community or your own community? Finally, actors and visitors were also asked to summarise their response to these questions in a dichotomous format at the end of the emotions questionnaire. Visitors were encouraged to engage in autophotography (40 still photographs and 1 video) and comment on the significance of the photography for them. One-to-one semi-structured interviews (11 personal, 13 emails and 1 mail) were conducted to focus on individuals and pertinent aspects of the research. Respondents for one-to-one interviews were selected based on emotional comments made during the focus group discussions or those who reserved comment in the focus groups and requested an opportunity to respond in private. Face-to-face interviews were restricted by the Covid-19 pandemic. Although semi-structured questions were prepared for the focus groups and one-to-one interviews, if visitors and actors strayed onto other subjects of relevance, the authors encouraged this to continue while the subject matter was relevant to the objectives of this research. Disclosed video recordings were made for analysis of nuanced emotional responses of visitors at all performances, focus group discussions and one-to-one interviews. The authors made coded verbatim transcripts for all research materials and conducted thematic analysis on all research sources.

MacCannell's (2012, p. 185) statement, 'a research domain without ethical reflexivity, that refuses to come to grips with itself and its subject matter in ethical terms is doomed to failure', focused our attention on the need for an ethical approach to this participatory and emotive qualitative research. Our emotions, privilege and positionality were continuously reflected upon throughout the research (Baker et al., 2019, p. 10). We recognise that our gender (men), religion (Catholic and Protestant), age (41 and 62), position of power as *researchers* and political affiliations conditioned the way our research could be conducted and was interpreted. As we played the lament at the UDR funeral our emotions had to be restrained, and we found ourselves with a smile on our faces as we played the music for the Waves of Tory. Even the pronouncement of this word had to be treated with care, as one community call it 'Tory' and the other 'Troy'. Pronunciation of words and letters were a cue to our religious/political backgrounds. The phrasing of our sentences and the emphasis on certain words could be interpreted in a tendentious manner. The location of the venue and our names at the introduction to the research indicated our affiliations and caused some participants to be circumspect of engaging in the research. Our aim was for a transparent process of data collection and rationalisation of qualitative methods. Most of the visitors attending accepted the performances as a place to gain knowledge and

Table 1. Living in the *Wake* of Rural Irish Troubles performances.

Performance	Location/Scene	Conducted By	Visitor Participation
Night Walk to the Wake House (Guided walk). Early 1970s setting.	400m walk on dark country lanes, setting the spatial and temporal scene for the following three performances. Booby trap and culvert bombs. UDR and IRA checkpoints.	Both authors in the role of local farmers & two actors as UDR soldier and IRA Volunteer.	Walking along lane, interacting with guides and negotiating booby traps. Interaction with UDR soldier and IRA volunteer at checkpoints
Performance 1 Wake of an IRA volunteer who had been shot by the British Army. Early 1970s setting.	Scene set in a cottage bedroom in a wake house. Coffin with cadaver bedecked in Republican regalia. Dialogue between parents of deceased, neighbours and American news reporter is used to explain the community's interpretation of events in Republican rural Irish history. Scene concluded with a Republican song.	7 Actors.	Visitors pay their respects to the relatives of the dead on arrival and are served Poitín during the wake.
Performance 2 Funeral of a Part-Time Ulster Defence Regiment (UDR) Soldier who was killed by an IRA bomb at his home. Early 1990s setting.	Outdoor funeral service with a coffin wrapped in the Union Jack is carried to the grave. Presentation of the flag to the relatives. Funeral tea afterwards in family home, furnished with Loyalist regalia. Dialogue between relatives and American news reporter, relates to events which have influenced the Protestant community's perception of history. Performance concludes with shouts of 'No Surrender'.	Both authors involved in playing music for the lament. 9 Actors.	Visitors carry the coffin and walk back to a room in the family home.
Performance 3 Mixed Marriage. 2019 setting. Bride and groom are descendants of the deceased in Performances 1 and 2.	Wedding Reception in Community Hall. A large, bright, warm room has been decorated with paintings by a local artist and wedding banners. Guests toast the bride and groom, and actors/visitors dance to traditional Irish music. The ceremony concludes with 'Auld Lang Syne'. Designed to encourage cross cultural visitors to participate through song and dance.	Both authors play ceili music. 15 actors enact the wedding celebration and interact with the visitors during the Ceili Dance.	Visitors served wedding cake and prosecco for toasts and invited to participated in the Ceili Dance with the actors. All hold hands and sing "Auld Lang Syne".

understanding about the other, however some groups/individuals choose not to attend or displayed displeasure with the performances. A victims' group, for example, did not accept the invitation to attend because to them the IRA volunteer was not a 'volunteer', but rather a 'terrorist', and they were not prepared to accept them being portrayed otherwise.

Analysis

Emotive tourism from sadness to inspiration

This analysis contributes to the SDG16 goals of promoting peace, peaceful societies and conflict resolution through the development of complex emotions via cross-cultural tourist participation. Psychologists have classified basic emotions such as grief, anger, joy, fear, disgust, surprise and sadness (Ekman, 1994) as the building blocks for 'more complex emotions that mediate human social interactions' like inspiration, empathy, and encouragement (Griffiths, 2002, p. 13). With reference to the tourist experience, Robinson (2016, p. 34) argues that 'we can detect some degree of hierarchy amongst the emotions.' Both communities relayed strong emotions of sadness and grief recalling the pointless loss of life representative in these statements: 'grief has no boundaries', 'equally sad and unnecessary loss of life on both sides of the community', and 'I shed real tears, both outside at the graveside (UDR funeral) and also when I witnessed the grief of the daughters and their mammoth loss'. This visitor also echoed her sense of grief at the IRA Wake 'great grief at seeing the body of the young man, even though he was of a different belief to myself'. Using emotions to facilitate the SDG16 conflict resolution, the sense of grief and sadness crossed religious/political boundaries, where visitors from opposite communities identified with the grief of the opposing communities.

Basic emotions of sadness, joy and anger are not less important than complex emotions such as empathy and inspiration but can establish a foundation for more reflective and complex emotions. Basic emotions provided an embryo for complex emotions to develop reconciliatory actions (Little, 2017). Intensive engagements that tackle violence and pain that communities feel or have engaged in can illuminate the humanity of others (Staub, 2001). During one-to-one interviews some visitors who had experienced sadness, grief and hope at the performances reflected upon the performances and made comments that suggest movement towards complex emotions and action. For example, some respondents stated: 'I felt sad, and then anger towards the politicians. We need to reflect this in our voting'; 'My experiences at the plays made me sad, but I felt joy and hope in the last scene. I now speak to my Protestant workmates'; and 'I felt uncomfortable with some of the scenes but I can now see beyond my prejudices towards the other side'. One Protestant man visitor from a rural area expanded:

> The performances certainly made me feel sad and grief for both communities and I have since conducted my own historical research which has caused me to empathise with the other community and my understanding of why they were involved in violence. I feel ashamed of some activities that my own community have been involved with in Ireland. Would I have supported violence over democracy if I had been a member of the Catholic community? If more people were stimulated to keep an objective mind towards our history there could be hope for the communities to work with one another.

This statement establishes three points that only became apparent through one-to-one discussion. First, the visitor originally expressed sadness and grief after the performances, but it inspired him to act after self-reflection that cultivated empathy. The emotive visitor experience has the potential to stimulate further thought and investigation, which is sympathetic towards the development of complex emotions and overall increases the potential for the emotive experience to develop reconciliatory praxis (Robinson, 2016). Second, the performances made him feel shame and hope that communities could work together. Tucker (2009, p. 254) recognises the deep connection between empathy and shame in that 'shame is brought on by heightened awareness of the self as seen through the eyes of the other'. This is significant if we are going to 'rethink how we wish to live in proximity to others' (Probyn, 2005, p. xiv). Tucker (2016) argues that empathy and hope are stimulated by imagination. From our findings, imagination was an empathy/hope generator for younger visitors, but older visitors used their life experiences along with imagination to generate emotions and specifically complex emotions. A senior Protestant woman who had experienced the conflict empathised with 'the remaining [Republican] family, as it could have been me, my dad was in the Royal Ulster Constabulary'. Remembering the circumstances her own family experienced focused her thoughts on others.

Finally, the quote raises the issue of praxis. Tucker refers to the 'healing powers of empathetic tourism' (2016, p. 38) and asks, does empathy 'catalyse some sort of action'? (2016, p. 41). We acknowledge the importance of empathy as one of the 'complex emotions and its role in peace through tourism' but caution the risk of overidentification because visitor's affective experiences may be beyond 'normal representational limits' (Rapson, 2016, p. 161). We proffer that a range of basic emotions can manifest themselves in complex emotions such as empathy and inspiration and therefore instigate reconciliatory actions through what we call emotive peace tourism.

Victims of fear and anger

Although Ekman (1994) and Parrott (2001) refer to fear and anger as basic and primary emotions respectively, there are 'numerous grey areas in attempting to categorise and name emotions' (Robinson, 2016, p. 33). Robinson (2016) acknowledges that certain emotions are more specific to tourism. In the light of detailed one-to-one interviews, we would argue that some basic emotions have an intricate involvement in emotive peace tourism. Powerful negative emotions were expressed by some visitors who empathised with the plight of their 'own' rather than 'others', describing the UDR funeral as highlighting the 'deliberate targeting of Protestant families to eliminate them from the area'. This was amplified through the emotion of anger at the IRA Wake with respondents perceiving it as 'hatred of all things British and using this to justify murder'. Put into context, these comments were made by visitors who during the Troubles lived in fear in rural areas. In subsequent one-to-one interviews (April 2020) these visitors qualified their statements: 'you don't know who is going to be next. You don't know who is watching you and you don't know what's going on. That sense of insecurity'. This visitor, drawing on a parallel with the Covid-19 pandemic, said nobody wanted to leave their house because we are all afraid of death:

> People were being forced continually to face the possibility of death, not knowing where it was coming from with no possibility of guarding against it. You analyse that in your head, it is a massive thing.

Fear affected the most mundane (in)actions, the visitor told us, including whether one displayed their car tax on the windscreen (signifying a non-Republican). The LWRIT visitor experience had stimulated 'dormant emotions', in this case the fear and anger that had been experienced by visitors close to victims in both communities. Such emotions were expressed forcefully during group discussions, but especially during one-to-one interviews. As we will discuss further below, such strong negative historical emotions were not uncommon, but in several instances one-to-one interviews became cathartic and even constructive spaces for emotional development.

The following account represents the views of a Catholic visitor (Visitor Q) who at a young age witnessed the impacts of a multiple shooting by loyalists in his community. The performances inspired him to share his emotions with the researchers through an emotional inter-personal interview.

> I felt really numb, for a long time. I was just 16. I just didn't understand why, why they would come and do that. You know the Troubles were always away in Belfast, I couldn't understand why this innocent man was brought into it.

Visitor Q went on to say that he would feel his views were quite sectarian and did not want to divulge them in the presence of other visitors at the focus groups. He also revealed that if he had been approached by the right people at that time he could have joined the IRA. Previous experience was an inner stimulus for emotions generated from the performances.

> When I saw my neighbour who was seventy, after the three-day funeral, three-day wake and funeral and stuff, she had no kids and her and her husband had been together for (x) amount of years. I was her surrogate son, she had a soft spot for me. She lived another 20 years after it and she never got to see the Police Ombudsman's report to say that there was collusion.

This statement refers to collusion between the security forces and loyalist paramilitaries and highlights a crucial aspect of peace and reconciliation. Access to justice is a central tenant of SDG16. Victims from both communities need to see that justice has been conducted as part of the emotive healing process.

With palpable emotion, the following concluding statement in relation to Visitor Q's referenced victim raises two significant points:

> So I had a very special relationship with her and [she] meant a lot to me in life. To see that person suffer, the way they did, that probably infested [instilled] a lot of insecurities and anger against the other community.

First, Visitor Q showed signs of affectual behaviour through discomfort, rapid blinking, twitching, periods of silence and signs of perspiring before making this statement. Analysis of the audio visual taken during the performances and focus groups demonstrated affectual intelligence. Facial expressions changed before the cognitive action took place. Second, an emotional inter-personal interview with the lead researcher developed into a deep emotionally rich dialogue. The lead researcher came from a rural Protestant community and shared personal experiences of how he felt about the unfolding of Irish history and politics in rural Northern Ireland. The ensuing text message received from Visitor Q's post-interview summarises the feelings and emotions both the interviewer and visitor could process:

> I am really pleased I participated in the experience and it without doubt reminded me of some of the fears and insecurities I had/have around those dark days in our history. The experience did remind me of some bigoted and sectarian feelings that I would struggle with morally. I don't want to portray myself as that type of person but I answered with the feelings I experienced. On a positive note overall it was a good experience and you actually acted like a counsellor during the interview by creating a safe and non-judgemental environment, allowing me to say and express feelings that perhaps I would not normally. Thank you for allowing me to take part. I hope we cross paths sometime in the future.

While gaining respect for others, Visitor Q retained negative emotions toward the Protestant community but sharing his long-standing emotional burden opened-up the possibility for reconciliation with others. Visitor Q valued 'a safe and non-judgemental environment' and these conditions allowed his affects and emotions to flow and form part of the healing process. The lead researcher also had an emotionally enriched experience and gained a much better understanding of the legitimate concerns of others that he would have dismissed in the past. Such self-reflective co-constitution involved participation towards the achievement of peace and conflict resolution.

Visitors who had lived through the Troubles were anxious to forget the past, and LWRIT stimulated some negative emotions. The experience also projected visitors back into a liminal world of what it was like to live daily with fear and anger and stimulated actions to avoid the reality of negative emotions. Buda's (2012; 2015a) research on danger-zone tourism in Palestine and Jordan maintains that fear 'possesses productive potentials in cases whereby it encourages us to confront dangers' (2015, p. 27). Solomon and Stone (2002, p. 420), however, see the emotion of fear as having the potential to 'propel us to remove ourselves from danger'. A retired police officer explained to us that when he was approached by an 'IRA guard' at the IRA Wake performance,

> the sudden appearance of the masked man speaking Irish from the front of the group instilled instant fear. Emotions welled up in me and I was frightened. I moved to the rear of the group in fear of being asked questions. I felt a sense of what it had been like for me 40 years ago.

While he affirmed, 'I never want to go back there', he went on to say that he wished his teenage sons had seen it so they could experience what it was like to live through the fear of conflict. Many other visitors who had experienced the Troubles and now revisited them through a sensual participant experience that invited them to carry coffins and shake hands with grieving relatives were adamant that 'I never want to return to that' and it reminded them of 'the quality of life with peace'.

For some, fear transitioned to anger at the apathetic representation from representative politicians and their lack of interest in attending this visitor experience which explored reconciliation and intercultural empathy. Fear can stimulate anger from which to confront 'the status quo and motivate pro-social action, defined as either direct helping or righteous behaviour in the interest of less fortunate other' (Henderson, 2008, p. 30). With caution not to overstate the impact of LWRIT, three Protestant visitors in one-to-one correspondence claimed that despite the negative emotions elicited by the LWRIT performances, they changed their vote in the 2019 United Kingdom General Election because of the emotive experiences elicited from the performances. One of the visitors stated, 'Yes, I changed my normal voting pattern because of a lack of interest shown by my politicians' and another declared she had crossed the political divide: 'I normally vote Alliance [liberal party] but changed to SDLP [Irish nationalist]'. Negative emotions may have motivated actions towards promoting peace and conflict resolution by way of peace through tourism. Such a finding highlights the potential importance of tourism to fulfill several SDG16 objectives.

Face to face in a liminal contact zone

Tourism is sui generis for Pratt's (2008, p. 8) 'contact zones' that 'invokes the space and time where subjects previously separated by geography and history are co-present, the point at which their trajectories now intersect'. The performances offered a unique opportunity for local residents to see their dissonant history through the lens of a tourist. Picard (2016, p. 15) sees 'emotions as being central … for dealing with transformations brought about through being a tourist'. Inspired in part by Levinas' (1969, p. 207) philosophy that 'the Other faces me and puts me in question and obliges me', the experience encouraged local people to come face-to-face with others in a liminally staged contact zone. The relationship between the face and others signals that we know ourselves through encounters with others and therefore we are indebted in alterity to others as they are indebted to us. Put another way, 'Face to Face is an exploration of how to inhabit the other' (Suchet, 2010, p. 158) and as such provides an opening for victims/oppressors to broker cultural and historical variance. This could be referred to as the receiving of the stranger, complete with their emotions as an unparalleled grounding in Pratt's contact zone. We rooted our research in the works of van Gennep (1960) and Turner's (1974) elaboration of this liminality to include space and time as part of the communitas. A diligent approach was adopted on the perspectives of rural community histories which were mirrored in the performances. The spatial and temporal setting for the histories began with a performative night walk to the venue, designed to immerse the visitors into an unknown liminal world. A former policeman commented,

> When we left the reception, my eyes had not adjusted to the dark, there was a fear of the unknown, the actor really set the scene, pointing out the border location, I felt uncomfortable, I had been there 40 years ago. What is going to happen next? Sticky [the actor] showed us how to avoid a culvert 'bomb', it was so real, I really felt I was jumping the culvert to avoid the bomb. We helped one another to get over the bomb.

This quote highlights the 'detachment from social norms' (Bristow & Jenkins, 2020, p. 220) and the transformation in time and space to future unknowns. Many other visitors shared similar comments about spatial and temporal immersion. The quote also exposes the fear of the unknown in this liminal world where the performative group tasks initiated a sense of communitas. This was further exemplified through the group joining together to sympathise with the relatives of the deceased IRA Volunteer and UDR Soldier and the pinnacle of communitas was Protestants and Catholics joining together to carry the coffin of the UDR soldier.

Autophotography was voluntary and few individuals engaged with it, but those who did produced fruitful data. A video was recorded by a Catholic woman of the lament being played and the carrying of the Protestant UDR soldier's coffin:

> Well R [Husband] was carrying the coffin and because he was carrying the coffin, it just reminds you of family funerals, where one of the marks of respect is to be one of the people carrying the coffin. For me,

> watching him carrying the coffin was very emotive, because it just reminds you of your family and again the person becomes more personal.

Significantly, her Catholic husband carried a coffin draped in the Union Jack, he had crossed community boundaries. An affectual example of a visitor being immersed in the liminal world is signalled by the actions of one woman who showed signs of rapid blinking, lips trembling and agitation for several minutes before interrupting the actors with her emotional language and gestures about the 1845 Irish Famine. The mood changed with signs of positive emotions, affects and communitas being expressed through smiles, laughter and group rhythmic movement during the celi dance at the Mixed Marriage performance.

The performances were an emotionally and affectually challenging bubble for both communities. They brought together people from different geographies and histories in an environment which transformed their 'gaze' from the 'everyday experience' (Urry & Larsen, 2011, p. 4) to examine their own histories through the lens of a tourist. The liminal world of the performances offered face-to-face contact with the other, but follow-up group discussions launched them into a more personal and intense face-to-face contact zone focusing on their experiences with others. A middle-aged Protestant man commented, 'I would not previously have had the opportunity of engaging in conversation with a "Republican" public representative before'. Acute passionate face-to-face discourse arose around subjects such as integrated education, the future of Northern Irish community relations, and the role of political parties. This plucked visitors 'out-of-place' from their liminal world back to their everyday world but focus group exchanges and one-to-one interviews showed transformations resulting from their liminal experiences. For example, a significant transformation in attitude occurred for one Protestant during his experience of the IRA Wake. He explained his emotions at the beginning of the scene:

> I looked at him in the coffin [IRA Volunteer] and couldn't help thinking; he looks like a real Republican. The dialogue had started, and it seemed to me like it was an IRA propaganda meeting, giving a biased story of how badly the Brits had treated the Irish. I felt bitter and angry that I was trapped and had to listen to this propaganda.

His attitude changed during his contact with the actors and the emotional surroundings of the zone:

> The soft voices used made it even more sincere and believable, and I began to feel embarrassed at what my ancestors may have been involved in. If my family had lived in Ireland and their land had been taken from them and given to the invaders, how would I have felt? History could have been reversed, this could have been my family and the young man could have been my son. I hadn't realised that Catholics had been forced to convert to Protestantism during the Famine. I asked myself, did my family "take the soup"? We were basically all the same, but history made us different. My views on Irish history had been challenged and I needed to research my own history. If it was true then I will need to review my thinking on Irish history and the role of my own community.

The visitor had opened his senses and emotions to what he had experienced through the liminal contact zone which signified a significant leap in his attitude towards others. The emotional contact zone provided a stage for objectives one and two, whereby cultural exchange extinguished the fear of others and contributed towards SDG16's objective of conflict resolution through the promotion of peaceful societies.

Peace and reconciliation through tourism

As a measure of reconciliation visitors and actors were asked to complete their emotions questionnaires as shown in Table 2. The results show a significant change in the attitudes of the actors and visitors towards the other community. The actor data shows more change than the visitors which is probably due to the actors being a mixed community group and having worked together for several months during the preparation and delivery of the performances. A young

Table 2. Emotions questionnaire results.

Did the performances influence your understanding of why the other community was involved in violence/defence?	Actors (15)	Visitors (89)
	53%	68%
Did the performances influence your attitude towards the other community in the following:		
Tolerance	69%	46%
Respect	69%	51%
Trust	54%	29%
Sympathy	55%	61%
Empathy	61%	59%

Catholic actor expressed his feelings as 'I felt really affected by the Protestant funeral more [sic] than I did by the Catholic scene, even though I come from a Catholic background. It was a side that I had never heard before'. Another view coming from an older Protestant actress who played a leading role in the IRA Wake said, 'I found it very intimidating, didn't want to be here but there is a belonging here. It would draw you in from a different background'.

The results illustrate two main changes. First, the figure is surprisingly low for change in trust. When visitors were asked about this they responded with surprise, 'I am shocked that we would be asked about trusting Catholics' and 'Why would we not trust Protestants'. It was the older generations who showed a reluctance to trust the other community. From discussions with older visitors this seems to be a deep routed historical problem linked to suspicions concerning members of the RUC (police force) and members of the IRA gathering intelligence, which put lives at risk. Second, the younger generations were more likely to change their understanding and attitude of the other. Older visitors and actors stated that history 'should be bottled' and forgotten. In line with Simone-Charteris et al. (2018), they didn't want to be reminded of the atrocities, although some did say 'lest we forget'. Younger Protestants and Catholics who had not lived through the Troubles were adamant that it 'is a real reminder of why something like this gives you an insight into how it happened and is a big reminder that something like that can never happen again and should never'. It was also recognised that young people should get a balanced view of their history 'not just one that they got from their parents'. One young Protestant during a discussion group affirmed that after experiencing the performances, 'I am going to end up in tears or something, but I think everybody in this country needs to experience that'. He expressed his views that some young people have listened to their parents and grandparents but need 'to see something like this in order for our community to move on, you have got to understand both sides'.

Some groups with more firm political views, for example, a Sinn Fein Member of Parliament and Community Group leaders from a Loyalist area, offered an invitation for the researchers to take the performances to their respective communities. The Sinn Fein MP confirmed, 'it would get people to look outside their own community and to see things from the other perspective' and the Loyalist Community Group Leader stated, 'this is forward thinking'. This offer would suggest that both communities considered it to be balanced, had value in understanding the liminal world of others, and therefore could function as a sustainable institution of peace towards the SDG16 goals.

Conclusion

There is debate about whether tourism brings peace or peace brings tourism. Pratt and Liu (2016) conclude that tourism benefits from peace rather than contributes to peace, and Khalilzadeh (2018) argues that tourism is more of a 'peacekeeper' than a 'peacemaker'. Our research demonstrates that an *in situ* liminal, emotive peace tourism experience, can generate tolerance, respect, trust, sympathy, and empathy towards others. We argue, then, that such emotional reconciliation is not only an important requisite for 'peacekeeping' but also essential for 'peacemaking'. This was acknowledged by an older Catholic visitor 'it is especially important for young people "lest they forget" and revert to conflict'. Our research has taken up the challenge of 'seizing tourism as a platform for imaginative and transformative action is unfinished work' (Boluk et al., 2019). This research, through out-of-place tourism, has used an innovative and imaginative approach to place tourism at the hub of SDG16's objectives that 'promote peaceful and inclusive societies for sustainable development, provide access to justice for all and build effective, accountable and inclusive institutions at all levels'.

There are four major areas where this research contributes towards reconciliation, peace, and the SDG16 objectives. First, all visitors experienced a number and variety of emotions during the performances. Basic emotions brought physically manifested reactions such as tears, but complex emotions brought lasting reactions, resulting in empathetic and even political actions. The second point relates to the generation of strong negative emotions which harkened back to tragic events during the Troubles and was an obstacle for older visitors with strong living memories. Conversely, a middle-aged Protestant stated that 'many middle-aged visitors had baggage from the past and benefited from the emotional experience'. Third, by taking cross-cultural visitors out of their everyday lives and face-to-face with others within an immersive sensual and participative experience, they were forced to examine their history through the lens of a tourist. This liminal cultural exchange instigated signs of breaking down the barriers between the two communities. A soupcon of sympathy, if not empathy, towards others was apparent.

Lastly, the contact zone allowed visitors and actors to immerse themselves in participatory activities such as the celi dance. Aussems (2016, p. 249) describes visitors 'listening to the story of the hosts allows them to reach the Other's reality and to better understand the former enemy's experience'. As such, UNWTO *Tourism in the 2030 Agenda* endorses 'tourism... can foster multicultural and inter-faith tolerance and understanding, laying the foundation for more peaceful societies. Sustainable tourism [can] prevent violence and conflict to take root and consolidate peace in post-conflict societies'. We advance the motivations of this quote through *in situ* emotive tourists (including day-trippers) in 'peacemaking' and 'peacekeeping'. We acknowledge the role of international tourists in peace keeping, but argue that peace could best be maintained and potentially assisted through local emotive peace tourism. Focus group discussions initiated dialogue between communities, some for the first time, and therefore opened new emotionally relational channels. Quantitative results confirmed a change in attitude towards others and qualitative data demonstrated that change was more welcome amongst the younger generation.

Our research indicates that reconciliatory changes took place at an individual level but as one visitor put it, 'one can feel shame for what one's community has done to the other. However, expressing this publicly can be difficult as there is still a feeling "out there" that one is letting one's own "side" down'. Some visitors, especially older ones are still living in the *wake* of rural Irish Troubles and are slow to show signs of emotional engagement, but the younger visitors were enthusiastic about their history and are emotionally enriched to make sure history does not repeat itself. Peace and reconciliation is a delicate process which cannot be imposed but can be stimulated, we argue, at a personal and through a participatory emotive space. This research, through the concepts 'emotive peace tourism' and 'out-of-place tourism', has demonstrated the positive impact of our self-reflection and empathetic emotive tourism experience within rural Northern Ireland. Moreover, we hope this research can be replicated in other post-conflict spaces as a toolkit and a means of implementing the objectives of SDG16 through the prism of tourism.

Acknowledgments

We would like to thank the three anonymous reviewers for helping us write a better, more nuanced, paper. Thanks also to Roberta Erwin who has supported us through the logistics of the research. We would also like to acknowledge all those who acted in, attended, and contributed to the research.

Disclosure statement

No potential conflict of interest was reported by the authors.

ORCID

Tristan Sturm http://orcid.org/0000-0001-5757-0117

References

Ahmed, S. (2004). *The cultural politics of emotion*. Routledge.

Allport, G. W. (1954). *The nature of prejudice*. Perseus Books.

Anderson, B. (2014). *Encountering affect: Capacities, apparatuses, conditions*. Ashgate.

Andrews, H., & Roberts, L. (2012). *Liminal landscapes: Travel, experience and spaces in-between*. Routledge.

Aussems, E. (2016). Cross-community tourism in Bosnia and Herzegovina: A path to reconciliation? *Journal of Tourism and Cultural Change, 14*(3), 240–254. https://doi.org/10.1080/14766825.2016.1169347

Baker, K., Eichhorn, M., & Griffiths, M. (2019). Decolonizing field ecology. *Biotropica, 51*(3), 288–292. https://doi.org/10.1111/btp.12663

Bigne, J. E., & Andreu, L. (2004). Emotions in segmentation: An empirical study. *Annals of Tourism Research, 31*(3), 682–696. https://doi.org/10.1016/j.annals.2003.12.018

Blanchard, L., & Higgins-Desbiolles, F. (2013). *Peace through tourism: Promoting human security through international citizenship*. Routledge.

Boast, R. (2011). Neocolonial collaboration: Museum as contact zone revisited. *Museum Anthropology, 34*(1), 56–70. https://doi.org/10.1111/j.1548-1379.2010.01107.x

Boluk, K. A., Cavaliere, C. T., & Higgins-Desbiolles, F. (2019). A critical framework for interrogating the United Nations Sustainable Development Goals 2030 Agenda in tourism. *Journal of Sustainable Tourism, 27*(7), 847–864. https://doi.org/10.1080/09669582.2019.1619748

Boyd, S W. (2019). Post-conflict tourism development in Northern Ireland: Moving beyond murals and dark sites associated with its past. In R. K. Isaac, E. Çakmak, & R. Butler, (Eds.), *Tourism and hospitality in conflict-ridden destinations* (pp. 226–239).Routledge.

Bristow, R. S., & Jenkins, I. S. (2020). Spatial and temporal tourism considerations in liminal landscapes. *Tourism Geographies, 22*(2), 219–228. https://doi.org/10.1080/14616688.2020.1725618

Buda, D. M. (2012). *Danger-zone tourism: Emotional performances in Jordan and Palestine. Doctoral Dissertation*. The University of Waikato.

Buda, D. M. (2015a). *Affective tourism*. Routledge.

Buda, D. M. (2015b). The death drive in tourism studies. *Annals of Tourism Research, 50*, 39–51. https://doi.org/10.1016/j.annals.2014.10.008

Caton, K. (2012). Thinking inside the box: Understanding discursive production and consumption in tourism. In I. Ateljevic, N. Morgan, & A. Pritchard (Eds.), *The critical turn in tourism studies: Creating an academy of hope* (pp. 121–134).Routledge.

Causevic, S., & Lynch, P. (2011). Phoenix tourism: Post-conflict tourism role. *Annals of Tourism Research, 38*(3), 780–800. https://doi.org/10.1016/j.annals.2010.12.004

Chard, C. (2002). Lassitude and revival in the warm south: Relaxing and exciting travel. In R. Wrigley & G. Revill (Eds.), *Pathologies of travel* (pp. 179–205). Rodopi.

Clifford, J. (1997). *Routes: Travel and translation in the late twentieth century*. Harvard University Press.

Cohen, E., & Cohen, S. A. (2012). Current sociological theories and issues in tourism. *Annals of Tourism Research, 39*(4), 2177–2202. https://doi.org/10.1016/j.annals.2012.07.009

Ekman, P. (1994). *The nature of emotion: Fundamental questions*. Oxford University Press.

Griffiths, P. E. (2002). *Basic emotions, complex emotions, Machiavellian emotions*. PhilSci-Archive, 1–43. http://philsci-archive.pitt.edu/604/.
Halpern, J. (2001). *From detached concern to empathy: Humanising medical practice*. Oxford University Press.
Henderson, V. L. (2008). Is there hope for anger? The politics of spatializing and (re)producing and emotion. *Emotion, Space and Society*, *1*(1), 28–37. https://doi.org/10.1016/j.emospa.2008.07.001
Higgins-Desbiolles, F. (2003). Reconciliation tourism: Tourism healing divided societies. *Tourism Recreation Research*, *28*(3), 35–44. https://doi.org/10.1080/02508281.2003.11081415
Hollan, D. (2011). Introduction. In D. Hollan & J. Throop (Eds.), *The anthropology of empathy* (pp. 1–21).Berghahn Books.
Joronen, M., and., & Griffiths, M. (2019). The moment to come: Geographies of hope in the hyperprecarious sites of occupied Palestine. *Geografiska Annaler, Series b: Human Geography*, *101*(2), 69–83. https://doi.org/10.1080/04353684.2019.1569473
Kelly, I., & Nkabahona, A. (2010). Tourism and reconciliation. In O. Moufakkir & I. Kelly (Eds.), *Tourism, progress and peace* (pp. 228–241). CAB International.
Khalilzadeh, J. (2018). Demonstration of exponential random graph models in tourism studies: Is tourism a means of global peace or the bottom line? *Annals of Tourism Research*, *69*(1), 31–41. https://doi.org/10.1016/j.annals.2017.12.007
Lévinas, E. (1969). *Totality and infinity: An essay on exteriority*. Duquesne University Press.
Light, D. (2017). Progress in tourism management: Progress in dark tourism and thanatourism research. *Tourism Management*, *61*, 275–301. https://doi.org/10.1016/j.tourman.2017.01.011
Little, A. (2017). Fear, hope and disappointment: Emotions in the politics of reconciliation and conflict transformation. *International Political Science Review*, *38*(2), 200–212. https://doi.org/10.1177/0192512115590635
MacCannell, D. (2012). Tourism research 2.0. *Annals of Tourism Research*, *39*(2), 883–907.
McCormack, D. (2003). An event of geographical ethics in spaces of affect. *Transactions of the Institute of British Geographers*, *28*(4), 488–507. https://doi.org/10.1111/j.0020-2754.2003.00106.x
Mesev, V., Shirlow, P., & Downs, J. (2009). The geography of Conflict and death in Belfast, Northern Ireland. *Annals of the Association of American Geographers*, *99*(5), 893–903. https://doi.org/10.1080/00045600903260556
Moufakkir, O., & Kelly, I. (2010). *Tourism, progress and peace*. CAB International.
Murtagh, B. (1998). Community, conflict and rural planning in Northern Ireland. *Journal of Rural Studies*, *14*(2), 221–231. https://doi.org/10.1016/S0743-0167(97)00067-3
Nagel, C., & Hopkins, P. (2010). Introduction: Spaces of multiculturalism. *Space and Polity*, *14*(1), 1–11. https://doi.org/10.1080/13562571003737650
Parrott, W. (2001). *Emotions in social psychology*. Psychology Press.
Pearce, P. (2012). The experience of visiting home and familiar places. *Annals of Tourism Research*, *39*(2), 1024–1047. https://doi.org/10.1016/j.annals.2011.11.018
Picard, D. (2016). Tourism, awe and inner journeys. In D. Picard & M. Robinson (Eds.), *Emotion in motion* (pp. 1–19). Routledge.
Pratt, M. L. (2008). *Imperial eyes: Travel writing and transculturation* (2nd ed.). Routledge.
Pratt, S., & Liu, A. (2016). Does tourism really lead to peace? A global view. *International Journal of Tourism Research*, *18*(1), 82–90. https://doi.org/10.1002/jtr.2035
Pratt, M. L. (1991). Arts of the contact zone. *Profession*, 33–40.
Probyn, E. (2005). *Blush: Faces of shame*. University of Minnesota Press.
Rapson, J. (2016). Emotional memory formation at former Nazi concentration camp sites. In D. Picard & M. Robinson (Eds.), *Emotion in motion* (2nd ed., pp. 161–178). Routledge.
Robinson, M. (2016). The emotional tourist. In D. Picard & M. Robinson (Eds.), *Emotion in motion* (2nd ed., pp. 21–46). Routledge.
Rodaway, P. (1994). *Sensuous geographies*. Routledge.
Royle, S. A. (2012). Beyond the boundaries in the island of Ireland. *Journal of Marine and Island Cultures*, *1*(2), 91–98. https://doi.org/10.1016/j.imic.2012.11.005
Schorch, P. (2013). Contact zones, third spaces, and the act of interpretation. *Museum and Society*, *11*(1), 68–81.
Shields, R. (2013). *Places on the margin: Alternative geographies of modernity*. Routledge.
Simone-Charteris, M. T., & Boyd, S. W. (2011). The potential for Northern Ireland to promote politico-religious tourism: An industry perspective. *Journal of Hospitality Marketing & Management*, *20*(3-4), 457–483. https://doi.org/10.1080/19368623.2011.562438
Simone-Charteris, M., Kirkpatrick, M., & McLaughlin, C. (2018). An investigation of the differences that exist between generations in relation to supporting dark tourism in Northern Ireland. *DBS Business Review*, *2*, 69–88. https://doi.org/10.22375/dbr.v2i0.26
Skinner, J. (2016). Walking the Falls: Dark tourism and the significance of movement on the political tour of West Belfast. *Tourist Studies*, *16*(1), 23–39. https://doi.org/10.1177/1468797615588427
Solomon, R. C., & Stone, L. D. (2002). On "positive" and "negative" emotions. *Journal for the Theory of Social Behaviour*, *32*(4), 417–435. https://doi.org/10.1111/1468-5914.00196

Staub, E. (2001). Individual and group identities in genocide and mass killing. In D. Richard, L. Jussim, & D. Wilder (Eds.), *Social identity, intergroup conflict and conflict reduction* (Vol. 3, pp. 159–184).Oxford University Press.

Stouraiti, A. (2012). Colonial encounters, local knowledge and the making of the cartographic archive in the Venetian Peloponnese. *European Review of History: Revue Europeenne D'histoire*, *19*(4), 491–514. https://doi.org/10.1080/13507486.2012.697874

Sturm, T. (2021). God's just Gaza war: Futurity foreclosed through evangelical apocalypse as orthogonal promise of Eretz Israel. *Geografiska Annaler: Series B, Human Geography*, 1–19. https://doi.org/10.1080/04353684.2021.1891557

Suchet, M. (2010). Face to face. *Psychoanalytic Dialogues*, *20*(2), 158–171. https://doi.org/10.1080/10481881003716248

Svasek, M. (2014). *Moving subjects, moving objects: Transnationalism, cultural production and emotions*. Berghahn.

Tourism NI Market Recovery Taskforce. (2020). *Northern Ireland Domestic Tourism Strategy* (pp. 1–14).

Tucker, H. (2009). Recognising emotion and its postcolonial potentialities: Discomfort and shame in a tourism encounter in Turkey. *Tourism Geographies*, *11*(4), 444–461. https://doi.org/10.1080/14616680903262612

Tucker, H. (2016). Empathy and tourism. *Annals of Tourism Research*, *57*, 31–43. https://doi.org/10.1016/j.annals.2015.12.001

Turner, V. (1974). Liminal to limonoid, in play, flow and ritual: An essay in comparative symbology. *Rice Institute Pamphlet-Rice University Studies*, *60*(3), 53–92.

Turner, V., & Turner, E. (1978). *Image and pilgrimage in Christian culture*. Columbia University Press.

UNWTO. (2018). *Tourism in the 2030 agenda*. https://www.unwto.org/tourism-in-2030-agenda

Urry, J., & Larsen, J. (2011). *The tourist gaze 3.0* (3rd ed.). Sage.

van Gennep, A. (1960). *The rites of passage*. University of Chicago Press.

Wilson, H. F. (2017). On geography and encounters: Bodies, borders, and difference. *Progress in Human Geography*, *41*(4), 451–471. https://doi.org/10.1177/0309132516645958

Beyond multicultural 'tolerance': guided tours and guidebooks as transformative tools for civic learning

Meghann Ormond and Francesco Vietti

ABSTRACT

In bringing people together that otherwise might have little more than passing contact with one another, tourism is appreciated for its potential to transform mindsets by fostering multi-perspectivity, a cornerstone of global citizenship education, among both 'tourists' and 'locals'. Hence, while tourism plays a significant role in marginalising and exploiting immigrants' bodies, labour and heritages, it also holds significant potential as a critical pedagogical tool for transcending the limits of multicultural tolerance discourse and combatting exploitation and xenophobia. In this article, we reflect on two Europe-based global citizenship-inspired initiatives bringing together migration and tourism in novel ways: Migrantour guided walking tours and the Roots Guide guidebook. They endeavour to rework guided tours and guidebooks, two of tourism's most common pedagogical tools, into 'good company' that supports the Arendtian practice of 'visiting' as a key mode of civic learning. In so doing, we explore the representational and structural opportunities and challenges that these two initiatives encounter as they seek to co-create multi-dimensional narratives and routes in ways that recognise guides' diverse experiences and perceptions of the places they call home, avoid stereotypical representations of 'communities' and hold space for the real-life frictions that accompany diversity.

Introduction

In bringing people together that otherwise might have little more than passing contact with one another, tourism has been appreciated for its educational potential to transform mindsets by exposing both 'visitors' and 'locals' to diverse perspectives on, and ways of living in, the world (Reisinger, 2013). Yet, while tourism may have the capacity to serve as a powerful vehicle for co-learning and emancipation, the superficial cultural and social exposure to and consumption of 'the Other' with which it is frequently associated can also serve as an instrument of oppression (Werry, 2008; Aytar & Rath, 2012). In this article, part of a special issue examining tourism's potential and limitations as a social force for peace, we call attention to the ways in which an ethics of 'tolerance' underpinning the construction and wielding of common 'tools' of tourism – like guided tours and guidebooks – perpetuates this 'tourism paradox' (Boluk & Carnicelli, 2019:

171) and examine strategies that deploy political theorist Hannah Arendt's (1977) notion of 'visiting' to move beyond it.

Political theorist Wendy Brown (2009) conceptualises 'tolerance' as a liberal democratic discursive de-politicisation of both the sources of political problems and the solutions to them, via the substitution of 'emotional and personal vocabularies' calling for justice and equality by political solutions that only hold space for uncomplicated celebration of 'diversity' and 'difference' (Brown, 2009: 16). Examples of this abound in tourism. Consider, for instance, the ways in which countless urban 'ethnic' neighbourhoods get sanitised as peaceful playgrounds of multicultural consumption for visitors (Shaw et al., 2004; Aytar & Rath, 2012), with immigrant and racialised 'Others' – spatially fixed and arranged into folklorised categories – rendered carriers of exotic scents, tastes and sounds that allow a privileged 'us' to reproduce and consume the variety of the world in 'our very own' cities. Conventional tourism practices' embrace of a de-politicised and containerised exposure to 'the Other' largely papers over social, economic and political inequalities while, at the same time, reinforcing, exacerbating and capitalising on them.

How, then, might critical tourism scholars and practitioners adjust the common 'tools' of tourism such that they can serve as pedagogical vehicles fostering not an ethics of multicultural tolerance but, rather instead, another kind of ethics that recognises our roles in 'the constitutive histories and powers organising contemporary problems and contemporary political subjects' (Brown, 2009: 16)? We examine this question through the lens of two Europe-based global citizenship education-inspired initiatives to which we are each closely tied in critical action research practice: Migrantour, a European network of intercultural guided urban walking tours, and Roots Guide, a guidebook of the Netherlands narrated by people with diverse migration backgrounds. Both initiatives seek to move beyond merely promoting 'our' tolerance of 'the migrant Other' and, instead, to foster the mutual recognition and dialogue among diverse social actors necessary to temper growing xenophobia and anti-immigrant sentiment. Yet Migrantour and the Roots Guide routinely encounter representational and structural challenges rooted in a discourse of multicultural tolerance that risk undermining the initiatives' objectives of re-politicising and challenging entrenched narratives that 'sometimes personalise, sometimes culturalise and sometimes naturalise' (Brown, 2009: 15) diverse perspectives and forms of inequality, contestation and conflict.

In the pages that follow, we first present this article's conceptual underpinnings. After observing how a multicultural politics of 'tolerance' gets implicitly reproduced through commonplace conventions and tools of tourism, we examine the ways in which a global citizenship-inspired politics of multi-perspectivity necessitates a fundamental transformation of such conventions and tools. We then discuss our guided tour and guidebook initiatives, Migrantour and the Roots Guide, exploring a set of representational and structural challenges we have encountered in working to make a shift towards fostering a more politically progressive form of engagement through 'visiting' (Arendt, 1992). We outline here the tactics we have developed to consciously and continuously disrupt 'tolerant' tourism narratives in order to foster greater civic recognition of and dialogue around real-life dissonances involving different social actors and interests. We conclude with brief reflections on how these experiences can be informative for practitioners developing similar initiatives and the scholars studying and/or contributing to them.

From touring to visiting

Recognising tourism's 'worldmaking' force as an 'interpretable and malleable carrier/creator/confirmer of being and becoming' (Hollinshead et al., 2009: 427), scholars like Belhassen and Caton (2011) and Higgins-Desbiolles and Blanchard (2010) have stressed the significance of incorporating a more critical moral pedagogy into formal tourism education, which has long been driven and heavily informed by free market-based moral logic, so that future tourism industry and policy actors – who are also engaged citizens – may consciously incorporate values of democracy, social justice and human rights into their practices.

Beyond the realm of formal tourism education, the values underpinning the production and use of tourism sites (e.g., museums, heritage sites, parks, etc.) and tools (e.g., guided tours and guidebooks) – powerful interpretive and performative vehicles for public pedagogy (Christie & Mason, 2003; Ablett & Dyer, 2009; Mair & Sumner, 2017) – also merit, and increasingly receive, considerable scrutiny for the uneven conditions they not only perpetuate but also engender between 'foreign' visitor and 'local' host as consumer and object of consumption, as spectator and performer of exotic difference (Urry, 2002). Cheng et al. (2010: 116) suggest that underpinning such relationships is a tacit co-enactment in tourism, first, of mutual 'strangerhood', 'whereby visitors and hosts treat each other as ideal types and objects' and, second, of service, 'whereby both parties understand that the tourist is at leisure and the host is there to serve' the tourist's needs and expectations, 'with no anticipation on either side that the tourist will adapt to the host context' (116). If tourism is 'one of the only contemporary sites outside of the education industry where explicitly designated, non-vocational learning about other times, places, and peoples takes place' (Werry, 2008: 17), then these fundamentally uneven conditions of 'strangerhood' and 'service' severely disable our capacity to learn about and from one another in a civically minded way.

Yet, in tourism's hegemonic narrative 'walls' that culturalise, naturalise and obscure uneven power structures (Brown, 2009), there are cracks through which light can enter. Wearing and Wearing (1996) refer to these cracks – moments and spaces in which mutual recognition and dialogue are possible – as 'chora', the 'space between being and becoming or the "space in which place is made possible" [… ,] given meaning by the people who make use of the space' (Grosz, 1995, in Wearing & Wearing, 1996: 233). It is on 'chora' – and the encounters that generate active 'chorasters' instead of flaneurs merely gazing through their own lenses upon the Other – that many educational, social and volunteer tourism initiatives focus, seeking to trouble both conditions of 'strangerhood' and 'service' in order to work towards more radically open, democratic and sustainable tourism practices.

In supporting the development of 'chorasters', Migrantour and the Roots Guide – explored later in this article – engage in a critical and reflexive form of global citizenship education (GCE) that embraces multi-perspectivity as a key tool with which to uncover and recognise the uneven processes that obscure difference, create inequalities, and maintain exploitation and our positions in these processes, 'with a view to changing them and their attendant assumptions, identities, attitudes and power relations' (Mannion et al., 2011: 452). A cornerstone of GCE, multi-perspectivity encompasses both the consciousness that one's 'view on the world is not universally shared, and others may differ profoundly' (Dill, 2013: 40) and the competence of 'think[ing] beyond the confines and particularities' (40) of one's own circumstances. It equips one to better grasp and navigate the range and dimensions of contemporary global interdependence and the social, economic, technological and environmental opportunities and challenges such interdependence poses.

The latest in a series of moral pedagogical movements concerned with the organisation and management of diversity over time and in space, GCE has emerged in the wake of assimilationism and multiculturalism. Whereas assimilationism sought social harmony through the reduction of diversity within the nation-state, multiculturalism sought it by placing emphasis instead on celebrating the societal gifts brought about through cultural diversity, while simultaneously abstracting 'culture' by obscuring internal heterogeneity and rendering it folklorised heritage that 'really need not be taken seriously' (Dill, 2013: 91), thus contributing to the de-politicisation inherent to tolerance discourse (Brown, 2009). GCE, by contrast, draws attention to our mutual imbrication in the narratives of one another's lives and how these 'multiple, overlapping enacted narratives are the sources of [our] allegiances, belonging and identity as an individual and a member of larger collectives' (Dill, 2013: 153). It has not escaped critique. GCE's acceptance of a de facto 'global world economy and society' (Mannion et al., 2011: 452) complicates framings of the nation-state as the privileged container of civic political engagement. Without a circumscribed sphere of mutual accountability, individuals' 'pre-political virtues' like care, respect and responsibility are held by some to be 'projected onto human problems unmediated by a [collective] political framing' (Butcher, 2017: 133-134) with little chance to effect political change. Yet, while complicating national belonging, GCE is increasingly appreciated for its framing of diversity as the intersectionality of forces that, at different scales and sites, shape us as individuals and our capacity to act (Nussbaum, 2002).

In our late-modern era where 'humans need to work together [...] to achieve common understanding and mutual recognition in a world without fixed truths' (Gillies, 2016: 152), an embrace of multi-perspectivity could attract critique for fostering ethical relativism (Wansink et al., 2018). Yet, for political theorist Hannah Arendt (1977) and philosopher and GCE proponent Martha Nussbaum (2002), both a plurality of perspectives and mutual recognition of these different perspectives are requisite to developing informed individual judgement and enabling adequate deliberation and collective action in democratic societies:

> Political thought is representative. [...] This process of representation does not blindly adopt the actual views of those who stand somewhere else, and hence look upon the world from a different perspective; this is a question neither of empathy, as though I tried to be or to feel like somebody else, nor of counting noses and joining a majority but of being and thinking in my own identity where actually I am not. The more people's standpoints I have present in my mind while I am pondering a given issue, and the better I can imagine how I would feel and think if I were in their place, the stronger will be my capacity for representative thinking and the more valid my final conclusions, my opinion.
>
> (Arendt, 1977: 241)

This 'enlarged mentality' (Arendt, 1977: 241) can be accomplished through what Arendt (1992) calls the practice of 'visiting'.

As a mode of civic learning, 'visiting' requires one, on the one hand, to pay 'considered attention' to and engage with 'stories of an event from each of the plurality of perspectives that might have an interest in telling it' and, on the other, to have one's sense-making interrupted and to reflexively embrace the ensuing disorientation that comes with grappling with 'how the world looks different to someone else' (Biesta & Cowell, 2012: 59). It is distinct from conventional notions of 'tourism' and 'empathy', which, Biesta and Cowell (2012: 50, citing Disch, 1997: 159) argue, tend to obscure and erase plurality as well as dismiss friction and distance: 'The former [tourism] does so by an objectivist stance that holds to "how we do things" as a lens through which different cultures can only appear as other. The latter [empathy] trades this spectatorial lens "to assume native glasses, identifying with the new culture so as to avoid the discomfort of being in an unfamiliar place"'.

With our judgment of the world around us and others in it mediated intersubjectively, not only 'our choice of company' (Arendt, 2003: 145-146) – i.e., the range of others' viewpoints with which we choose to engage and surround ourselves – matters but so do the forms or vehicles through which we are exposed to our company's viewpoints. For Arendt, the key to enlarging one's mentality is 'not to see through the eyes of someone else, but to see with your own eyes from a position that is not your own' (Biesta, 2016: 187; cf. Gatta, 2014). Yet the dialogical exchange of perspectives that this requires has been poorly facilitated by tourism's conventional pedagogical vehicles, which are instead largely used to sustain the earlier-mentioned objectifying conditions of 'strangerhood' and 'service' (Cheng et al., 2010). Cohen (1985: 13, 15) argues that guided tours and guidebooks derive their utility from translating 'the strangeness of a foreign culture into a cultural idiom familiar to the visitors' in ways that both integrate tourists 'into the visited setting as well as insulat[ing them] from that setting', ultimately themselves becoming the 'company' or 'bosom companions' (Bhattacharyya, 1997) with which visitors most closely engage. If this is indeed so, then to what extent can guided tours and guidebooks effectively be used to move beyond the treatment of one another as ideal types and objects and instead to foster 'visiting' as a form of civic learning? To what extent can they themselves be adapted to become *'good* company' (Gillies, 2016: 154, our italics) for both tourists-as-visitors *and* their hosts? Guidebooks, by and large, have functioned unreflexively as Orientalist texts, 'secur[ing] ethnocentric perspectives, producing other spaces and subjects for Western consumption, and caricatur[ing] entire histories and cultures according to a predetermined set of values that reinforce European Superiority' (Lisle, 2008: 163; Bhattacharyya, 1997). Likewise, guided tours are frequently dominated by heritage narratives that erase or marginalise the presence and claims of subaltern groups (Modlin et al., 2011; Dolezal & Gudka, 2019). Indeed, Cheng et al. (2010) argue that guided tours, guidebooks and tourism maps actively train their users' gazes 'away from thinking about either structures of inequality or the power of collective action and social movements', noting that chorastic opportunities tend to come about instead more through the accidental 'slippages and fractures' (Cheng et al., 2010: 116) in users' consumption of them.

Yet, over the last decades and around the world, civic and political activist groups increasingly have appropriated guided tours and guidebooks, recognising them as powerful vehicles of public pedagogy 'containing possibilities for both reproduction [of] and resistance' (Sandlin et al. 2010: 2) to dominant narratives (see, e.g., Cheng et al., 2010; Counter-Cartographies Collective et al. 2012; Dolezal & Gudka, 2019). Such social and responsible tourism initiatives seek to counter-map traditionally marginalised people's (e.g., homeless, Black, LGBTQ, etc.) narratives onto the ways in which both 'tourists' and 'locals' have framed and experienced them and the places relevant to them. These initiatives' attention to 'dissonant heritages' (Ashworth & Tunbridge, 1996) not easily reconciled with dominant values and experiences can be understood as efforts to re-politicise and enlarge public discourse by 'empower[ing] voices that are marginalised or ignored in authorised [heritage] discourses about places and the people within them' (Campos-Delgado 2018: 490). This enables a shift away from an emphasis on 'tolerance' (Brown, 2009) and towards acknowledgement of what anthropologist Anna Tsing (2005: 4) calls 'friction' – 'the awkward, unequal, unstable, and creative qualities of interconnection across difference' –, a disorientation fundamental to Arendtian (Arendt, 1977) 'visiting' as civic learning. Yet there is little work examining *how* such initiatives endeavour to accomplish these objectives (but see, e.g., Christie and Mason (2003) on critically reflexive guide training). In other words, beyond swapping out old messages for newer, more progressive ones, what are the additional mechanics necessary to fundamentally transform tourism's conventional pedagogical vehicles in such a way that they embrace and embody a different kind of ethical engagement with difference?

Contextualising Migrantour and the Roots Guide

In this article, we reflect on our shared experience of being in pivotal roles in two initiatives to which we ourselves have contributed to creating and developing over many years. Both of us are university scholars engaged in critical action research (Kemmis et al., 2013). Meghann, the Roots Guide's editor/curator and scientific coordinator since 2018, is a cultural geographer and a migrant with both American and Portuguese citizenship living in the Netherlands. Francesco, Migrantour's scientific coordinator from 2009-19, is an Italian social anthropologist. We share a focus on collaborating with practitioners and civil society in the fields of mobility and heritage to contribute to public debate on migration and tourism (Pink, 2005; Beck & Maida, 2013). For the following discussion of challenges and opportunities that we and those with whom we have collaborated in the development of our respective initiatives have encountered, we draw on qualitative interviews with production team members and guides as well as our own ongoing observations of decision-making dynamics within our respective initiatives. The type of self-reporting in which we engage here aligns with critical action research practice, our embedded participation offering privileged insight into our initiatives' practices, our understanding of our initiatives' intentions and practices, and the conditions under which those practices are formed (Kemmis et al., 2013: 188).

Migrantour and the Roots Guide are countermapping initiatives focused specifically on migrant heritage. Often categorised as 'foreign' guests, people with migration backgrounds have long been excluded from participation and demonised in public discourse as well as marginalised and excluded in the authorised heritage discourses (Smith, 2006) of their receiving contexts. Where people with migration backgrounds and their tangible and intangible heritages have been incorporated, they frequently have been exoticised, objectified and celebrated by 'locals' and 'tourists' alike as representative 'hosts' of the faraway countries and cultures from which they or their ancestors have come (Rath, 2007; Aytar & Rath, 2012). This 'touristification' of cultural diversity has come about, note Aytar and Rath (2012: 2), in a 'globalizing world – where local difference and place identity are increasingly important – [such that] heritage and cultural diversity have become crucial components of the cultural capital of post-industrial societies'. Specific expressions of immigrant heritages, hence, become conveniently 'tolerated' in Brown's (2009) sense so long as they maintain their worth as cultural capital capable of attracting visitors, employment and investment. Indeed, as a tool of public pedagogy, tourism has played a significant role in fostering and celebrating multiculturalism (Dill, 2013).

The potential of people with migration backgrounds to transcend 'strangerhood' (Cheng et al., 2010: 116) and acquire what we refer to here as 'localhood' – in other words, to becoming themselves recognised hosts of the places in which they live – requires the enfolding of the experiences and 'memories of migrants in their diversity and specificity' (Gouriévidis, 2014: 1) into dominant discourse and recognition of immigrants' interwovenness into the fabric of their places of residence. However, it is only in recent decades, and within the framework of multicultural policy, that migration, claims Gouriévidis (2014: 1), has been able to escape its status as a *'non lieu de mémoire'* ('non-place of memory'). She notes that sites of public pedagogy, like museums, are increasingly 'encouraged to reflect the socio-cultural implications of [...] the increasingly plural face of the populations composing modern states' (Gouriévidis, 2014: 1). Yet, while gradually gaining visibility in the processes and places where the collective identity of local and national communities is built (Basso Peressut & Pozzi, 2012), migration-related stories, memories and representations by and large continue to occupy a peripheral place in dominant historical, cultural and social narratives (Hintermann & Johansson, 2010). It is in this context that Migrantour and the Roots Guide have developed.

Migrantour

In line with the Council of Europe's (2005) Convention on the Value of Cultural Heritage for Society's (the Faro Convention) advocacy for recognising and strengthening 'heritage communities', Migrantour fosters the expansion of the 'right to cultural heritage' among people with migration backgrounds such that they not only have 'the right to benefit from the existing heritage, but also the right to take part in the selection of new cultural expressions aimed at belonging to the notion of cultural heritage' (Zagato, 2015: 147). To do this, Migrantour focuses on the development and use of intercultural guided walking tours as a tool for local community-building and to advance 'dissonant' (Ashworth & Tunbridge, 1996) narratives about the role of migration in processes of urban transformation and heritage-making.

Originating in Turin, Italy, Migrantour is a decade-old network that brings people with first- and second-generation migration backgrounds together with tour operators, local institutions and NGOs to co-create intercultural guided walking tours in 15 European cities (Turin, Milan, Genoa, Florence, Rome, Marseille, Paris, Lisbon, Brussels, Ljubljana, Naples, Bologna, Cagliari, Catania and Pavia), with new cities currently joining. To date, and thanks in part to European Commission funding, network partners have been able to train more than 600 people with first-

and second-generation migration backgrounds as guides (or 'intercultural companions'), co-create 40 intercultural guided walking tour routes, and reach approximately 30,000 walking tour participants (ACRA., 2016, 2020). While accessible to international tourists, the year-round tours are undertaken primarily by local residents and school children from the cities in which they run. In this article, we focus specifically on Migrantour's development in Italian cities.

Roots Guide

Like all modern nation-states, the Netherlands is supported by heritage narratives that are officially authorised and shared to forge belonging to an 'imagined community' (Anderson, 1984/2006; Smith, 2006). However, while more than one-fifth of the Netherlands' population has a first- or second-generation migration background, contemporary national authorised heritage discourse largely has obscured and marginalised their contributions and heritages, with attention to the country's slavery and colonial heritage notably absent (Witte, 2019). In light of this, Roots Guide has explicitly worked to develop a public pedagogical tool targeting domestic tourists in the Netherlands that draws on written and visual storytelling, guides' meaningful places and experiential travel techniques to enable users to undertake both inner and outer journeys that inspire them to use migration as a lens through which to revisit what they think they know about themselves and life in the Netherlands, starting at their own doorsteps.

Roots Guide is an interactive, reflexive guidebook of the Netherlands in its final stages of development at the time of writing, with a planned launch in mid-2021. The project emerged in 2016 as a collaboration between a Netherlands-based non-profit organisation – which offers storytelling workshops and events for and by people with diverse backgrounds – and a small group of artists and academics. The resulting guidebook features personal stories and personally significant places of 34 people with a diverse array of internal and international migration experience who draw on their own perspectives and experiences to guide users around 'our Netherlands'. Arranged not around specific regions or cities like a conventional guidebook but rather in global citizenship-inspired thematic clusters, the guides' stories come together to foster users' cognitive and socio-emotional engagement with themes like peace and conflict, identity and belonging, and community participation (UNESCO, 2015).

Co-creating narratives and routes

Because people with migration backgrounds are frequently exoticised (Aytar & Rath, 2012), 'tolerated' (Brown, 2009) and/or marginalised (Gouriévidis, 2014) in dominant heritage discourse and the tools of tourism that propagate it, Migrantour and the Roots Guide seek to de-centre mainstream narratives by valuing diverse and dissenting voices. To do this, both social tourism-based initiatives have made use of participatory storytelling and mapping techniques with their guides to ensure that guides have a significant sense of ownership in both narrative content and route development. In acknowledging that mapping and storytelling are always 'a situated, political process with a social context, purpose and effects' (Counter-Cartographies Collective et al., 2012: 441; Campos-Delgado, 2019), the initiatives explicitly recognise their guides as locals that – regardless of whether they have resided in or intend to reside in these places for a few months or their whole lives – contribute to and intimately know the places in which they live, possessing the authority to speak about these places and the lived and inherited successes, defeats, joys and struggles with which they associate them.

With Migrantour, network partners throughout Europe each implement a roughly standardised route co-creation and guide training programme informed by critical pedagogy and designed as an emancipatory praxis of co-learning, based on principles of citizenship and agency and aimed at strengthening co-learners' socio-political critical reflexivity and creating new knowledge that will challenge dominant discourses, ideologies and practices (Boluk & Carnicelli, 2019). First, a preliminary study is carried out on their respective city's history, focusing on connections between different migratory flows, the transformations of the places in which migrants have come to live, and the tangible and intangible cultural heritages in these places that have emerged as a result. Then, in response to an open call disseminated by network partners, people with first- and second-generation migration backgrounds interested in becoming part-time guides – most often members of cultural associations, cultural mediators, former teachers and tour guides – participate in a free training course. The training course (24-30 hours in length) involves contributions both from academic perspectives (e.g., anthropologists, sociologists, geographers and historians) on topics related to immigration in urban contexts and from professional tour guides on group management and storytelling techniques.

Migrantour guides engage in a walking tour route and narrative co-creation process with network partner staff, fellow guides-in-training, and the academic and professional coaches from the training course. This process begins with guides-in-training drawing a series of mental maps (Lynch, 1960) through which they express their experiences of and relationships with the city in which they live. These maps show their daily routes, the places they usually visit and landmarks significant to them. With their own mental maps based on experiences prior to joining Migrantour in hand, guides-in-training then exchange information with one another, overlaying their personal routes to develop a shared map of significant places. Once this common base has been generated, guides-in-training carry out ethnographic fieldwork – engaging in participant observation, undertaking qualitative interviews with residents and creating a visual archive documenting the changes that have taken place in their residential areas over time.

Finally, the guides' personal narratives and the fieldwork material are rooted to specific places in the area covered by the guided walking tour in which they can most effectively resonate with each other. For example, Rome's Esquilino market has become a stop on a Migrantour walking tour route because it offers an opportunity for guides not only to explain some aspects of the market's history as a meeting place for generations of the city's immigrants but also for guides to recount their personal memories about markets frequented in the places they lived before moving to Rome and to arrange encounters with the market's migrant workers open to describing aspects of their jobs and goods on sale to tour participants. The resulting tangible and intangible heritage presented during the guided walking tours is therefore always the outcome of an intersection of narratives, with no one narrative being privileged over another (see Harrison, 2010). Some Migrantour guides experience 'friction' (Tsing, 2005) when merging these different degrees of narration because they must shift from their own experience and point of view to assume the responsibility of sharing others' stories and points of view. This sometimes requires challenging their own convictions, as V., a Genoa Migrantour guide with a Romanian background, observes:

> Meeting with the other guides also helped me to question myself. For example, before attending the training course, I admit that I had a lot of prejudice towards Roma people, a negative feeling I had developed in Romania and that I brought with me to Italy. Basically, I did not want to be confused with them. During the training I had to work on this attitude and, only thanks to this path that I have taken, now, when I talk about the Roma community during the walks, I'm sincere and credible, starting with admitting that I also had prejudices and telling about how I've tried to overcome them.

The Roots Guide guidebook's narrative development involves prospective guides – identified both through an open call and selective sampling to ensure representation across the country's provinces – volunteering their time to participate in a series of in-depth interviews and photoshoots with the storytelling facilitating team. Through these interviews and photoshoots, guides acquire the skills to develop and reflect upon a story expressing a specific message or question that they wish to share with 'the Dutch public', which they are free to interpret as they wish. Guides' first-person narratives are then crafted by the storytelling facilitation team from the interview material, after which they are reviewed, edited and sometimes completely rewritten by each guide until they are satisfied personally with the result. This months- – sometimes years- – long dialogical co-creation process seeks to ensure that the written and visual stories faithfully reflect the guides' desired message or question as well as their individual voice. Many guides' emotion-infused, hand-drawn point-of-view sketches (Hawthorne et al. 2015) of their meaningful places also are included. Clusters of these written and visual stories then get framed by a set of intra- and inter-personal experiential travel and creative reflection activities co-designed by the production team and guides to help guidebook users engage with the guides' stories and the global citizenship themes that emerge from them as well as to develop users' own stories.

Unlike with Migrantour, Roots Guide's guides did not themselves initially shoulder the responsibility to represent the Roots Guide's broader narrative. Due to the Roots Guide's conceptual evolution over time from a loosely narrated collection of individual stories to a more tightly narrated interactive guidebook, guides' personal narratives were developed in intense collaboration with the storytelling facilitation team but in relative isolation from other guides. As Roots Guide is still in development, guides are at the time of this article's publication beginning to gain a fuller overview of how their personal narratives relate to fellow guides' narratives and to co-develop the user-focused activities that thematically frame their stories. Hence, guides' experience of Roots Guide to date has been mediated largely by their experience with the storytelling facilitation process as well as by regular newsletters and gatherings (e.g., shared meals and guide-initiated workshops and tours).

Guides interviewed perceive the co-creation process as intensely personally empowering. Those with international migration experiences have come to recognise that their perspectives and voices matter. Others with internal migration experiences, meanwhile, have begun to see themselves as migrants, sometimes for the very first time reflecting on the impoverished ways in which migration and privilege are discussed in mainstream Dutch society. One guide, M., a dairy farmer and cheesemaker adopted as a baby from South Korea by white Dutch parents, for example, found the dialogical narrative development process deeply transformative, helping her to better understand and articulate herself in personal, civic and professional realms: 'It would be nice', she observes, 'if everyone could go through this experience once in their lives as part of the process of learning how to be a citizen'. While invited to participate in the Roots Guide initiative as much as they would like, guides demonstrate a varied sense of overall project ownership: some talk about the Roots Guide as 'ours', while others are hesitant to 'impose' themselves, referring to it as 'yours' when engaging with the production team. Therefore, the production team has sought, first, to be very careful to not silence guides' voices where they are most present, such as when editing their stories to adhere to a standardised story length or to ensure narrative coherence and clarity for readers and, second, to ensure that the guidebook's broader narrative framing and activities complement and enhance guides' individual messages.

Avoiding stereotypical 'community' representations

Mutual objectification and stereotyping condition relations between hosts and guests in tourism encounters, oftentimes leading to superficial consumption of the Other instead of meaningful dialogue and exchange (Urry, 2002; Cheng et al., 2010). For Migrantour and the Roots Guide to foster the practice of 'visiting' (Arendt, 1992) as an alternative to the gazing flaneur's practice of tourism (Wearing & Wearing, 1996), our initiatives have had to be continuously alert to the ways in which they both can and do fall into the trap of reproducing multiculturalist representations of 'diversity' and 'community'.

Migrantour network members and guides must constantly reflect on the risks of simplifying and trivialising, even unconsciously, the themes dealt with during the tours. Throughout different points of Migrantour's initial development and in the initiative's spread throughout different European cities, network partners and guides have experimented with the delivery of content during the walks that has produced a range of results and sometimes provoked criticism by walking tour participants and researchers analysing Migrantour's impact. For example, early in the initiative's history, Migrantour network members often opted to engage in 'strategic essentialism' (Spivak, 1990) – the provisional acceptance and tactical use of essentialist foundations for identity categories as a strategy for collective representation to advance a clear political goal. Although sensitive to the challenges posed by this strategy, Migrantour guides were encouraged to bring together a collection of symbolic foods, crafts, photographs, anecdotes and so on that would be useful to evoke the stories gathered during their training course. These objects materialised and reified cultural differences attributed to different immigrant communities during tours, first, by mediating the encounter between the guides and tour participants, making the experience manageable and acceptable for both parties and, second, by mobilising a counter-narrative emphasising migration's role in enriching cities' cultural heritage in the face of dominant discourses stigmatising 'ethnic neighbourhoods' as notorious and dangerous places. In some cases, however, the effects of such choices ran contrary to intentions, effectively consolidating new stereotypes and folklorising cultural differences in ways that led some walking tour participants to perceive themselves as 'looking at immigrants' as if they were in a 'human zoo', rather than having opportunities for mutual dialogue and exchange.

Roots Guide faced similar representational challenges. Its production team initially struggled to avoid reproducing conventional guidebook conventions that exotify, tokenise and commodify others (Lisle, 2008). While its storytelling facilitators were focused on drawing out individual contributors' stories as organically and faithfully as possible, the initial overall framing of the collection of narratives was problematic. In seeking to trouble notions of 'Dutch' identity, several individuals' personal stories were linked to a set of narratives about iconic objects in the Netherlands with foreign origins. For instance, the co-created personal story of N., a Dutch-Turkish shop owner selling Turkish handicrafts, was paired up with a brief history of the iconic 'Dutch' tulip's own Central and West Asian origins that, while inspired by N.'s own reflections, were penned only by the storytelling facilitators. Likewise, the production team developed their own place recommendations accompanying guides' stories with a view to help guidebook users 'travel faraway at home' through places and events linked to the guides' national, cultural and religious origins. Two years into the project, the production team recognised that this first iteration had largely gone the path of Collins and Castillo's' (1998) guidebook celebrating the multicultural character of Sydney, Australia, in the way it categorised guides primarily as representatives of their or their parents' faraway countries of origin. In uncritically celebrating multicultural diversity and exoticising its guides' 'non-Dutch' otherness, the team was essentially eclipsing, and hence de-politicising (Brown, 2009), the Roots Guide's guides' personal everyday lived experiences within the Netherlands.

To turn Roots Guide into a guidebook that would not essentialise, 'other' or exclude, the production team chose to fundamentally trouble and twist elements conventionally deemed fundamental to a guidebook by consciously adopting a 'heritage-from-below' (Robertson, 2012) counter-mapping approach that frames guides as local hosts with the authority to speak about their past and present experiences in the Netherlands on their own terms. Consequently, the places featured in the guidebook now are chosen by guides themselves to highlight personally meaningful places associated with their stories (e.g., the Amsterdam metro line that internal migrant A. appreciates for the great diversity it showcases in contrast with her tiny hometown; the steel factory that reminds L., a second-generation descendent of Spanish guestworkers, of her father; or the suburban McDonald's fast-food restaurant cherished by S. as a meaningful spot for connecting with new friends after moving to the Netherlands from Palestine). This marked departure in approach has required guides to not only think about their stories as contributing to the enrichment of the Netherlands' heritage but also the everyday places significant to them as characters in those stories.

Roots Guide's guides' stories are now strictly about them as individuals, with no 'community' representation taken for granted by storytelling facilitators. The guides themselves decide how they wish to represent themselves and they, by and large, opt to frame themselves as individuals, talking about issues that personally matter to them (e.g., love, education, inequality, legacy, etc.). Though guides are informed that they will be read by 'the Dutch public' as part of an educational tool seeking to 'reframe how migration is thought about in the Netherlands', few have sought to use their stories as platforms from which to make declarations about how 'migrants or refugees are like this or that'. This de-emphasis on framing guides as representatives of specific 'communities', however, has also conflicted with the production team's desire to have greater representative diversity in terms of national origin, religious beliefs, gender, profession and sexual orientation among the guides, acknowledging that Roots Guide is also intended as a space for supporting the voices of groups traditionally marginalised in guidebooks, history books and mainstream media.

Over time, Migrantour also came to dismiss its initial multiculturalist focus on celebrating cultural diversity. In seeking to offer an expanded, more complex heritage narrative, Migrantour began to see their guides not as ambassadors presenting or selling 'their' heritage but rather as storytellers sharing their own views and interpretations of particular parts of a city (Bryon, 2012). Furthermore, in the wake of critical reflection (Mellino & Vietti, 2019; Pozzi & Ceschi, 2019), Migrantour network members opted to foreground ethnicity's intersections with gender, age, social class, length of residence and other variables in examining the outcome of integration processes and inter-group relations between 'majorities' and 'minorities' (Pastore & Ponzo, 2016). They are increasingly committed to avoiding the production of normalised, pacified and univocal representations of the neighbourhoods featured in the walking tours, discussing with guides how to hold more space for the emergence of contradictions, conflicts and resistance in their narratives. Consequently, inequalities and relationships of subordination and domination related to the neoliberal distribution of power and that concern tourism's (and Migrantour's) impacts on people and the cities in which they live are highlighted more explicitly:

> I live in the Porta Palazzo district and mostly accompany school groups on visits to the district. Using language appropriate to their age, I always try to explain to the children why there are houses in the area like the one where I live, with small apartments, lots of people and no lift, and others are all renovated [...] I also explain to them why there are certain messages written on the walls, why they see so many police around and why some people would like to remove poor people from the neighbourhood in order to open new shopping centres and trendy clubs. I don't use the word 'gentrification' with them, but I give them concrete examples of how changes in the neighbourhood affect my life as a citizen with a migration background.
>
> (M., a Turin Migrantour guide)

In the last three years, the Migrantour network has cautiously expanded, seeking partners in new cities that will not manage Migrantour walks as a stand-alone activity but, rather, as part of a broader commitment to social cohesion in their respective areas. In the Italian city of Catania, for example, Migrantour walks are organised by an organisation that works with residents with both migration and non-migration backgrounds to implement actions aimed at regenerating the city's San Berillo neighbourhood 'from below'. In this way, Migrantour distinguishes itself from 'slum tourism' by committing to not making diversity, poverty and marginality the object of the tourist gaze (Frenzel, 2020). Instead, by deconstructing migration-related stigma and stereotyping as well as embedding 'immigration issues' within broader socio-economic forces (e.g., unemployment and housing dynamics) that bring about societal transformation, the tours are intended to function as vehicles for residents 'to understand the everyday places where they live, work, shop and socialise', providing them 'with a basis for fighting proposed changes that often destroy the centres of social life, erase cultural meanings and restrict local participatory practices' (Low, 2011: 391) and addressing the poverty-tourism nexus in urban areas by involving marginalised and vulnerable groups in a process of empowerment (Dolezal & Gudka, 2019).

Fostering conditions for 'visiting'

Migrantour and the Roots Guide have chosen the guided walking tour and guidebook, respectively, as public pedagogical tools to enhance participants' and users' perspective consciousness within the places in which they live. While the initiatives share a focus on integrating migrants and migration into local and national heritage discourse, their differing modes of delivery have unique parameters, offering different possibilities for participants and users to engage with the initiatives' objectives. Guided walking tours offer physical proximity and potential for dialogue with guides and fellow tour participants, with real-time updating and tailoring of content to tour participants' needs and interests. Guidebooks, on the other hand, enable users – individuals or in small groups of family or friends – to turn to them when and how they like. Yet they communicate unidirectionally to a specific audience and their content can quickly become out-dated. Guided walking tours necessitate tangible engagement with heritage, while guidebook narratives can compensate for potential lacks in visible heritage with the emotional pull of their stories and descriptive detail. Furthermore, walking tours require an itinerary of experiences concentrated within time and space, while a guidebook does not need to be used in a linear or continuous fashion. Here, we draw attention to some ways in which these formats' diverse characteristics impact how we have developed content to foster conditions conducive to the practice of 'visiting' (Arendt, 1992).

Migrantour's initial focus was on national and international tourists interested in discovering neighbourhoods not usually included in conventional city tours, guidebooks and tourist maps. Over time, however, most network partners realised that the guided walking tours mainly attracted local residents wishing to know some parts of their city better and school-aged children on educational excursions. This increasingly led guides to see themselves as facilitating access to everyday places in neighbourhoods frequently perceived by other city residents as alien and inhospitable, potentiating a more invested exchange reminiscent of Jane's Walks, an initiative inspired by urban activist Jane Jacobs's work championing the voices of everyday people in planning and civic involvement. This is evidenced in an account from A., a Milan Migrantour guide with an Italo-Moroccan background:

> The most important stop on the Via Padova walking tour, in my opinion, is when we go to the House of Muslim Culture. It always happens that none of the participants has ever entered an Islamic space of worship in Italy. So, we explain that we must take off our shoes, then the imam – someone who has lived in Milan for 50 years and knows everything about the neighbourhood – welcomes us, we sit in a circle, and we talk for at least half an hour, answering all the tour participants' doubts and questions. When the meeting ends, people have a completely new idea of what Islamic prayer spaces in Italy are like and how they work.

In physically uniting local hosts capable of making these places accessible and understandable to tour participants living in the same areas who have imagined them through the filter of popular stereotypes and prejudices, guides like A. are actively 'opening the city' (Sennett, 2018) to render urban borders more porous within an ever-more xenophobic political context.

While Migrantour's guided walking tours provide the opportunity for immediate encounter and dialogue, Roots Guide – which targets residents of all ages in the Netherlands with some travel experience that are open to engaging in a different kind of interaction and dialogue – focuses on developing connection between its local guides and users differently. In harnessing the emotive power of storytelling to connect people, its interactive format foregrounds not only *guides'* personal narratives and places significant to them but also the recognition and development of *guidebook users'* own stories and the constellation of places meaningful to them. In disrupting the guidebook format's conventional unidirectional gaze, Roots Guide ultimately seeks to map both its guides' and guidebook users' heritages onto the contemporary landscape and put them into dialogue, troubling constructions of 'Dutchness', 'strangerhood' and 'localhood' through this juxtaposition. It does this via exposing guidebook users to first-person stories thematically clustered and arranged in a narrative arc that brings attention to the various drivers for migration, the challenges of setting up one's life in a new place, and the ways in which a sense of home and connection get forged. Each cluster of guides' stories is supported by a suite of 'inner' and 'outer' journeys comprising interactive reflexive activities designed to enhance users' emotional, cognitive and physical connections to the guides' stories and places associated with them as well as with users' own everyday landscapes and companions. These 'inner' and 'outer' journeys include: 1) questions that foster users' dialogic engagement with the guides' stories and broader global citizenship concerns (e.g., various forms of privilege, stereotypes, constructions of home, practices of community, etc.); 2) guides' personally meaningful places linked to their stories that users can visit; 3) experiential travel activities that draw on mindfulness and psychogeographical techniques to spark mindsets more familiar to users when they travel far away from home than in their everyday lives (e.g., spontaneity, curiosity, appreciation and reflexivity); and 4) pages with prompts and dedicated writing space on which users are encouraged to build up the layers of their own stories.

Conclusion

For Arendt (1977), the ability to develop informed individual judgement and engage in adequate deliberation and responsible collective action in democratic societies necessitates being able to think 'representatively' – in other words, a multi-perspectivity that both entails being conscious that perspectives beyond one's own exist and actively recognising the legitimacy of a range of these different perspectives. Critiquing both 'tourism' and 'empathy' as modes of engaging with others, Arendt (1977) advances the notion of 'visiting' as a preferred mode of civic learning through which to develop 'a certain kind of modesty: living one among many, engaged in a world that does not mirror oneself[, enabling] richness of meaning rather than clarity of meaning' (Sennett, 2018: 302). With this article, we have embraced 'visiting' (Arendt, 1992) as an alternative ethical practice to conventional tourism practices that have long served as vehicles for perpetuating a multicultural discourse of 'tolerance' (Brown, 2009; Dill, 2013) that abstracts, naturalises and depoliticises oppression, inequality and conflict. In so doing, we have sought to engage with what Blanchard and Higgins-Desbiolles (2013: 2) call 'the invisible elephant in the room': '[t]he essential tension between the multiple dimensions of peace (as a process) and the confining dimensions of tourism (as a product)' by suggesting that those working in tourism not only face the urgent task of replacing oppressive content with more progressive messages but also that some of the most common pedagogical vehicles of tourism – like guided tours and guidebooks – need to be fundamentally retooled in order to adequately convey those messages and support dialogic civic encounters with the diverse people voicing them.

Because simply being exposed to and immersed in the stories and points of view of others does not, in and of itself, necessarily foster the disorientation necessary for personal transformation (Reisinger, 2013) or lead to meaningful cross-cultural dialogue, especially in short visits, it becomes even more imperative for guided tours and guidebooks to adapt in ways such that they become 'good company' (Gillies, 2016) capable of fostering the discovery of 'intersection points' between hosts and visitors 'from which to intervene and struggle together' (Counter-Cartographies Collective et al. 2012: 447; Christie & Mason, 2003). In this article, we used Migrantour and the Roots Guide – two Europe-based initiatives aimed at 'mobilising hospitality' (Germann-Molz & Gibson, 2007) to resist dominant discourse associating 'the host with home, territory, stability, and ownership on one side, and [...] the guest with mobility, estrangement and un-belonging on the other' (Germann-Molz & Gibson, 2007: 16) – to examine how we might trouble and twist conventional guided tours and guidebooks into pedagogical tools offering 'good company' that enable 'visiting' (Arendt, 1992) in different European contexts where xenophobia and anti-immigrant sentiment are on the rise. To offer 'good company', Migrantour and the Roots Guide have focused on co-creating multi-dimensional tour narratives and routes in dialogic ways that recognise guides' diverse experiences and perceptions of the places they call home, avoid stereotypical representations of 'communities' and hold space for the real-life frictions that accompany diversity and its management. Doing so has required explicit engagement with three different but overlapping challenges: narrative ownership, representation and participant/user engagement.

First, in contrast with conventional tools of tourism that frequently advance 'tolerant' narratives *about* 'the (migrant) Other', Migrantour and the Roots Guide have explicitly embraced a 'nothing about me without me' approach by recruiting people with diverse migration backgrounds as guides and by foregrounding guides' involvement in the development of narrative content and routes. Participatory storytelling and (auto-)ethnographic mapping techniques have been essential to accomplish this, with guides themselves being stimulated individually and collectively to develop both the consciousness and competence of multi-perspectivity, strengthening guides' critical reflexivity (Christie & Mason, 2003). In this way, even before reaching the general public, the initiatives' public pedagogical objectives get enacted during the process of narrative and route co-creation.

Second, unlike conventional tools of tourism that often represent 'the (migrant) Other' as members of 'exotic' communities distinct and divorced from 'our own', Migrantour and the Roots Guide have sought to recognise the 'localhood' of people with migration backgrounds and their corresponding authority to contest and broaden local and national heritage discourses that may be dissonant with essentialist constructions of 'culture' and 'community' typical within multicultural 'tolerance' discourse (Brown, 2009; Dill, 2013). A key challenge for both Migrantour and the Roots Guide has been to avoid reproducing and reifying superficial multicultural tropes through which both guides and tour participants/guidebook users are used to engaging with each other. In order to open up and hold space for the discomfort for all involved to acknowledge and engage with 'the constitutive histories and powers organising contemporary problems and contemporary political subjects' (Brown, 2009: 16), both initiatives have adopted a global citizenship-aligned ethic of framing people both with and without migration backgrounds first and foremost as individuals with intersectional collective identities, allowing for 'immigrant issues' to be unpacked in relation to gender, age, class and broader socio-economic forces. Recognition of the significance of holding space for emotional discomfort, an essential component of transformative learning, is increasing among scholars concerned with pedagogy of/as/and tourism and strengthening tourism's potential for bringing about societal change (Werry, 2008; Reisinger, 2013; Walker & Manyamba, 2020).

Finally, Migrantour and the Roots Guide have retooled conventional disembodied guided tour and unidirectionally narrated guidebook formats to enable 'visiting' (Arendt, 1992) by embracing an interactive, dialogic approach that recognises the significance not only of the authority of guides' interpretations but also of visitors' own interpretations and of bringing these in relation to one another in ways that facilitate deeper, more reflexive learning (Ablett & Dyer, 2009). This dialogic approach holds potential to undermine the tacit co-enactment in tourism described by Cheng et al. (2010: 116) of mutual 'strangerhood' and of service.

While a straightforward path to facilitating progressive social change may be lacking, much can be done within tourism to strengthen providers' and users' awareness of its profound pedagogical potential, of what kinds of politics conventional tourism tools and practices are already inculcating, and of the urgency for consciously troubling and carefully twisting these in ways that work towards achieving more humane ways of living together in an ever-more diverse world.

Acknowledgements

We would like to sincerely thank our article's three anonymous reviewers, Ingeborg Mehus and Chih-Chen Trista Lin for very helpful comments that have strengthened the article's message and honed its contribution to the field.

Disclosure statement

Meghann Ormond has been, since 2018, the curator/editor and scientific coordinator of Roots Guide, a project legally owned by the non-profit organisation Stichting Pocket Stories, registered at the Kamer van Koophandel in the Netherlands under registration number 66886090 with ANBI status from 21/06/2017. Her position is unpaid and voluntary. Francesco Vietti was the scientific coordinator for the Migrantour project from 2009-19. He continues to collaborate with the social tourism cooperative Viaggi Solidali (cooperative registration number: A161747 of 05/01/2005), a Migrantour network partner. The authors do not financially benefit from this research or its publication.

ORCID

Meghann Ormond http://orcid.org/0000-0002-4056-9619
Francesco Vietti http://orcid.org/0000-0002-5603-3392

References

Ablett, P. G., & Dyer, P. K. (2009). Heritage and hermeneutics: towards a broader interpretation of interpretation. *Current Issues in Tourism*, *12*(3), 209–233. https://doi.org/10.1080/13683500802316063

ACRA. (2016). *Migrantour: A European network of migrant-driven intercultural routes to understand cultural diversity: Final report*. Fondazione ACRA.

ACRA. (2020). *New Roots – Migrantour intercultural walks building bridges for newcomers' active participation: Final Report*. Fondazione ACRA.

Anderson, B. (2006). *Imagined communities: Reflections on the origin and spread of nationalism*. Verso Books. (Original work published 1984)

Arendt, H. (1977). Truth and politics. In *Between past and Future* (pp. 227–264). Penguin.

Arendt, H. (1992). *Lectures on Kant's political philosophy*. University of Chicago Press.

Arendt, H. (2003). *Responsibility and judgment*. Random House.

Ashworth, G., & Tunbridge, J. (1996). *Dissonant heritage: The management of the past as a resource in conflict*. Wiley.

Aytar, V. & Rath, J. (Eds.). (2012). *Selling ethnic neighbourhoods: The rise of neighbourhoods as places of leisure and consumption*. Routledge.

Basso Peressut, L. & Pozzi, C. (Eds.). (2012). *Museums in an age of migrations*. Politecnico di Milano.

Beck, S. & Maida, C.A. (Eds.). (2013). *Toward engaged anthropology*. Berghahn.

Belhassen, Y., & Caton, K. (2011). On the need for critical pedagogy in tourism education. *Tourism Management*, *32*(6), 1389–1396. https://doi.org/10.1016/j.tourman.2011.01.014

Bhattacharyya, D. P. (1997). Mediating India: An analysis of a guidebook. *Annals of Tourism Research*, *24*(2), 371–389. https://doi.org/10.1016/S0160-7383(97)80007-2

Biesta, G. (2016). Reconciling ourselves to reality: Arendt, education and the challenge of being at home in the world. *Journal of Educational Administration and History*, *48*(2), 183–192. https://doi.org/10.1080/00220620.2016.1144580

Biesta, G., & Cowell, G. (2012). How is community done? Understanding civic learning through psychogeographic mapping. *International Journal of Lifelong Education*, *31*(1), 47–61. https://doi.org/10.1080/02601370.2012.636587

Blanchard, L.A. & Higgins-Desbiolles, F. (Eds.). (2013). *Peace through tourism: Promoting human security through international citizenship*. Routledge.

Boluk, K., & Carnicelli, S. (2019). Tourism for the emancipation of the oppressed: Towards a critical tourism education drawing on Freirean philosophy. *Annals of Tourism Research, 76*, 168–179. https://doi.org/10.1016/j.annals.2019.04.002

Brown, W. (2009). *Regulating aversion: Tolerance in the age of identity and empire*. Princeton University Press.

Butcher, J. (2017). Citizenship, global citizenship and volunteer tourism: a critical analysis. *Tourism Recreation Research, 42*(2), 129–138. https://doi.org/10.1080/02508281.2017.1295172

Bryon, J. (2012). Tour guides as storytellers–from selling to sharing. *Scandinavian Journal of Hospitality and Tourism, 12*(1), 27–43.

Campos-Delgado, A. (2018). Counter-mapping migration: irregular migrants' stories through cognitive mapping. *Mobilities, 13*(4), 488–504. https://doi.org/10.1080/17450101.2017.1421022

Cheng, W., Barraclough, L., & Pulido, L. (2010). Radicalising teaching and tourism: A People's Guide as active and activist history. *Left History, 15*(1), 111–127.

Christie, M. F., & Mason, P. A. (2003). Transformative tour guiding: Training tour guides to be critically reflective practitioners. *Journal of Ecotourism, 2*(1), 1–16. https://doi.org/10.1080/14724040308668130

Cohen, E. (1985). The tourist guide: The origins, structure and dynamics of a role. *Annals of Tourism Research, 12*(1), 5–29. https://doi.org/10.1016/0160-7383(85)90037-4

Collins, J., & Castillo, A. (1998). *Cosmopolitan Sydney: A guide to Sydney's multicultural people and places*. Pluto Press.

Council of Europe (2005). Council of Europe framework convention on the value of cultural heritage for society, Faro, 27.X.2005, Series - No. 199.

Counter-Cartographies Collective, Dalton, C., & Mason-Deese, L. (2012). Counter (mapping) actions: Mapping as militant research. *ACME, 11*(3), 439–466.

Dill, J. S. (2013). *The longings and limits of global citizenship education*. Routledge.

Disch, L. (1997). Please sit down, but don't make yourself at home': Arendtian 'visiting' and the prefigurative politics of consciousness-raising. In Calhoun, C. & McGowan, J. (Eds.), *Hannah Arendt and the meaning of politics* (pp. 132–165). University of Minnesota Press.

Dolezal, C., & Gudka, J. (2019). London's 'Unseen Tours': Slumming or Social Tourism?. In Smith, A. and Graham, A. (Eds.), *Destination London: The expansion of the visitor economy* (pp. 141–163). University of Westminster Press.

Frenzel, F. (2020). Touring poverty in townships, inner-city, and rural South Africa. In Rogerson, J.M. & Visser, G. (Eds.), *New directions in South African tourism geographies* (pp. 167–181). Springer.

Gatta, G. (2014). Visiting or house-swapping? Arendt and Jaspers on empathy, enlarged mentality and the space between. *Philosophy & Social Criticism, 40*(10), 997–1017. https://doi.org/10.1177/0191453714548501

Germann-Molz, J. & Gibson, S. (Eds.). (2007). *Mobilizing hospitality*. Ashgate.

Gillies, D. (2016). Visiting good company: Arendt and the development of the reflective practitioner. *Journal of Educational Administration and History, 48*(2), 148–159. https://doi.org/10.1080/00220620.2016.1144576

Gouriévidis, L. (Ed.). (2014). *Museum and migration. History, memory and politics*. Routledge.

Grosz, E. (1995). *Space, time and perversion: Essays on the politics of bodies*. Routledge.

Harrison, R. (2010). Heritage as social action. In West, S. (Ed.), *Understanding heritage in practice* (pp. 240–276). Manchester University Press.

Hawthorne, T. L., Solís, P., Terry, B., Price, M., & Atchison, C. L. (2015). Critical reflection mapping as a hybrid methodology for examining socio-spatial perceptions of new research sites. *Annals of the Association of American Geographers, 105*(1), 22–47. https://doi.org/10.1080/00045608.2014.960041

Higgins-Desbiolles, F., & Blanchard, L. (2010). Challenging peace through tourism: Placing tourism in the context of human rights, justice and peace. In Moufakkir, O. & Kelly. I. (Eds.), *Tourism progress and peace* (pp. 35–47). CABI.

Hintermann, C., & Johansson, C. (2010). *Migration and memory: Representations of migration in Europe since 1960*. StudienVerlag.

Hollinshead, K., Ateljevic, I., & Ali, N. (2009). Worldmaking agency–worldmaking authority: The sovereign constitutive role of tourism. *Tourism Geographies, 11*(4), 427–443. https://doi.org/10.1080/14616680903262562

Kemmis, S., McTaggart, R., & Nixon, R. (2013). *The action research planner: Doing critical participatory action research*. Springer.

Lisle, D. (2008). Humanitarian travels: Ethical communication in Lonely Planet guidebooks. *Review of International Studies, 34*(S1), 155–172. https://doi.org/10.1017/S0260210508007845

Low, S. M. (2011). Claiming space for an engaged anthropology: spatial inequality and social exclusion. *American Anthropologist, 113*(3), 389–407. https://doi.org/10.1111/j.1548-1433.2011.01349.x

Lynch, K. (1960). *The image of the city*. MIT Press.

Mair, H., & Sumner, J. (2017). Critical tourism pedagogies: Exploring the potential through food. *Journal of Hospitality, Leisure, Sport & Tourism Education, 21*, 195–203. https://doi.org/10.1016/j.jhlste.2017.06.001

Mannion, G., Biesta, G., Priestley, M., & Ross, H. (2011). The global dimension in education and education for global citizenship: genealogy and critique. *Globalisation, Societies and Education, 9*(3-4), 443–456. https://doi.org/10.1080/14767724.2011.605327

Mellino, M., & Vietti, F. (2019). L'antropologia applicata tra 'tecniche di mercato' e 'pratiche politiche. Riflessioni Sui Migranti, Migrantour e Noi. *Antropologia Pubblica, 5*(1), 123–132.

Modlin, E. A., Alderman, D. H., & Gentry, G. W. (2011). Tour guides as creators of empathy: The role of affective inequality in marginalizing the enslaved at plantation house museums. *Tourist Studies, 11*(1), 3–19. https://doi.org/10.1177/1468797611412007

Nussbaum, M. (2002). Education for citizenship in an era of global connection. *Studies in Philosophy and Education, 21*(4/5), 289–303. https://doi.org/10.1023/A:1019837105053

Pastore, F. & Ponzo, I. (Eds.). (2016). *Inter-group relations and migrant integration in European cities*. Springer.

Pink, S. (2005). *Applications of anthropology*. Berghahn.

Pozzi, G., & Ceschi, S. (2019). L'antropologia applicata tra 'tecniche di mercato' e 'pratiche politiche. Riflessioni Sui Migranti, Migrantour e Noi. *Antropologia Pubblica, 5*(2), 127–144.

Rath, J. (Ed.) (2007). *Tourism, ethnic diversity and the city*. Routledge.

Reisinger, Y. (Ed.). (2013). *Transformational tourism: Tourist perspectives*. CABI.

Robertson, I. J. M. (2012). *Heritage from below*. Ashgate.

Sandlin, J. A., Schultz, B. D., & Burdick, J. (2010). Understanding, mapping and exploring the terrain of public pedagogy. In Sandlin, J. A., Schultz, B. D. and Burdick, J. (Eds.), *Handbook of public pedagogy* (pp. 1–6). Routledge.

Sennett, R. (2018). *Building and dwelling: Ethics for the city*. Penguin.

Shaw, S., Bagwell, S., & Karmowska, J. (2004). Ethnoscapes as spectacle: Reimaging multicultural districts as new destinations for leisure and tourism consumption. *Urban Studies, 41*(10), 1983–2000. https://doi.org/10.1080/0042098042000256341

Smith, L. (2006). *Uses of heritage*. Routledge.

Spivak, G. C. (1990). *The post-colonial critic: Interviews, strategies, dialogues*. Routledge.

Tsing, A. L. (2005). *Friction: An ethnography of global connection*. Princeton: Princeton University Press.

UNESCO (2015). Global citizenship education: Topics and learning objectives. Paris: UNESCO. http://unesdoc.unesco.org/images/0023/002329/232993e.pdf

Urry, J. (2002). *The tourist gaze*. Sage.

Walker, J., & Manyamba, V. N. (2020). Towards an emotion-focused, discomfort-embracing transformative tourism education. *Journal of Hospitality, Leisure, Sport & Tourism Education, 26*, 100213. https://doi.org/10.1016/j.jhlste.2019.100213

Wansink, B., Akkerman, S., Zuiker, I., & Wubbels, T. (2018). Where does teaching multi-perspectivity in history education begin and end? An analysis of the uses of temporality. *Theory & Research in Social Education, 46*(4), 495–527. https://doi.org/10.1080/00933104.2018.1480439

Wearing, B., & Wearing, S. (1996). Refocussing the tourist experience: The flaneur and the choraster. *Leisure Studies, 15*(4), 229–243. https://doi.org/10.1080/026143696375530

Werry, M. (2008). Pedagogy of/as/and tourism: Or, shameful lessons. *The Review of Education, Pedagogy & Cultural Studies, 30*(1), 14–42. https://doi.org/10.1080/10714410701566207

Witte, M. (2019). Black citizenship, Afropolitan critiques: vernacular heritage-making and the negotiation of race in the Netherlands. *Social Anthropology, 27*(4), 609–625. https://doi.org/10.1111/1469-8676.12680

Zagato, L. (2015). The notion of 'Heritage Community' in the Council of Europe's Faro Convention. Its impact on the European legal framework. In Adell, N. (Ed.), *Between imagined communities and communities of practice. Participation, territory, and the making of heritage* (pp. 141–168). Universität Verlag.

One stone, two birds: harnessing interfaith tourism for peacebuilding and socio-economic development

Dagnachew Leta Senbeto

ABSTRACT
This study examines the importance of interfaith tourism to peace and socio-economic development and assesses factors that influence interfaith tourism vis-à-vis its role in peacebuilding and socio-economic development. The researcher adopted an exploratory research design and conducted in-depth semi-structured interviews and media analysis. The findings revealed that interfaith tourism constitutes a powerful social force that can reinforce peace and socio-economic development. However, challenges related to low tourism development, poor engagement, and collaboration among tourism stakeholders limit interfaith tourism benefits. This research contributes to the scarce literature concerning the socio-political aspects of tourism and the importance of interfaith tourism for peace and local development. It is argued that interfaith tourism supports the United Nation Sustainable Development Goal (UN SDG) 16 in maintaining peace and social justice in developing nations like Africa. Practically, the findings benefit policymakers and tourism stakeholders in nurturing peace, inclusive societies, and unanimity using interfaith tourism development.

Introduction

The world continues to face several challenges that stem from various issues such as conflict, racism, social unrest, and the COVID-19 pandemic. Hence, mutual understanding, sustainable wellbeing, and peacebuilding constitute an important agenda for countries across the globe to alleviate social crises. Studies have suggested that tourism can be an instrument to reinforce peace and tolerance, social justice, and mutual understanding among societies in the aftermath of the COVID-19 pandemic (Higgins-Desbiolles, 2020; Sigala, 2020). In his pilgrimage to Mecca, Malcolm X (the prominent African American freedom advocator) experienced racial indifference, mutual understanding, and dialogue between diverse constituencies. As stated in his autobiography, Malcolm X noted that he became open to human and social conditions regardless of racial and religious backgrounds (Malcolm, 1965). Despite the traditional view of tourism as an economic activity, tourism is also a relevant social force. Consequently, research interest in the social and political aspects of tourism has increased in the last decade (Higgins-Desbiolles, 2006; Litvin et al., 2020). Among several tourism segments, interfaith tourism has additional potential to maximize cross-cultural and mutual understanding among diverse constituencies. Owing to the limited number of studies on interfaith tourism in the tourism literature, there is no clear definition of this concept. Interfaith tourism, one aspect of religious tourism, is practiced by visitors of multiple faiths, including followers and non-followers of a particular religion (McIntosh et al., 2019; Öter & Çetinkaya, 2016). Interfaith religious studies are concerned with enhancing dialogue among religious leaders and religious coexistence and harmony among people of different faiths.

Drawing on social resilience and dynamic capability approaches and assessing the interactive features among Christians and Muslims in Ethiopia, I extend the debate on the importance of tourism as a powerful social force for peacebuilding and the common good. More importantly, this paper is initiated in light of Senbeto's (2020) discussion concerning the potential and challenges of interfaith tourism in Ethiopia. The aim of this study is two-fold: 1) assess interfaith tourism potential for peacebuilding and socio-economic development, and 2) explore factors that influence interfaith tourism in peacebuilding and socio-economic development. This paper contributes to theory and practice in several ways. Theoretically, it contributes to the debate on the relevance of social resilience and dynamic capabilities of interfaith tourism to uphold peace and socio-economic development. Considering academic discourses related to the role of tourism in safeguarding the fine balance between the commercial benefit of religious tourism and the larger agenda of peacebuilding (Sharma, 2013), this paper demonstrates the reality of interfaith religious harmony in Ethiopia and the lived scenario of collaboration at the grassroots level and its importance for peacebuilding and socio-economic development. The findings contribute to the United Nations Sustainable Development Goals (UN SDG) 16 in fostering peace, social justice, and community cohesion through the pursuit of interfaith tourism. Practically, the insights adduced in this study will be beneficial to destination management organizations, policymakers, and concerned stakeholders by drawing their attention to the view that interfaith tourism can contribute to peace and socio-economic development.

Literature review

Religion, tourism, and peace

There is a relationship between religion and tourism in that the former is a motivator, and the latter is an originator for visitor attraction. Religious tourism is one of the earliest forms of tourism. According to some accounts, it is now among the fastest-growing tourism segments in terms of visitor number and generating tourism receipt (Olsen & Esplin, 2020; Terzidou et al., 2018). Venerated places such as churches, mosques, monasteries, temples, and cathedrals attract tourists worldwide. Values (i.e. natural or cultural) play a crucial role as far as the relationship between religion and tourism is concerned; hence, Hughes et al. (2013) define religious tourism "as tourism to sites of current and/or previous religious significance" (p. 210). The current religious tourism literature associates the motive to visit sacred sites with cultural tourism features in which people are attracted to religious sites for not only sacredness, spirituality, or ritual activities, but also reasons related to tradition, education, society, or culture (Albayrak et al., 2018; Buzinde, 2020). In addition to these attractions, artefacts and storytelling in and around religious tourist destinations fascinate tourists.

Tourism is known for promoting knowledge and cross-cultural understanding among individuals and societies. Although the religious tourism literature has primarily focused on the economic aspects driven by the traditional orientation of economic benefits, the association between tourism and socio-political issues like peace is gaining attention (Bandyopadhyay, 2019; Collins-Kreiner, 2020; Timothy, 2019). Tourists develop a sense of belonging, tolerance, and empathy with other people by visiting new places and interacting with different cultures and lifestyles. The proposition that tourism contributes to peace and tolerance is essential to social justice, global understanding, and conflict resolution, and it helps to build a more harmonious world. The concept of peace through tourism was first introduced in 1929 in a conference themed "Travel for Peace" organized by the British Travel and Holidays Association (Wohlmuther & Wintersteiner, 2014). However, the 1980 Manila Declaration on World Tourism described tourism as a "vital force for peace and international understanding" was the trailblazer in establishing a nexus between tourism and peace (UNWTO, 1980). Subsequently, the 1985 Tourism Bill of Rights and Tourist Code, and the 1995 Charter for Sustainable Tourism explained the importance of tourism to global peace and mutual understanding (UNWTO, 1985). Yet, scholars have noted that a deeper understanding of the mechanisms of how tourism promotes peace is lacking. In relation to this, previous studies have suggested more research focusing on the value of tourism to peacebuilding, cross-cultural understanding, and community cohesion (Chien & Ritchie, 2018; Farmaki, 2017; Higgins-Desbiolles, 2006). Hence, the present paper explores interfaith religious tourism and its impact on peacebuilding and local economic development.

Interfaith tourism, peace and socio-economic development: Social resilience and dynamic capabilities approaches

Interfaith features are revealed through mutual understanding established on the harmonious coexistence of culture, people, and religions or demonstrated by interreligious dialogue among religious leaders. The literature on interfaith religious tourism is a subset of the religious tourism literature. Interfaith tourism is associated with a faith other than one's own and is considered an opportunity to express traditional, cultural, and religious beliefs and practices. A major feature of interfaith tourism is that visitors belonging to several faiths or atheists (no faith) travel to a religious tourism destination for recreational, cultural, or educational values (McIntosh et al., 2019; Öter & Çetinkaya, 2016). Interfaith issues have been studied in several social science disciplines and fields of study, mainly in sociology and anthropology. For example, building on the religious anthropology perspective in Lebanon, Haddad (2014) asserted the benefits of interfaith features in facilitating spontaneous and natural dialogue among leaders and members of different religions. In addition, Farra-Haddad mentioned that pilgrimage to interfaith tourism destinations enhances mutual understanding and cross-cultural understanding among tourists and local communities. In a similar vein, Fartacek (2012) notes how religious sites in Syria received considerable respect by several religious groups, including non-members of a particular religious group like the Sunnites, Christians, Alawites, and Druzes, who pave the way for respect and understanding. However, compared to other social science disciplines like sociology and anthropology, less research has been carried out on interfaith issues in tourism literature.

Previous studies have noted that social resilience is used as a lens to understand the socio-cultural attributes in managing crises and conflict resolution (Espiner et al., 2017; Kutzner, 2019). According to Adger (2000), social resilience is defined as "the ability of groups or communities to cope with external stresses and disturbances as a result of social, political, and environmental change" (p. 347). While the economic role of tourism has been extensively examined, the social and political power of tourism is relatively neglected (Chilufya et al., 2019; Mostafanezhad & Norum, 2019). In addition, studying dynamic capabilities in specific local contexts helps to understand how locally driven competence and treasures can support sustainable development and propel the process of creating values and reinforce positive implications to socio-economic development (Salvato & Vassolo, 2018; Teece, 2009). Drawing from resource-based theory (Barney, 2001), dynamic capability underlines a diverse constituency of resources, including intangible treasures such as beliefs, norms, and traditions. Past studies have explained dynamic capability as the ability or potential to inherently integrate, shape, and reconfigure internal and external resources or experiences to address or deal with changing or challenging situations (Ferreira et al., 2020; Nonaka et al., 2016). Bearing this in mind, adopting social resilience and dynamic capability approaches in the present study helps understand the importance of inter-faith tourism to peacebuilding and socio-economic development.

Method

Research context

Bordering the Red Sea in the northeastern part of Africa, Ethiopia has a population of over 100 million (World Population Review, 2020). Ethiopia is home to ancient civilizations and is known for safeguarding its sovereignty from major foreign aggression including colonialism. Archaeological and paleontological evidence state that Ethiopia is the "cradle of humankind." (Brunet, 2010; Hilton-Barber, 2004). The country is recognized as the origin of coffee and the Blue Nile (the world's longest river). Ethiopia is known for being a "land of extremes" with its diverse topographic and landscape features. Ethiopia's mosaic of cultures and civilizations is its greatest gift with its diversified culture, tradition, ethnicity, and language (Senbeto, 2020). Bearing such uniqueness and diversity, Ethiopia owns several natural and cultural attractions of which nine of them are listed as world heritage sites by UNESCO (United Nation Educational, Scientific, and Cultural Organization). Although Ethiopia is endowed with rich religious and cultural treasures and inbound tourism growing from time to time, tourism is currently an under-developed economic segment and its contribution to the national economy has remained limited. According to the World Bank report, annual tourist arrivals in Ethiopia are less than 1 million (849,000) with tourist's expenditure growth of 46% (World Bank, 2020) (Figure 1).

Religion and cultural diversity are the primary attributes of tourism in Ethiopia. The name "Ethiopia" is featured in the history of the two main world religions, Christianity and Islam (Haggai, 2010). In the Bible, the name "Ethiopia" is mentioned more than 42 times. In 330 C.E., Ethiopia was the second nation after Armenia to adopt Christianity when King Ezana was baptized (Ray, 2012). The country has played a crucial role in the history of Islam. For example, Ethiopia saved Prophet Mohammed's followers from the hardships they encountered at home. According to the national census conducted in 2007, Ethiopia has over 32 million followers of the Ethiopian Orthodox Church (43.5%) and 25 million (33.9%) followers of Islam. 18 million (20.6%) people identified as Protestant, and the remaining 2% adhered to traditional religions (Ethiopian Central Statistical Agency, 2007).

Figure 1. World heritage sites in Ethiopia.
Source: Google map (2020)

As the report shows, Christianity and Islam are the two major religions in Ethiopia; both religions have numerous tourism and heritage resources. Churches, monasteries, religious festivals, and precious objects attract local and international tourists. The Ethiopian Orthodox Church, for example, possesses several historical, cultural, and architectural assets in both material and non-material forms. Besides, in many different parts of Ethiopia, mosques, old Islamic objects, and shrines draw substantial tourism potential. Building on the interfaith features of Christian-Muslim religions, this study examines the importance of interfaith tourism to peace, stability, and socio-economic development. The paper also explores the challenges that restrain interfaith tourism potential and development.

Data collection and procedures

This paper examines the potential of interfaith tourism in promoting peace and socio-economic development. The researcher employed an interpretive/constructivist approach (Guba & Lincoln, 1994) given the viability of this method to explore outlooks drawn from shared understanding, realities, experiences, and backgrounds. The realities of interfaith religious tourism result from various social interactions, experiences, and mental constructions. Unlike a positivist approach that relies on theory and quantitative hypothesis testing, the constructivist researcher pursues qualitative data collection and analysis mechanisms to understand lived experiences and phenomena in context. A multi-source data collection method was employed to collect and analyze both primary and secondary qualitative evidence (Saunders et al., 2009). For primary data collection, semi-structured interviews were conducted with 16 informants composed of government officials, community and religious leaders, visitors, and experts, as shown in Table 1.

A purposive sampling technique was utilized on predetermined criteria. Informants were selected based on their experience, knowledge, participation, and engagement in interfaith activities and tourism. Informants from the community and religious leaders were selected considering their experience of interfaith features and knowledge and observation of people's religious coexistence. Hence, interview questions for such informant groups were focused mainly on examining interfaith features and religious coexistence. Experts (in culture, tourism, and heritage) and government officials were found to gather information and opinion about the potential and challenges of interfaith tourism development for peacebuilding and socio-economic development. The interview questions for such informants stressed interfaith tourism development challenges, possible mechanisms and challenges for peacebuilding and socio-economic development. In addition, to understand observations and experience visitors, three informants who visited places known for interfaith features and interfaith tourist destinations were

Table 1. Background information of the informants.

Informants code	Category	Gender	Age
I1	Government officials	M	35
I2	Government officials	M	42
I3	Community	F	45
I4	Community	M	39
I5	Community	M	52
I6	Community	F	41
I7	Religious figures	M	47
I8	Religious figures	M	38
I9	Religious figures	M	43
I10	Religious figures	M	56
I11	Expert	F	32
I12	Expert	M	31
I13	Expert	M	29
I14	Visitor	M	28
I15	Visitor	M	43
I16	Visitor	F	31

interviewed. The interview protocol and format (see in the Appendix) were partly informed by previous studies (see Öter & Çetinkaya, 2016; Sharma, 2013), and each session lasted around 30 min. Some of the interviews were conducted face-to-face, whereas others were conducted via email using English and Amharic (working language in Ethiopia). In keeping with ethical considerations, respondents were informed of the objectives of the study and their consent was sought coupled with the guarantee of their anonymity and confidentiality.

In supporting the primary interview data, media analysis was pursued to strengthen the research findings. Some of the main media sources were websites and social media outlets of government and non-government organizations like the Inter-Religious Council of Ethiopia, The International Dialogue Centre (KAICIID), Universal Peace Federation, Land of Origin, and Ministry of Tourism and Culture. The researcher consulted news reports, press releases, and reports about interfaith potential in Ethiopia from major international and national media outlets such as The Washington Post, Ethiopian Press Agency, Addis Standard, and AllAfrica.com. Data analysis was conducted following exploratory qualitative data analysis techniques. Exploratory qualitative data analysis allows for the amalgamation of data from several sources to draw generalizations and generate understanding from observations, experiences, and opinions (Cox, 2017; Young, 1981). After manually transcribing the interviews, the transcribed data were coded and analyzed according to concepts and indicators of clearly defined issues. In addition, qualitative content analysis was performed to extract relevant ideas from the media sources.

Results and discussion

Interfaith tourism: the bright side for peacebuilding and socio-economic development

This research shows evidence to suggest that interfaith religious coexistence and harmony in Ethiopia are beneficial to tourism development and society's wellbeing. Informants and media sources revealed that interfaith destinations and religious coexistence supports peacebuilding and socio-economic development. According to informants (I7 and I9), both religious and non-religious people have participated on a pilgrimage to Kulubi Gabriel (a monastery located in Eastern Ethiopia) and Zekuala (a monastery situated around central Ethiopia), especially in earlier times. Kulubi Gabriel is located in the southeastern part of Ethiopia. The monastery is named after the Angel Gabriel, "the favoured messenger of God" as stated in the sacred literature of Christians, Muslims, and Jews. Pilgrimage to Kulubi Gabriel is known as one of Ethiopia's primary religious tourism destinations with vast gatherings, especially during the biannual celebrations held in July and December. At this time, people walk up the hill to accomplish their vow and then provide gifts to the monastery. One can observe Christians expressing their thankfulness or providing a "silent" (vow) to the church, but many Muslims also participate alongside them. In supporting such experiences, informant 8 stated that:

> The monastery welcomes people of different religions for different reasons. For example, Muslims (who call Gabriel' Jubril' as written in the Holy Qur'an) visited the monastery. In addition, traditional religion followers visited the monastery, as there was a tree which is seen as a ritual symbol that stood in the place of the monastery.

The informants added that non-believers also participated in the pilgrimage by accompanying their family and friends or providing services to pilgrims during a religious celebration. Adding to this, previous literature noted that pilgrimage to Kulubi Gabriel has both religious and non-religious dimensions (Lester, 2019; Levine, 2017). The northeastern part of Ethiopia is known for interfaith religious coexistence and harmony between Christians and Muslims. According to informants (I3, I5, and I6), Christians and Muslims actively participate in the building and maintenance of churches, mosques, and monasteries. Pilgrimage to Gishen Mariyam monastery is one of the leading interests in religious tourism, especially during annual celebrations in September.

The monastery is famous for religious teaching and scholarship and holds a high volume of books concerning the True Cross of Christ (The African Report, 2011). According to informants, many people, including non-Christians, visit the monastery for healing and vows. Such interfaith tourism collaboration extends to service provision during the pilgrimage and religious celebration, supporting local development and livelihood at the community level. In this regard, informant 4 said:

> By the way, if you go to Gishen Mariam (a monastery located in Northeastern Ethiopia), the service provider and the guard of the tourists, as well as the heritage, have mostly executed by the Muslims people, and I think the same is true for other religious resources and sites.

Peaceful coexistence between Christians, Muslims, and other traditional religious followers has been demonstrated in different parts of the country. For example, the monastery in Zekuala named after the Egyptian hermit Abune Gebre Menfes Kidus, attracts pilgrims of all faiths. The hermit was one of nine saints to enter Ethiopia. He found a suitable place for prayer at a crater lake on the mountain top. According to his hagiography, the saint came to Ethiopia by God's order and lived in Zekuala for 265 years. The landscape where the monastery is located offers an outstanding view over the great East African Rift Valley. In addition, the Crater Lake features unique vegetation, as well as birds and monkeys. Pilgrimage to the Zekuala's monastery is held twice yearly in July and October. Non-believers also participate in this pilgrimage and accompany their family and friends. Most importantly, Zekuala is a sacred place where Christianity has been peacefully practiced alongside other traditional religions for centuries, and people of different religious backgrounds have lived together under mutual understanding and community cohesion (according to I8 and I2).

Muslim sacred sites also draw a large number of visitors for pilgrimage. Dirre Sheik Hussein, for example, is a prominent Muslim pilgrimage site located in the Southeastern part of Ethiopia. Dirre Sheik Hussein welcomes people from different faith groups and adherents of indigenous religions, especially during the pilgrimage season. In addition, several Mosques and Shrines across different parts of the country attract non-Muslims (according to I1, I10, and I12). Since the early medieval period, the site has featured impressive groups of buildings, monumental tombs, and courts demonstrating Islamic architecture and building structure. Dirre Sheik Hussein is famous for pilgrimage and has attracted pilgrims from different parts of Ethiopia as well as the Middle East since the 10th century. Besides, people of other religions including traditional religious followers, have lived together peacefully in and around the Shrine for centuries. Al-Najashi mosque, located in Ethiopia, draws tourists' attention regardless of religious position. Informants assert that the Mosque is one of the most ancient Mosques in the world and is believed to be the first Islamic settlement in Africa by Muslim refugees who escaped from mistreatment in Mecca (I4, I6, and I8). Such historical significance of the Mosque is the main reason for attracting both Muslim and non-Muslim visitors. With rich historical and cultural attributes and the composition of 110 mosques and 102 shrines, Harar is referred to as the fourth-holiest city in Islam (Ethiopian Press Agency, 2020). Likewise, the walled city of Harar, situated in the eastern part of Ethiopia, is famous for the concentration of historical Mosques and Shrines and attracts visitors of different religions. Sacred Muslim heritage sites are the major interfaith tourism destination that could support peacebuilding, mutual understanding, and socio-economic development through tourism.

In support of this, studies have noted that Mosques, Shrines, and other Muslim sacred sites have experienced a long period of interfaith tourist flow that promote peace and social harmony between societies despite religious convulsions that affect such relationships (Levine, 2017; Zeleke, 2012). Given such propositions, informants suggested that interfaith tourism contributes positively to economic development as it focuses on the physical environment and society's wellbeing (I11, I12, and I13). In such regard, informant 13 noted that:

> Interfaith tourism could be a potential contributor to economic regeneration, and it may assist the diversification of tourism resources. The social interaction, events, and festivals, and the well-established relationship between different religions could draw tourists' attention and support tourism development at the national level.

As with other major global pilgrimage destinations like Karbala in Iraq, where Christian, Zoroastrian, and Yezidi pilgrims join with their Muslim brothers and sisters to honor the Shi'a Islam martyrs, the above-mentioned Muslim and Christian religious sites in Ethiopia are receptive to members of other faiths. Buttressing the present findings, past studies assert that interfaith pilgrimage tourism creates a sense of togetherness. Christians and other non-Muslims feel welcome to participate in a ritual in which Muslims offer their prayers at a Muslim shrine (Abbink, 2011; Braukamper & Braukämper, 2002). Events and festivals are also the main features of interfaith tourism in which followers and non-followers of different religions celebrate together, according to informants 14 and 15. Some of these festivals and events include Epiphany, Meskel (the founding of the True Cross), Eid al-Fitr, and Eid al-Adha. In addition to Christian and Muslim festivals and events, Irreechaa, the Oromo thanksgiving ritual event, is one of the main indigenous practices that attract visitors and followers of different religions (I15). Previous studies support this view that inter-faith tourism is initiated to obtain personal transformation, meaning, and fulfillment instead of solely traveling for pleasure (Collins-Kreiner, 2020; Öter & Çetinkaya, 2016; Sharma, 2013). Thus, the findings support the argument that interfaith tourism is often encouraged by an intrinsic drive for personal and social conditioning, cultural exchange, and community cohesion.

Social attributes

The long-standing Christian–Muslim coexistence retains peace and collaboration. Social activities such as weddings, funerals, holiday celebrations, and gatherings strengthen interfaith potential and religious coexistence. Informants indicate that Christian–Muslim harmony has resulted in interfaith marriage beyond the customary collaboration and coexistence between different religions (I3, I5, and I6). Interfaith marriage and interfaith romance are widely reflected in performing arts and folklore, which express the long-lived religious coexistence. A news article about Christian–Muslim coexistence in Ethiopia noted that "Muslims and Christians here have lived in the same neighborhoods, celebrated each other's holidays, intermarried and blended religions with indigenous beliefs" (McCrummen, 2007). In line with this, literature stated that interfaith marriage is the best indicator of community cohesion and social integration, thereby it supports conflict resolution and peacebuilding (Gaines et al., 2015; Sherkat, 2004; Van Niekerk & Verkuyten, 2018). The relationship between Christians and Muslims has resulted in the sharing of customs and traditions, social gathering, peace, and humanitarian missions. Many Christians, including the informants, believe that their Muslim brothers and sisters possess positive attributes such as confidentiality and honesty. Similarly, Muslims recognize the truthfulness and brotherliness of the Christians in their community. For instance, informant 16 said:

> Traditional organizations such as 'iqub' (a traditional means of saving), 'idir' (a self-help traditional association similar to life insurance), and other related organization managed by followers of different religions could play a proactive socio-economic role as well as it improves the value of interfaith religious harmony at local level.

The Inter-Religious Council of Ethiopia and Ministry of Tourism and Culture's official websites mentioned the importance of religious coexistence and interfaith features. The former Ministry of Federal Affairs noted that interfaith resources would "exploit the long-standing heritage of peaceful coexistence, religious respect and tolerance in Ethiopia to consolidate peace, democracy and development" (Addis Standard, 2013). The results of this study support previous arguments that social interaction between different groups and constituencies help to enhance peace and community wellbeing (Akinola & Uzodike, 2018; Ani, 2017; Appiah-Thompson, 2020). In summary, the interview and media analysis results indicate that interfaith tourism, supported by people's religious coexistence, and the resources of interfaith religious tourism, could encourage peace-building and socio-economic development.

Factors affecting the potential of interfaith tourism

This paper argues that fostering interfaith religious coexistence helps alleviate tension and advance interaction between diverse faith groups, thereby assisting socio-economic development. Nevertheless, several challenges hinder the contribution of interfaith tourism in improving tourism's socio-political and economic power. Despite the government's recent attention to tourism development, the evolving challenges related to limited tourism development, stakeholder engagement, and collaboration borders the potential of interfaith tourism. According to informants, little effort has been made to enhance interfaith tourism resources through promotions, advertisements and other marketing strategies, and basic tourism amenities (i.e. accommodation and transportation) are extremely limited (I14 and I15). For example, few tourism brochures and publicity folders explain complex sites like Zekuala or Dire Sheihk Hussein, which feature harmonious interfaith relationships coupled with other indigenous practices. Despite recent conservation and preservation efforts initiated by the government, an absence of viable conservation schemes hampers further development of interfaith tourism destinations. Compared to religious coexistence and harmony developed at the community level, interfaith promotion and development at the macro-level among stakeholders (government and non-government organization) remain at its infancy. Despite the efforts made by some organizations and associations like The Inter-Religious Council of Ethiopia and the Interfaith Forum for Development, Dialogue, and Action (EIFDDA), stakeholder consciousness and holistic efforts to enhance the importance of interfaith tourism have remained limited.

The peaceful coexistence or collaboration between Christians and Muslims in Ethiopia experience several challenges from emerging thoughts and social tensions. According to classical sociology literature, social tension is defined as a disintegration process caused by problems that result in negative attitudes, pressure, or aggressive dispositions (Artemov et al., 2017; Charles et al., 2009). Likewise, the majority of informants revealed that emerging religious thoughts, conflicts and extremism are perceived as the greatest threat to the development of interfaith tourism. In line with this, Semu (2016) noted that "due to the proliferation of extremist teachings of the Salafists and Wahhabists, attacks on pilgrims and marginalization of adherents have been strengthening (p. 1)". Terrorism and related stresses from neighboring countries also pose a challenge to Christians and Muslims' coexistence. There is a general fear of the emergence of modernity and acculturation-related challenges to sacred sites. According to informants (I8, I9, I10, and I15), during most pilgrimage site or religious celebration, people entertain themselves with modern and secular music, not spiritual songs, and they tend to ignore religious rules like dressing style, silence in and around sacred sites, and focus on spiritual values. In summary, the findings indicate that tourism development and management, facility and infrastructure, marketing, and stakeholder collaborations are the main factors that hinder the potential of interfaith tourism.

Conclusion and implications

This study examined the potential of interfaith tourism to ensure peace and socioeconomic development. Looking back at previous studies, research has been conducted on interfaith features from cultural and sociological perspectives (e.g. McIntosh et al., 2019; Öter & Çetinkaya, 2016; Sharma, 2013). Yet so far, there is a paucity of research on understanding interfaith tourism from peacebuilding and local development settings. In response to this, the present study found that religious coexistence and harmony are the main drivers of interfaith tourism, contributing to communal wellbeing and socio-economic development. As informants from community groups

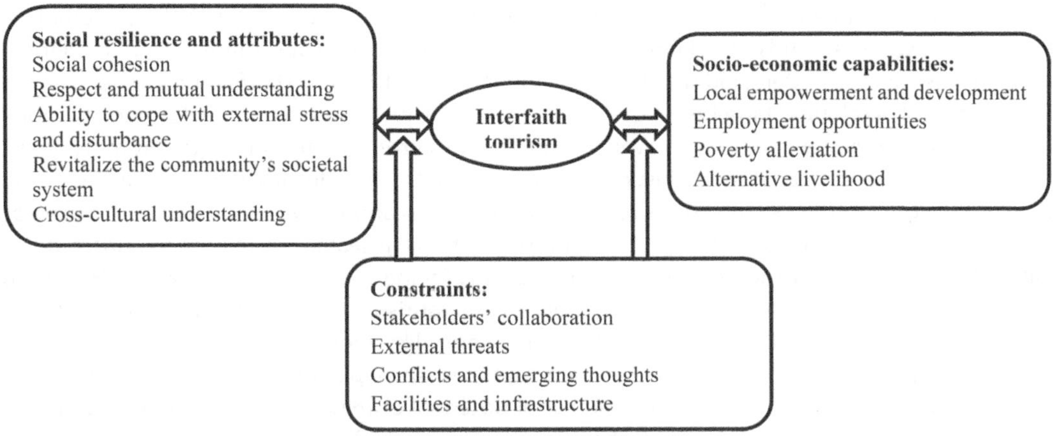

Figure 2. Interfaith tourism features.

and religious leaders noticed, interfaith religious coexistence helps to maintain peace and mutual understanding. In addition, government officials and experts underline the necessity of interfaith tourism for socio-economic development by considering tourism planning, marketing facility development schemes, and stakeholder engagement and collaboration. Drawing on social resilience and dynamic capability approaches, Figure 2 summarizes a comprehensive outlook of interfaith tourism potential and highlights the fine balance between the commercialization of religious tourism resources and the larger agenda of cooperation and peacebuilding.

From a social resilience point of view, the findings show that interfaith tourism can play a crucial role in developing a resilient society and organization necessary to manage internal tensions, competition, and conflicts. Scholars have argued that social resilience is measured through cognitive and structural dimensions (Cutter et al., 2014; Paton et al., 2008). In line with cognitive, social resilience, beliefs, values, and attitudes drawn from interfaith tourism attributes could encourage cross-cultural understanding and social cohesion. Regarding structural social resilience, interfaith tourism embraces personal and psychological wellbeing as visitors travel to sacred sites for healing purposes and escape stress and disturbances. The findings correspond with tourism's role in peace dividend perspectives (Butler & Baum, 1999). Interfaith tourism supports peacebuilding efforts via the interaction of locally based people instead of excessive military cost, diplomatic or peacebuilding projects. From the dynamic capabilities setting, the findings indicate that interfaith tourism can drive local development through alternative livelihood, empowerment, and poverty alleviation at the grassroots level. Given that the notion of dynamic capability is an inward-looking approach that emphasizes how to incorporate sustainable development (Nagano, 2020; Surmeier, 2020), I argue that the possession and employment of interfaith tourism provides dynamic microeconomic capabilities. Most importantly, since tourism thrives on peace and security, the contribution of interfaith tourism to peace can ensure the stability of other tourism segments and economic activities. Yet, proper tourism planning and marketing and stakeholder collaboration are necessary to encourage interfaith tourism potentials.

Theoretical and practical implications

This study offers further theoretical implications to the limited literature related to socio-political aspects, social justice, and peacebuilding. More specifically, the study responds to the scarce literature that focuses on the importance of interfaith tourism for peacebuilding and socio-economic development. Examining the social and political aspects of tourism, assessing the interfaith phenomenon, and exploring some of the attractive promises that tourism holds, the findings shed light on tourism's role at the societal and global community level. By revealing the socio-political role of interfaith tourism in fostering peace and tolerance at the societal level, the current study extends the knowledge of the scope of religious tourism and it helps to advance the current understanding of the economic role of tourism. The findings underscore the importance of exploring tangible (e.g. monasteries, mosques, and monuments) and intangible (e.g. people's coexistence, culture, events, and festivals) tourism resources to promote peace, community wellbeing, and social justice at the grassroots level. In addition, the present study provides insights into interfaith tourism, peace, and socio-economic development in the developing country setting. Previous studies have suggested more research on religious tourism potential and contributions in the developing world (Collins-Kreiner, 2020; Lee et al., 2020). Hence, the findings offer an understanding of the role interfaith tourism resources hold for peacebuilding efforts and socio-economic livelihood in the global south like Africa.

Practically, the study will be useful to destination managers, planners, policymakers, and practitioners as it draws their attention to the powerful potential of religion and tourism in maintaining common social and economic goals and confronting violence. The current study provides a constructivist perspective to raise the value of peace and cross-cultural exchange in interfaith tourism, which can positively impact people's perception of humanity and global understanding. Given the prevalence of social crises in various parts of the world fueled by injustice and inequality, this research's findings shed light on the importance of interfaith tourism in promoting tolerance, mutual understanding, and a sense of belonging. It is often the case that international organisations' interventions are necessary to resolve conflicts such as civil war, political instability, and social unrest in less developed territories like Africa. Nevertheless, the success rate of conflict resolution by such international intermediaries has been minimally coupled with the excess budget spent on diplomatic relations (Akinola & Uzodike, 2018; Ani, 2017). Against this backdrop, my study offered insights into the role of interfaith tourism for peacebuilding and a bottom-up approach to encourage peace and stability. The results also support developmental goals, particularly sustainable development goal 16 as a mechanism to reinforce peace and mutual understanding (UN, 2018). Destination managers and policymakers need to extend the widely practiced interfaith religious coexistence at the micro-level. Thus, efforts to formulate and develop stakeholder collaboration, proper tourism planning, infrastructure development, and management and marketing mechanisms are necessary to conserve and properly utilize the potential of inter-faith tourism. In relation to this, tourism marketers in Ethiopia need to consider the value of interfaith tourism in terms of adding value to the unique cultural, historical, and religious heritage of the country, since Ethiopia's tourism competitors depend mostly on natural/safari tourism in Africa (Fang, 2020).

Limitations and future research

While the present study contributes to the existing body of knowledge on religious tourism and the tourism literature in general, it still has certain limitations. First, my research is limited to Ethiopia's religious perspectives; hence, the findings may not be generalized to different contexts and religious backgrounds. Additionally, using a small sample size and a qualitative research design may limit the conclusions drawn from this study. Therefore, future studies can explore the different religions' role and their visitors' intention and engagement toward interfaith tourism perceptions and experiences. Quantitative studies are also encouraged to extend the current attempt by examining specific interfaith tourism factors that promote peace and local development using a larger sample size that includes the insights of tourism organizations, host communities, tour operators, and destination management organizations. Measuring the economic advantage of tourism, tourist arrival, and tourist behavior in and around interfaith destinations is necessary to formulate strategies and draw an approximate understanding of the effect of interfaith tourism on socio-economic development. Exploratory studies are needed to understand the value of interfaith tourism from different sustainability perspectives. For example, future studies may explore the role of interfaith visitors' intentions and engagement toward climate change and environmental sustainability. Replication of the current findings in different geographical settings will also have momentous implications. Comparative studies on interfaith tourism across developed and developing countries are appreciated. Furthermore, given the lack of conceptual understanding of interfaith tourism in the tourism literature, future studies can establish a general framework for the practice of interfaith tourism.

Acknowledgements

The author would like to thank the Institute of International Business and Governance, established with the substantial support of a grant from the Research Grants Council of the Hong Kong Special Administrative Region, China (UGC/IDS16/17), for its support.

Disclosure statement

No potential conflict of interest was reported by the author.

ORCID

Dagnachew Leta Senbeto http://orcid.org/0000-0002-8743-3348

References

Abbink, J. (2011). Religion in public spaces: Emerging Muslim–Christian polemics in Ethiopia. *African Affairs*, *110*(439), 253–274. https://doi.org/10.1093/afraf/adr002

Addis Standard. (2013). Ethiopia undertakes three days interreligious conference in hopes of bringing about peace and tolerance. https://addisstandard.com/ethiopia-undertakes-three-days-interreligious-conference-in-hopes-of-bringing-about-peace-and-tolerance/.

Adger, W. N. (2000). Social and ecological resilience: Are they related? *Progress in Human Geography*, *24* (3), 347–364. https://doi.org/10.1191/030913200701540465

Akinola, A. O., & Uzodike, U. O. (2018). Ubuntu and the quest for conflict resolution in Africa. *Journal of Black Studies*, *49*(2), 91–113. https://doi.org/10.1177/0021934717736186

Appiah-Thompson, C. (2020). The concept of peace, conflict and conflict transformation in African religious philosophy. *Journal of Peace Education*, *17*(2), 161–185. https://doi.org/10.1080/17400201.2019.1688140

Ani, N. C. (2017). Re-empowering indigenous principles for conflict resolution in Africa: Implications for the African Union. *Journal of Pan African Studies*, *10*(9), 15–36.

Albayrak, T., Herstein, R., Caber, M., Drori, N., Bideci, M., & Berger, R. (2018). Exploring religious tourist experiences in Jerusalem: The intersection of Abrahamic religions. *Tourism Management*, *69*, 285–296. https://doi.org/10.1016/j.tourman.2018.06.022

Artemov, G., Aleinikov, A., Abgadzhava, D., Pinkevich, A., Abalian, A. (2017). Social tension: The possibility of conflict diagnosis (on the example of St. Petersburg). *Economics & Sociology*, *10*(1), 192–208.

Barney, J. B. (2001). Resource-based theories of competitive advantage: A ten-year retrospective on the resource-based view. *Journal of Management*, *27*(6), 643–650. https://doi.org/10.1177/014920630102700602

Bandyopadhyay, R. (2019). Volunteer tourism and "The White Man's Burden": Globalization of suffering, white savior complex, religion and modernity. *Journal of Sustainable Tourism*, *27*(3), 327–343. https://doi.org/10.1080/09669582.2019.1578361

Brunet, M. (2010). Short note: The track of a new cradle of mankind in Sahelo-Saharan Africa (Chad, Libya, Egypt, Cameroon). *Journal of African Earth Sciences*, *58*(4), 680–683. https://doi.org/10.1016/j.jafrearsci.2010.08.010

Buzinde, C. N. (2020). Theoretical linkages between wellbeing and tourism: The case of self-determination theory and spiritual tourism. *Annals of Tourism Research*, *83*, 102911–102920. https://doi.org/10.1016/j.annals.2020.102920

Braukamper, U., & Braukämper, U. (2002). *Islamic history and culture in southern Ethiopia: collected essays* (Vol. 9). LIT Verlag Münster.

Butler, R. W., & Baum, T. (1999). The tourism potential of the peace dividend. *Journal of Travel Research*, *38*(1), 24–29. https://doi.org/10.1177/004728759903800106

Collins-Kreiner, N. (2020). Religion and tourism: A diverse and fragmented field in need of a holistic agenda. *Annals of Tourism Research*, *82*, 1–22. https://doi.org/10.1016/j.annals.2020.102892.

Charles, S. T., Piazza, J. R., Luong, G., & Almeida, D. M. (2009). Now you see it, now you don't: Age differences in affective reactivity to social tensions. *Psychology and Aging*, *24*(3), 645–653. https://doi.org/10.1037/a0016673

Chien, P. M., & Ritchie, B. W. (2018). Understanding intergroup conflicts in tourism. *Annals of Tourism Research*, *72*(C), 177–179. https://doi.org/10.1016/j.annals.2018.03.004

Chilufya, A., Hughes, E., & Scheyvens, R. (2019). Tourists and community development: corporate social responsibility or tourist social responsibility? *Journal of Sustainable Tourism*, *27*(10), 1513–1529. https://doi.org/10.1080/09669582.2019.1643871

Cox, V. (2017). *Exploratory data analysis. Translating statistics to make*. Apress.

Cutter, S. L., Ash, K. D., & Emrich, C. T. (2014). The geographies of community disaster resilience. *Global Environmental Change*, *29*, 65–77. https://doi.org/10.1016/j.gloenvcha.2014.08.005

Ethiopian Central Statistical Agency. (2007). Ethiopian census. First draft. https://web.archive.org/web/20120604045256/http://www.csa.gov.et/pdf/Cen2007_firstdraft.pdf.

Haggai, E. (2010). *Islam and Christianity in the Horn of Africa*. Lynne Rienner Publishers.

Espiner, S., Orchiston, C., & Higham, J. (2017). Resilience and sustainability: A complementary relationship? Towards a practical conceptual model for the sustainability–resilience nexus in tourism. *Journal of Sustainable Tourism*, *25*(10), 1385–1400. https://doi.org/10.1080/09669582.2017.1281929

Fang, W. T. (2020). Cultural tourism. In *Tourism in emerging economies* (pp. 75–101). Springer.

Farmaki, A. (2017). The tourism and peace nexus. *Tourism Management*, *59*, 528–540. https://doi.org/10.1016/j.tourman.2016.09.012

Farra-Haddad, N. (2014). Dismantling religious boundaries by sharing the Baraka through pilgrimages in Lebanon. *DISKUS*, *15*(0), 54–73. https://doi.org/10.18792/diskus.v15i0.10

Fartacek, G. (2012). Rethinking ethnic boundaries: Rituals of pilgrimage and the construction of holy places in Syria. In G. Kiliánová, C. Jahoda, & M. Ferencová (Eds.), *Ritual, conflict and consensus: Case studies from Asia and Europe* (pp. 119–130). Verlag der Österreichischen Akademie der Wissenschaften.

Ferreira, J., Coelho, A., & Moutinho, L. (2020). Dynamic capabilities, creativity and innovation capability and their impact on competitive advantage and firm performance: The moderating role of entrepreneurial orientation. *Technovation*, *92-93*, 102018–102061. https://doi.org/10.1016/j.technovation.2018.11.004

Gaines, S. O., Clark, E. M., & Afful, S. E. (2015). Interethnic marriage in the United States: An introduction. *Journal of Social Issues*, *71*(4), 647–658. https://doi.org/10.1111/josi.12141

Guba, E. G., & Lincoln, Y. S. (1994). Competing paradigms in qualitative research. In N. K. Denzin & Y. S. Lincoln (Eds.), *Handbook of Qualitative Research*. (pp. 105–117). Sage.

Higgins-Desbiolles, F. (2006). More than an "industry": The forgotten power of tourism as a social force. *Tourism Management*, *27*(6), 1192–1208. https://doi.org/10.1016/j.tourman.2005.05.020

Higgins-Desbiolles, F. (2020). Socializing tourism for social and ecological justice after COVID-19. *Tourism Geographies*, *22*(3), 610–623. https://doi.org/10.1080/14616688.2020.1757748

Hilton-Barber, B. (2004). *Field guide to the cradle of humankind: Sterkfontein, Swartkrans, Kromdraai & Environs World Heritage Site*. Struik.

Hughes, K., Bond, N., & Ballantyne, R. (2013). Designing and managing interpretive experiences at religious sites: Visitors' perceptions of Canterbury Cathedral. *Tourism Management*, *36*, 210–220. https://doi.org/10.1016/j.tourman.2012.11.022

Kutzner, D. (2019). Environmental change, resilience, and adaptation in nature-based tourism: conceptualizing the social-ecological resilience of birdwatching tour operations. *Journal of Sustainable Tourism*, *27*(8), 1142–1166. https://doi.org/10.1080/09669582.2019.1601730

Lee, H. T., Jan, F. H., & Lin, Y. H. (2020). How authentic experience affects traditional religious tourism development: evidence from the Dajia Mazu Pilgrimage, Taiwan. *Journal of Travel Research*, 1–18. https://doi.org/doi/0047287520921240.

Levine, D. N. (2017). *Social theory as a vocation: Genres of theory work in sociology*. Routledge.

McIntosh, I. S., Haddad, N. F., & Munro, D. (Eds.). (2019). *Peace journeys: A new direction in religious tourism and pilgrimage research*. Cambridge Scholars Publishing.

Malcolm, X. (1965). *The Autobiography of Malcolm X*. Grove Press.

Lester, M. (2019). *Sacred journeys: Your guide to the world's most transformative spaces, places, and sites*. Simon and Schuster.

Litvin, S. W., Smith, W. W., & McEwen, W. R. (2020). Not in my backyard: Personal politics and resident attitudes toward tourism. *Journal of Travel Research*, *59*(4), 674–685. https://doi.org/10.1177/0047287519853039
McCrummen, S. (2007, May 14). Christians, muslims coexistence in Ethiopia shows signs of fraying. *The Washington Post*. https://sudantribune.com/spip.php?page=imprimable&id_article=21853http://news.bbc.co.uk/2/hi/talking_point/debates/african_debates/574023.stm.
Mostafanezhad, M., & Norum, R. (2019). The anthropocenic imaginary: Political ecologies of tourism in a geological epoch. *Journal of Sustainable Tourism*, *27*(4), 421–435. https://doi.org/10.1080/09669582.2018.1544252
Nagano, H. (2020). The impact of knowledge diversity: Integrating two economic perspectives through the dynamic capability approach. *Managerial and Decision Economics*, *41*(6), 1057–1070. https://doi.org/10.1002/mde.3159
Nonaka, I., Hirose, A., & Takeda, Y. (2016). "Meso" – Foundations of dynamic capabilities: Team-level synthesis and distributed leadership as the source of dynamic creativity. *Global Strategy Journal*, *6*(3), 168–182. https://doi.org/10.1002/gsj.1125
Olsen, D. H., & Esplin, S. C. (2020). The role of religious leaders in religious heritage tourism development: The case of the Church of Jesus Christ of Latter-day Saints. *Religions*, *11*(5), 222–256. https://doi.org/10.3390/rel11050256
Öter, Z., & Çetinkaya, M. Y. (2016). Interfaith tourist behaviour at religious heritage sites: House of the virgin mary in Turkey. *International Journal of Religious Tourism and Pilgrimage*, *4*(4), 1–18.
Paton, D., Smith, L., Daly, M., & Johnston, D. (2008). Risk perception and volcanic hazard mitigation: Individual and social perspectives. *Journal of Volcanology and Geothermal Research*, *172*(3-4), 179–188. https://doi.org/10.1016/j.jvolgeores.2007.12.026
Ray, W. D. (2012). *Tasting Heaven on Earth: Worship in Sixth-Century Constantinople*. Eerdmans Publishing.
Saunders, M., Lewis, P., & Thornhill, A. (2009). *Research methods for business students*. Pearson Education.
Sherkat, D. E. (2004). Religious intermarriage in the United States: Trends, patterns, and predictors. *Social Science Research*, *33*(4), 606–625. https://doi:10.1016/j.ssresearch.2003.11.001 https://doi.org/10.1016/j.ssresearch.2003.11.001
Sigala, M. (2020). Tourism and COVID-19: impacts and implications for advancing and resetting industry and research. *Journal of Business Research*, *117*, 312–321. https://doi.org/10.1016/j.jbusres.2020.06.015
Salvato, C., & Vassolo, R. (2018). The sources of dynamism in dynamic capabilities. *Strategic Management Journal*, *39*(6), 1728–1752. https://doi.org/10.1002/smj.2703
Senbeto, D. L. (2019). Interfaith tourism in Ethiopia: an opportunity for socio-economic development and peace-building?. In McIntosh, I. S., Haddad, N. F., & Munro, D. (Eds.), *Peace journeys: A new direction in religious tourism and pilgrimage research* (pp. 80–90). Cambridge Scholars Publishing.
Senbeto, D. L. (2020). Rock art tourism in Ethiopia and cultural heritage management challenges. In Novelli, M., Adu-Ampong, E. A. & Ribeiro, M. A. (Eds.), *Routledge handbook of tourism in Africa* (pp. 370–380). Routledge.
Semu, K. T. (2016). Dynamics of the cult of Sheik Hussein of Bale, Ethiopia: Its course and curse of the extremists, a historical perspective. *Ethiopian Journal of the Social Sciences and Humanities*, *12*(1), 63–101.
Surmeier, A. (2020). Dynamic capability building and social upgrading in tourism-Potentials and limits of sustainability standards. *Journal of Sustainable Tourism*, *28*(10), 1421–1498. https://doi.org/10.1080/09669582.2020.1744615
Sharma, V. (2013). Faith tourism: For a healthy environment and a more sensitive world. *International Journal of Religious Tourism and Pilgrimage*, *1*(1), 15–23. https://doi.org/10.21427/D7772J
Teece, D. J. (2009). *Dynamic capabilities and strategic management: Organizing for innovation and growth*. Oxford University Press on Demand.
Terzidou, M., Scarles, C., & Saunders, M. N. (2018). The complexities of religious tourism motivations: Sacred places, vows and visions. *Annals of Tourism Research*, *70*, 54–65. https://doi.org/10.1016/j.annals.2018.02.011
The African Report. (2011). Ethiopian Christians celebrate the discovery of the True Cross. https://www.theafricareport.com/8302/ethiopian-christians-celebrate-the-discovery-of-the-true-cross/
Timothy, D. J. (2019). Tourism, border disputes and claims to territorial sovereignty. In Isaac, R. K., Çakmak, E., & Butler, R. (Eds.), *Tourism and hospitality in conflict- ridden destinations*. Routledge.
Van Niekerk, J., & Verkuyten, M. (2018). Interfaith marriage attitudes in Muslim majority countries: A multilevel approach. *The International Journal for the Psychology of Religion*, *28*(4), 257–270. https://doi.org/10.1080/10508619.2018.1517015
UN. (2018). *The sustainable development goals report 2018*. Available at: https://unstats.un.org/sdgs/report/2018.
UNWTO. (1985). Tourism bill of rights and tourist code. http://www.world-tourism.org/sustainable/concepts.
UNWTO. (1980). *Manila declaration*. World Tourism Organization.
Wohlmuther, C., & Wintersteiner, W. (Eds.). (2014). *International handbook on tourism and peace.*
World Bank. (2020). International tourism, number of arrivals – Ethiopia. 2020 from https://data.worldbank.org/indicator/ST.INT.ARVL?locations=ET.
World Population Review. (2020). Ethiopian population. http://worldpopulationreview.com/countries/ethiopia-population/.
Young, F. W. (1981). Quantitative analysis of qualitative data. *Psychometrika*, *46*(4), 357–388. https://doi.org/10.1007/BF02293796
Zeleke, M. (2012). *Pilgrimage at Islamic Sufi Shrines in North Eastern Ethiopia.* https://epub.unibayreuth.de/206/1/BIGSAS%20WORKS%202%20Oct%2019%202012.pdf#page=104.

Appendix. Interview protocol

1. Experiences, perceptions, and opinions regarding interfaith features and religious coexistence.
    a. What are the main features of religious coexistence and harmony that have been developed among people?
    b. Which places are mainly known by interfaith tourism destination and religious coexistence?
    c. Would you think that interfaith features support peacebuilding and local development opportunities?
    d. Is there challenges that hinder interfaith and people's religious coexistence in ensuring peace and socio-economic development?
2. Potentials, challenges, and possible ways related to interfaith tourism.
    a. How do interfaith tourism support peacebuilding and socio-economic development?
    b. What are the main challenges that influence interfaith tourism potentials?
    c. Would you suggest possible ways to enhance the importance of interfaith tourism for peacebuilding and socio-economic development?

"Don't look back in anger". War museums' role in the post conflict tourism-peace nexus

Fabio Carbone

ABSTRACT

This article explores the role of war related attractions as an infrastructure for peace, particularly in post conflict areas. Emphasis is given to war museums and their narratives. The article questions the view based on the implicit causal relation between the representation of war and the promotion of peace. An alternative approach is proposed by linking theories from tourism, museum and peace studies. The application of 'the forgiveness model' to give rise to a different kind of war museum narrative is put forward, along with a theoretical model on tourism and peace based on the conception of war related attractions as local infrastructures for peace.

Crafting peace is a skilled work:

it requires passion, patience, experience and tenacity.

Blessed are those who sow peace (Francesco, 2015)

Introduction

Data disclosed by the Institute for Economics and Peace (2018) about a decline in global peace, confirm the common perception of living in a less stable and secure world. Designing and implementing strategies aimed at cultivating inter- and intra-state peaceful relations is thus one of the greatest challenges and priorities of our time. In this context, the United Nations Agenda 2030 for Sustainable Development and its 17 Sustainable Development Goals (SDGs) represents a shared blueprint for peace and prosperity worldwide (UN, 2015a). Goal 16 specifically addresses the creation of a more peaceful and just world. The entire civil society and each sector are called to get involved and seek effective ways to contribute to the achievement of this and the other SDGs.

The tourism sector, despite being highly vulnerable to conflicts, violence and instability, does not seem to engage consistently or directly with the promotion of peace, despite a vibrant academic debate. Interestingly, the United Nations World Tourism Organisation (UNWTO) commits to contribute specially for the achievement of the SDGs 8, 12 and 14 (respectively, 'Decent Work and Economic Growth', 'Responsible Consumption and Production', and 'Preservation of Life below Water'). It is curious that the SDG 16, 'Peace, Justice and Strong Institutions', is excluded from the specific commitment assumed by the highest global representative of the tourism sector. One of the possible explanations for this significant omission might be the lack of concrete theoretical and practical guidelines.

Against this backcloth, this article explores new theory-based paths to make better sense of the association between tourism and peace. The debate on the causal relation between tourism and peace is indeed fascinating and controversial (see, e.g., Becken & Carmignani, 2016; Burnett & Uysal, 1990), and mostly based on the common assumption that travel eases negative prejudices and reduces sociocultural stereotypes through a direct experience of a different culture. The study here challenges such assumptions and demonstrates this is a major theoretical weakness of the peace through tourism debate. To this end, the work proposes an interdisciplinary approach that takes greater account of peace and conflict studies, in order to avoid simplifications and bias that would compromise the impact of any theoretical extension applicable to the real world. Namely, three main fields of study converge:

1. tourism studies;
2. peace and conflict studies;
3. heritage and museum studies.

The author reaffirms the potential contribution that tourism can make to the creation of a "peace culture" (Boulding, 2008, p. 1452) whilst responding to the urgent need to define ways of developing this potential in the context of peace education. The latter still represents a missing element in the formation of contemporary women and men (van den Dungen, 2017), and it gives added salience to the present work. The paper further develops the study by Packer et al. (2019) on the potential impact of the interpretation of war heritage in the process of reconciliation by examining the role of war related attractions, namely war museums, in the context of post-conflict tourism and peace.

This is a challenging choice as the centre of interest and analysis since a vast literature exists about the more immediate association between peace, peace education and peace museums (see, among others, Apsel, 2015; Barrett, 2012; Luli & Kawano, 2020; Muzaini, 2008), while the debate on war museums as promoters of peace is scarcely explored. This article, therefore, is not concerned with anti-war or peace museums, whose advocacy and purpose is well defined and uncontroversial. The focus is circumscribed to the role of war museums and the message they do and could convey. War museums' political connotation as an element of national identity and pride is opportunely considered in the study, as well as the debate on the management of the so-called 'difficult heritage'.

Cohen (2011) identifies the visits to conflict-related attractions with 'dark tourism', a form of tourism that focuses on showing tourists how the effects of violence and conflict have impacted society via sites related to death, tragedies and disasters (Lennon & Foley, 2000; Seaton, 1996). Dark tourism remains controversial as it is perceived as supporting the commodification of tragedy and human death. Although 'post-conflict tourism' is also *"usually conceptualised under dark tourism scholarship"* (Causevic & Lynch, 2011, p. 780), the present paper does not focus on the association between tourism, war and peace with the aim of debating on a specific definition, but rather on the role given to tourism in the process of peace education and reconciliation, conflict transformation and social renewal of a destination, following in the footsteps of Causevic and Lynch (2011).

In reflecting on the role of museums as vehicles for promoting peace through tourism, and more specifically on the function of war museums, it should be noted that in the past museums used to promote themselves as mere tourist attractions in order to increase profitability (Kirshenblatt-Gimblett, 1998). The different objectives and functions of tourism and museums - one being commercial and the other non-profit - was one of the most significant limitations to the establishment of a more fruitful and lasting interaction between them. Indeed, this alliance used to be undermined by an underlying flaw, namely, the distance between the respective missions. In what follows, the evolution of the roles of tourism and museums will be made to converge to pave the way for innovative collaborations towards an effective role in the context of the revisited concept of peacebuilding as a local, bottom-up process.

Tourism and peace: a critical synthesis of the academic debate

Despite the existence of a conspicuous literature on tourism and peace, this subject still appears fuzzy, resulting in different interpretations of the topic itself. In this regard, the paper identifies three broad approaches through which previous studies have engaged with the tourism-peace nexus:

1. an abstract approach;
2. a case study-based approach;
3. a complex approach.

The abstract approach is at the very origin of the debate. It helps makes sense of the association between tourism and peace by emphasising the unique opportunity given by travel and tourism to enter directly in contact with different cultures, and establish a positive, peaceful host-visitor interaction.

This approach can be contextualised in the global aspiration for shared responsibility (Bauman, 2018) as per the vision within the United Nations SDGs, as opposed to the pervasive sense of acceptance of violence and misery as inevitable societal features (Duffield, 2014). A sort of neo-humanism alerts to the urgency to regain the human dimension by revaluing areas such as philosophy (Fitch, 2018) and rediscovering its strong political vocation to inspire public policies as an alternative to the constraints of neo-liberalism (Di Cesare, 2018). The idea that tourism could make a positive contribution to peacebuilding could be traced back to the post-World War II period (Hugues, 2004), and D'Amore (1988) responded to this philosophical call by launching the motto "Tourism. Industry of Peace" and the idea of a "Higher Purpose of Tourism". These ideas have been also at the basis of the establishment, by Louis D'Amore himself, of the International Institute for Peace through Tourism, to engage conversations about the ways in which to understand the notion of 'peace through tourism'.

This discourse is mainly based on the assumption that travelling reduces cultural and psychological distances among people (Nyaupane et al., 2008) and as a result it positively influences international politics and fosters world peace (Edgell et al., 2008) by encouraging cooperation among nations (Askjellerud, 2003; Causevic, 2010). Nevertheless, this discourse does not define the causal relations underlying the process described. Indeed, objections have been raised about the chain of reasoning underlying this assumption of a sort of peace by tourism-induced osmosis (see, e.g., Brown, 1989; Carbone, 2019).

For example, it has been argued that this process assumes that contacts and interactions produced by travel result in the formation of personal meaningful relationships with locals who are perceived as individuals, rather than in a stereotypical way (Suvantola, 2018). Nevertheless, the assumption of tourism as a vehicle for socio-cultural encounter (Saarinen & Manwa, 2008) promoting positive contacts between different cultures has been challenged - totally or partially - by empirical evidence from studies on host-tourist social dynamics (Fan et al., 2020; Tasci & Severt, 2017).

If, on the one side, in terms of sustainable tourism development the evidence suggests that the residents' attitudes are positively affected by host–tourist interaction (Eusébio et al., 2018), and intergroup social contact can nurture mutual understanding (Litvin, 2003; Li & Wang, 2020), on the other side several studies have shown that tourism can have the effect of reinforcing cultural stereotypes about the places visited, rather than dissipating them. Bruner (1991), for instance, denounces the fact that developing countries' cultural displays serve as a mirror for Western fantasies, performances reflecting what tourists desire and expect to find. In this sense, the market itself capitalises on the existence of stereotypes (Walle, 1993).What is conveyed about different cultures within tourism marketing is intentionally structured around those stereotypes, somehow feeding the Western discourse on the "exotic other" (Suvantola, 2018). Although some forms of tourism are unquestionably more responsible and respectful than others (Maoz, 2006), the study carried out by Sirakaya-Turk et al. (2014) shows that tourist experience can paradoxically reinforce visitor's prejudices and stereotyped representations of host communities, rather than promoting a more careful and respectful approach towards local cultures. The study by Yang (2011) on ethnic tourism and cultural representation, which shows how stereotypical conceptions of minority people are sometimes reinforced by representations in ethnic tourism, corroborates the point.

The major limitation of this 'abstract discourse' on tourism and peace seems thus to stem from the misleading assumption that tourism *per se* is a promoter of peace. The speculative reasoning of a causal link between tourism and peace appears rather naïve and weakened by the lack of a theoretical framework. The simplistic formula proposed by Lisle (2000) "war causes tourism, tourism causes peace" is an example of this approach. Back in the 1990s, Hall (1994, p. 91) had already argued that tourism seen as a force for peace is a "simplistic interpretation of the complexities of tourism and international relations". More recently, new criticisms have been raised against such argument that rests on the contact hypothesis (see, for instance: Carbone, 2017, 2019; Farmaki, 2017).

A conspicuous amount of case study literature exists generically linked to the debate on tourism and peace (see: Causevic & Lynch, 2011; Cho, 2007; Moufakkir & Kelly, 2010; Pearse, 2016; Scott, 2012; Wohlmuther & Wintersteiner, 2014). Such studies deal with how tourism can become (or not) a catalyst for change and development. Nevertheless, taken as a whole, this body of work illustrates the lack of a consensus over what the tourism-peace nexus actually is. If unanimity exists about the fact that tourism prospers in stability and peace and cannot flourish in conflictual environments (Farmaki, 2017; Fyall et al., 2006; Khalilzadeh, 2018), the actual nexus between tourism and the creation of peace remains fuzzy. Different case studies are thus based on different possible perspectives and interpretations of the tourism-peace nexus.

A third strand of related literature is the one that is classified here as the 'complex approach' to the study of the tourism-peace nexus. These contributions are distinguished by employing - implicitly or explicitly - the paradigm of complexity to the discussion on the causal relation between tourism and peace. The concept of complexity is used with reference to the holistic and transdisciplinary character of the debate on tourism and peace, through which the connections between and interdependence of the areas of research involved are emphasised. Several authors try to look beyond the area of tourism by establishing links with other social, political and human aspects that could eventually represent a prior condition to promote peace-through-tourism activities. The association between tourism and peace is complemented by subjects such as sustainability (De Villiers, 2014), globalisation (Reid, 2003), cultural heritage (Carbone & Oosterbeek, 2021), human rights (Higgins-Desbiolles & Whyte, 2015), international relations (Tavares da Silva & Breda, 2021; Carbone, 2017), social innovation (Malek & Costa, 2015) and academic mobility (Monteiro et al., 2021).

The search for new semantics and new lexical formulas is a feature of these works. Wohlmuther and Wintersteiner (2014), for instance, suggest the concept of "peace sensitive tourism" as an alternative to "tourism and peace". Instead, Isaac and Hodge (2011) examine the idea of "justice tourism", described as a form of purposeful travel directly aiming to meet criteria of social, political and environmental sustainability. In turn, Tucker (2016) coins the notion of "empathy in tourism", as an emotional pre-requisite for cross-cultural understanding. Table 1

Table 1. Tourism and peace: a critical overview on the academic debate.

	Abstract approach	Case study-based approach	Complex approach
Main characteristics	Account of works containing intellectual speculations on the causal link between tourism and peace; emphasises the ethical, philosophical connotation of the topic.	Conspicuous literature provides evidence about how tourism can eventually be a catalyst for change and development (Causevic & Lynch, 2011; Cho, 2007; Moufakkir & Kelly, 2010; Pearse, 2016; Scott, 2012; Wohlmuther & Wintersteiner, 2014)	This approach applies - implicitly or explicitly - the paradigm of complexity to the discussion on the causal relation between tourism and peace.
Contribution to the field	Bauman (2018) observed the need for ethics and philosophy within the current debates in social sciences, and Fitch (2018) echoes by arguing the need to set on being Human, revaluing areas such as philosophy. In this context, this approach to the peace-through-tourism debate provides a decisive momentum to "Review tourism's capacity to contribute to human well-being, human rights recognition, conflict resolution, and to the worldwide enhancement of justice and a culture of peace" (Pearse, 2016, p. 87), within a sort of neo-humanism in which the strong political vocation of philosophy and its capacity to inspire public policies is being debated	Different perspectives are provided by this corpus. Some analyse from a political, and intra or interstate conflict point of view specific forms of cultural tourism, (or even dark tourism) carried out in places that, for whatever reason, lead back to the topic of peace: tourist activities at Hiroshima (Selmi et al., 2012; Yoshida et al., 2016), Auschwitz (Miles, 2002) or the beaches of Okinawa (Figal, 2012), just to name a few examples.	A look beyond the area of tourism by establishing and exploring links with other social, political and human aspects that could eventually represent a prior condition to actually promote peace through tourist activity (Carbone, 2016, 2017; De Villiers, 2014; Reid, 2003). Tucker (2016) focuses on the specific topic of "empathy in tourism". Other areas are linked to the debate, such as cultural heritage management (Carbone et al., 2012) , social innovation (Malek & Costa, 2015), sustainability (Oosterbeek & Pollice, 2014), geopolitics (Issac, 2011) and the potential causal relationships between cultural heritage management, public participation and cultural diplomacy within the process of promotion of intercultural dialogue and peace through tourism (Carbone, 2017).
Main limitations	Not totally free from naïve bias; lack of a precise theoretical framework; misleading assumption that tourism *per se* promotes and spread peace.	This literature reflects the fundamental lack of unanimous agreement about what the research subject of tourism and peace actually is.	It represents the latest approach within the literature, and it needs to be further developed. It opens the path to outline a less vague and more consistent object of study, but should receive improved attention from the scholars.
Further remarks	Sirakaya-Turk et al. (2014) prove that tourists can, in certain conditions, strengthen their prejudicial beliefs towards the host community, instead of reducing their distance. These findings strongly contribute to end the momentum of simplistic approaches to the debate on the tourism-peace nexus.	The number of case studies is growing, keep reflecting the vast range of possible perspectives and interpretations.	New semantics and new lexical formulas are explored, with the aim to provide a more holistic theoretical approach. Wohlmuther and Wintersteiner (2014), for instance, suggest the idea of "peace sensitive tourism" as alternative to "tourism and peace". Isaac and Hodge (2011) examine instead the idea of "justice tourism".

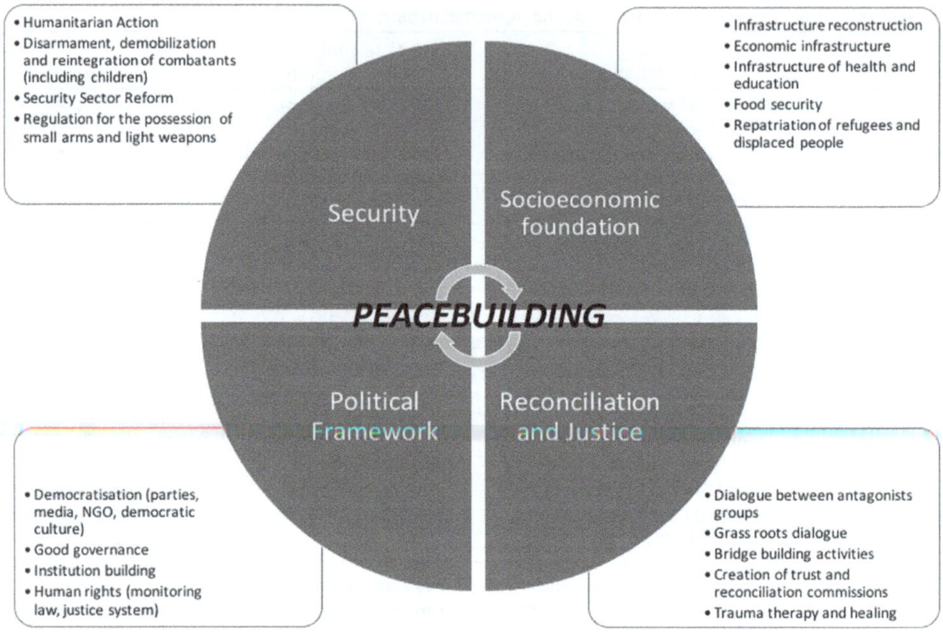

Figure 1. The Peacebuilding Palette. Source: Adapted from Smith (2004, p. 28).

summarises the classification described in this section, by type of approach, reporting identified characteristics, strengths and limitations of each.

'Building peace' - A conceptual overview

Post-conflict peacebuilding is the process of identifying and supporting structures which will tend to strengthen and solidify peace in order to avoid a relapse into conflict (Boutros-Ghali, 1992). The creation of peace is now approached as also a bottom-up process, in which a variety of peace-building agents intervene in order to promote capacity building, reconciliation and social transformation (Öjendal et al., 2017). Chandler (2017) argues that we are witnessing a transformation from peacebuilding as an international project based on universalist assumptions, to the understanding of peace as a necessarily indigenous process based on plural and non-linear understandings of differences.
According to the categorisation elaborated by Smith (2004, p. 10), peacebuilding activities aim to reach four main goals:

1. to provide security;
2. to establish the socio-economic foundations of long-term peace;
3. to establish the political framework of long-term peace;
4. to generate reconciliation, a healing of the wounds of war and justice.

The interaction of these dimensions, the so called "Peacebuilding Palette", is illustrated in Figure 1.

The idea of "sustaining peace" represents an improved architecture of the United Nations Peacebuilding, reflecting the UN's wider view of the role peacebuilding and prevention of violent conflict activities will play moving forward. It is, for instance, recognised that the "critical determinant of success is fostering inclusive national ownership", which entails increased involvement of domestic actors (UN, 2015b, p. 8).

The United Nations (UN) acknowledges that they have not invested enough on addressing root causes of conflicts (UN, 2015c). According to Richmond (2011, p. 58) "what is needed is an exploration of culture, needs and welfare-based approaches to peacebuilding that are context sensitive". This specific call opens the door to new proposals about innovative partnerships and practices in the context of peace creation.

Post-conflict tourism

Tourism is capable of consolidating peace in post-conflict societies, as claimed by the UNWTO (2020). In this sense, tourism represents not only a means to recover and consolidate economic stability and to expand infrastructure and services, but also a catalyst for social change (Ashley et al., 2007). Hence, tourism could potentially be channelled to contribute to reconciliation processes and, more generally, to the promotion of a culture of peace. Attempts to explore such a potential are increasing, as demonstrated by the increasing number of reports on the positive effects of tourism in post-conflict areas. For example, Alluri (2009) discusses the role of tourism in post-conflict peacebuilding in Rwanda, where local communities find socio-economic relief in the presence of tourists. Upadhayaya et al. (2013) report on the experience of Pokharona, in Nepal, arguing that the promotion of peace-through-tourism should be considered part of the Corporate Social Responsibility (CSR) agenda in the tourism sector, thus complementing the considerations already made by Nicolau (2008) on the CSR in tourism.

The academic debate on post-conflict tourism was inaugurated by Smith (1998) with a paper on tourist activities after the World War II (WWII). The author stated: "*WWII was a milestone in tourism: it created the technology for fast, efficient global transit; it generated the motivations for mass tourism and (…) favoured tourism development*" (Smith, 1998, p. 204). Although this analysis can be considered today too basic when compared with the more recent debate on the transformative power of tourism (Kirillova et al., 2017; UNWTO, 2016) and the latest developments of the debate on tourism and peace, it marked a milestone. Since then, in fact, post conflict tourism was approached in a more systematic way and from different perspectives. From the international relations point of view, for example, tourism was associated to the dynamics of international cooperation and cultural diplomacy (Carbone, 2017; Edgell et al., 2008). In the specific case of post conflict societies, tourism was also associated to the process of transitional justice and state-building, as a driver of social and economic recovery, for the (re)construction of contested national identities and development of nation branding (Dinnie, 2008). It is also worth mentioning the debate linking tourism to the promotion and protection of human rights (Higgins-Desbiolles & Blanchard, 2010; Higgins-Desbiolles & Whyte, 2015).

The distinctive period of post-conflict tourism development was theorised and categorised as "phoenix tourism", in the context of social renewal of the destination and its people (Causevic & Lynch, 2011). Other studies shed light on the process of transformation of war inherited sites into cultural tourist attractions (Packer et al., 2019). The importance of the intercultural dialogue and public participation in the process of heritage management have been highlighted as key aspects of the peace-through-tourism debate (Carbone et al., 2020; Malek et al., 2017; Nogués-Pedregal, 2012; Nyaupane et al., 2015).

The new museum

Museums have traditionally been places where "*society can celebrate its past and constitute a sense for its cultural identity*" (Tufts & Milne, 1999, p. 614). The contribution that museums, cultural institutions and cultural heritage could make to peacebuilding, has been largely ignored so far (Walters, 2017). However, an intense debate about the social functions of the museum could be an opportunity to treat the museum institution among the catalysts for the creation of peace.

At the end of the last century a debate inaugurated by Vergo (1989) introduced what has since then been known as *new museology*, and in the last two decades concerns for equality, diversity, social justice and human rights moved from the margins of museum research, policy and practice, to its very core, to actively inform contemporary public debates (Carbone, 2018; Dudley & Rodrigues, 2020; Sandell & Nightingale, 2012) . The proposal discussed - not without controversies - at the General Assembly of the ICOM (Kyoto, 2019, September 9th) to opt for a new definition of the museum which better reflects its social role is indicative of this current transitional moment.

The potential of museums as key intellectual and civic resources for social change has been recently emphasised by Janes and Sandell (2019) who pioneered the notion of *museum activism*. Challenging perspectives on the cultural and political significance of contemporary museums and curatorial activism places emphasis on the "dissatisfaction of the new museology towards the old one, judged to be too much about museums *methods* and too little about the purposes of museums" (Vergo, 1989, p. 3). A growing number of cultural institutions are concerned about constructing new narratives representing a plurality of lived experiences, histories and identities aimed at supporting more progressive and ethically informed ways of looking at societal issues. Specifically, museums, galleries and heritage organisations are engaging with global social challenges such as the need to construct fairer and more just societies, amid shifting politics of identity at global, national and local levels. Nevertheless, some authors highlight the fact that such a momentum is slowing down or, in some contexts, even reversing (Sandell & Nightingale, 2012).

War museums as a local infrastructure for peace

As noted earlier, museums as 'local infrastructures for peace' can act as vehicles for 'peace formation' (Richmond, 2009). It is striking that there is no legal or academic definition for military and war museums, or war memorials. This aspect has already been highlighted by the War Memorials Trust practitioners (2017) in the UK and by academics such Hacker and Vining (2013, p. 42), who indicate as a possible reason for such a gap "the well-known academic distaste for military studies". It seems quite inappropriate the indiscriminately interchangeable use of the terms, *military* museum and *war* museum.

In the name of scientific rigour, this article therefore proposes the following operational definition of the term war museum:

A war museum is an institution open to the public aiming to preserve and transmit, for different possible purposes, the memory of one (or more) specific armed conflict from any point in history to the present day. It should showcase material remains of military operations and/or objects that give an insight into both combatants and non-combatants' experience of war. It should offer physical and/or virtual reconstructions, sensory trails and/or other technological means to provide visitors with a suggestive, powerful experience.

Among the war museums to reference it is possibly uncontroversial to identify the Imperial War Museums (UK) exploring conflicts from WWI to the present day. Other interesting experiments exist nowadays, such as the *Museo Diffuso della Resistenza, della Deportazione, della Guerra, dei Diritti e della Libertà* (Widespread Museum of Resistance, Deportation, War, Rights and Freedom) in Italy. Among the war museums related to more recent conflicts, it is worth mentioning the *Tunel Spasa*, an underground tunnel that served as a lifeline for the city of Sarajevo in 1993 to find shelter from Serbian sniper fire and mortar fire, turned in a war museum on the Siege of Sarajevo (1992 – 1996). The Islamic Revolution and the Holy Defence Museum, in Tehran (Iran), focusing on the Iran-Iraq war (1980 – 1988), is today another outstanding example of war museum, thanks to the massive employment of new technologies that guarantee a touching and remarkable experience. It is finally worth mentioning the recent project of the Afghanistan Centre for Memory and Dialogue, in Kabul, aiming at promoting peace and reconciliation in a country currently affected by an intra-state, violent conflict.

Despite having already acknowledged the potential value of the association between tourism and war heritage within processes of reconciliation (Jansen-Verbeke & George, 2013; Packer et al., 2019), defining a war museum as a local infrastructure for peace is not devoid of controversy. First and foremost, this is because the narratives of war related attractions and, in particular, war museums, is often the result of a "process of selective appropriation" (Somers & Gibson, 1994, p. 60).

War museums are often used as a political space for influencing and celebrating national identity (Sternfeld, 2018). After all, the museum itself is the tool of the authorized discourse, representing the attitude of governments toward war. Especially in countries where the conflict is still fresh and unresolved, these institutions often become "like armies, an instruments and means of politics" (Muchitsch, 2013, p. 12). In these cases, they reflect the mythopoetic ability of a state, that is, "the capacity of a state as a geopolitical subject to tell its story in order to create cohesion within its borders and respect in the international arena" (Caracciolo, 2020, p. 7). War heritage is in this sense a controversial, 'dissonant heritage' conveniently mobilised in remembering war, reconstructing landscapes, political systems and identities after the conflict (Gegner & Ziino, 2012).

Wellington (2017) observes a range of different exhibitions in Britain, Canada and Australia, and reveals complex imperial dynamics in the ways these countries developed diverging understandings of World War I, even despite their cultural, political and institutional similarities. It is, therefore, necessary to overcome somehow the arbitrariness of the dissonant heritage. The use of 'dissonant heritage' for tourism purposes is a topic of debate among academics. For instance, while Šešić Dragićević and Mijatović Rogač (2014) view this practice as an opportunity to overcome inter-regional conflicting interpretations of common cultural heritage, Battilani et al. (2018) state that a proper cultural policy and public debate are paramount especially in places where dissonant heritage exists and it is used for tourism, in order to avoid visitor-resident conflict.

There are across the global landscape of war museums those that have begun to move towards the presentation of war from a multinational perspective through narratives that contribute to overcoming the use of images which depict the other side as implacable enemies. For instance, new war museums, especially about World War I, present that war from a multinational perspective with the participation of historians and museum makers of several countries involved in the war. *The Historial de la Grande Guerre* in *Péronne*, France, is a case in point. Yet, such museums still constitute an exception. By and large, the political use of war-related museums as places of celebration of identity continue to hold the potential to promote violence and conflict such as in the stimulation of feelings of anger, revenge, fear and persecution (Barber et al., 2005; McCullough, 2008). Considering war museums as a place of discussion on reconciliation and peace, requires the abandonment of the above described use of war museums. This approach tends to close the communities in themselves rather than open them to the world and, most importantly, in post-conflict areas, it is not conducive to reconciliation with former opponents. Reconciliation and peacebuilding are, according to Bar-Tal and Vered (2016, p. 249), "a long process of major societal change involving building a new socio-psychological repertoire that allows reaching an agreement with the adversary and [...] an allied relationship based on mutual trust and acceptance, cooperation, and consideration of mutual needs". It is, therefore, necessary to think about what Somers and Gibson (1994, p. 63) refer to as metanarratives or master-narratives, in which we are embedded as 'contemporary actors', in this case aiming to reconcile and promote a culture of peace.

A second challenging issue that emerges in proposing war museums as a centre of promotion of peace is related to the necessary acknowledgment and respect for the time required by a post conflict society to be psychologically and emotionally ready to deal with the wounds of conflict. As Tizzoni (2013) pointed out, the practice to promote tourism as a socio-political instrument of dialogue among former enemies started especially after World War II. Nevertheless, the visits to the symbolic places of the World War II started to be a tourist practice only later, after a long and painful process of elaboration of one of the most tragic and controversial periods in history. Only in the 1990s, almost fifty years after the end of the war, war museums and memorials started to be established abandoning the commemorative logic, conceived with didactic functions (Tizzoni, 2013; Vierregg, 2000). This historical observation alerts us to the fact that, in practice, the conscious use of tourism and war museums in post-conflict areas must necessarily be conceived as a delayed, long-term practice, to be implemented in full respect of the will and the process of elaboration of the communities previously affected by the conflict.

An additional point is worth raising. Although Suzuki (2009, p. 250) states that war museums "envision what they do as a form of public education and peace activism", in reality, this intended message of peace remains, to date, both in theory and practice, mostly implicit. A demonstration is given by the work elaborated by Ribbens and Captain (2011), in which, along the over 40 pages used to describe themes, perspectives and scope of war museums and memorial centres in the Netherlands, the world *peace* appears only once. The view that war museums' visitors "learn from the horrific past and go home with renewed determination to build peace" (Suzuki, 2009, p. 250), is questionable. And so is the view that war heritage's purpose is to convey the message for universal peace "by visually reminding visitors of the extreme devastation that a war can cause" (Luli & Kawano, 2020, p. 514). This is because both views are based on the belief that the "identification with pain, a universal true feeling, would lead to structural social change and create a peaceful world where there is no more pain" (Berlant, 2000: 45, 59). According to the historical materialist school of thought, a tragedy such as an armed conflict is remembered with both an acute sense of irreversibility and an immense sense of regret that compels us to imagine proactively other alternative courses of history (Yoneyama, 1999). This debate, so far scarcely considered in the literature, indicated the value of strategies that overcome ongoing controversies and also offer narratives able to transform a memory of war into a clear and explicit message of peace.

Let us consider, for instance, the hostile visions and ethnic divisions strongly persisting in post-war Bosnia and Herzegovina, especially among war victims. In intergenerational terms, youth born after the end of the war are exposed to routines performed by parents, educators, and media tending to dehumanise the rival communities. These young people are even more predisposed to resist reconciliation efforts than their parents, fearing that reconciliation means accepting the suffering their parents endured during the war (Omer, 2017). This example goes a long way in demonstrating the need to involve the population, especially young people, in practices that explicitly promote intercommunity dialogue grounded in peace education, enabling a realistic adjustment to the post war reality, with the ultimate purpose of reconciliation.

One of the most congenial spaces for this process to take place is a war museum, as this is the place where the conflict is remembered, interpreted and processed. This is where a participatory process where locals and tourists are involved could develop reflective capacities for achieving and maintaining peace, by proposing specific activities such as those suggested by peace education scholars such as Harris (2002) and Reardon (2002), namely:

1. understanding the richness of the concept of peace;
2. examining fears, gaining knowledge about security systems;
3. understanding violence;
4. developing intercultural understanding;
5. promoting social justice with peace;
6. encouraging respect to the living;
7. ending violence.

The bridging of theories

Three different areas of research have been highlighted above: peacebuilding, post-conflict tourism, and the changing role of museums within society. In view of the above, it seems plausible to suggest the opportunity for a renewed theoretical and practical association between tourism and museums in the context of the peace-through-tourism debate, also considering the reviewed concept of peacebuilding. This opportunity is afforded by the convergence of the change towards a boosted engagement of museums with social concerns and the current trend in tourism to overcome the dominance of the neoliberalist approach. The latter has been said to lead *"to consider pre-eminently the financial market potential of the sector, disregarding almost completely its capacity to contribute to human well-being, human rights recognition, conflict resolution, and to the worldwide enhancement of justice and a culture of peace"* (Pearse, 2016, p. 87).

The tourism sector is indeed experiencing a phase characterised by more responsible and conscious practices aiming to produce positive changes in societies. The concept of transformative (or transformational) tourism, for instance, emphasises the positive change that travel and tourism can have on human behaviour and, consequently, on society (Kirillova et al., 2017; Reisinger, 2013). In this sense, the model proposed by Cooper and Hall (2016, p. 24) "holiday travel for leisure" could also include an idea of tourism with civic commitment such as, the reframing implicit in 'responsible tourism' or 'eco-tourism'. In particular, the present work advocates for the establishment of a stronger link between tourism and museums.

The role of tourism and museums described above, paves the way for innovative collaborations towards an effective role in the context of the revisited concept of peace building as a local, bottom-up process. Destinations, local communities and tourist experiences might be managed accordingly, by integrating systematically the tourism sector in the field of education and international relations, and by making explicit the engagement with the construction of a culture of peace to replace the traditional idea of leisure. To this end, the tourism sector is enabled to consider one of its main attractions, museums, as a centre for peace education, dialogue and - where required - reconciliation. Museums have the potential to sustain and enrich society at times marked by insecurity and increasing conflict, by stimulating a curiosity that is vital to understanding (Thomas, 2016). By doing so, tourism systems would practically and effectively start to include what in the context of peace and conflict studies Richmond (2013, p. 271) calls "Local Infrastructures for Peace", defined as those institutions, mechanisms, resources, and skills used so that different groups can constructively interact with one another to address potential causes of actual or potential tension (socio-economic, political, religious, ethnic or religious differences, for instance).

Figure 2 depicts a rudimental, system-thinking driven blueprint of the dynamics underlying the association between peacebuilding and post-conflict tourism, but one where the details of the ecological (spatial and temporal) patterns of causes and consequences of identified relationships, are admittedly absent as they are inevitably context-dependent and hard to predict. Nevertheless, building on the four dimensions of peacebuilding depicted in Figure 1, the basic blueprint displayed in Figure 2, highlights that it is possible to link tourism - as an economic sector - not only to the process of socioeconomic recovery in post conflict societies, but also to the process of promotion of reconciliation and peace. This is made possible by the use of some components of the tourist system, namely tourist attractions such as museums, in general, and war-related attractions and war museums, in particular, as local infrastructures for peace. These would be spaces where both locals and tourists are explicitly encouraged to engage in relevant discussions for the active promotion of reconciliation and peace through curatorship solutions (see next subsection) and specific activities, in line with the idea of New Museum explored in the next section. So, these components of the tourism system become the bridge between tourism activity and the peacebuilding process (in Figure 2, the area in which the two rectangles overlap).

By reframing war museums as spaces for peace education and dialogue between former opponents based on a bottom-up process of reconciliation, a contribution is made to support a form of peacebuilding that avoids top-down oriented or teleologically driven peace strategies. Such a proposal aligns to the current emphasis placed on local forms of peacebuilding and,

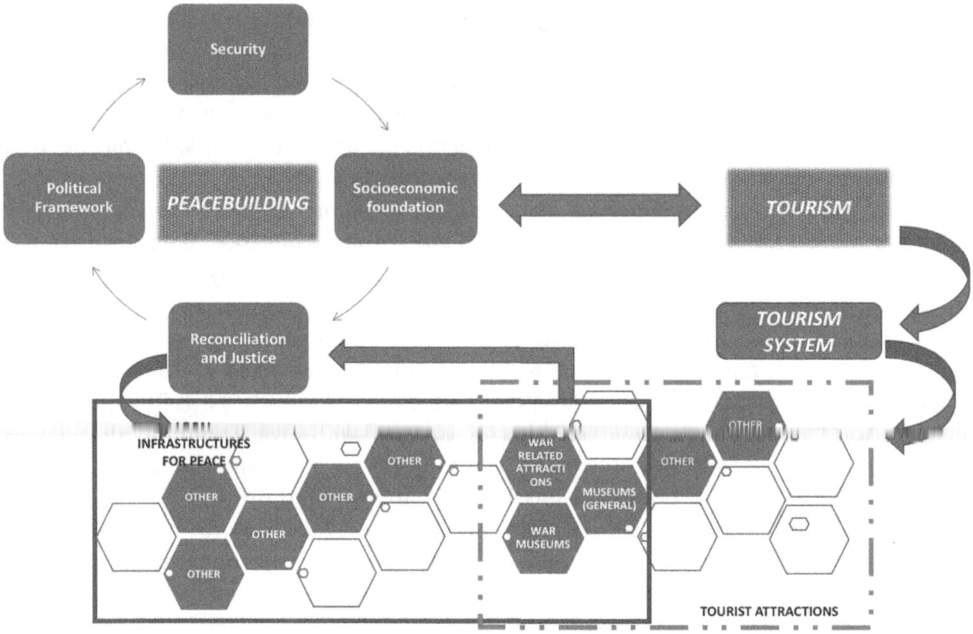

Figure 2. Dynamics underlying the association between tourism and peace through the use of museums as local infrastructure for peace. Source: The author.

particularly, to the criticisms levelled by Richmond and Franks (2007) at the concept of liberal peace. The latter is based on the idea that "the threefold transformation to peace, democracy and market economy is a self-strengthening process leading to sustainable development" (Kurtenbach, 2007, p. 6) and assumes that "democratisation and market liberalisation are themselves sources of peace" (Sriram, 2007, p. 579). But the imposition of such concepts might have counterproductive effects or might simply not match local characteristics, culture and priorities (Sriram, 2007, p. 580). In this sense, liberal peace not only disregards existing socio-political structures, but also fails to adequately appreciate local context. The 'local' and 'local people' are denied agency and depoliticised at the same time (Bräuchler, 2015; Richmond, 2009, 2011).

Proposal for an internationally shared war museums' narrative

Transforming a memory of war into a message of peace requires the review of the current practices of locals and tourists' engagement in the museum experience, starting with the reconsideration of the narratives offered by war museums. Developing a metanarrative, a shared vision, involves a process and a product. Using a war museum to display the process through which a community passes from the harm inflicted by the war to the reconciliation with former opponents can represent a valid alternative to the misleading belief according to which displaying the atrocity of war implicitly involves a message of peace.

As Bräuchler (2015, p. 9) states: "In the end, what it is all about is reconciliation, about how people in divided societies can live together again and realise positive peace". The concept of reconciliation appears in Hegel's 1807 Phenomenology of Spirits as moment of transition. In this sense, reconciliation does not stand on its own, it rather follows an offence, or a wound, and a forgiveness that has to be "unconditional, that is, must not demand or expect repentance in exchange" (Petersson, 2013, pp. ix-x; 163). Even though "there is no such thing like a blueprint for reconciliation" (Bräuchler, 2015, pp. 9-10) and there is no set method to educate on forgiveness and reconciliation, academics and practitioners have developed a variety of processes for people to explore on their own forgiveness journeys (Enright, 2008; Luskin, 2002; Tutu & Tutu,

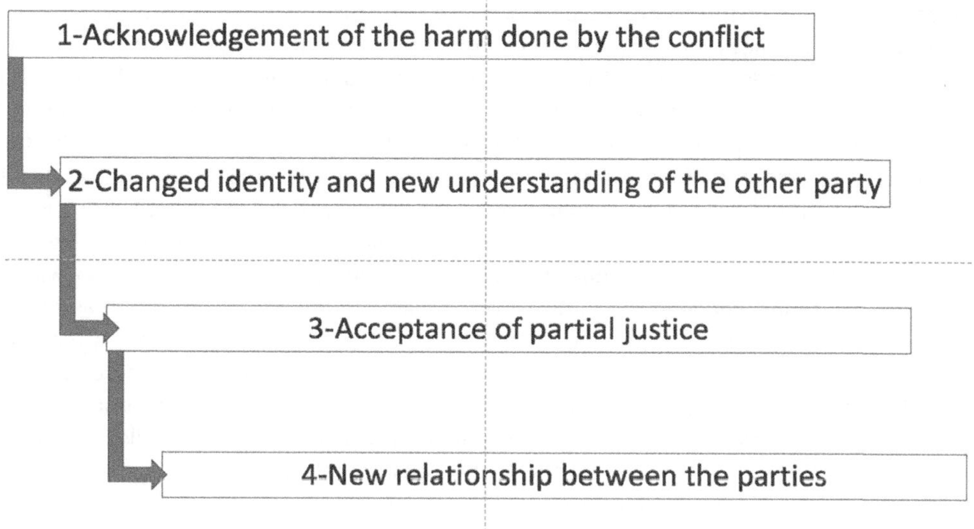

Figure 3. Synthesis of the Long and Brecke (2003) forgiveness model.

2014). Yet, as noted by (Bräuchler, 2015, p. 10), "reconciliation is a multidimensional process that takes place on different levels: interpersonal, intracommunal, and intercommunal, between communities and the state, among states", a complexity that is best dealt with by framing forgiveness and reconciliation processes contextually.

While in the past reconciliation has been perceived by peacebuilders as an outcome, today it is rather perceived as a long-term, broad, inclusive and voluntary process (Bloomfield, 2003) that "goes beyond the agenda of formal conflict resolution to changing the motivations, goals, beliefs, attitudes, and emotions of the great majority of the society members regarding the conflict, the nature of the relationship between the parties, and the parties themselves" (Bar-Tal & Bennink, 2004, p. 12). In this context, the present study identifies in the widely accepted forgiveness model by Long and Brecke (2003) the most appropriate reference model for the creation of a war museums' narrative. The model effectively describes the entire behavioural and psychological process leading communities and individuals to reconcile after a conflict (Figure 3).

According to Long and Brecke (2003), reconciliation is achieved through a process where former opponents accept the harm inflicted by the conflict (1) and overcome the feelings of anger and revenge. It allows them to build a renewed, post-conflict identity (2). In this phase, former opponents start to see themselves "as something more than a victim in the relationship, and to see the other as something other than an enemy" (Kaufman, 2004, p. 145), and it is essential to accept an unavoidable partial justice (3). At the end of this process, the parties that were once in conflict can finally establish new positive relations in a spirit of reconciliation and mutual trust (4).

The most significant criticism to this model has been moved from the area of evolutionary psychology. Ren (2014, p. 167) wondered what could motivate the public to hold certain attitudes towards reconciliation and follow certain patterns of behaviour. However, the study conducted by Packer et al. (2019) shows that a raised awareness of the importance of peace is among the most significant effects on the visitor of the war-related tourist attraction they analysed. These findings "are encouraging as they suggest that war heritage interpretation can lead visitors to adopt a more inclusive perspective, rather than a narrow focus on national pride or patriotism" (Packer et al., 2019, p. 113).

It is, therefore, reasonable to argue that the potential described by those authors can be further developed and subsequently expressed through a clear reference to the reconciliation process by following the forgiveness model to build war museums' narrative. It is important to better define the role of the tourist as part of these dynamics. The presence of tourists is complementary to the promotion of peace and reconciliation. International tourists coming from previously opposed countries (this could be the case of Iraqi tourists in Iran, or Serbians in Bosnia) could be directly involved in a participatory process of setting up the museum, sharing their own view about the happenings, following the same path made available to locals (the forgiveness process). The presence of international tourists from countries outside the conflict, will bring their experiences and visions of the conflict and, by doing so, they will represent in the debate the international public opinion. How astonishing, imagine if in a few years, it will be possible for a European tourist to visit a war museum in Yemen, or in Syria, for instance. What will be, for instance, the gap found between the information available in the country of origin and the story told in the museum? Such an experience promises to be remarkable, even from a more purely touristic point of view, a formative and eye-opening experience about the importance of an active global citizenship. The host community will perceive their harm as internationally acknowledged, an important step to overcoming the trauma and heal the wound, following the steps proposed by the museum and based on the forgiveness model. Activities facilitating transformative experiences on personal and inter-personal levels, involve:

- logical thinking;
- deep critical reflection;
- challenging of assumptions;
- learning lessons;
- creativity, innovation and reimagining;
- deep conversations;
- inclusive solutions,
- shared responsibility,
- honesty.

Concluding discussion

For tourism to effectively contribute to peace, it is necessary to link theories and practices from different fields of study. Theoretically, the use of concepts from fields of research other than tourism helps to understand and make better sense of the complex dynamics that the peace-through-tourism discourse underlies. This knowledge should then inform detailed guidelines to be shared internationally within the sector. With this aim in mind, this article linked theories from tourism, museum and peace and conflict studies, aiming to fill the gap related to the lack of clarity about how post-conflict tourism could contribute to the process of reconciliation. The study drew from the three areas considered likely to allow for both theoretical convergence and a practical connection through which museums bridge the gap between tourism and peacebuilding. Reciprocal and complementary relations between the transformational role of tourism and the renewed social role of museums is thus established. In this context, museums, namely war museums, if considered as local infrastructures for peace, represent the platform binding together tourism and peacebuilding. It is suggested that the model be shared and applied internationally in order to be effective, alongside the application of the forgiveness model to inspire war related attractions' narratives in inclusive, participatory experiences aimed at locals and tourists, particularly when the latter come from countries that took part in the conflict as opponents of the former.

Whilst the article goes some way in debunking the claim that showing the horrors of war alone is enough to persuade audiences of the need for peace, it must be acknowledged that in the absence of concrete examples where the proposed reinventing of the war museum has been trialled, the sceptical reader might legitimately note that we have, as yet, no way of knowing whether this approach would work or not. But as per the common destiny of theory-based explorative studies such as this, it is at least hoped that the propositions put forward in the present study will stimulate further debate and be adopted at least experimentally by museums keen on breaking new ground in finding new and better ways for promoting peace. As for the success or otherwise of the underlying propositions, we will inevitably have to wait for them to be tested empirically in future research.

Another caveat worth acknowledging is the forcefully normative and prescriptive thrust of this article in encouraging museums to give an unambiguously clear and explicit message of peace, which is both the premise and corollary of the thesis of this article. Whilst some might argue that it is not the purpose of museums to give explicit messages or advocate fixed moral positions that might challenge democratic notions of the museum as a space for dialogue and negotiation, the article makes no apologies for its modicum of healthy, ethically-driven didacticism in promoting the advocacy of peace.

Acknowledgments

I am gratefully acknowledging the most helpful advice and suggestions by Freya Higgins-Desbiolles, Lynda Blanchard and Yoko Urbain, Special Issue Editors 'Peace Through Tourism', and by the four anonymous reviewers on earlier versions of this paper; my colleagues at the Centre for Trust, Peace and Social Relations - RG "Peacebuilding and Conflict Transformation" (Coventry University, UK), for their insights. A special thanks to Glauco de Vita, for his fraternal support and his revisions. This work is dedicated to Manar, and the people of Syria. To all those who struggle, affected by armed conflicts.

Disclosure statement

In accordance with Taylor & Francis policy and my ethical obligation as a researcher, I am reporting that there are no financial or non-financial competing interests to report.

ORCID

Fabio Carbone http://orcid.org/0000-0002-6711-4892

References

Alluri, R. M. (2009). *The Role of Tourism in Post-Conflict Peacebuilding in Rwanda*. Swisspeace.

Apsel, J. (2015). *Introducing Peace Museums*. Routledge.

Ashley, C., De Brine, P., Lehr, A., & Wilde, H. (2007). *The role of the tourism sector in expanding economic opportunity. Corporate Social Responsibility Initiative, MA*. Harvard University.

Askjellerud, S. (2003). The Tourist: A Messenger of Peace? *Annals of Tourism Research*, *30*(3), 741–744. https://doi.org/10.1016/S0160-7383(03)00049-5

Barber, L., Maltby, J., & Macaskill, A. (2005). Angry memories and thoughts of revenge: The relationship between forgiveness and anger rumination. *Personality and Individual Differences*, *39*(2), 253–262. https://doi.org/10.1016/j.paid.2005.01.006

Barrett, C. (2012). Sadako or Al Qaeda? A peace museum approach to counter terrorism. In C. Barret & J. Apsel (Eds.), *Museums for peace: transforming cultures* (pp. 71–87). The International Network of Museums for Peace.

Bar-Tal, D., & Bennink, H. G. (2004). The nature of reconciliation as an outcome and as a process. In Y. Bar-Simav-Tov (Ed.), *From conflict resolution to reconciliation* (pp. 11–38). Oxford University Press.

Bar-Tal, D., & Vered, S. (2016). Peace and Reconciliation. In H. S. Friedman (Ed.), *Encyclopedia of Mental Health* (Second Edition) (pp. 249–257). Academic Press.

Battilani, P., Bernini, C., & Mariotti, A. (2018). How to cope with dissonant heritage: a way towards sustainable tourism development. *Journal of Sustainable Tourism*, *26*(8), 1417–1436. https://doi.org/10.1080/09669582.2018.1458856

Bauman, Z. (2018). *Di nuovo soli*. Castelvecchi.

Becken, S., & Carmignani, F. (2016). Does tourism lead to peace? *Annals of Tourism Research*, *61*, 63–79. https://doi.org/10.1016/j.annals.2016.09.002

Berlant, L. (2000). The subject of true feeling: pain, privacy, and policy. In J. Dean (Ed.), *Cultural Studies and Political Theory* (pp. 42–62). Cornell University Press.

Bloomfield, D. (2003). Reconciliation: an introduction. In D. Bloomfield, T. Barnes, & L. Huyse (Eds.), *Reconciliation after violent conflict: a handbook* (pp. 10–18). International Institute for Democracy and Electoral Assistance (IDEA).

Boulding, E. (2008). Peace Culture. In L. Kurtz (Ed.), *Encyclopedia of Violence, Peace, & Conflict. (*Second Edition*)* (pp. 1452–1465). Academic Press.

Boutros-Ghali, B. (1992). *An agenda for peace: preventive diplomacy, peacemaking and peace-keeping.* United Nations-Security Council.

Bräuchler, B. (2015). *The cultural dimension of peace. Decentralization and reconciliation in Indonesia.* Palgrave Macmillan.

Brown, F. (1989). Is tourism really a peacemaker? *Tourism Management, 10*(4), 270–271. https://doi.org/10.1016/0261-5177(89)90001-0

Bruner, E. M. (1991). Transformation of self in tourism. *Annals of Tourism Research, 18*(2), 238–250. https://doi.org/10.1016/0160-7383(91)90007-X

Burnett, G. W., & Uysal, M. (1990). On the nature of peace in relationship to tourism: Three cases. *The Tourist Review, 45*(1), 2–7. https://doi.org/10.1108/eb058035

Caracciolo, L. (2020). Tutti i miti portano a Roma, Limes- Il potere del mito, 2/20. *Limes - Rivista Italiana di Geopolitica, 2,* 7–34.

Carbone, F. (2016). Cultural Heritage Quality Management: Analysis of archaeological heritage managers' perception. *European Journal of Tourism Research, 14,* 114–118.

Carbone, F. (2017). International Tourism and Cultural Diplomacy: A new conceptual approach towards global mutual understanding and peace through tourism. *Tourism. An International Interdisciplinary Journal, 65*(1), 61–74.

Carbone, F. (2018). Post-multicultural challenges for cultural heritage managers and museums in the age of migrations. *Museum Management and Curatorship, 34*(1), 1–22.

Carbone, F. (2019). Tourism and peace: a tool for global integration and peace. In L. Oosterbeek & L. Caron (Eds.), *Resilience and Transformation in the territories of low demographic density: integrated methodologies of human and social sciences for integrated cultural landscape management* (pp. 149–165). Instituto Terra e Memoria.

Carbone, F., & Oosterbeek, L. (2021). Peace: A Roadmap for Heritage and Tourism. In J. Tavares, Z. Breda, & F. Carbone (Eds.), *Role and Impact of Tourism in Peacebuilding and Conflict Transformation* (pp. 1–15). IGI.

Carbone, F., Oosterbeek, L., & Costa, C. (2012). The educational and awareness purposes of the Paideia approach for heritage management. *Natural Hazards and Earth System Sciences, 12*(6), 1983–1986. https://doi.org/10.5194/nhess-12-1983-2012

Carbone, F., Oosterbeek, L., Costa, C., & Ferreira, A. M. (2020). Extending and adapting the concept of quality management for museums and cultural heritage attractions: A comparative study of southern European cultural heritage managers' perceptions. *Tourism Management Perspectives, 35,* 100698 https://doi.org/10.1016/j.tmp.2020.100698

Causevic, S. (2010). Tourism which erases borders: An introspection into Bosnia and Herzegovina. In O. Moufakkir & I. Kelly (Eds.), *Tourism, Progress and Peace* (pp. 48–64). CABI.

Causevic, S., & Lynch, P. (2011). Phoenix tourism. Post-conflict tourism role. *Annals of Tourism Research, 38*(3), 780–800. https://doi.org/10.1016/j.annals.2010.12.004

Chandler, D. (2017). *Peacebuilding: The Twenty Years' Crisis, 1997-2017 (Rethinking Peace and Conflict Studies).* Palgrave Macmillan.

Cho, M. (2007). A re-examination of tourism and peace: The case of the Mt. Gumgang tourism development on the Korean Peninsula. *Tourism Management, 28*(2), 556–569.

Cohen, E. H. (2011). Educational dark tourism at an in populo site: The Holocaust Museum in Jerusalem. *Annals of Tourism Research, 38*(1), 193–209. https://doi.org/10.1016/j.annals.2010.08.003

Cooper, C., & Hall, C. M. (2016). *Contemporary tourism. An international approach.* (3rd ed.). Goodfellow.

D'Amore, L. (1988). Tourism - The World's Peace Industry. *Journal of Travel Research, 27*(1), 35–40. https://doi.org/10.1177/004728758802700107

De Villiers, D. (2014). Cornerstones for a Better World: Peace, Tourism and Sustainable Development. In C. Wohlmuther & W. Wintersteiner (Eds.), *International Handbook on Tourism and Peace* (pp. 78–86). Drava.

Di Cesare, D. (2018). *Sulla vocazione politica della filosofia.* Bollati Boringhieri.

Dinnie, K. (2008). *Nation Branding: Concepts, Issues, Practice.* Elsevier/Butterworth-Heine-Mann.

Dudley, L., & Rodrigues, E. (2020). What is a museum today? *Museological Review,* (24), 4–6.

Duffield, M. R. (2014). *Global Governance and the New Wars: The Merging of Development and Security.* (3rd ed.). *London.* & Zed Books.

Edgell, D. L., Allen, M. D., Smith, G., & Swanson, J. R. (2008). Political and foreign policy implications of tourism. In D. L. Edgell, M. D. Allen, G. Smith, & J. R. Swanson (Eds.), *Tourism Policy and Planning* (pp. 141–180). Butterworth-Heinemann.

Enright, R. (2008). *Forgiveness is a choice*. American Psychological Association.
Eusébio, C., Vieira, A. L., & Lima, S. (2018). Place attachment, host-tourist interactions, and residents' attitudes towards tourism development: the case of Boa Vista Island in Cape Verde. *Journal of Sustainable Tourism*, *26*(6), 890–909. https://doi.org/10.1080/09669582.2018.1425695
Fan, D. X. F., Qiu, H., Jenkins, C. L., & Lau, C. (2020). Towards a better tourist-host relationship: the role of social contact between tourists' perceived cultural distance and travel attitude. *Journal of Sustainable Tourism*, 1–25.
Farmaki, A. (2017). The tourism and peace nexus. *Tourism Management*, *59*(Supplement C), 528–540. https://doi.org/10.1016/j.tourman.2016.09.012
Figal, G. (2012). *Beach heads. War, Peace, and tourism in post-war*. Okinawa, London: Rowman & Littlefield.
Fitch, C. (2018). *Globalography: our interconnected world revealed in 50 maps*. White Lion.
Francesco (2015). Holy Mass Homily of the Holy Father. Apostolic journey of His Holiness pope Francis to Sarajevo (Bosnia and Herzegovina), Koševo stadium Saturday, 6 June 2015. Retrieved from http://www.vatican.va/content/francesco/en/homilies/2015/documents/papa-francesco_20150606_omelia-sarajevo.html
Fyall, A., Prideaux, B., & Timothy, D. J. (2006). War and tourism: an introduction. *International Journal of Tourism Research*, *8*(3), 153–155. https://doi.org/10.1002/jtr.564
Gegner, M., & Ziino, B. (2012). *The Heritage of War*. Routledge.
Hacker, B. C., & Vining, M. (2013). Military Museums and Social History. In W. Muchitsch (Ed.), *Does War Belong in Museums? The Representation of Violence in Exhibitions* (pp. 41–59). Creative Commons.
Hall, C. M. (1994). *Tourism and politics: Policy, power, and place*. Wiley.
Harris, M. I. (2002). Conceptual underpinnings of peace education. In G. Salomon & B. Nevo (Eds.), *Peace education: The concept, principles, and practices around the world* (pp. 15–25). Lawrence Erlbaum Associates Publishers.
Higgins-Desbiolles, F., & Blanchard, L. (2010). Challenging peace through tourism: placing tourism in the context of human rights, justice and peace. In O. Moufakkir & I. Kelly (Eds.), *Tourism, Progresss and Peace* (pp. 35–47). CABI.
Higgins-Desbiolles, F., & Whyte, K. P. (2015). Tourism and human rights. In M. C. Hall, S. Gossling, & D. Scott (Eds.), *The Routledge Handbook of Tourism and Sustainability*. (pp. 105–116). Routledge.
Hugues, G. (2004). Tourism, Sustainability, and Social Theory. In A. A. Lew, C. M. Hall, & A. M. Williams (Eds.), *A Companion to Tourism* (pp. 498–509). Blackwell.
Institute for Economics and Peace (2018). *Global Peace Index 2018*. Institute for Economics and Peace.
Isaac, R. K., & Hodge, D. (2011). An Exploratory Study: Justice Tourism in Controversial Areas. The Case of Palestine. *Tourism Planning & Development*, *8*(1), 101–108. https://doi.org/10.1080/21568316.2011.554048
Janes, R. R., & Sandell, R. (2019). *Museum Activism. Lodon and* Routledge.
Jansen-Verbeke, M., & George, W. (2013). Reflections on the Great War centenary: From warscapes to memory-scapes in 100 years. In R. Butler & W. Suntikul (Eds.), *Tourism and War* (pp. 273–294). Routledge.
Kaufman, S. J. (2004). Peace means saying you are sorry - And meaning it. *International Studies Review*, *6*(1), 145–147. https://doi.org/10.1111/j.1079-1760.2004.00386.x
Khalilzadeh, J. (2018). Demonstration of exponential random graph models in tourism studies: Is tourism a means of global peace or the bottom line? *Annals of Tourism Research*, *69*, 31–41. https://doi.org/10.1016/j.annals.2017.12.007
Kirillova, K., Lehto, X., & Cai, L. (2017). What triggers transformative tourism experiences? *Tourism Recreation Research*, *42*(4), 498–511. https://doi.org/10.1080/02508281.2017.1342349
Kirshenblatt-Gimblett, B. (1998). *Destination Culture: Tourism, Museums, and Heritage*. University of California Press.
Kurtenbach, S. (2007). *Why is libearl peace so difficult? Some lessons from central America*. GIGA.
Lennon, J., & Foley, M. (2000). *Dark tourism: The attraction of death and disaster*. Continuum.
Li, F., & Wang, B. (2020). Social contact theory and attitude change through tourism: Researching Chinese visitors to North Korea. *Tour Manag Perspect*, *36*, 100743 https://doi.org/10.1016/j.tmp.2020.100743
Lisle, D. (2000). Consuming Danger: Reimagining the War/Tourism Divide. *Alternatives: Global, Local, Political*, *25*(1), 91–116. https://doi.org/10.1177/030437540002500106
Litvin, S. W. (2003). Tourism and understanding. *Annals of Tourism Research*, *30*(1), 77–93. https://doi.org/10.1016/S0160-7383(02)00048-8
Long, J. W., & Brecke, P. (2003). *War and Reconciliation. Reason and Emotion in Conflict Resolution*. MIT Press.
Luli, v d D., & Kawano, N. (2020). Online tourist reviews and accidental conveyors of memories of the atomic bomb. *Journal of Tourism and Cultural Change*, *18*(5), 514–531. https://doi.org/10.1080/14766825.2019.1702048
Luskin, F. (2002). *Forgive for good*. HarperCollins.
Malek, A., Carbone, F., & Alder, J. (2017). Community Engagement, Rural Institutions and Rural Tourism Business in Developing Countries. In A. Oriade & P. Robinson (Eds.), *Rural Tourism and Enterprise* (pp. 145–157). CAB International.
Malek, A., & Costa, C. (2015). Integrating Communities into Tourism Planning Through Social Innovation. *Tourism Planning & Development*, *12*(3), 281–299. https://doi.org/10.1080/21568316.2014.951125
Maoz, D. (2006). The mutual gaze. *Annals of Tourism Research*, *33*(1), 221–239. https://doi.org/10.1016/j.annals.2005.10.010
McCullough, M. E. (2008). *Beyond revenge. The evolution of the forgiveness instinct*. Jossey-Bass.

Monteiro, A., Lopes, S., & Carbone, F. (2021). Academic Mobility: Bridging Tourism and Peace Education. In J. Tavares da Silva, Z. Breda, & F. Carbone (Eds.), *Role and Impact of Tourism in Peacebuilding and Conflict Transformation* (pp. 275–301). IGI.

Moufakkir, O., & Kelly, I. (2010). *Tourism, progress and peace*. CABI.

Muchitsch, W. (2013). *Does War Belong in Museums?. The Representation of Violence in Exhibitions*. Creative Commons.

Muzaini, H. (2008). Peace Education: Peace Museums. In L. Kurtz (Ed.), *Encyclopedia of Violence, Peace, & Conflict (*Second Edition*)* (pp. 1488–1498). Academic Press.

Nicolau, J. L. (2008). Corporate social responsibility: Worth-creating activities. *Annals of Tourism Research*, *35*(4), 990–1006. https://doi.org/10.1016/j.annals.2008.09.003

Nogués-Pedregal, A. M. (2012). Tourism in the Muslim World. *Annals of Tourism Research*, *39*(1), 511–513. https://doi.org/10.1016/j.annals.2011.11.007

Nyaupane, G. P., Teye, V., & Paris, C. (2008). Innocents abroad: attitude change toward hosts. *Annals of Tourism Research*, *35*(3), 650–667. https://doi.org/10.1016/j.annals.2008.03.002

Nyaupane, G. P., Timothy, D. J., & Poudel, S. (2015). Understanding tourists in religious destinations: A social distance perspective. *Tourism Management*, *48*, 343–353. https://doi.org/10.1016/j.tourman.2014.12.009

Öjendal, J., Leonardsson, H., & Lundqvist, M. (2017). *Local Peacebuilding – challenges and opportunities*. The Expert Group for Aid Studies (EBA).

Omer, A. (2017). Bridging ethnoreligious divides in Bosnia and Herzegovina: choosing peace together through interreligious action for conflict transformation. *Contending Modernities*. https://contendingmodernities.nd.edu/theorizing-modernities/bridging-ethnoreligious-divides-bosnia-herzegovina-choosing-peace-together-interreligious-action-conflict-transformation/

Oosterbeek, L., & Pollice, F. (2014). Cultural heritage and local development. Local communities through heritage awareness and global understanding. *Territori Della Cultura*, (18)

Packer, J., Ballantyne, R., & Uzzell, D. (2019). Interpreting war heritage: Impacts of Anzac museum and battlefield visits on Australians' understanding of national identity. *Annals of Tourism Research*, *76*, 105–116. https://doi.org/10.1016/j.annals.2019.03.012

Pearse, A. (2016). Peace through tourism: promoting human security through international citizenship (Book review). *Journal of Policy Research in Tourism, Leisure and Events*, *8*(1), 87–89. https://doi.org/10.1080/19407963.2014.977507

Petersson, D. (2013). *The art of reconciliation*. Palgrave Macmillan.

Reardon, B. (2002). Human rights and the global campaign for peace education. *International Review of Education/ Internationale Zeitschrift fr Erziehungswissenschaft/ Revue Inter*, *48*(3/4), 283–284. https://doi.org/10.1023/A:1020369813176

Reid, G. D. (2003). *Tourism, globalization and development*. Pluto.

Reisinger, Y. (2013). *Transformational Tourism. Tourist Perspectives*. CABI.

Ren, L. (2014). *Rationality and Emotion: Comparative Studies of the Franco-German and Sino-Japanese Reconciliations*. Springer.

Ribbens, K., & Captain, E. (2011). *Exhibiting the war. The Future of World War II Museums in the Netherlands*. NIOD, Institute for War, Holocaust and Genocide Studies.

Richmond, O. (2009). Becoming liberal, unbecoming liberalism: liberal local hybridity via the everyday as a response to the paradoxes of liberal peacebuilding. *Journal of Intervention and Statebuilding*, *3*(3), 324–344. https://doi.org/10.1080/17502970903086719

Richmond, O. P. (2011). *A Post-Liberal Peace*. Routledge.

Richmond, O. P. (2013). Peace formation and local infrastructures for peace. *Alternatives: Global, Local, Political*, *38*(4), 271–287. https://doi.org/10.1177/0304375413512100

Richmond, O., & Franks, J. (2007). Liberal Hubris? *Security Dialogue*, *38*(1), 27–48. https://doi.org/10.1177/0967010607075971

Saarinen, J., & Manwa, H. (2008). Tourism as a socio-cultural encounter: host-guest relations in tourism development in Botswana. *Botswana Notes and Records*, *39*, 43–53.

Sandell, R., & Nightingale, E. (2012). *Museums, Equality and Social Justice*. Routledge.

Scott, J. (2012). Tourism, civil society and peace in Cyprus. *Annals of Tourism Research*, *39*(4), 2114–2132. https://doi.org/10.1016/j.annals.2012.07.007

Seaton, A. V. (1996). Guided by the dark: From thanatopsis to thanatourism. *International Journal of Heritage Studies*, *2*(4), 234–244. https://doi.org/10.1080/13527259608722178

Selmi, N., Tur, C., & Dornier, R. (2012). To what extent may sites of death be tourist destinations? The case of Hiroshima in japan and Struthof in France. *Asian Business Management*, *11*(3), 311–328.

Šešić Dragićević, M., & Mijatović Rogač, L. (2014). Balkan dissonant heritage narratives (and their attractiveness) for tourism. *American Journal of Tourism Management*, *3*(1B), 10–19.

Sirakaya-Turk, E., Nyaupane, G., & Uysal, M. (2014). Guests and hosts revisited: prejudicial attitudes of guests toward the host population. *Journal of Travel Research*, *53*(3), 336–352. https://doi.org/10.1177/0047287513500580

Smith, D. (2004). *Towards a strategic framework for peacebuilding: Getting their act together (Evaluation Report 1/ 2004).* Royal Norwegian Ministry of Foreign Affairs.
Smith, V. L. (1998). War and tourism. *Annals of Tourism Research, 25*(1), 202–227. https://doi.org/10.1016/S0160-7383(97)00086-8
Somers, M. R., & Gibson, G. D. (1994). Reclaiming the epistemological "other": narrative and the social constitution of identity. In C. Calhoun (Ed.), *Social Theory and the Politics of Identity* (pp. 37–99). Blackwell.
Sriram, C. L. (2007). Justice as peace? liberal peacebuilding and strategies of transitional justice. *Global Society, 21*(4), 579–591.
Sternfeld, N. (2018). *Das radikaldemokratische Museum.* De Gruyter.
Suvantola, J. (2018). *Tourist's Experience of Place.* Routledge.
Suzuki, T. (2009). Narrating war and peace at battle ruins: Okinawan tourism-activism discourses. In A. Hodges (Ed.), *Discourses of War and Peace* (pp. 250.). Oxford University Press.
Tasci, A. D. A., & Severt, D. (2017). A triple lens measurement of host–guest perceptions for sustainable gaze in tourism. *Journal of Sustainable Tourism, 25*(6), 711–731. https://doi.org/10.1080/09669582.2016.1225746
Tavares da Silva, J., Breda, Z., & Carbone, F. (2021). *Role and impact of tourism in peacebuilding and conflict transformation.* New York: IGI.
Thomas, N. (2016). *The Return of Curiosity: What Museums are Good for in the Twenty-First Century.*
Tizzoni, E. (2013). Turismo di guerra, turismo di pace: sguardi incrociati su Italia e Francia. *Diacronie, 15*(N° 15, 3), 3. http://journals.openedition.org/diacronie/430 https://doi.org/10.4000/diacronie.430
Tucker, H. (2016). Empathy and tourism: Limits and possibilities. *Annals of Tourism Research, 57*, 31–43. https://doi.org/10.1016/j.annals.2015.12.001
Tufts, S., & Milne, S. (1999). Museums: A supply-side perspective. *Annals of Tourism Research, 26*(3), 613–631. https://doi.org/10.1016/S0160-7383(99)00024-9
Tutu, D., & Tutu, M. (2014). *The book of forgiving.* HarperCollins.
UN (2015a). *Transforming our world: the 2030 Agenda for Sustainable Development.* United Nations.
UN (2015b). *The challenge of sustaining peace. Report of the advisory group of experts for the 2015 review of the United Nations Peacebuilding architecture.* United Nations.
UN. (2015c). *Uniting our Strengths for Peace - Politics. Partnerships and People, Report of the High Level Independent Panel on UN Peace Operations.* United Nations.
UNWTO (2016). *Global Report on the Transformative Power of Tourism: a paradigm shift towards a more responsible traveller.* (Vol. XIV). UNWTO.
UNWTO (2020). Tourism for SDGs. Retrieved from http://tourism4sdgs.org/sdg-16-peace-justice-institutions/
Upadhayaya, P. K., Khatiwada, S. S., Upreti, B. R., & Sapkota, T. (2013). *Corporate Social Responsibility Guideline for Promoting Peace in Tourism Sector in Pokhara.* Pokhara Tourism Council, Nepal Tourism Board, NCCR North-South.
van den Dungen, P. (2017). The heritage of peace: the importance of peace museums for the development of a culture of peace. In D. Laven, D. Walters, & P. Davis (Eds.), *Heritage and Peacebuilding* (pp. 7–16): Boydell & Brewer.
Vergo, P. (1989). *The new museology.* Reaktion.
Vierregg, H. (2000). Rappresentare la storia contemporanea nei musei. In F. Di Valerio & V. Paticchia (Eds.), *Un futuro per il passato. Memoria e musei nel terzo millennio* (pp. 131–150). Clueb.
Walle, A. H. (1993). Tourism and traditional people: forging equitable strategies. *Journal of Travel Research, 31*(3), 14–19. https://doi.org/10.1177/004728759303100303
Walters, D. (2017). Can museums build peace? The role of museums in peacebuilding and internationalism. In D. Laven, D. Walters, & P. Davis (Eds.), *Heritage and Peacebuilding.* (pp. 39–52): Boydell & Brewer.
War-Memorials-Trust (2017). *Definition of a war memorial for funding purposes.* WMT.
Wellington, J. (2017). *Exhibiting War: The Great War, Museums, and Memory in Britain, Canada, and Australia.* Cambridge University Press.
Wohlmuther, C., & Wintersteiner, W. (2014). *International handbook on tourism and peace.* DRAVA.
Yang, L. (2011). Ethnic tourism and cultural representation. *Annals of Tourism Research, 38*(2), 561–585. https://doi.org/10.1016/j.annals.2010.10.009
Yoneyama, L. (1999). *Hiroshima Traces: Time, Space and the dialectics of memory.* University of California Press.
Yoshida, K., Bui, H. T., & Lee, T. J. (2016). Does tourism illuminate the darkness of Hiroshima and Nagasaki. *Journal of Destination Marketing & Management, 5*(4), 333–340.

Dances with despots: tourists and the afterlife of statues

Elizabeth Carnegie and Jerzy Kociatkiewicz

ABSTRACT

In times of liquid modernity, when human lifespan often exceeds that of grand political structures, monumental statues continue to be built and celebrated as symbols of enduring ideological triumphs. In their apparent permanence, these statues often outlive the political systems they were designed to glorify, creating a dilemma of how to exhibit their ambiguous or disgraced presence. In this article, we argue that the heritagization of political figures and pasts is central to the reframing of such narratives and that tourists have a key, if sometimes unwitting, role to play in the shaping of the emerging political imaginaries. Focusing on statue parks in Central and Eastern Europe showcasing communist-era sculptures, we examine strategies of exhibition and tourist responses to the multivalent presence of the monuments of past regimes. We identify four approaches of destruction, delegitimization, decontextualization, and depoliticization, each tied to a particular political moment and rhetorical goal. Examining these shifting modes of preservation, presentation and interpretation, we query the tourists' role as participants in the processes of stabilization and peace-building, proposing that in times of global re-evaluation of the symbolism of past monuments, these sites can serve to guide much needed analysis and reflection.

Introduction

Bauman (2000) famously described our current condition as that of liquid modernity: times of rapid change and unstable social structures. In times of such flux, individual actors tend to enjoy longer lifespans than the institutions governing their lives (Bauman, 2004). While this has not stopped the practice of glorifying public triumphs (or aspirations) and the political status quo through the erection of imposing statues and monumental buildings, it does imply that such landmarks are likely to, relatively quickly, witness dramatic revalorisation(s) of their status and significance. However, their very solidity and apparent importance mean that such revalorisation is symbolically charged and can lead to a problematic and disputed afterlife. Wells (2007, p. 139) notes that

> because the purpose of monuments is often to commemorate a particular moment in a government's ascendancy and at the same time to insist on the permanence or atemporality of its rule, they are, of course, difficult to dispose of.

Our article is an attempt to take stock of the variety of fates of disgraced monuments and to investigate their possible roles in forming a basis for sustainable tourism and peace-oriented heritage (Wallis, 2019). We contend that tourists explore these sites under consideration here at the point where past and present meet and merge with their own expectations of place.

Our main insights come from a study of recent history museum and memory sites in Central and Eastern Europe, particularly those including (or consisting of) collections of communist-era monuments. Consequently, our focus concerns the fate of communist monuments in post-communist Central and Eastern Europe, but we believe the relevance of our analysis stretches much further, and thus we highlight some of the parallels to events and reconfigurations touching on other ideologies and political shifts. Taken as a paradigm case (Flyvbjerg, 2006) for the wider processes of reconsidering meaning of past monuments, our research sites offer a "mature" case study which benefits from a long-term view on the ways in which monuments and statues of former political leaders are treated before, during and, particularly, after a significant sociopolitical shift. The main conundrum at the heart of this article is the juxtaposition of the solidity and apparent immutability of statues and their changing interpretations, as reflected in shifts in their treatment and in varying reception over time, and at different sites, by their local and international viewers. Consequently, our key aim in this article is to work towards an understanding of the complex ways that nations, communities and localities deal with these once powerful and now symbolic tangible remnants of the recent past in order to shape and, indeed, consolidate the political present.

The contribution of our study lies in providing a field study-derived structuring for the reflection on the afterlife of monumental statues. The study follows a story starting with collapse of the political and rhetorical regime supporting monuments in their original context (Dresser, 2007; Poria et al.,2014). We explore how the various approaches to destruction, banishment and subsequent redisplays and re-contextualisation of the material icons of the recent political past are used in attempts to build and support peace, stability and sustainability over the period of thirty years following seismic political change throughout the region. We examine how these tangible, often monumental, objects become subject to being physically moved, damaged or destroyed (Goodrich & Bombardella, 2016; Hautamäki & Laine, 2020). The complexity lies in the myriad of ways that individual and collective local, national and tourist remembering differs, colludes and collides with the formal narratives of contemporary place-making. We determine that the various approaches to these problematic and tangible remnants work towards the goal of maintaining and ideally ensuring political and historical sustainability of the present within the liquid modern world of rapidly changing contexts.

Construction of place-based memory in the political present

Crane (2000, p. 2) defines memory as the "act of thinking about things in their absence" and notes that memories can be triggered consciously by the desire to remember and subconsciously and even unwelcome where there is the desire to forget. Museums and heritage sites function as containers of memory (Assmann & Czaplicka, 1995; Carnegie, 2006, 2012) and may be shaped to promote national or local narratives of social and collective remembering, yet as individual memories are unique to that individual, so too will be their responses to such objects and displays (Bennett, 2003). Constructed visions and versions of the past may fit the need of the political present, but can clash with memories triggered by the places and objects encountered through museum and heritage visits. We argue that changing the context under which objects (here monumental statuary) are viewed and thus understood gives them new meaning in the present but does not necessarily succeed in destroying or changing the range of personal memories and associations that deny national narratives (Tunbridge & Ashworth, 1996).

Drozdzewski et al. (2016) follow Hill (2013, p. 381) in asserting that memory is enmeshed in everyday corporality through "practices of incorporation" and "practices of inscription," and argue that memory is articulated, felt, enacted, and experienced through the body, and the body is thus a place of memory. Violence facilitates the construction of "remembrance-scapes" that also serve to locate memory's absent presence, and people who have suffered loss or trauma or political change often return to geographical locations associated with these events; the sites remain central to how individuals and nations maintain memories. Memories of place are both spatially constituted in concrete or physical form, and constructed through narrative and memories of events. Consequently, places in memory can function as positioning tools for building, anchoring, and maintaining identities. As Zhang et al. (2018, p. 116) note, "heritage production, (re)presentation and consumption are closely connected to place promotion power relations, whilst identity (re)construction is often seen as a negotiation between dynamic and contested heritage discourses".

Despite the existence of real sites of memory and mourning associated with events, governments often chose to "place memory" by constructing sites of mourning such as war memorials in strategic locales within cities (Drozdzewski et al., 2016). Formal remembering is encouraged as the new sites become the symbolic spaces of memory even when they do not have an actual relationship to war, atrocity or political change that is being remembered or, in the case of this article, consciously disremembered. Pivotal to memory's power is its politics and social usefulness (Bell, 1997), stressing its importance in embodying and sustaining value systems and stability.

Here, we examine the opposite process, showing how destruction or banishment of the objects of memory constitutes an attempt to eradicate the memories themselves, or at least to change the context for remembering. The aim is not easily achieved: objects continue to maintain ghostly presence at original sites in the memories of people interviewed for our study. As Beardslee (2016) points out, intangible heritage does not necessarily require physical contact with sites of commemoration.

Our interest began with the movement of bodies through heritage sites, and the movement of memories: embodied, carried, and emerging as narratives which complement, define or challenge the contemporary readings of space. Statues, the focus of our particular attention, are non-human actors who mimic bodies and their movement, whether in celebration, or, as is the case in the studied sites, in the violence of banishment, mockery, or destruction. Their destruction or displacement is a political act rife with meaning and only the seeming opposite of the erection of new memorial spaces for collective remembering and mourning.

Yet the significance of such actions reaches beyond the grand political gesture. The incorporeal bodies, the "unsettling" representations of "ghosts of place" are removed from symbolic and important spaces where they have been landmarks in the daily lives of people beyond their political associations to spaces where they do not belong. These symbols of a shared national and indeed local history are not erased by the symbolic destruction of their likenesses; they acquire new meaning, possibly supporting the political present yet still provide powerful, possibly painful and even whimsical links to the past.

Even when partially destroyed, as Feldman et al. (2006, p. 261) argues in relation to memories of the Holocaust, object fragments still retain power to "conjure images of the social body and in their ability to reanimate the social body". This paradox of power and powerlessness is also evident in the monumental statues of the deposed, dangerous as symbols of a political past and yet more readily destructible than the memories of associated deeds and impact on the community. The destruction of images, the rendering them down to size as splintered broken bodies and the banishment of "whole bodies" out of context or within a carefully curated setting to walled and heritagized spaces, changes their meaning in the present but not their past. Feldman et al. (2006) discusses the missing body problem and the impossibility of embodying experience in traditional museum settings. Yet as tourists bring their bodies, as local visitors bring their memories, they people these displays and re-animate them, in agreement or tension with the ways intended by the organisers and funders.

Monuments as constructed spaces

Williams Williams, Williams (2008, p. 5) describes how the establishment of monuments to WWI events created "social practices of visitations", and how continued visits and touristic appreciation rendered them sites "of cultural significance". He sees visitations to sites of memory and mourning as part of the long tradition of pilgrimage and funerary rites, and argues that they necessitate some form of reflection on what the visitor has learned: a physical as well as cognitive engagement linked to place. Edson (2017, p. 21) claims that people are "better informed today about world events in part due to technology" and are more sophisticated and demanding. The mostly international visitors to statue parks are both knowledgeable and largely interested in learning more through the actual engagement with symbolic spaces, albeit those constructed as a consequence of the act of banishment and revisionism.

For Milton (2011, p. 163), violent acts on sites and monuments cannot simply be termed violence of iconoclasm: such acts are a way of aggressive rewriting of a narrative of the past that contrasts with or challenges the objects' or sites' meaning in the present. Herscher (2011, p. 151) discusses how states, "post-conflict or otherwise", seek to preserve places, mobilising heritage to serve their own aims, as in the case of statues mobilised to populate new heritagized spaces away from their previous, often symbolic and outwardly political, contexts. Many forms of remembering and forgetting as present concurrently and, we contend, elicit varied emotional responses from tourists and locals alike. However, political sustainability requires the formal narratives to dominate through and despite the myriad of competing "voices" and memories (Tunbridge & Ashworth, 1996; Ashworth & Graham, 1997).

Key questions remain: what purpose do these spaces serve in sustaining "appropriate" or politically useful visions and versions of regimes? Can they freeze the memories in the way the statues are frozen, suspended in artificial groupings and forced into a single representation of their role and significance in complex pasts? Who might these museums be for, as opposed to why they were created in the early spirit of change? We attempt to tackle these questions by examining who the actual local and tourist visitors are and how they remember or fit the depicted past within their own frames of reference and expectations of place. In short, we seek to determine how memories or viewpoints are sustained or changed by the encounter, and to question how visitors, in particular international visitors support the use of such spaces to reinforce messages of peace in the political present (Light, 2001; Higgins-Desbiolles & Blanchard, 2010; Farmaki, 2017).

We are concerned with timelines and with temporal and relational distance to past events affecting the responses to and treatment of symbolic figureheads and political monuments of the past. Time alone does not fix narratives or ensure that they fade from memory in linear ways (Healy & Tumarkin, 2011). Indeed, in an era lacking stable structures, extant narratives and representations continually risk becoming uninteresting, unhelpful, or subversive when viewed against emerging consensuses and newly dominant values (Ashworth & Graham, 1997; Rivera, 2008). Recent contestations of colonialism, structural racism, and heteronormativity underscore precariousness of triumphal celebrations of past successes (Goodrich & Bombardella, 2016).

While formal narratives strive to solidify political stability in the present, memories ebb and flow and are subject to revisionism, generational shift and can be triggered by political unrest or citizen dissatisfaction. Our study suggests that heritagization of political pasts positively engages both the local population and the largely foreign visitors, and thus supports sustainable political strategies. These, in turn, are influenced by the expectations of tourist audiences wittingly or unwittingly supporting visited representations, with heritagized spaces shaped for tourist audiences or at least owing their continued existence to outside interest (Capitello et al., 2017).

Methodology

The fieldwork material of our study consists of ethnographic observations and interviews with visitors and curatorial staff carried out in museums and memorial sites in Estonia, Germany, Hungary, Latvia, Lithuania, Russia, and Poland between 2016 and 2019, as part of a project investigating relationships between visitors and recent history exhibitions in Central and Eastern Europe. In line with most common ethnographic practice (Clifford & Marcus, 1986; Goodall, 2000), our fieldwork was oriented towards understanding the field rather than towards finding answers to any pre-specified research question. Consequently, only small amounts of the material we collected turn out to be relevant to the topic of the article: the afterlife of statues representing the communist past. For this reason, in the following discussion we draw primarily on our research at three sites focused specifically on preservation and exhibition of political art of the past regimes: Memento Park near Budapest in Hungary, Grūtas Park near Druskininkai in Lithuania, and a statue park at the Museum of National History in Tallinn, Estonia. Our collected material consists of field notes, interviews, and discussion notes. We agree with Erll (2011, p. 110) that "memories are robustly plural" and with Tucker and Carnegie (2014) who stress the impossibility of plurality within dominant narratives in the political present. Thus, our summary is an abstraction of the multiple and multifaceted ways in which tourists construct and shape their own knowledge and experience of the contexts being remembered. Indeed, these experiences are often reflected on in the months and years after the visits, helping to fix or at least disseminate contemporary and shaped readings of the past.

While the study is not autoethnographic, our observations are necessarily influenced by our own backgrounds and viewpoints. In particular, we draw on our conflicting and complementary stances: while we are both academics working in Western Europe, one of us brings discipline knowledge of heritage and museum studies, and the other the personal experience of growing up and continued political engagement in the region coupled with a sociological perspective. Consequently, our observations include our own responses as insider/outsider, academics and tourists, observers and witnesses to cultural change.

However, as the aim of this article is to present a generalizing argument rather than a report on the field, we present our findings in a more summarizing vein than is usual in ethnographic and ethnography-inspired studies. With concision in mind, we have restrained from presenting the sties through direct quotations from interviews and field notes, integrating the description with our discussion of the concepts we find important.

The field study

In the sections, we present concise descriptions of three sites important to our argument: Memento Park in Hungary, Grūtas Park in Lithuania, and the statue park of the Estonian History Museum. Each consists of a grouping of monumental statues, but the exposition, arrangement, and information provided differs widely. We present them not just as research sites, but as exemplars of different ways of presenting the monumental remnants of previous regimes, and of the ways in which tourists engage with such presentations. We also discuss one more strategy of dealing with troublesome statuaries which does not involve creating a heritage site: that of destruction.

The monumental statues of our study, created under a now delegitimized political system, form "a memory of waste" (Crane, 2000, p. 28) and now present as discarded objects of past regimes and toppled despots. All museums and heritage sites are based on "fragmented, dismembered, isolated, de-figured and disjoined objects" (ibid., p. 34); they are deliberately disfigured and displaced as if to confuse the ghosts of memory. The parks examined here offer a curated reading of this disposal, of the disposed/deposed and disowned. Heritagization of the stone bodies as survivors of the first wave of destruction being put out to grass transforms them even more fully into memory objects. Their solid and giant forms have been mobilised to their new site beyond the everyday gaze, just as the bodies of tourists and locals who come to see them move to the monuments. Cutting the statues "down to size" in the various approaches under consideration here, takes away or diminishes or manages the potentially strong emotions surrounding memories (Wulff, 2007) such as anger, grief, admiration and fear. The taming or erasing of both political and personal threat is clearly important for sustaining peace in the political present and also for making tourist attractions and spaces comfortable for visitors, even as the emotional life of other cultures and past political regimes sparks interest, curiosity and potentially empathy in tourists (Heelas, 2007; Tucker, 2016).

It is a truism to note that the scale and the solidity of public monuments serve to create an aura of permanence and imperviousness to any change enacted in a human timeframe. Their actual lifespan and continued significance is more problematic. A number of strategies have been employed to deal with outmoded or rejected messages these statuaries or other public monuments might be understood to carry, and to find meanings and contemporary context able to accommodate them. The four item typology presented below does not constitute all the possible responses, nor does it cover all of the nuances within each strategy, but can serve as a preliminary analytical tool for understanding and discussing the broad range of possible fates befalling once potent symbols of the rejected past.

Destruction

The first, and most obvious, strategy is destruction: physical obliteration of a monument can serve both to actualize rejection of the originally ascribed meaning, and to prevent unwelcome message from further propagation (continued physical presence might imply continued rhetorical presence). Acts of destruction and defacement often predate political change, or even signal arrested attempts at promoting such change. Thus, the statue of Soviet secret police founder Feliks Dzierżyński in Warsaw was defaced multiple times in the communist era. More recently, many statues connected to racial discrimination, from the slave merchant Edward Colston in Bristol, United Kingdom, through multiple Confederacy-linked figures in the United States, to white abolitionist Victor Schoelcher in Fort-de-France, Martinique have been toppled or defaced in acts of protest.

More significantly, revolutions or dramatic political shifts are marked by symbolic destruction of previous regime's cherished monuments. The demolition of the same statue of Dzierżyński in November 1989, of Lenin statue in Kiev in 2013, and of the monument to Saddam Hussein in Baghdad in 2003 are all celebrated as significant symbolic moments witnessing political change, commemorated through widely disseminated photographs and mediated and re-lived as part of the reinforcement of change. Contemporary global audiences and subsequent generations access these celebratory images and continue to be influenced by them.

In some cases, the act of destruction might itself be felt to warrant commemoration beyond graphic depiction. Thus, during the 1956 Hungarian Revolution, a monumental statue of Joseph Stalin in Budapest (erected only three years earlier) was torn down, leaving only a pair of bronze boots on a ten-metre plinth. These were later removed together with the base of the statue, but, in 2006, a new monument depicting a pair of boots atop a tall pedestal was erected in Memento Park, Budapest, to commemorate the event. We believe it is the only permanent monument created specifically to celebrate the act of statue destruction.

Outside of the revolutionary moment, however, the destruction of statues and monuments is much more problematic (Poria et al., 2014). Public discourse more commonly celebrates conservation rather than destruction (cf. the rapidly growing list of currently 1121 UNESCO World Heritage sites). There is a tendency for conservative bodies to advocate for conservation (Poria et al., 2014) which implies sustainability of material evidence of pastness. There is a point to be made that conservation can be, and in the case of UNESCO often is de-politicizing at least in intent. Heritage, and world heritage in particular, offers a political overlay of imagined value beyond the political and beyond the local context, making the sites conserving politically contentious or subversive objects worthy of particular attention.

Removal or demolition of monuments commonly encounters protests and condemnation, not necessarily from proponents of their original rhetorical interpretation. Global protests are often about destruction in the abstract, not really the local story. Ultimately unsuccessful international protests regarding the destruction of Bamiyan Buddha statues by the Taliban regime in 2001 remain the most obvious exemplar, but acts such as the removal of the Monument to Brotherhood in Arms (depicting Polish and Soviet World War II soldiers) in 2011 also merited strong opposition.

Delegitimization

Consequently, revolutionary fervour rarely brings about the effacement of all monumental traces of the previous regime, and the strategy of destruction is much less commonly followed in times of political stability. Instead, a second strategy, that of delegitimization, offers another way of dealing with markers of the recently rejected past. It involves diverse activities undermining the perceived rhetorical thrust originally espoused by the monument. At its simplest, an explanatory plaque can add a contemporary commentary while retaining original framing. Usually, more elaborate changes are deemed necessary.

Kattago (2009) chronicles the changes to the Bronze Soldier, a large statue in Tallinn with changing official designation and legitimacy. Erected in 1947 and designated the "Monument to the Liberators," the statue commemorated the taking of Tallinn by Soviet troops in 1944. It was later adorned with a continually burning torch. After Estonia's secession from the Soviet Union in 1991, the dominant discourse shifted to understand the event as occupation rather than liberation. The monument was retained as it marked a burial site of Soviet soldiers. However, it was renamed "To Those Killed in the Second World War," the torch was removed, and some of the inscriptions were changed. Nevertheless, the statue remains controversial and any rise in tensions between Estonia and Russia rekindles discussion of whether further changes are needed. Political sustainability calls for decisions to be taken that seem to serve the present.

Large political monuments tend to be situated in significant and prestigious settings and to dominate and organize their spatial surroundings, with framing often chosen to enhance rhetorical appeal. Consequently, delegitimization tends to involve removal to less salubrious locations. The Bronze Soldier statue was itself relocated to less prominent location in 2007, sparking further protests.

Memento Park in Hungary, a statue park situated on the outskirts of Budapest, serves as an emblematic case. Established in 1993, it was originally envisioned as an exhibition of totalitarian art from Nazi and Soviet regimes. Ultimately, only half of the original project was completed, and the park showcases only communist-era monuments. It largely follows the original layout and presentation, with minor subsequent changes. The flat exhibition area, screened by a brick wall and hedges, collects some three dozen sculptures and sculpture groups (the site's own sources count 42 artworks). Statues are placed without consideration to original positioning or intended presentation: many of the monumental sculptures, in particular, are placed without plinths at ground level, nullifying any relation of awe. Indeed, the disrupted sense of scale appears to invite playfulness, even mockery. Identifying descriptions for the statues themselves are sparse, but framing is provided through other means. The park's website (subtitled: "the biggest statues of the cold war") offers a mixture of different tones and possible interpretations. These include photos of tourists posing with statues (Light, 2000a), commendatory quotations from Hungarian political figures and historians praising the park's value and presentation ("an example for solving a controversial problem in an intelligent and elegant manner"), and sample lesson plans for pupils of different ages (only in Hungarian). Topics range from charting the rise of communism to identifying art styles. and.

Tourist photos open up a theme of mockery: people mimicking the statue poses, jumping around, pretending to high-five the giant monuments. Dancing with despots becomes a legitimate visitor activity, and the appropriate response to the spent rhetorical force of the exhibited statues. The jocular approach is continued by the reception/shop display on the site itself: a mash-up poster shows Lenin giving the site a facebook "like" (and invites the reader to do likewise), and items for sale include light-hearted parodies of communist slogans and socialist-realist imagery.

Nevertheless, the largely foreign visitors we have observed and interviewed treated the visit much more seriously: as a learning experience, immersion in the region's past, broadening their horizons. While a few struck relatively timid poses for photos with the statues, for most tourists a visit consisted of walking among the statues, reading plaques and gazing at the monuments.

Thus, the strategy of deligitimization, which presents statues as rhetorical objects of the political milieu in which they originated, but with the validity of their message is denied, subverted, or explained away, is only moderately successful in co-opting tourists to produce a sustained narrative of the past. Mockery relies on strong relationship with the object of mockery, and at the present distance from the fall of communism, most visitors approach it as a largely unfamiliar period in history, tinged with generalized respect for the past suffering of the local population (Light, 2000b).

Decontextualization

At the time of the park's creation, the interpretation of statues primarily as carriers for rhetorical and political messages was the predominant one, at least for the local populace. It is never the only possible reading, though, and its pervasiveness can wane over time. The park's visitors whom we interviewed between 2016-19 showed interest in statues as works of art (or, at least, as representatives of a particular artistic style) and emblems of a historical epoch rather than as political creations.

Vabamu Museum of Occupations and Freedom, a recent history museum in Tallinn, originally opened in 2003. In its original staging of the permanent collection, the curators placed two large statues of Soviet era dignitaries in the underground passage leading to the lavatories, effectively positioning them in the symbolically degrading role of toilet guards. During a 2017 renovation, these statues were removed, and the general emphasis on countering or debunking communist propaganda was lessened throughout the exhibition.

This change is consistent with the third strategy for past monuments: decontextualization. It involves stripping of the original meanings and associations from the monuments, without any corresponding attempt to fix alternative readings. It requires no shared understanding, and no common political standpoint uniting the curators and the visitors, but provides little in the way of guidance as to the value of the presented works, be it artistic or didactic. Decontextualized presentation can accommodate difference in viewpoints among the visitors or conflicted feelings regarding previous political regime(s).

Grūtas Park, a statue park near the small town of Druskininkai in Lithuania exemplifies it well. Created in 2001 by a local entrepreneur, the park exhibits 86 statues (mostly life-size or larger) in a landscaped forest. It also houses a small zoo, with some animals wandering freely around the statues. The site's website proclaims its aim to

> provide an opportunity for Lithuanian people, visitors coming to our country as well as future generations to see the naked Soviet ideology which suppressed and hurt the spirit of our nation for many decades (Grūtas Park, n.d.)

However, if this was indeed the intention, there is little evidence of it in the presentation of the statues themselves. Explanatory plaques describe, in very neutral terms, the subject of each sculpture, as well as the name of the sculptor. The leafy surroundings and secluded location contribute to the peaceful atmosphere of the place, while grazing alpacas set the space apart from the everyday. Grūtas Park is not a popular destination, and during our one day of fieldwork at the site in May 2016 we did not encounter any local visitors. The few international tourists we have interviewed reported bafflement by lack of context or explication. Thus, the strategy of decontextualization should probably be viewed as transitory, providing space and time for new contexts and new interpretations to arise (Light, 2000b). Although both Memento Park and Grūtas Park potentially serve the same aims and likely audience, both were clearly created with different agendas. Memento Park denies or defies the logic of time and place through striking juxtapositions of its statues, but it is a carefully structured and contained site. Grūtas Park, in contrast, seems deliberately eccentric and it is not clear whether the elements such as the zoo animals and play parks were added to make the park a "good day out for all the family" or as a conscious statement about the political past. Information boards at the entrance to the park introduced the owner and presented the park as his project, but provided little guidance regarding his curatorial aims. As visitors ourselves from different cultures, countries and political backgrounds we, like the few tourists viewing the site, walked out unsure and confused.

Depoliticization

The final strategy, that of depoliticization, represents an emergence of a new interpretive scheme. It focuses on presenting statuaries as artworks or historical markers rather than as symbols of a political ideology. The focus shifts back onto the objects, rather than on the political system enabling their creation. Such presentation can be found in the sculpture garden of the Estonian History Museum in Tallinn. The relatively modest space, located behind the museum itself, houses 21 different Soviet era sculptures, arranged in a rough circle. Descriptive plaques focus on the sculptors and art styles, similar to how other works of art are presented elsewhere in the museum context. This is by far the smallest of the sites we have viewed, but in many ways the most informative—its compactness allows visitors to compare and contrast different iterations of the Soviet era socialist-realist style . It is also the newest site: laid out in its present form only in the 2010s, and the least politically-charged: the sculptures appear primarily as artistic creations of individual sculptors, albeit working in a specific time and place.

The visitors exhibited a variety of behaviours, from hurried passage, through thoughtful appreciation, to playfulness or mockery; there was no clear approved script for them to follow. Like the interviewees we spoke to in Memento Park, visitors here viewed the statues as object from a bygone era rather than as participants in an ongoing political struggle.

Discussion

Our interest here is twofold. Through examining how nations and communities deal with symbolic structures of monumental statuaries to move forward politically and emotionally from the events of recent past, and to support individual, collective and national remembering allowing for sustainability within the political present and future. Multiple viewpoints and potentially contentious and competing memories and narratives are not readily known or accessible and our fieldwork frames the studied sites as meeting points for remembering and forgetting both within and beyond the heritagized spaces. We also query the role and impact of tourists in shaping the heritagized visions and version of the past within the sites and the extent to which the studied sites have been shaped to encourage tourism supporting the fixing of such narratives despite a seeming lack of interpretation for tourist audiences.

Who visits?

All of the above key sites are essentially outward-facing. They actively encourage tourists and seem specifically marketed as tourist attractions. Lehrer and Milton (2011, p. 5) rightly challenge terms such as dark tourism being applied to visits to sites associated with painful pasts as they note that while such visits "raise concerns about voyeurism and crass commercialisation, they may just as often draw people earnestly seeking to mediate on peace, imagine common futures, and even forge these through dialogue or political action".

An interesting issue is how locals and tourists, often coming from different backgrounds, counties and cultures, perceive sites such as Memento and Grūtas Park set up in peripheral zones to contain the figureheads of past regimes. These are displayed and curated as both collectively banished and creatively presented. This act of banishment is the one thing they share in the present. Collectively, they are offered up as political waste in a political wasteland (Wallis, 2019).

Feldman reframes Clifford's notion of "museums as contact zones" (Clifford, 1997, p. 192) arguing to substitute conflict for contact: zones where inequalities meet; this (re)reading forces a reassessment of heritage spaces. Analogically, places like Memento and Grūtas Park constitute meeting sites where tourists and local visitors alike have to negotiate these constructed placeless places.

Lehrer and Milton (2011, p. 7) argue that representation of recent past constitutes "attempts to curate difficult knowledge" within the framework of "transitional justice," drawing on symbolic objects and spaces of the past to render the present less contentious and more acceptable. States, nations or communities work to reclaim and reject ideologies and their symbolic markers both within the geographical sphere – landscapes of memory - and as objects associated with such landscapes. Hartmann (2016) notes that formal heritage spaces can support the erasure of collective remembering and Mitchell (1992, p. 30) adds that the pulling down (and we might add, re-placing) of public art "is as important to its function as its putting up". The framing of statues in themed spaces merges the forms of museum (generally emphasizing material culture and the "real"), heritage constructions as politic (Tunbridge & Ashworth, 1996) and the wholly experiential "theme park" spaces created rather than curated as other worlds. In the latter, sense-making is framed by inauthenticity and also potentially edutainment (Balloffet et al., 2014; Babic, 2016), setting the atmosphere for the experience (Bonn et al., 2007).

Visitor responses: correct or expected?

Visitors respond to sites that present difficult knowledge according to their own understanding of and relationship to the past (Carnegie & Kociatkiewicz, 2019). Such sites reinforce understanding of the past and offer continuity in terms of historical "truths" or at least narratives of events and place. As Lehrer and Milton (2011, p. 9) note, both tourists and local community visitors' emotional responses to exhibitions change over time as people "unevenly engage with the processes of 'working through' in relation to their communal tragedies". Historic events and their mediated representations or exhibitions that follow become solidified as accepted formal versions of the past shaped for contemporary audiences (Silverman, 1995). Individual and community remembering and tourist knowledge may challenge perceived constructions of past events focusing on significant points that shape or aim to define formal remembering as opposed to the lived experience of individuals or the formal teachings received by tourists in another political context. How these representations are received and perceived depends on where, how and importantly when visitors developed their knowledge and values of the past being depicted. Importantly for this study, we, as curators, locals, visitors and academics, witness the emergence of images and objects from their challenging original context into the "visible, material touchstones of new experiences and narratives" (Lehrer & Milton, 2011, p. 17).

One of the key ways of rendering objects harmless is to change the context for their remembering. Simon (2011, p. 207) discusses the need for a "pedagogy of witnessing" where the past and present can be represented "without reducing one to the other, or dictating the terms on which this is to be accomplished". This call for ethical or moral curating denies, obviously, the political, social and economic acts that brought about the changes in society either through violent acts or by peaceful means. Individuals do not curate their own memories although they may build up collective memory banks within their localities, but only where it is safe or desirable to do so. Otherwise, memories which challenge the "desired" formal narratives or are too painful go underground or are subject to selective self-banishment.

Museums and heritage spaces collect and reframe objects in order to give them sense in the present. This suggests that curators present objects in a way they understand as best suited to contemporary audiences. Social, cultural and political changes, fashion and geography all influence these interpretations. Notable examples include the "New Museology" debates largely focused around accessibility representations, and more recent attempts to frame the post-modern museum as a more knowing and known space (Stam, 1993; Dwyer & Alderman, 2008; Zhang et al., 2018). As Watson (2007, p. 4) argues, all curated sites have some sense of being part of a local or regional community or of serving national narratives.

Museums and heritage site representations follow certain curatorial approaches; these tend towards chronological, geographical, thematic and aesthetic or a mixture of these (Simpson, 1996: Bennett, 2003; Stogner, 2009). Visitors need to understand curatorial logic to be able to negotiate spaces and effectively engage with them (Hennes, 2010). We contend that Memento and Grūtas Parks defy the logic of association with place and are chronologically and literally "all over the place". Sensemaking is therefore limited to an understanding of the relationship of each of the figureheads depicted in the statues to one another. What they all have in common is that they reflect a past political regime. This deliberate juxtaposition reduces them to that of just past political figures although some have no geographical link to the localities they are now in and some were clearly important political forces such as Lenin. Some were the works of the best artists and sculptures of the period, others more roughly cast where the representation and not "artistic merit" was required. The works in Memento Park are displayed as in a temple, in Grūtas Park a grotto within a zoo. This deliberate loss of usual curatorial sensemaking and the seemingly provocative and confusing display strategies, are at once playful and mocking and immersive in their creation of imaginary. They result in presentations that reflect Bencard's Bencard (2014) world of "history without stories".

Williams (2008, p. 8) defines a monument as "a sculpture, structure or physical marker designed to memorialise". He claims that "a memorial is seen to be, if not apolitical at least safe in the refuge of history", as it capitalizes on respect our culture demands to be given to all dead. But disrespect, as we have shown, can also appear as an early, engaged response to political change, a way of moving beyond the all-or-nothing decisions on how to treat markers of significant suffering, with the usual options being obliteration or sanctification (Williams, 2008, p. 185). The monumental statues forming the core of our study are not, generally, witnesses to atrocities or immediate markers of suffering. They are seen, however, as glorifying, directly or indirectly, an oppressive regime whose vanquishing forms the founding narrative of the current political system, and the first three strategies for dealing with disgraced statues invite tourists to participate in creating or emphasizing the point of rupture at the end of the previous regime.

Yet, as Nadkarni et al. (2003, p. 196) notes, the demise of socialism "did not represent revolution, but the end of the age of revolutions" and came about largely peaceably despite representing a point of rupture. She adds that the "communist monuments had inspired not only anger, but also indifference, irony or affection" (ibid., p. 198). The initial argument for Memento Park embodied a compromise between proponents of immediate destruction of communist-era and those willing to keep them in situ. Statues' preservation was deemed an "emblem of democracy" which allowed insights into "aesthetics, ideology and historical politics of the previous era" (ibid.,

p. 198). She notes that some protestors considered the park a "human zoo" and an anti-democratic act of dismissing varied contexts and significance of different statues. In particular, the statue of the Soviet soldier Osztapenko, erected on a route leading to the popular holiday spots, functioned primarily as a place marker and not a political symbol. The same example was brought up (without prompting) fourteen years later by one of the local visitors to Memento Park whom we interviewed. Nadkarni further noted that while "countless Lenins proved the fact of Soviet occupation, it was perhaps even more pressing to remove Osztapenko, who called attention to the ways forty years of Socialism had become cosy and familiar" (ibid.).

Memento Park's opening in 1993 coincided with the second anniversary of the removal of communist troops from Hungary, and Nadkarni notes the irony in the way the event was reported in the Western press as a money-making venture, presenting communist statues as reformed servants of the new capitalist economy. This, and the site's marketing appealing to international tourists through socialist realist kitsch, provided for the early visitors an interpretive framework that "offered proof of both the oppressed Socialist past that feared and hated these statues and the democratic present that is free to laugh at them" (ibid., p. 203).

However, contestation does not necessarily come from "conflicting accounts of what actually happened in the past… as much as the question of who or what is entitled to speak for that past in the present" (Hodgkin and Radstone, 2003, p. 1). Instead, the "past is constituted in narrative, always representation, always construction" (ibid., p. 2). Heritage spaces constitute a particular form of formal remembering, selectively framing the past to suit the political needs of the present: the past represented through disgraced statues is used to highlight ruptured continuity and momentousness of political change. But personal memories are contradictory, flawed and complex with the potential to challenge, corroborate, and intertwine with the formal narratives of place. The Budapest locals whom we have interviewed in relation to Memento Park still associate the individual statues in their original setting, as part of their own memories: as former meeting points and other memory markers.

Memorials embody a historical "return of the oppressed" (Kapralski, 2011, p. 187), and thus invite unambiguous demarcation of the categories of the oppressed and the victims. But, intricateness and contradictions of history and of memory give monuments irreducible complexity. Sites such as Memento Park or Grūtas Park defy easy characterization as a failure, commercial enterprise, or a tourist site; they function as places where memories are gathered interpreted differently by different stakeholders. That they are also places that many locals we spoke actively choose not to visit is yet another of their roles.

Tourists as supportive peacemakers

This leads us to consider on-site tourist behaviours and how they support the wider aims of the parks in the present. Are their responses in keeping with the overall tone of the parks which is more mocking than reverential, chaotic in the mix of statues from time and place, artistic quality and local significant? Our initial thoughts on our first visit to Memento Park fixated on the purely human response to the scale of the works there, as we witnessed some of the visitors climb the monuments, pose and "dance" with them, leading to our titular metaphor of dancing with despots. We thought of tourists performing as proxies for local visitors who may chose not to visit. But the act of mocking evident in the dancing and general playfulness, they were giving an appropriate repose, disrespecting the statues and therefore participated in creating rupture from the ideology and regime that they stood for. We still hold this interpretation as valid, but also as incomplete.

Formal narratives and representations of the political past confined and contained as a heritage park are relatively straightforward. Meanwhile, the range of emotions felt by those for whom they formed part of the everyday lived experience and the subsequent generations for whom the park represents Bell's (1997, p. 827) "unsettling ghosts of place" remains complex and unruly. Tourists may indeed have an imagined ideal role in sustaining the present through seemingly mocking the past, but only in the sense that by appearing to do so they may serve formal political narratives.

Having conducted fieldwork at a ten different sites as part of the project (including the three described in more detail in this text) we argue that tourist responses, and in particular Western international tourists, are also complex, and influenced but not determined by the context of the space, and the desire to learn and understand. Through interaction with sites tourists sustain and challenging their own "knowledge bank" of what they learned about this period of time and what they can then witness. We argue that a more thoughtful tourism develops to reflect this complexity.

However, reflexive engagement with heritage sites can be an uncomfortable, even painful experience, depending on the visitors' relationship to the presented past (Poria et al., 2014; Carnegie & Kociatkiewicz, 2019). Moreover, complexity and ambiguity are both difficult to convey to visitors unfamiliar with the context, and often problematic for the curatorial teams socially, ideologically, and politically engaged in potentially unstable milieus. We would also recognise that in conducting our fieldwork, we actively invited visitors to reflect on both their understanding of the context of the parks and their feelings about them. This can have served to evoke feelings and insights which would have not been part of their experience without us present, but the thoughtful responses we received also suggest that reflexivity can be a welcome part of the visitor experience.

Tourists interviewed as part of this project came as thoughtful visitors, aware of the key moments of recent history. However, they did not necessarily know how to "read" the site in terms of expected behaviours. The ticket booth at Memento Park sells ice-cream and souvenirs which are parodistic and aimed at Western tourists (a tin that contained the "last breath of communism", ironic posters). Similarly, Grūtas Park is a confusing space, part zoo, part heritage park where animals graze among statues. The inference is that sculptures are now part of a human zoo, tamed, controlled, if not endangered. This show of levity gives visitors permission to mock, to joke and as the title of this article suggests, to dance among the despots.

In many ways, this challenges the idea that tourist to such sites are necessarily deferential, yet it does not equate to a deliberately mocking or disengaged form sometimes associated with dark tourism. Rather, the confused response should be recognized as the desired behaviour for encountering and assimilating the complexity of history. As North (1990, p. 861) notes, "changing the nature of the art means changing the role of the audience as well … as the aesthetic shifts from the object to the experience it provokes".

Conclusion

Armada (2010) argues that when one memory is prioritised, other memories are executed. However, our study suggests memory is complex and nether linear nor permanently erased or erasable just as the symbolic structures of memory – the monumental statuary can be erased yet remain in popular and public memory (Dresser, 2007; Goodrich & Bombardella, 2016). The parks may form a bridge between the past and present values, or may be consciously reframed in the present. Our research determines that the absence of interpretation within public spaces should not be viewed as forgetting or even as deliberate attempts to hide the past (Armada, 2010). And Grūtas Park shows that seeming neutrality in interpretation does not necessarily or ever suggest neutrality of behalf of the organiser.

Less political and arguably more reflective ways of treating the material have been emerging more recently: the art-historical responses of the strategy of depoliticization (Drayton, 2019; Hautamäki & Laine, 2020). The same statues previously displayed in Tallinn under the name of a statue graveyard are now presented as a floodlit outdoor art gallery without an overt call to a mocking or condemnatory response. Thus, sustaining the political present becomes both easier and harder over time – easier as memories become less acute, harder because a new form of disillusionment about current regimes and nostalgia comes as a consequence of passing of time

(Light, 2000a). Political swings equally threaten the political status quo, and both retention and removal of past symbols can create different resonances in different contexts.

North (1990, p. 861) argues that "sculpture becomes public by taking the spatial experience of its audience as a subject." The resulting "cultural space" becomes a meeting ground for both agitation (Tunbridge & Ashworth, 1996; Hicks, 2020; Hautamäki & Laine, 2020) and potentially peace making (Higgins-Desbiolles & Blanchard, 2010; Hautamäki & Laine, 2020). As Farmaki (2017, p. 538) notes, peace building is not a "static but a continuous process" and tourism can support or moderate positive peace building. The statue parks we studies have evolved into predominantly tourist-facing spaces in which tourists contribute to and support the narrative of achieved peace and historical justice. Visitor-tourists are eager to understand and partake in experiencing historical milieus, ready to accept and internalize curated visions of the past.

Memento Park is an established part of the tourist offering in Budapest and although its peripheral siting attracts only those committed to making the journey, it has become a 'known space' popularised through and for tourism and as such has a wider reach than its peripheral siting would suggest. The statue park at the Museum of National History in Tallinnwas already a tourist attraction when a 'statue graveyard' – described as a symbol of rejection, decay and conscious neglect. The move to a formal display heralds the recognition of the importance of tourism in framing the recent past and its reception with external audiences.

Our findings determine that thoughtful tourists have a real and relevant role to play in forming and indeed solidifying narratives of the past. In this we argue, that while representations of the recent past are not always consciously shaped for external and tourist audiences, tourists often form a key audience. At the same time, the local people we interviewed still remember the statutes and where they stood when they pass the actual spaces where they were once erected. As time passes these memories become more symbolic and will sooner or later become folk memories (Hartmann, 2016).

This study of statues as monuments to cultural change and changing cultures is topical and timely. Statues are often seen as both focus and pretext for discussion of key figures' role in history (Drayton, 2019). At present, statue parks are both places of fragmented memory and forgetting. Folk memories blur timelines and can become both a threat to and evidence of acculturation as narratives change and shift over time. Managing these processes of change becomes part of the successive governments' necessary strategic agendas to maintain the political status quo (Ashworth & Graham, 1997; Hicks, 2020). In this sense, tourists' engagement with statue sites helps support and maintain a consensus fixing the narratives of the past, and thus contribute, steadily rather than spectacularly, to stability and peace through tourism.

ORCID

Elizabeth Carnegie http://orcid.org/0000-0003-3912-0627
Jerzy Kociatkiewicz http://orcid.org/0000-0001-6449-1185

References

Armada, B. J. (2010). Memory's execution: (Dis)placing the dissident body. In G. Dickinson, C. Blair, & B. L. Ott (Eds.), *Places of public memory: The rhetoric of museums and memorials* (pp. 216–237). University of Alabama Press.

Ashworth, G., & Graham, B. (1997). Heritage, identity and Europe. *Tijdschrift Voor Economische en Sociale Geografie, 88*(4), 381–388. https://doi.org/10.1111/j.1467-9663.1997.tb01632.x

Assmann, J., & Czaplicka, J. (1995). Collective memory and cultural identity. *New German Critique, 65*(65), 125–133. https://doi.org/10.2307/488538

Babic, D. (2016). Bridging the boundaries between museum and heritage studies. *Museum International, 68/1-2*, 15–28.

Balloffet, P., Courvoisier, F. H., & Lagier, J. (2014). From museum to amusement park: The opportunities and risks of edutainment. *International Journal of Arts Management, 16*(2), 4–18.

Bauman, Z. (2000). *Liquid modernity*. Polity Press.

Bauman, Z. (2004). *Identity: Conversations with Benedetto Vecchi*. Polity Press.

Beardslee, T. (2016). Whom does heritage empower, and whom does it silence? Intangible cultural heritage at the Jemaa el Fnaa, Marrakech. *International Journal of Heritage Studies, 22*(2), 89–101. https://doi.org/10.1080/13527258.2015.1037333

Bell, M. M. (1997). The ghosts of place. *Theory and Society, 26*(6), 813–836.

Bencard, A. (2014). Presence in the museum: On metonymies, discontinuity and history without stories. *Museum and Society, 12*(1), 29–43.

Bennett, T. (2003). Stored virtue: Memory, the body and the evolutionary museum. In K. Hodgkin, & S. Radstone (Eds.), *Regimes of Memory* (pp. 40–52). Routledge.

Bonn, M. A., Joseph-Mathews, S. M., Dai, M., Hayes, S., & Cave, J. (2007). Heritage/culture attraction atmospherics: Creating the right environment for the heritage/cultural visitor. *Journal of Travel Research, 45*(3), 345–354. https://doi.org/10.1177/0047287506295947

Capitello, R., Agnoli, L., Charters, S., & Begalli, D. (2017). Exploring place image from visitors' expected and lived experiences: A discrete choice approach. *Journal of Customer Behaviour, 16*(1), 19–34. https://doi.org/10.1362/147539217X14909732699462

Carnegie, E. (2006). It wasn't all bad': Representations of working class cultures within social history museums and their impacts on audiences. *Museum and Society*, 4(2), 69–83.

Carnegie, E. (2012). Curating people? Museum mediated memories and the politics of representation. In E. Boesen, F. Lentz, M. Marque, D. Scuto, & R. Wagener (Eds.), *Grand Narratives and Peripheral Memories* (pp. 143–160). Histoire.

Carnegie, E., & Kociatkiewicz, J. (2019). Occupying whateverland: journeys to museums in the Baltic. *Annals of Tourism Research, 75*, 238–247. https://doi.org/10.1016/j.annals.2018.09.010

Clifford, J. (1997). Routes: Travel and translation in the late twentieth century. Harvard University Press.

Clifford, J., & Marcus, G. E. (Eds.). (1986). *Writing Culture: The poetics and politics of ethnography*. University of California Press.

Crane, S. A. (2000). *Museums and Memory*. Stanford University Press.

Drayton, R. (2019). Rhodes must not fall? *Third Text, 33*(4–5), 651–666. https://doi.org/10.1080/09528822.2019.1653073

Dresser, M. (2007). Set in stone? Statues and slavery in London. *History Workshop Journal, 64*(1), 162–199. https://doi.org/10.1093/hwj/dbm032

Drozdzewski, D., De Nardi, S., & Waterton, E. (2016). The significance of memory in the present. In D. Drozdzewski, S. De Nardi, & E. Waterton (Eds.), *Memory, place and identity: Commemoration and remembrance of war and conflict* (pp. 1–17). Routledge.

Dwyer, O. J., & Alderman, D. A. (2008). Memorial landscapes: Analytic questions and metaphors. *GeoJournal, 73*(3), 165–178. https://doi.org/10.1007/s10708-008-9201-5

Erll, A. (2011). *Memory in Culture*. Palgrave Macmillan.

Farmaki, A. (2017). The tourism and peace nexus. *Tourism Management, 59* (2017), 528–540. https://doi.org/10.1016/j.tourman.2016.09.012

Feldman, J. D. (2006). Contact points: Museums and the lost body problem. In E. Edwards, C. Gosden, R. B. Philips (Eds.), *Sensible objects: Colonialism, museums and material culture* (pp. 245–268). Bloomsbury Academic.

Flyvbjerg, B. (2006). Five misunderstandings about case-study research. *Qualitative Inquiry, 12*(2), 219–245. https://doi.org/10.1177/1077800405284363

Goodall, H. L. Jr. (2000). *Writing the New Ethnography*. Altamira.

Goodrich, A., & Bombardella, P. (2016). What are statues good for? *Koers - Bulletin for Christian Scholarship, 81*(3), 1–10. https://doi.org/10.19108/KOERS.81.3.2272

Grūtas Park (n.d). http://grutoparkas.lt/en_US/.

Hartmann, K. M. (2016). Fragmentation and forgetting: Sarajevo's Vijećnica. *International Journal of Heritage Studies, 22*(4), 312–324. https://doi.org/10.1080/13527258.2016.1138317

Hautamäki, R., & Laine, S. (2020). Heritage of the Finnish Civil War monuments in Tampere. *Landscape Research, 45*(6), 742–757. https://doi.org/10.1080/01426397.2020.1749580

Healy, C., & Tumarkin, M. (2011). Social memory and historical justice: Introduction. *Memory Studies, 4*(1), 3–12. https://doi.org/10.1177/1750698010382156

Heelas, P. (2007). Emotion talk across cultures. In H. Wulff (Ed.), *The emotions: A cultural reader* (pp. 31–36). Berg.

Hennes, T. (2010). Exhibitions: From a perspective of encounter. *Curator: The Museum Journal, 53*(1), 21–33. https://doi.org/10.1111/J.2151-6952.2009.00005.x

Herscher, A. (2011). Points of No Return: Cultural Heritage and Counter-Memory in Post-Yugoslavia. In E. Lehrer, C. Milton, M. Patterson (Eds.), *Curating difficult knowledge: Violent pasts in public places* (pp. 147–160). Palgrave Macmillan.

Hicks, D. (2020). *The Brutish Museums. The Benin Bronzes, Colonial Violence and Cultural Restitution*. Pluto Press.

Higgins-Desbiolles, F., & Blanchard, L. (2010). Challenging peace through tourism. In O. Moufakkir, & I. Kelly (Eds.), *Tourism, progress and peace* (pp. 35–48). CABI.

Hill, L. (2013). Archaeologies and geographies of the post-industrial past: Landscape, memory and the spectral. *Cultural Geographies*, *20*(3), 379–396. https://doi.org/10.1177/1474474013480121
Hodgkin, K., & Radstone, S. (2003). Introduction: Contested pasts. In K. Hodgkin, & S. Radstone (Eds.), Contested pasts: The politics of memory (pp. 1–22). Routledge.
Kapralski, S. (2011). (Mis)representations of the Jewish Past in Poland's Memoryscapes: Nationalism, religion, and political economies of commemoration. In E. Lehrer, C. Milton, & M. Patterson (Eds.), *Curating difficult knowledge: Violent pasts in public places* (pp. 179–192). Palgrave Macmillan.
Kattago, S. (2009). War memorials and the politics of memory: The Soviet War Memorial in Tallinn. *Constellations*, *16*(1), 150–166. https://doi.org/10.1111/j.1467-8675.2009.00525.x
Light, D. (2000a). An unwanted past: Contemporary tourism and the heritage of communism in Romania. *International Journal of Heritage Studies*, *6*(2), 145–160. https://doi.org/10.1080/135272500404197
Light, D. (2000b). Gazing on communism: Heritage tourism and post-communist identities in Germany, Hungary and Romania. *Tourism Geographies*, *2*(2), 157–176. https://doi.org/10.1080/14616680050027879
Light, D. (2001). Facing the future': Tourism and identity building in post-socialist Romania. *Political Geography*, *20*(8), 1053–1074. https://doi.org/10.1016/S0962-6298(01)00044-0
Lehrer, E., & Milton, C. (2011). Introduction: Witnesses to witnessing. In E. Lehrer, C. Milton, & M. Patterson (Eds.), *Curating difficult knowledge: Violent pasts in public places* (pp. 1–19). Palgrave Macmillan.
Mitchell, W. J. T. (1992). The violence of public art: *Do the right thing*. In W. J. T. Mitchell (Ed.), *Art and the public sphere* (pp. 29–48). The University of Chicago Press.
Nadkarni, M., (2003). The death of socialism and the afterlife of its monuments: making and marketing the past in Budapest's Statue Park Museum. In K. Hodgkin, & S. Radstone (Eds.), *Contested pasts: The politics of memory* (pp. 193–207). Routledge.
North, M. (1990). The public as sculpture: From Heavenly City to mass ornament. *Critical Inquiry*, *16*(4), 860–879. https://doi.org/10.1086/448564
Poria, Y., Ivanov, S., & Webster, C. (2014). Attitudes and willingness to donate towards heritage restoration: An exploratory study about Bulgarian socialist monuments. *Journal of Heritage Tourism*, *9*(1), 68–74. https://doi.org/10.1080/1743873X.2013.778266
Rivera, L. A. (2008). Managing "spoiled" national identity: War, tourism, and memory in Croatia. *American Sociological Review*, *73*(4), 613–634. https://doi.org/10.1177/000312240807300405
Silverman, L. (1995). Visitor meaning-making in museums for a new age. *Curator: The Museum Journal*, *38*(3), 161–170. https://doi.org/10.1111/j.2151-6952.1995.tb01052.x
Simon, R. I. (2011). Afterword: The turn to pedagogy: A needed conversation on the practice of curating difficult knowledge. In E. Lehrer, C. Milton, & M. Patterson (Eds.), *Curating difficult knowledge: Violent pasts in public places* (pp. 193–209). Palgrave Macmillan.
Simpson, M. (1996). *Making representations: Museums in the post-colonial era*. Routledge.
Stam, D. (1993). The informed muse: The implications of 'The New Museology' for museum practice. *Museum Management and Curatorship*, *12*(3), 267–283.
Stogner, M. B. (2009). The media enhanced museum experience: debating the use of media technology in cultural exhibitions. *Curator: The Museum Journal*, *52*(4), 385–397. https://doi.org/10.1111/j.2151-6952.2009.tb00360.x
Tucker, H., & Carnegie, E. (2014). World heritage and the contradictions of 'universal value. *Annals of Tourism Research*, *47*, 63–76. https://doi.org/10.1016/j.annals.2014.04.003
Tucker, H. (2016). Empathy and tourism: Limits and possibilities. *Annals of Tourism Research*, *57*, 31–43. https://doi.org/10.1016/j.annals.2015.12.001
Tunbridge, J. E., & Ashworth, G. (1996). *Dissonant Heritage: The management of the past as a resource in conflict*. John Wiley & Sons.
Wallis, J. (2019). *Difficult histories and positive identities*. Cumberland Lodge.
Watson, S. E R. (2007). *Museums and their communities*. Psychology Press.
Wells, K. (2007). The material and visual culture of cities. *Space and Culture*, *10*(2), 136–144. https://doi.org/10.1177/1206331206298544
Williams, P. (2008). *Memorial Museums: The rush to commemorate atrocities*. Berg.
Wulff, H. (2007). Introduction: The cultural study of mood and meaning. In H. Wulff (Ed.), *The emotions: A cultural reader* (pp. 1–16). New York.
Zhang, C. X., Honggen, X., Morgan, N., & Ly, T. P. (2018). Politics of memories: Identity construction in museums. *Annals of Tourism Research*, *73*(1), 116–130. https://doi.org/10.1016/j.annals.2018.09.011

Part III

Radical Peace Tourism in Practice

Making waves: Peace Boat Japan as a model of sustainable peace through tourism

Lynda-ann Blanchard, Sumiko Hatakeyama and Akira Kawasaki

ABSTRACT

Linking peace to tourism has led to key questions in the academic literature. Is tourism a generator of peace or simply its beneficiary? Can tourism be a social force or is it merely an industry? What kind of pedagogical principles can tourism utilise to promote learning for peace? This article offers a case study of Peace Boat, a Japan-based non-government organisation that contributes to human rights literacy and peace to reflect on some of these critical questions. It is based on a qualitative and collaborative research partnership that harnesses the power of stories of peace encounters to better understand the practices, pedagogies and potentials of peace tourism. We argue that when we utilise peace studies' concepts such as "positive and negative peace" and "peace with justice", we illuminate how dialogue in contested spaces is core to peace through tourism approaches. This study reveals a radical pedagogy and enables a more reflexive approach to reclaiming tourism in the interests of securing sustainability, human rights, justice and peace. In analysing the work of Peace Boat through its global voyages, we show how it fosters global solidarity, providing evidence that tourism can indeed contribute to sowing the seeds for global peace.

Prologue: Peace Boat 83 as a creative student protest

Advertised as "Peace Boat 83", this first voyage was organised by a group of students from Waseda University in Japan, as a creative response to the Japanese government's censorship regarding Japan's past military aggression in the Asia-Pacific.

So, on September 2, 1983, a small passenger ship *Nippon Maru* set sail on a 13-day voyage across the Pacific. During the voyage, 159 passengers, mainly in their 20 s and 30 s, visited Ogasawara, Iwo Jima, Guam, and Saipan. At each of these places, participants learned first-hand about World War II from those who experienced it.

Tatsuya Yoshioka, one of the founding members of Peace Boat, was initially disappointed that the regional political situation at that time did not allow the voyage to visit China or South Korea. Yet, the first voyage fundamentally defied his belief. The legacy of World War II and Japanese occupation was "so visible and real" in Saipan: taxi drivers spoke Japanese; artilleries and anti-aircraft guns were left abandoned. The voyage also highlighted that peace and reconciliation in the region is not just about the history and legacy of world wars. In Guam, meeting with the Indigenous Chamorro people and the visit to the U.S. Andersen Air Force Base revealed to Peace Boat participants structures of oppression, inequality, and dominance. At the same time, Yoshioka was also shocked by how many tourists, including those on their honeymoon, were innocently enjoying the "tropical paradise," with no contemplation about the historical, social, and political aspects of the place they are visiting.

"That's when Peace Boat became somehow different from a study tour. The first voyage convinced me that tourism must and can become a tool for peace and reconciliation". (Quoted from Tatsuya Yoshioka, co-founder Peace Boat interviewed by the authors, October 12, 2020).

Introduction: seeking sustainable peace through tourism—the case of Peace Boat

The prologue sets the scene for this research article—as primarily an investigation into the Peace Boat model of tourism for sustainable peace. Indeed, key tensions and conundrums are exposed in this opening story that constitute contested spaces within the peace tourism scholarship. Firstly, the industry imperative that "peace" is a necessary precursor for tourism; secondly, that tourism as "a tropical paradise" leisure industry can sit beside tourism as a socio-political force; and thirdly, the capacity for tourism to engage both "negative" and "positive" peace, key concepts in peace studies as discussed below. At heart of the theoretical discussion here, is that final tension. Harnessing the peace and conflict studies literature enables a sophisticated and enlightening analysis of Peace Boat practice of tourism, revealing a radical pedagogy for peace education and a reflexive positioning around sustainable tourism. Peace Boat from its very beginnings, disrupts the conventions of tourism, whilst also growing into a thriving business over nearly four decades.

Peace Boat is a Japan based international non-government organisation (NGO) promoting peace, human rights, and sustainability. It now holds Special Consultative Status with the Economic and Social Council (ECOSOC) of the United Nations (UN).[1] As an NGO, Peace Boat carries out its main activities through a passenger ship that travels the world. Every year, Peace Boat organises three global voyages and two East Asia regional voyages. The global voyages are three-month round-the-world cruises, each visiting around 20 countries and carrying 1000 participants. In its thirty-seven years of operation and up until January 2020, Peace Boat has organised close to 70 around-the-world voyages, visiting more than 200 ports. However, Peace Boat is also unique as a cruise liner because voyages do not fit the conventional model of luxury cruises. Working in partnership with the UN Sustainable Development Goals (SDGs) Action Campaign, Peace Boat's activities onboard and in port aim to empower participants, strengthen local capacity for sustainability and build people-to-people cooperation beyond borders. The voyages, run on a social business model and blend sustainable tourism practice, lifelong learning and friendship activities with educational programmes, cooperative projects, and advocacy.

The stories from Peace Boat in this article present a case study that reveals a unique model for "sustainable peace through tourism". One that engages aspects of tourism as an industry, in terms of a cruise ship with large numbers of staff and clientele visiting hundreds of destination communities around the globe; but that also pursues a radical pedagogy to enable a more reflexive approach to reclaiming tourism within the context of sustainability, human rights, justice and peace. There is an obvious tension to begin. The cruise industry is noted in the literature for the environmental harms that are problematic in terms of sustainable tourism (Genc, 2016), particularly as a significant growth sector of tourism (Ruiz-Guerra et al., 2019). What is more encouraging are cruise industry efforts to address sustainability (Johnston, 2020); and indeed, Peace Boat negotiates this tension in reflexive practice daily, and furthermore has been working on the design and development of a world first sustainable Ecoship (see Figure 1).

What does sustainable peace through tourism look like for Peace Boat? Firstly, as elaborated in the upcoming discussion of peace studies literature, the starting point for Peace Boat is "values-explicit" and "reflexive". In actively seeking sustainable peace, the tourism practice engages with the definition of "peace" as more than the absence of war; that includes contested spaces; and values critical thinking. The stories evidenced here demonstrate that as participants are exposed to various local and global realities and contexts—including issues of human rights, inequality and injustice—there is discussion and learning that highlights the complexities of indirect and structural violence. Similar to Jim Ife's (2009) conception of human rights education, Peace Boat's pedagogy and practice concerns "learning from below"—about the hopes, needs and experiences of human beings in specific contexts, communities and countries.

Secondly, for Peace Boat, "sustainable peace" challenges tourism as a whole to reflect on its mission. Here, tourism is not seen solely as an industry but also as an important social force. Specifically, tourism is seen as having the potential to create the conditions for more ethical relations in human rights and peace practice in two distinct ways. To enable relationships in which listening to (human rights or justice) issues identified by "communities" and/or "marginalised others" is key; and, in shifting locations from formal (touristic) spaces to also include informal spaces for the sharing of stories and the creation of more equal learning (tourism) relationships (Blanchard & Nix, 2019). Put simply, Peace Boat is not framed by the tourism industry, but by the tourism space.

Finally, to engage sustainable peace through tourism, it is imperative to apply a peace education model to tourism practice. This then necessitates and creates a radical pedagogy, which is centred around the principles of active community and cooperative learning. The approach is in keeping with Freya Higgins-Desbiolles' (2020) recent work on "socialising tourism". Each component of the enterprise serves a particular function in the learning experience (or tourism experience) on which the learning community (or tourism space) is built. The excursions permit participants to have some direct experiences with the destination country, its landscapes and the issues of justice and peace that its citizens are addressing. A reflexive approach within tourism spaces is at the heart of the learning and community-building relationships and partnerships.

From its beginnings as a modest student protest, to the development of Global University (Yoshioka, 2008, 2013); and, from hosting over 100 Hibakusha—survivors of atomic bombing of Hiroshima and Nagasaki—to contributing to the International Campaign to Abolish Nuclear Weapons (ICAN) Nobel Peace Prize 2017 (Nahory, 2017); there are several stories to tell and listen to. What we discover is that "peace through tourism" is not just about plain sailing and cannot avoid contested spaces; indeed, this is where it flourishes. It is in seeking creative ways to work to address social, economic, political and environmental (in)justices, that Peace Boat utilises the cruise ship voyage as a social force for global peace. In this article we ask, in what ways Peace Boat is more than just another example of a niche or boutique market model where "peace tourism" is simply packaged as a product for sale? To address this research question, we will first situate our case study in review of the scholarship developed within peace studies and tourism studies .

Peace through tourism: can tourism generate peace?

> Peace studies must take on the task of nurturing new modes of thought (Reardon, 2015, p. 113).

Linking peace to tourism has led to key questions in the academic literature: Is tourism a generator of peace or simply the beneficiary of peace (Litvin, 1988)? Can tourism be a social force or is tourism simply an industry (Higgins-Desbiolles, 2006)? On one hand, there is the perspective linking tourism, progress and peace (Moufakkir & Kelly, 2010) to notions of sustainable tourism. Yet, if we harness the peace studies scholarship—such as concepts of "positive and negative peace" (Galtung, 1996) or "peace with justice" (Rees, 2003)—to shed light on the link, the promise of promoting peace through tourism is revealed in contested spaces (Blanchard & Higgins-Desbiolles, 2013).

Unmasking the rhetoric of peace, in the words of peace educator Betty Reardon, "does not exclude conflict… it governs and guides it to become a source of growth and change rather than harm and destruction" (2015, p. 115). Johan Galtung (1996) calls this "positive peace", where the often hidden, conflicted structural and cultural dimensions of life must be addressed as potential sources of indirect violence in any claims about promoting peace. Thus, structural violence includes inequality, dispossession and poverty. Whereas cultural violence includes patriarchy and racism. This broader perspective is also referred to as "peace with justice" (Rees, 2003, p. 20–21). Elsewhere peace scholars call to advance a philosophy of peace (Webel & Galtung, 2007, pp. 4–5) and a paradigm of/for peace (Reardon, 2015, p. 113). This call for critical awareness resonates with the values-explicit and self-reflexive nature of peace studies (e.g. Galtung, 1996; Reardon, 1988). Other notions such as "ecological civilization" imply a civil society that operates within the limits of the ecosystems of which humanity is a part (Gare, 2010, p. 31).

These concepts of "ecological civilization" and "peace with justice" underpin "sustainable development". Beginning with these ideas within peace studies gives us the cue to linking "sustainable peace" to tourism.

The founder of the International Institute for Peace Through Tourism (IIPT) envisioned over three decades ago, "[t]hrough tourism, we can come to appreciate the rich human, cultural, and ecological diversity of the world mosaic and to evolve" (D'Amore, 1988, p. 270). This vision is reiterated a decade later in the "Global Code of Ethics for Tourism" which states that "through the direct, spontaneous and non-mediatised contacts it engenders between ... different cultures and lifestyles, tourism presents a vital force for peace and a factor of friendship and understanding among the people of the world" (UNWTO, 1999, p. 1) . Yet, tourism scholars may interpret this in positivistic and uncritical ways such as evoked in the title *Tourism, Progress and Peace* (Moufakkir & Kelly, 2010). According to Werner Wintersteiner and Cordula Wohlmuther's understanding, what they call "peace-sensitive tourism" starts "not only with a criticism ... the negative effects of tourism in economic, social, ecological and cultural matters, but ... [e]qually, it has to denounce the hypocrisy of a tourism "new speak", using the language of alternative and responsible tourism or the rhetoric of peace without any substantial changes" (2014, p. 50).

Broadening that focus requires reflexive thinking about tourism as an industry and as a social force. In the seminal article "More than an Industry: The forgotten power of tourism as a social force", Higgins-Desbiolles (2006) posits the notion of tourism as a "social force". In it she argues that an industry focus in the theory and practice of tourism, has led to the "marketisation" model that has dominated and overshadowed the "transformative capacity" model that gives equal weight to social and ecological effects or impacts of tourism. This conceptualisation has been extended in her most recent scholarship that critically engages sustainable tourism. In introducing the concept of "socialising tourism", Higgins-Desbiolles (2020) argues the global health pandemic has not only spotlighted the dominance of the economic imperatives of tourism—in bringing the global industry to almost total collapse—but has offered an opportunity for transformation towards a more sustainable tourism. She states:

> The question this raises is: what balance will be struck between supporting tourism multinationals, tourism small business, vulnerable tourism workers and destination communities in an environment that challenges the very foundations of current economic systems? ... Despite [OECD] efforts to enable a return to "business as usual" as soon as possible, some have identified this as a transformational moment opening up possibilities for resetting tourism. (2020, p. 3).

"Resetting tourism" brings into focus claims made by Wintersteiner and Wohlmuther, as discussed above, about being wary of both tourism "new speak" and using the language of "responsible tourism" to engage rhetoric without any attention to systemic changes. Higgins-Desbiolles makes this exact point after quoting the United Nations World Tourism Organisation (UNWTO) as "seizing on the rhetoric":

> Tourism has the potential to recover and once again establish itself as a part of national economies and of the wider sustainable development agenda ... This crisis may also offer a unique opportunity to shape the sector to ensure it not only grows, but it grows better with inclusivity, sustainability and responsibility prioritized. Furthermore, to build for the future, special attention should be placed on building resilience and on promoting sustainability at all levels. (UNWTO, 2020, p. 33 cited in Higgins-Desbiolles, 2020, p. 3).

Within the recent literature on cruise ship tourism, there is research concerned with future "cruise tourism as an option for development", as in the case study research in Oman where "recommendations for policy makers were suggested on how to increase the role of cruise tourism in strengthening the economy" (Aziz et al., 2020, p. 95) . However, Julia Eva Peters is more circumspect in making a specific link between sustainability and cruise contexts when she argues that: "[i]n the pursuit of sustainability in tourism, a rethink is necessary ... This is particularly true in those areas that are experiencing high growth rates ... as can currently be observed in the cruise segment" (2020, p. 111).

In the broader scope of tourism literature, the resistance within the tourism industry for reflexive and critical practice is writ large as "[the] commercial perspective portrays tourism as an unalloyed blessing to populations worldwide" (Higgins-Desbiolles et al., 2013, p. 1). There is also a body of tourism scholarship that begins with promoting the industry as necessarily beneficial to "progress" for community development with critique concerned with "the ability of tourism to contribute to harmonious relationships among peoples of the world" (Moufakkir & Kelly, 2010, p. xix). This view appears at odds with research that details a very different picture. Martin Mowforth and Ian Munt, for example, note that:

> [o]f all the problems experienced by local communities facing tourism development schemes, the most harrowing involve accounts of people being displaced. Such events normally reflect the distribution of power around the activity of tourism and highlight the powerlessness of many local communities. And it seems rare that displacement and subsequent resettlement of displaced people result in more even and equal development (2003, pp. 236–237).

This propensity for the industry to emphasise the positive economic benefits while ignoring the social and ecological costs that accompany it, is clearly documented in the scholarship—bringing into sharp focus human rights and justice issues (Scheyvens, 2002; Smith & Duffy, 2003; Fennell, 2006; Higgins-Desbiolles & Blanchard, 2010; Jamal & Higham, 2020). Surely, this must be at root of any claims about sustainable tourism. In response, the tireless work and contribution of NGOs indicates vigilance in raising awareness of the adverse impacts of the tourism industry—in concert with government policy and practice—on local livelihoods, communities and lands (Stancliffe, 2013; EQUATIONS 2013). Arguably, this work has paid off, influencing tourism practice in terms of raising broader awareness of damaging industrial practice and likely contributing to positive growth in the flourishing niche markets for tourism. Yet, a more circumspect and critical approach argues for a view of sustainable tourism, as more than simply products packaged in suitable niche markets such as eco-tourism, pro-poor tourism, volunteer tourism, small group tourism or peace tourism— the latter, somewhat narrowly defined in the existing literature as:

> [involving] visits to places, at home and abroad, which are significant because of their association with such notions as peace-making, peaceful conflict resolution, prevention of war, resistance to war, protesting war, non-violence and reconciliation (Van den Dugen, 2014, p. 62).

Engaging a broader approach requires investigation into the ways in which sustainable tourism meets sustainable peace. Taking seriously the linkages between sustainability, peace and tourism, we might take the lead from peace scholar, activist and poet, Stuart Rees who states that while "structural violence cements injustice" (2015, p. 13), we must work towards addressing social, economic and ecological justice to achieve sustainable peace:

> [p]eace with justice is characterized by an absence of violence, whether it is direct and observable or indirect and invisible. A just peace would be marked not only by an end to hostilities between peoples, but also by fairness in social, economic and political arrangements. It would be a fairness characterised by equality and opportunity and by associated influences in the building of just societies… (Rees, 2003, pp. 20–21) .

The tourism literature includes promising evidence of such reflexive critical thinking with regard to the notion of sustainable tourism. Karla Boluk et al. (2017) argue critical thinking is needed to realise sustainability in tourism systems and utilise the UN SDGs as the roadmap. Furthermore, Boluk et al. (2019) propose a pedagogical framework for the development of the "critical tourism citizen". Similarly, Yaniv Belhassen and Kellee Caton (2011) discuss the need for critical thinking in tourism education. Critical and creative thinking is exactly what is needed to negotiate sustainable peace, according to Joseph Camilleri and Deborah Guess (2020) in their new book *Towards a Just and Ecologically Sustainable Peace*.

We might also learn from peace educator Betty Reardon who says the major obstacle in moving to a peace system can be seen to "lie in the system and the thinking that sustains it" (2015, p. 123). This is useful in an application in linking peace with sustainable tourism. Unless there is a rethinking in terms of the "values" underpinning mass tourism—the exploitative aspects of neo-liberal market imperatives and, socio-political power inequalities—the promise of peace through tourism will remain unrealised (Higgins-Desbiolles et al., 2019). Since we have been stopped in our tracks by this global pandemic, perhaps it is the reflexive moment needed to rethink sustainable peace through tourism (Lew et al., 2020).

For example, a reflexive approach to this casts new light on the major thrust in the tourism literature equating "peace" to "post-conflict" tourism. The notion of *sustainable* peace through tourism is much more interested in, and concerned with, conflicted spaces: seeking to understand structural, cultural and environmental conflict. Borrowing from Reardon (1990; 1993; 1999), educating for peace requires us to unmask indirect and structural violence and develop capacity to create possibilities for nonviolent social change in the promotion of human and environmental security. If we pause to reflect on the prologue with which this article began, we can depict such conflicted spaces in tourism spaces:

> In Guam, meeting with the Indigenous Chamorro people and the visit to the U.S. Andersen Air Force Base revealed to Peace Boat participants structures of oppression, inequality, and dominance. At the same time, … many tourists, including those on their honeymoon, were innocently enjoying the "tropical paradise", with no contemplation about the historical, social, and political aspects of the place they are visiting.

In referencing the prologue here, we note tourism literature on mass tourism that draws attention to the fact that tourists are happy to remain oblivious to the local struggles. This is depicted in Jamaica Kincaid's (1988) post-colonial narrative *A Small Place*. Engaging conflicted spaces in tourism spaces is perhaps the key to *sustainable* peace through tourism. This enables examination into the contradictions of tourism policy and practice; interrogation of the "new speak" in global standards of tourism and their application; and a diversity of individual and corporate approaches to "responsible" tourism. The case study of Peace Boat represents a willingness to use the foundations of a mass tourism cruise ship for radical peace pedagogies—for critical and reflexive practices that embody the possibilities of peace through tourism.

Methodology: bridging academia and praxis

Norman Denzin and Yvonna Lincoln (2008) theorise the research act as having deeply disciplined dimensions enshrined within academe. Therefore, it is useful to be somewhat reflexive of the research process when introducing interdisciplinary work, such as bringing "peace studies" together with "tourism studies"—with the former recognised as essentially interdisciplinary at its core (see Galtung, 2010). The literature review offered in the former section was intended to engage both "disciplines" in this regard. Furthermore, we acknowledge in the tourism literature "reflexivity" is discussed in terms of critical, engaged research practice "which, while alive in other disciplines and fields, has received rather limited attention within tourism studies" (Ateljevic et al., 2005, p. 9; also see pp. 13–14). Reflexive practice is a key idea in this research as we note from the methodology literature:

> [R]eflexivity as a concept and practice is widely recognised and acknowledged in qualitative social science research … [as] reflexivity unveils … how the researcher(s) arrive(s) at certain positions during the knowledge construction process (Subramani, 2019, p. 1).

In this study, the authors come from different perspectives: as an academic of peace studies and as staff members of Peace Boat. This collaboration enabled a dialogic allowing a cross-referencing of the scholarship on peace studies—including reference to peace education, human rights, nonviolence and sustainable development—together with the implementation of these ideas and experiences based on the practices of the "hybrid" tourism enterprise of Peace Boat. As the co-authors have collaborated on Peace Boat's Global University for over a decade, we draw on this relationship as a key component of our research methodology for this project. In conversation, the authors have selected the case study stories (see the example below) to form the basis of evidence from which to develop our theoretical ideas. Furthermore, as this research project is primarily qualitative, new ideas of "storytelling as research" (Davis, 2004) are engaged—via stories about the practice of Peace Boat. In this approach, we also follow insights from tourism studies:

> [W]e argue narrative inquiry offers a research method for understanding tourism as a dynamic, reflexive process negotiated by tourists and hosts within social environments. In synthesising tourism literature, we [note] how a shift in methodological focus, from quantitative to qualitative narrative inquiry, shifts research focus towards issues and research questions of great pertinence in contemporary liminal tourism spaces. (Nelson et al., 2017, p. 1)

In this qualitative inquiry, narrative data reveal the purpose, processes and impacts Peace Boat presents. For example, the case study story from Peace Boat as told in the prologue, provides a springboard for key research themes and the narrative example in the text box below points to reflexive practice -. As Peace Boat's pedagogy is one that is centred on experiential learning and intercultural communication, such narrative data presents key evidence in this case study.

> In 1984, Peace Boat again organised an expedition, whose itinerary this time included Hong Kong and Shanghai. In the latter port, an excursion was organised for the group to visit Nanjing, the site of a massacre during WWII. In this voyage, participants had a chance to hear from an old lady, who miraculously survived the Nanjing Massacre. A young participant, who took part in the tour, was deeply moved by her testimony. In the letter she wrote to Peace Boat upon her return, she said:
>
> "I am probably not going to become a peace activist. But I do not think I will ever forget what I heard from that lady in Nanjing. I will be doing what I can do in my capacity to contribute to peace because I have a responsibility as someone who was fortunate enough to hear first-hand testimony about the violence she had gone through"
> (As recalled by Tatsuya Yoshioka, co-founder Peace Boat in interview with the authors, October 12, 2020).
>
> "A focus on hearing testimonies from those who are affected by the issue at stake had a powerful impact on Peace Boat participants, as recorded in post excursion comments and letters. After the first two voyages, the founding members were convinced that such activity constituted a "radical pedagogy" in promoting peace— and tourism would pave the way."
> (Quoted from director, Peace Boat Global University, Akira Kawasaki, in interview with the authors, October 12, 2020).

Case study approaches are valuable when research questions are concerned with qualitative dimensions such as when the research questions "... require an extensive and "in-depth" description of some social phenomenon" (Yin, 2014, p. 4). Although there are limitations to this method of a case study approach, such as not being robust in terms of drawing generalisations or broad comparisons, its strength lies in the textured context-specific descriptive insights and understandings that can be used to build tentative hypotheses to help inform future research (Flyvbjerg, 2006; Lean, 2009). Case study methodology has been essential here to try to build deep understanding of the interface between tourism and peace practice as undertaken by Peace Boat. This project draws on the nearly four decades of global tourism practice of Peace Boat and entails analysis of archival written records, documentation from Peace Boat's website, participant observation and semi-structured qualitative interviews with Peace Boat leadership. This offers plausible evidence through triangulated data (Decrop, 1999). The collaboration between the authors has spanned over a decade and this has supported such critically engaged and collaborative work together analysing Peace Boat praxis and impacts in linking peace, sustainability and tourism.

The positionality of researchers is important to acknowledge in qualitative research processes (Wilson et al., 2020). This research is based on engaged collaboration between an involved academic and two practitioners. One author works in academic peace and human rights education and has been an invited guest educator with Peace Boat's Global University on board programme. The other two authors are staff of Peace Boat. One of whom has served as

International Coordinator of Peace Boat and travelled to over 20 countries with "Hibakusha"—atomic bomb survivors who testify to the humanitarian consequences of nuclear weapons. The other co-author is a member of the Executive Committee of Peace Boat, serves on the International Steering Group of the International Campaign to Abolish Nuclear Weapons (ICAN), is director of Peace Boat Global University and since 2008 has coordinated "Global Voyage for a Nuclear-Free World: Peace Boat Hibakusha Project". This will be an ongoing research relationship extending beyond this preliminary study as we have a long-term commitment to working together in developing ideas concerned with "peace tourism" and investigating action for sustainable tourism. Such longitudinal research collaboration is rare in tourism studies and offers novel and rich insights into tourism's possibilities (see Cole, 2004; Ritchie, 2005). This has allowed the building of this rich case study exploring the pedagogies of peace that can inform tourism practices, to which we now turn.

The Peace Boat model of tourism

In this case study, we explore in detail below the ways Peace Boat is more than just another example of a niche or boutique market model where "peace tourism" is simply packaged as a product for sale. The Peace Boat case attests that while tourism and cruises in particular are often considered to be for "leisure" and are simply the beneficiary of peace, they can also serve as generators of peace. Crucial to this point is the role of the ship as a "space of tourism", not merely as a mode of transportation. Peace Boat was quick in realising that the vessel can provide a uniquely safe environment for dialogue. Likened elsewhere as a "floating peace village" (Yoshioka, 2008), the onboard community becomes amenable to informal spaces—such as the cafeteria, shared cabins, or even pools and gyms—becoming "spaces" where frank conversations take place and issues can be discussed. Even inconveniences, such as being an environment where the internet is not easily accessible, becomes an opportunity for participants to "disconnect" from their former lives and recreate a "presence" in an onboard community. As one participant shared:

> We were more connected because we were in a space where we did not have the Internet... We were dealing with each other, spending more time engaging with each other. We got to know (and talk) more about the issues of our own countries, own cultures, heritage, food.... I did not expect to find a *family* on Peace Boat but I did (Batac, 2020).

This space of tourism is often contested, where tensions and conundrums arise, and difference of opinions and values become visible. Yet, Peace Boat stories below underline how these contested spaces can also be key to tourism as a generator of peace.

Cultivating regional and global citizenship

In 2001, during Peace Boat's 33rd Global Voyage, the International Students Programme (ISP) was launched. It gathered students and peace activists from regions in conflict for half the voyage—from Tokyo, Japan to Las Palmas, Spain. Among the participants were Rami Nasser from Palestine and Keren Assaf from Israel, who hardly talked to each other when they met for the first-time onboard Peace Boat. They did not shake hands. For both it was the first time to meet somebody in person from the "other side" of the conflict. Yet, with the time allowed during the time at sea, during the sessions, events, and workshops onboard and in ports, Rami and Keren built up a friendship. Reflecting on the experience, Keren said: "I think this voyage gave me a good perspective and proportion, to step outside and examine myself and my situation and clear my head." Post-voyage, Rami and Keren decided to establish a programme in their region, similar to one onboard Peace Boat, giving the opportunity to other young Palestinians and Israelis to meet, to listen to each other's stories, and to talk about the future—within safe spaces. Rami today testifies:

> Peace Boat has given me so many ideas for the work of my organisation in Palestine. If Peace Boat continues to work as it does, through cultural exchange, information sharing, and solidarity with oppressed people, I believe Peace Boat can help to change a generation of young people—and sooner or later we will have peace (Rami Nasser as quoted in Peace Boat, 2020a).

Peace Boat has organised ISP on several subsequent global voyages, bringing together students and activists from Palestine, Israel, the Balkans, Northern Ireland, Cyprus, Pakistan, India, Sri Lanka, China, Taiwan, the USA and Colombia (Peace Boat, 2020b; 2020c; 2020d). The ISP has generated educational and exchange opportunities for students and youth from all over the world, as one of the core peace education programmes on Peace Boat.

Another initiative is a similarly ambitious attempt to cultivate regional and international citizenship—bringing together participants from both Japan and South Korea. Initiated in 2005, Peace & Green Boat voyages are run in collaboration with the Green Foundation, South Korea. Since 2005, over the course of fifteen years, this collaborative initiative has organised more than ten joint voyages, bringing together over 10,000 participants from Korea and Japan in journeys of dialogue, mutual understanding and trust— and for joint action towards building a more informed East Asian Community. The Green Foundation's founder and representative Choi Yul stated in 2017 on the occasion of the tenth Peace & Green Boat excursion:

> At the time we first set sail together in 2005, tensions between Korea and Japan were extremely heated. We have experienced various difficulties each and every year. However, we have continued to create these voyages together, making great efforts to overcome these difficulties. This is because we believe that the seeds for conflict resolution lie in people. I feel that finally, the fruits of our dialogue are starting to come to the surface. Even more because of the current climate, we must continue more and more exchange in order to nurture the tree of peace and sustainability in East Asia. Each and every one of us holds a great responsibility for the peace and environment of our region. (Yul as quoted in Peace Boat, 2020e).

Cultivation of regional and global citizenship requires careful and active participation. Enabling contested spaces also requires providing safe spaces—a minimum requirement for open dialogues that necessitates careful consideration. Peace Boat values engaging controversial issues. Yet, when dealing with historical and political issues that are often complex and sensitive, coordinators must know how to navigate the "contested spaces". They must know how to reconcile conflicting perspectives and protect those whose position is most vulnerable. This is especially so given that Peace Boat is becoming increasingly diverse—with approximately 30% of the participants coming from countries and regions other than Japan—underscoring the essential role skilled coordinators play in shaping tourism as a generator of peace.

Ship voyage for active communal learning

What makes Peace Boat a "floating peace village" is not simply the tourism spaces it provides. Many Peace Boat projects explicitly harness pedagogical principles that resonate with peace education—such as, learning with and from one another; nurturing spaces for active communal learning; and treatment of learners as agents of change (Reardon, 1988). Peace Boat's Global University—which began in 2000, first as a Japanese-language peace education project and subsequently developed into an international tertiary accredited peace and sustainability course—in particular embraces these principles in its curriculum design.

The ship voyage allows for a particularly dynamic approach to learning about global issues. Voyaging through different countries enables participants to bear witness to a wide range of realities and testimonies that they otherwise would have no chance to encounter. In Singapore, for example, participants engage with a local NGO—that supports migrant workers—to learn about the harsh and inhumane conditions workers face. In Myanmar, the same participants hear stories from families who willingly send their members abroad to seek economic opportunities. Onboard, invited guest educators share possibilities and limitations of international policies and regulations concerning workforce migration and discuss international dimensions of human slavery. Such exposure to different standpoints challenges participants to consider what it means to come up with a solution that substantially rejects structural and direct violence; whether there is a contested solution; and whose voice must be heard when deciding steps to address injustice. Participants come to realise that seeking "peace with justice" does not involve "simple" solutions but lies in contested spaces of interrelated issues. This is illustrated in these two comments from former Global University participants:

> I thought I knew stuff. And I knew. But what Peace Boat did was to open things further for me. I understood things from a wider perspective. There were a lot of simulation exercises that helped us understand the diversity of thought processes. People think differently and people feel differently. People have different perspectives on different issues. That is important when it comes to peacebuilding. You are not always right. You are not always neutral. I understood that deeply during my time on Peace Boat (Zain Deen, 2020).

> That experience shaped the kind of peacebuilding that I would do .... The kind of peacebuilding that looks at the side of those who are erased and marginalised (Batac, 2020).

In Peace Boat's Global University programmes, participants are active learners. Their own educational, cultural, political, and religious backgrounds contribute to a pool of knowledge in discussions that take place. There is no one expert voice. For example, participants have visited the Killing Fields in Cambodia, where hundreds of human skulls are displayed in a glass shrine. As they stood in front of the shrine, one student was deeply moved by the display, saying that the event should never be forgotten and repeated again. Another participant was surprised at such a reaction—claiming that displaying human skulls as if they were to attract tourists was an insult to the dead. These differing opinions led to post-tour reflexive thinking on what it means to memorialise war and violence (in tourism spaces) and from whose perspective such a question should be considered. As one participant observed:

> Most of the discussions we had during these sessions were very deep discussions. I think this kind of deep conversation should be implemented in our societies also: we learn from each other and expand knowledge on what peace is, what real reconciliation is, what conflict resolution is all about (Peace Boat, 2014, p. 12).

Another participant highlighted how learning with and from one another made many issues "real":

> We learned about Tamils and Sinhalas, Rohingyas and surrounding religious issues in Bangladesh, the issues of ethnic conflict in Myanmar, migrant workers in Singapore, genocide in Cambodia, legacy of WWII impacting the relation between Japan and its neighbours, etc. We got to learn more about these things because we were learning about each other's countries. There is a face now when we talk about Cambodia or Sri Lanka, when we talk about Japan (Batac, 2020).

A comment by Fauziah Mohd Hasan, a physician and humanitarian activist who has joined Global University as a guest (educator) navigator underlines the value of such a learning community:

> The programme nurtures the bonds, human compassion, and solidarity that can unite us across the divides of race and religion, gender and geography. This is absolutely critical for a collective ambition for our world (Peace Boat, 2018, p. 10).

Finally, Global University puts emphasis on raising participants' capacity to take actions, as an action-oriented component is incorporated in the curriculum, where participants devise and implement their own ideas into an onboard project. The projects vary—for example: a photo exhibition on transnational migration; presentation on gender discrimination in the workplace; a talk show on atomic bombings; and puppetry as a way to tell Hibakusha testimony. Experiences of taking an action, however small the action may be, empower participants, as the feedback evidence shows:

> When you look at the macro-level, it becomes hard to see how we as youths can create peaceful and inclusive societies, but I believe the solution is in each and every one of us (Quoted in an archival essay as participant observation from 2017).

A comment by John Gee, founding member of Transient Workers Count Too (TWC2), highlights the long-term implication of such a learning experience:

> The students' active interest was inspiring and so was the commitment they showed in finding effective ways to share something of what they learned with the passengers on the ship. I'm sure that they will carry what they learned and experienced with them into their careers and studies in the years to come (Peace Boat, 2018, p. 10).

Measuring the impact of projects like the above is an important next step for Peace Boat, as this is also about how to transform the transient nature of travel and bring about a longer engagement with "tourism spaces" (such as elaborated in some of the transformative tourism scholarship, e.g. Lew et al., 2020). The difference for Peace Boat tourism practice lies in the relationship-building aspects of ongoing collaboration with participants, NGOs and communities. Some of these "tourism" relationships and partnerships have lasted for more than twenty years. The ongoing COVID-19 pandemic also pushes Peace Boat to reflect on these "tourism" collaborations/partnerships and rethink how to leverage online tools and platforms for active communal learning, including "virtual tourism"—a move which seemingly contradicts its core pedagogy in harnessing the power of direct people-to-people exposure and building "in port" community relationships.

Making waves for nuclear disarmament

On October 24, 2020, the UN Treaty on the Prohibition of Nuclear Weapons (TPNW) secured the required 50 states parties for its entry into force.[2] The following day, Peace Boat organised a 12-hour live broadcast to celebrate this historical moment (Peace Boat, 2020f). Appearing in the broadcast were almost 70 people from over a dozen countries. Connecting Peace Boat's domestic and international partners, as well as Hibakusha living in Japan and abroad, the broadcast illustrated how Peace Boat has built a constituency of advocates for peace through its tourism endeavours. Nuclear disarmament is one area where Peace Boat has been particularly effective in engaging tourism as a social force. Through Peace Boat Hibakusha Project, or Global Voyages for a Nuclear-Free World, approximately 180 Hibakusha—not only Japanese Hibakusha living in Japan but also Korean Hibakusha as well as Hibakusha living overseas—have travelled around the world, giving personal testimonies about the effects of the atomic bombs on humanity and ecology. In around 100 cities in more than 60 countries, the Hibakusha have given testimonies to schools, civil society organisations, government-related institutions and parliaments to create personal connections through stories revealing the profound humanitarian consequences of the use of nuclear weapons (Hatakeyama & Kawasaki, 2020).

In this process, the project has also empowered Hibakusha. Setsuko Thurlow, a Hiroshima survivor living in Canada, for example has written about her experiences on one of the Peace Boat voyages she has joined:

> [T]he journey was an opportunity for many Hibakusha to reposition our own views and reflect on our experiences of the atomic bombings in a broader perspective. By visiting many countries, listening to people from the local communities and learning from guest educators, we were able to revisit our own ideas, or misconceptions, about the war itself....By providing a community for healing and learning, reflecting on and revisiting our pain, the Global Voyage for a Nuclear-Free World transformed many Hibakusha from storytellers into activists for nuclear abolition (Thurlow, 2020, pp. ix–x).

As the Peace Boat Hibakusha Project illuminates, it is in linking education to advocacy that Peace Boat achieves harnessing tourism as a social force. Peace Boat Hibakusha Project has developed hand in hand with Peace Boat's advocacy work on nuclear disarmament at the UN and other disarmament fora. In particular, Peace Boat has been one of the key organisations of ICAN, which was awarded the Nobel Peace Prize in 2017 for its work leading up to the adoption of TPNW. Peace Boat made a special effort to organise high-level meetings with officials in countries that were not supportive of the ban. Meeting directly with official, Hibakusha told their harrowing stories and urged local councils to convince their governments to participate in the negotiations to support TPNW (Nahory, 2017, pp. 5–6). In the words of Elayne Whyte Gómez, who chaired the U.N. negotiations that established the Treaty: "The international community heard the Hibakushas' cry and adopted a norm that illegitimises nuclear weapons" (Whyte Gómez as quoted in Hiroshima Peace Media Center, 2017). Beatrice Fihn, Executive Director of ICAN, has also highlighted the role of Hibakusha as having shown the campaigners "the kind of brave action to win their struggle against nuclear weapons" (Fihn, 2020).

Further enhancing Peace Boat's approach to tourism as a social force for sustainable peace is the perspective of uniting different "nuclear" victim populations. Peace Boat works with survivors of nuclear testing, former workers at uranium mining sites and nuclear weapons factories, residents of lands polluted by nuclear power plant breaches, as well as communities impacted by the history and legacy of nuclear violence—such as Indigenous peoples in Australia and Tahiti. Connecting stories of survivors of direct and indirect "nuclear" violence from different parts of the world elucidates structures of inequality and discrimination that have underpinned our world threatened by the accumulation of nuclear weapons. This connection and realisation lead not only to solidarity among civil society organisations across borders, but also lead to a powerful, international, advocacy network to advocate for the abolition of nuclear weapons. In generating this kind of global solidarity, tourism can sow seeds for global peace.

Building the world's greenest ship

Peace Boat has attempted to demonstrate—through trials and errors—that making an explicit commitment to environmental and planetary sustainability offers added value to the tourism industry. In 1990, alerted by the acceleration of environmental degradation, Peace Boat collaborated with Professor Koya Ogino of Kyoto University to conduct a project called "Earth's Health Check-up". Throughout the Global Voyage, the team led by Professor Ogino measured water and air quality, investigated soil contamination and looked into waste issues at different ports of call (Peace Boat Environment Team, 1992). In 1992, the Peace Boat delegation attended the Earth Summit in Rio de Janeiro, highlighting the need for world leaders to take actions. In the words of Peace Boat co-founder Tatsuya Yoshioka, "[b]eing indifferent to environmental degradation and the negative impact tourism may have on environmental sustainability is as if we are willingly eroding resources that we need to maintain our business." (Quote from interview with the authors, October 12, 2020). Yet the most troubling issue for Peace Boat had been the negative environmental impact of the cruise industry itself. The average cruise ship carrying 3,000 passengers and crew generates 80,000 litres of sewage a day and many cruise lines use out-dated filter systems, resulting in minimally treated sewage being dumped into the water (Peace Boat, 2021). Moreover, most ocean liners use bunker fuel to power their engines, which is the heavy, residual oil left over after gasoline and diesel have been extracted from crude oil and can emit dangerous levels of sulphur dioxide (Peace Boat, 2021).

The Great East Japan Earthquake and Tsunami that caused the Fukushima Nuclear Power Plant accident in March 2011, gave Peace Boat the one last push that was necessary to declare a most ambitious plan: to build the world's greenest cruise ship. Peace Boat thus launched the Ecoship Project in 2014 to challenge the status-quo and has set itself a radical set of targets to demonstrate what is possible, and necessary for the industry to exist in a sustainable world. A key Ecoship deliverable is an estimated 40% CO2 reduction in comparison with a cruise ship with conventional propulsion built before 2000; and an estimated 30% reduction compared to a good current design. These results will be achieved through the combination of the propulsion efficiency, hull forms, accommodation efficiency, hull air bubbles and renewable technologies, as well as by route, speed and management measures. Once set sail, the Ecoship will be the future platform for Peace Boat's global voyages hosting exhibitions on green technology in up to 100 ports and serving as a floating laboratory contributing to research on the ocean, the climate and technology (Peace Boat, 2021; also see Figure 1).

Peace Boat has always ambitiously combined tourism and activism to create a social force. The Ecoship is the next stage of Peace Boat's journey to radically embody its philosophy, that tourism can be a positive social force that engages tourism standards and practices to inspire people to make the world a better and more sustainable place. Furthermore, as vividly shown by the current COVID-19 pandemic, tourism is particularly vulnerable to crises. Unlike mega companies with different sources of income, small organisations like Peace Boat are easily affected whenever events affecting ship travel—such as, terrorism, extreme weather events, and pandemics—occur. For Peace Boat, a radical embracing of environmental sustainability also gives Peace Boat a unique added-value and an advantage within the tourism industry, which would in turn ensure its business sustainability: "For us, it was always obvious that environmental sustainability is directly linked to our business sustainability", says founding director of Peace Boat Tatsuya Yoshioka. (Quote from Tatsuya Yoshioka, in interview with the authors, October 12, 2020).

Figure 1. Industry innovation for sustainable peace through tourism—Peace Boat Ecoship.
Copyright: Peace Boat, permission granted to reprint in Journal of Sustainable Tourism.

Conclusion: sustainable peace tourism praxis–harnessing pedagogical principles

In her search for a peace paradigm, Reardon (2009) stresses the role of peace education and a critical pedagogy of human rights learning. This is based on a philosophy and practice in which the social purposes are oriented toward social change and transformation. Three key tenets of the model are: community—and the notion of learning with and from each other; nurturing spaces for authentic communal learning; and learners as active agents (Jenkins, 2008).

As depicted in the stories of Peace Boat praxis above, engaging such principles is a productive measure of sustainable peace through tourism.

We also propose pedagogical principles from human rights education. In his book *Human Rights from Below* (2009), Ife appeals for us to consider that thinking about human rights does not start with the UN Declaration of Human Rights. Although a template for a common humanity for peace with justice, the document is aspirational rather than practical. For Ife, we need to begin with "human rights from below", to begin with how the most marginalised and vulnerable individuals, communities and lands can access rights. In an application to tourism, this principle may act as a measure for sustainable peace. For example, starting with considerations of the meaning and value of the SDGs to "community" and "ecology", in essence, involves translating aspirational policy documents drafted by bodies such as the UNWTO into accountability measures in business vis-à-vis the development and implementation of effective tourism policy and practice.

Lastly, we propose radical pedagogy. Following radical educators such as Paulo Freire (1970, 1998) and bell hooks (2010), this is a kind of critical pedagogy in which learners engage actively with lived experience. In his book *Passion for Peace*, as a radical peace educator Rees suggests for us to find ways to challenge the status quo—in theorising the creative use of power, his concern is to discredit the use of "one-dimensional" power in all its forms, whether social/cultural or economic/political, as direct and structural violence:

> [Q]uestioning the status quo of power distribution is likely to occur among those who not only identify injustice … but the advance of civil liberties and the building of civil societies have depended on the courage and vision of those who "talked back" (2003, p. 76).

Human rights educator and scholar Audrey Osler is also a radical when she suggests that praxis should be "an emancipatory or transformatory form … where learners get to study justice/injustice and equality/inequality … to support greater justice and equality in learners' lives and to promote solidarity with others to achieve these ends in the wider society" (2015, pp. 30–32). The application of such principles to sustainable peace through tourism are elaborated throughout this article. Indeed, stories from Peace Boat curated here provide ample evidence of radical pedagogy in tourism spaces.

In bringing peace studies and tourism studies together, the theoretical shift is in the contribution peace theory can make to an analysis of peace through tourism. The argument here is that tourism as a social investment for societies is a powerful industry to affect positive change. Is this too good to be true? The Peace Boat experience may be perceived as simply an educational experience or as a "peace tourism" niche market product packaged for sale. Some may view the language and practice of Peace Boat as naïve and dismiss the "journeys of dialogue", the 'floating peace village", and stories of "mutual understanding and trust" as too idealistic for the business of tourism. However, this analysis has explained the language and practice of nonviolence *is* visionary, creative, optimistic, values explicit and impactful in the real world. We would argue that any legitimate claims about "peace through tourism" must take seriously the language and practice of nonviolence. We note this is also recently being theorised in the tourism literature. In their recent article "Tourism for the Emancipation of the Oppressed", Karla Boluk and Sandro Carnicelli (2019) draw on Freirean philosophy to forward their idea of critical tourism studies. As mentioned earlier in the literature review, a pedagogical framework for the cultivation of the "Critical Tourism Citizen" is being developed by tourism scholars (see Boluk et al., 2019), which presents five strategies for fostering critical tourism citizenship—topics, dialogue, positionality, reflection, praxis—all of which are clearly evident in this research and case study of Peace Boat. Such work provides us with scope for further research ideas regarding critical and compassionate thinking towards others' circumstances and a sense of shared citizenship that can be linked also with reference to recent work on justice through tourism (e.g. Jamal &

Higham, 2020). Although there was limited scope to elaborate these ideas in this article this plants seeds for an intriguing set of ideas to further link peace, sustainability and tourism.

The case of Peace Boat demonstrates how tourism can lay a foundation for sustainable peace. In presenting a challenge to claims that industry imperatives must prevail and that render tourism to be a passive beneficiary of peace, Peace Boat exemplifies the capacities of tourism as a social force. In moving beyond the recent debates in peace through tourism (see Salazar, 2006) this analysis of Peace Boat provides evidence that tourism can generate peace through a radical pedagogy that engages "positive peace" such as engaging social and ecological justice. This tourism model consciously harnesses pedagogical principles of peace education: as participants create a learning community, negotiate "positive" and "negative" peace and inspire each other to become change agents. Moreover, the reflexive business practice sees value in "buying into" human rights, justice, sustainability and a common humanity. In presenting this analysis of the work of Peace Boat through its global voyages, we show how it fosters global solidarity, providing evidence that tourism can indeed contribute to sowing the seeds for global peace.

Notes

1. Special consultative status is granted to NGOs which have a special competence in, and are concerned specifically with, certain fields of activity covered by the ECOSOC. With this status, Peace Boat can participate in, present written contributions to and make statements with all ECOSOC subsidiary bodies, attend international conferences convened by the United Nations and organize side events. It provides access to not only ECOSOC, but also to its many subsidiary bodies, to the various human rights mechanisms of the UN, ad-hoc processes on small arms, as well as special events organised by the President of the General Assembly.
2. The TPNW is the first ever treaty that completely and comprehensively bans the development, possession and use of nuclear weapons – making them finally illegal under international law. The treaty entered into force on January 22, 2021; 90 days after the 50th state ratified the treaty.

Disclosure statement

No potential conflict of interest was reported by the authors.

References

Ateljevic, I., Harris, C., Wilson, E., & Collins, F. L. (2005). Getting 'entangled': Reflexivity and the 'critical turn' in tourism studies. *Tourism Recreation Research, 30*(2), 9–21. https://doi.org/10.1080/02508281.2005.11081469

Aziz, H., El-Said, O., & Bontenbal, M. (2020). An analytical approach to cruise tourism as an option for development: A case study of the Sultanate of Oman. *Tourism in Marine Environments, 15*(2), 95–109. https://doi.org/10.3727/154427320X15828792518754

Batac, M. (2020). Online Event "The World is Our Classroom: Peace Boat's Global University." August 21, 2020. https://www.facebook.com/watch/?v=587350338624999.

Belhassen, Y., & Caton, K. (2011). On the need for critical pedagogy in tourism education. *Tourism Management, 32*(6), 1389–1396. https://doi.org/10.1016/j.tourman.2011.01.014

Blanchard, L., & Higgins-Desbiolles, F. (Eds.) (2013). *Peace through tourism: Promoting human security through international citizenship*. Routledge.

Blanchard, L., & Nix, M. (2019). Creating spaces for radical pedagogy in higher education. *Human Rights Education Review, 2*(2), 64–83. https://doi.org/10.7577/hrer.3363

Boluk, K. A., & Carnicelli, S. (2019). Tourism for the emancipation of the oppressed: Towards a critical tourism education drawing on Freirean philosophy. *Annals of Tourism Research, 76*, 168–179. https://doi.org/10.1016/j.annals.2019.04.002

Boluk, K., & Cavaliere, C. T., & Duffy, L. N. (2019). A pedagogical framework for the development of the critical tourism citizen. *Journal of Sustainable Tourism, 27*(7), 865–881. https://doi.org/10.1080/09669582.2019.1615928

Boluk, K., Cavaliere, C. T., & Higgins-Desbiolles, F. (2017). Critical thinking to realize sustainability in tourism systems: Reflecting on the 2030 sustainable development goals. *Journal of Sustainable Tourism, 25*(9), 1201–1204. https://doi.org/10.1080/09669582.2017.1333263

Camilleri, J. and Guess, D. (Eds.). (2020). *Towards a just and ecologically sustainable peace: navigating the great transition*. Palgrave Macmillan.
Cole, S. (2004). Shared benefits: Longitudinal research in eastern Indonesia. In L. Goodson & J Phillimore (Eds.), *Qualitative research in tourism: Ontologies, epistemologies and methodologies* (pp. 292–310). Routledge.
D'Amore, L. J. (1988). Tourism—a vital force for peace. *Tourism Management*, *9*(2), 151–154.
Davis, L. (2004). Risky stories: speaking and writing in colonial spaces. *Native Studies Review*, *15*(1), 1–20.
Decrop, A. (1999). Triangulation in qualitative tourism research. *Tourism Management*, *20*(1), 157–161. https://doi.org/10.1016/S0261-5177(98)00102-2
Denzin, N., & Lincoln, Y. S. (2008). *The landscape of qualitative research*. Sage.
EQUATIONS. (2013). Of peoples and places: Tourism and zones of conflict in India. In L. Blanchard & F. Higgins-Desbiolles (Eds.), *Peace through tourism: Promoting human security through international citizenship* (pp. 116–134). Routledge.
Fennell, D. A. (2006). *Tourism ethics*. Cromwell Press.
Fihn, B. (2020). *We all owe our lives to Hiroshima survivors |opinion, Newsweek*. Retrieved February 18, 2021, from https://www.newsweek.com/hiroshima-we-owe-hibakusha-world-1523387.
Flyvbjerg, B. (2006). Five misunderstandings about case study research. *Qualitative Enquiry*, *12*(2):219–245. https://doi.org/10.1177/1077800405284363.
Freire, P. (1970). *Pedagogy of the oppressed*. Continuum.
Freire, P. (1998). *Pedagogy of freedom*. Rowman & Littlefield.
Galtung, J. (1996). *Peace by peaceful means: Peace and conflict, development and civilisation*. Sage.
Galtung, J. (2010). Peace studies and conflict resolution: The need for transdisciplinarity. *Transcultural Psychiatry*, *47*(1), 20–32. https://doi.org/10.1177/1363461510362041
Gare, A. (2010). Toward an ecological civilization: The science, ethics and politics of ecopoiesis. *Process Studies*, *39*(1), 5–38. https://doi.org/10.5840/process20103912
Genc, R. (2016). Sustainability in cruise ship management. *International Journal of Social Science Studies*, *4*(6), 76–83.
Hatakeyama, S., Kawasaki, A. (2020). *Navigating disarmament education: The Peace Boat model*. New York: United Nations. https://www.un.org/disarmament/civil-society-and-disarmament-2020/.
Higgins-Desbiolles, F. (2006). More than an industry: Tourism as a social force. *Tourism Management*, *27*(6), 1192–1208. https://doi.org/10.1016/j.tourman.2005.05.020
Higgins-Desbiolles, F. (2020). Socialising tourism for social and ecological justice after COVID-19. *Tourism Geographies*, *22*(3), 610–623. https://doi.org/10.1080/14616688.2020.1757748
Higgins-Desbiolles, F., & Blanchard, L. (2010). Challenging peace through tourism: Placing tourism in the context of human rights, justice and peace. In O. Moufakkir & I. Kelly (Eds.), *Tourism, progress and peace* (pp. 35–47). CABI.
Higgins-Desbiolles, F., Carnicelli, S., Krolikowski, C., Wijesinghe, G., & Boluk, K. (2019). Degrowing tourism: Rethinking tourism. *Journal of Sustainable Tourism*, *27*(12), 1926–1944. https://doi.org/10.1080/09669582.2019.1601732
Higgins-Desbiolles, F., Whyte, K. P., & Tedmanson, D. (2013). Tourism and environmental justice. In K. A. Schwab & D. L. Dustin (Eds.), *Just leisure: Things that we believe in*. (pp. 91–101). Sagamore-Venture Publishing.
Hiroshima Peace Media Center. (2017). *Elayne Whyte Gómez visits Hiroshima for U.N. conference on disarmament issues*. November 30, 2017. http://www.hiroshimapeacemedia.jp/?p=79627.
hooks, b. (2010). *Teaching critical thinking: Practical wisdom*. Routledge.
Ife, J. (2009). *Human rights from below: Achieving rights through community development*. Cambridge.
Jamal, T., & Higham, J. (2020). Justice and ethics: Towards a new platform for tourism and sustainability. *Journal of Sustainable Tourism*, *29*(2–3), 143–157. https://doi.org/10.1080/09669582.2020.1835933
Jenkins, T. (2008). The international institute on peace education: Twenty-six years modelling critical, participatory peace pedagogy. *Journal of Peace Education and Social Justice*, *2*(2), 166–174.
Johnston, B. (2020). *How the cruise industry is sailing towards a sustainable future*. https://www.afr.com/life-and-luxury/travel/how-the-cruise-industry-is-sailing-towards-a-sustainable-future-20200128-p53vio, accessed 14 February 2021.
Kincaid, J. (1988). *A small place*. Farrar, Straus & Giroux.
Lean, G. (2009). Transformative travel: inspiring sustainability. In R. Bushell & P. J. Sheldon (Eds.), *Wellness and tourism: Mind, body, spirit, place*. Cognizant.
Lew, A., Cheer, J. M., Haywood, M., Brouder, P., & Salazar, N. B. (2020). Visions of travel and tourism after the global COVID-19 transformation of 2020. *Tourism Geographies*, *22*(3), 455–466. https://doi.org/10.1080/14616688.2020.1770326
Litvin, S. (1988). Tourism: The world's peace industry? *Journal of Travel Research*, *37*(1), 63–66. https://doi.org/10.1177/004728759803700108
Moufakkir, O. & Kelly, I. (Eds.). (2010). *Tourism, progress and peace*. CABI.
Mowforth, M., & Munt, I. (2003). *Tourism and sustainability: Development and new tourism in the third world* (2nd ed.). Routledge.

Nahory, C. (2017). Education: A weapon for peace—The case of the nuclear ban treaty. In United Nations Office for Disarmament Affairs (UNODA) (Ed.), Celebrating 15 years of disarmament and non-proliferation education. New York: United Nations. https://s3.amazonaws.com/unoda-web/wp-content/uploads/2018/01/op31.pdf

Nelson, K., Ragusa, A., Black, R. (2017). *The storied self in tourism spaces: Qualitative methodology as a framework for understanding self-identity in Niseko, Japan.* https://researchoutput.csu.edu.au/ws/portalfiles/portal/22018082/CAUTHE_2018_Unpacking_host_narratives_Resubmission_11Dec2017ATR_1_.pdf

Osler, A. (2015). Human rights education, postcolonial scholarship and action for social justice. *Theory & Research in Social Education*, *43*(2), 244–274. https://doi.org/10.1080/00933104.2015.1034393

Peace Boat Environment Team. (1992). *Is the earth still blue?: Peace Boat's earth's health check-up through a global voyage*. Daisan Shokan.

Peace Boat. (2014). *Global University 2014: Programme Report.* https://peaceboat.org/wordpress/wp-content/uploads/2015/04/UNIV_84_SpecialProgramme-_Report_Eng_FINAL.pdf

Peace Boat. (2018). *Global University 2018: Programme Report.* https://peaceboat.org/english/assets/media/images/GU-2018.pdf.

Peace Boat. (2020a). *Looking back: How did the International Student (IS) Programme for students from conflict areas start?* May 22, 2020. https://peaceboat.org/english/news/looking-back-IS-programme.

Peace Boat. (2020b). *Looking back: The beginnings of cooperation with the Tehran Peace Museum*. June 12, 2020. https://peaceboat.org/english/news/beginnings-of-cooperation-with-tehran-peace-museum.

Peace Boat. (2020c). *Peace Boat's 20th Anniversary and sailing with the Topaz. July 2, 2020.* https://peaceboat.org/english/news/looking-back-20th-anniversary

Peace Boat. (2020d). *Looking back: Dealing with the past in Germany and Japan - Peace Boat in cooperation with the University of Tübingen, Germany. June 5, 2020.* https://peaceboat.org/english/news/looking-back-tuebingen.

Peace Boat. (2020e). *Looking back: Peace & Green Boat – A peaceful and sustainable East Asian community. July 15, 2020.* https://peaceboat.org/english/news/looking-back-peace-green-boat.

Peace Boat. (2020f). *Online Live Streaming Peace Boat's 12-hr TV Celebrating Confirmation of TPNW's Entry into Force. October 25, 2020*, https://youtu.be/PAZ_gZiBMtg.

Peace Boat. (2021). *Design for a Paradigm Shift. Peace Boat Ecoship.* http://ecoship-pb.com/technology/.

Peters, J. E. (2020). Nakreu: evaluations of cruise passengers' attitudes towards sustainability in cruise contexts. *Tourism in Marine Environments*, *15*(2), 111–121. https://doi.org/10.3727/154427320X15833193080731

Reardon, B. A. (1988). *Comprehensive peace education: Educating for global responsibility*. Teachers College Press.

Reardon, B. A. (1990). Feminist concepts of peace and security. In P. Smoker (Ed.), *A reader in peace studies: Educating for global responsibility*. Teachers College Press.

Reardon, B. A. (1993). *Women and peace: Feminist visions for global security*. State University of New York Press.

Reardon, B. A. (1999). *Peace education: A review and projection*. Malmo University.

Reardon, B. A. (2009). *Human rights learning: Pedagogies and politics of peace* [Paper presentation]. This Paper is Based on a Lecture Delivered for the UNESCO Chair for Peace Education Master Conference at the University of Puerto Rico,. University of Puerto Rico, UNESCO, April 15th.

Reardon, B. A. (2015). Toward a paradigm of peace. In B. A. Reardon & D. T. Snauwaert (Eds.), *Betty A. Reardon: a pioneer in education for peace and human rights* (pp. 109–120). Cham: Springer International.

Rees, S. (2003). *Passion for peace: Exercising power creatively*. UNSW Press.

Rees, S. (2015). Consensus and conflict over peace. In L. Blanchard & H. Middleton (Eds.), *Conversations in peace* (pp. 7–13). University of Sydney Press.

Ritchie, B. W. (2005). Longitudinal research methods. In B. W Ritchie, P. Burns & C. Palmer (Eds.), *Tourism research methods: Integrating theory with practice* (pp. 131–149). CABI.

Ruiz-Guerra, I., Molina-Moreno, V., Cortes-Garcia, F., & Nunez-Cacho, P. (2019). Prediction of the impact on air quality of the cities receiving cruise tourism: The case of the Port of Barcelona. *Heliyon*, *5*(3), e01280. https://doi.org/10.1016/j.heliyon.2019.e01280

Salazar, N. B. (2006, July–December). Building a "Culture of Peace" through tourism: Reflexive and analytical notes and queries. *Universitas Humanística*, *62*, 319–333.

Scheyvens, R. (2002). *Tourism for development: Empowering communities*. Prentice-Hall.

Smith, M., & Duffy, R. (2003). *The ethics of tourism development*. Routledge.

Stancliffe, A. (2013). Tourism concern: putting human rights principles into practice. In L. Blanchard & F. Higgins-Desbiolles (Eds.) *Peace through tourism: Promoting human security through international citizenship* (pp. 75–88). Routledge.

Subramani, S. (2019). Practising reflexivity: Ethics, methodology and theory construction. *Methodological Innovations*, *12*(2), 205979911986327. https://doi.org/10.1177/2059799119863276

Thurlow, S. (2020). Prologue. In S. Hatakeyama and A. Kawasaki (Eds.), *Navigating disarmament education: The Peace Boat model* (pp. ix–xi). United Nations.

UNWTO (2020). *Supporting jobs and economies through travel and tourism*. Retrieved October 26, 2020, from https://webunwto.s3.eu-west-1.amazonaws.com/s3fs-public/2020-04/COVID19_Recommendations_English_1.pdf

UNWTO (1999). *The global code of ethics for tourism*. Retrieved October 26, 2020, from https://www.unwto.org/global-code-of-ethics-for-tourism

Van den Dugen, P. (2014). Peace tourism. In C. Wohlmuther & W. Wintersteiner (Eds). *International handbook on tourism and peace* (pp.62–77). Drava.

Webel, C., & Galtung, J. (Eds.), (2007). *Handbook of peace and conflict studies*. Routledge.

Wilson, E., Mura, P., Sharif, S. P., & Wijesinghe, S. N. R. (2020). Beyond the third moment? Mapping the state of qualitative tourism research. *Current Issues in Tourism*, *23*(7), 795–810. https://doi.org/10.1080/13683500.2019.1568971

Wohlmuther, C., & Wintersteiner, W. (Eds) (2014). *International handbook on tourism and peace*. Drava.

Yin, R. (2014). *Case study research: Design and methods*. Thousand Oaks, CA: Sage Publications Inc.

Yoshioka, T. (2008). Experience, action, and the floating peace village. In J. Chan (Ed.), *Another Japan is possible: New social movements and global citizenship education (Part VIII 47)* (pp. 317–322). Stanford University Press.

Yoshioka, T. (2013). The floating peace village: An experiment in nonviolence. In L. Blanchard and F. Higgins-Desbiolles (Eds.), *Peace through tourism: Promoting human security through international citizenship* (pp. 135–144). Routledge.

Zain Deen, R. (2020). *Online Event The world is our classroom: Peace Boat's Global University, August 21, 2020*. https://www.facebook.com/watch/?v=587350338624999

A diverse economies approach for promoting peace and justice in volunteer tourism

Phoebe Everingham, Tamara N. Young, Stephen L. Wearing and Kevin Lyons

ABSTRACT

The extent to which volunteer tourism can contribute to peace promotion through intercultural exchange and understanding is a central theme in this paper. We present the tensions of situating volunteer tourism within a broader tourism industry that promotes hyper-consumption and homogenous development discourses. As volunteer tourism becomes increasingly commodified and co-opted by a profit-driven industry that perpetuates neoliberal capitalism and neocolonial development, the benefits of volunteer tourism to local communities are questioned. The differences between commodified and decommodified volunteer tourism are highlighted, with the heterogeneity of volunteer tourism through a diverse economies lens considered. Through a case study of a not-for-profit volunteer organisation in Ecuador, we examine volunteer tourism as a force for peace and social justice, outlining how volunteer programs can facilitate encounters that foster mutual understanding and respect for different cultures and knowledge systems. When volunteer tourism operates outside of hyper-consumption and neocolonial development aid models, peace can be (re)centred through intercultural learning and exchange.

Introduction

Volunteer tourism has the potential to be peacebuilding through intercultural exchanges benefiting both the travelled and travelling cultures (Wearing, Stevenson, and Young 2010). These peer-to-peer exchanges place volunteer tourism as alternative tourism (Higgins-Desbiolles 2009; Wearing 2001; Wearing and McGehee 2013), which is more ethical, responsible, and capable of promoting peace and justice for destination host communities (Higgins-Desbiolles 2008; Higgins-Desbiolles 2009). Volunteer tourism researchers emphasise these intercultural exchanges as peace-promoting when mutual respect, tolerance and understanding between the volunteers and the communities they visit is central (Lyons et al., 2010). In these contexts, volunteer tourism creates possibilities for encounters that can foster mutual understanding and respect across cultural difference (Everingham 2015; Everingham and Motta 2020; Wearing, Young, and Everingham 2017; Wearing and Grabowski 2011; Wearing, Stevenson, and Young 2010). However, for peace to be foregrounded in volunteer tourism, research and practice need to move beyond privileging the volunteers' experience and focus on mutually beneficial exchanges *between* hosts and guests through two-way learning exchanges (Everingham 2015, 2016; Everingham and Motta 2020). This facilitation of mutually beneficial cultural exchange and understanding needs to be deliberately fostered and facilitated by organisations, and actively work towards subverting neo-colonial stereotypes of 'others' (Everingham 2015, 2016; Everingham and Motta 2020; Raymond and Hall 2008).

While volunteer tourism has the potential to promote peace through providing opportunities to learn about others, these ideals have been challenged through the increasing commodification of the industry – where the consumption desires of volunteers are prioritised at the expense of host communities. Moreover, volunteer tourism has been criticised for perpetuating neoliberal and neocolonial agendas (Coren and Gray 2012; Gray and Campbell 2007; Higgins-Desbiolles 2008; Lyons et al. 2012; Mostafanezhad 2014; Vrasti 2013). While volunteer tourism was initially promoted as a form of alternative tourism, it has primarily come to mirror the unjust and unsustainable impacts of mass tourism (Higgins-Desbiolles 2008). This co-option of volunteer tourism into an "inequitable, unjust, and unsustainable" mainstream tourism industry (Higgins-Desbiolles 2008, p. 347) diverts it from fulfilling its capacities as an alternative form of tourism. From this perspective, volunteer tourism can reinforce economic inequity and cultural inequalities between tourists and host communities (Jafari 1989; Pratt and Liu 2016), thus operating as an incubator for structural violence when dominated by tourists from the Global North travelling to the Global South (McGregor 2003).

To address these inequalities perpetuated by tourism and commodified volunteer tourism, we see the need for new ways to examine how justice might be promoted in and through tourism. The commodifying and neoliberalising effects on the volunteer tourism industry, and problematic links to hyper-consumption and development (all of which perpetuate structural violence), require attention. To this end, in this article, we 'read for difference' (Gibson-Graham 2006; 2008) to examine *existing* ways in which volunteer tourism can work towards peace and justice. We present the volunteer tourism industry as diverse: including commercialised programs, not-for-profits and non-government organisations, some of which are small-scale and community-oriented. Thus to 'read for difference' in volunteer tourism, we must move away from homogeneous singular conceptions of 'volunteer tourism', 'development' and 'capitalism.' We argue volunteer tourism must (re)align its goals towards social justice agendas. In doing so, volunteer tourism can go some way to promote positive peace: a form of peace that evolves through addressing structural violence (Galtung 1969).

Taking a postcapitalist diverse economies lens provides a language and framework to contest hyper-consumption forms of profit-driven volunteer tourism and neocolonial, neoliberal forms of development and demonstrate how volunteer tourism can act as a form of collaborative consumption that is beneficial for *both* tourists and host communities. To illustrate how decommodified volunteer tourism can lead to collaborative consumption through facilitating intercultural communication – a key pillar in moving towards more socially just and peaceful worlds - we draw on a case study of a not-for-profit volunteer tourism organisation in Ecuador, South America. The diverse economies postcapitalist perspective that we engage in this article provides room for academic debate on volunteer tourism to move away from a simplistic and reductionist analysis that often accompanies critical approaches. Our approach in this paper centres around peace and justice in volunteer tourism, foregrounding a politics of possibility, one that can "uncover the multiplicity of dynamic forms of actually existing justice present in the here and now" (Williams 2017, p. 2219). A diverse economies approach opens up the imagination to how postcapitalist economies can be brought into being (Gibson-Graham 2008). Using the lens of collaborative consumption (Dredge and Gyimóthy 2015), we provide insights into the importance of intercultural exchange and sharing in volunteer tourism – benefitting both the host communities and volunteers as well as demonstrating the ways in which more peaceful worlds can be brought into being.

Hyper-consumption, neocolonial and neoliberal development as structural violence

Despite impressive foreign exchange earnings, transnational tourism corporations incur high economic leakage rates and government revenue losses through investment incentives that result in limited contributions to local communities in rural communities and developing countries (Scheyvens, Banks, and Hughes 2016). Mass tourism underpinned by hyper-consumption models leads to host communities too often receiving minimal financial benefit through the creation of poorly paid menial jobs with the bulk of tourism-related revenue leaking out of the destination community back to capitalist investors (Fletcher et al. 2019; Higgins-Desbiolles et al. 2019). In this way, tourism in the free-market economy represents the commercialisation of the human desire to travel and exploit natural and cultural resources for profit accumulation, thus, exemplifying the hyper-consumption model of tourism. Transnational tourism organisations pursue the ideals of a laissez-faire system of government and regulation, and operate efficiently in this market system (Wearing and Wearing 1999). Communities seeking to diversify their economy through tourism development can experience it as another extractive industry, with many negative consequences on societies, cultures, and the natural environment (Smith and Krannich 1998; Higgins-Desbiolles et al. 2019).

In contrast, volunteer tourism emphasises holidays "that might involve aiding or alleviating the material poverty of some groups in society … while furthering knowledge and awareness of these issues" (Wearing 2001, p. 1). This definition framed volunteer tourism as social development, providing formal modes of sharing and consumption that would ideally enrich the host community and the tourist. However, while volunteer tourism was initially positioned as a positive alternative to the hyper-consumption economic model of mass tourism, it has been increasingly co-opted by rampant hyper-consumption (Lyons et al. 2012; Wearing, Young, and Everingham 2017). Hyper-consumption is the process by which "the consumption of non-essential goods is supercharged so that consumers are encouraged to select items they did not know they needed or wanted" (Ritzer and Miles 2019, p. 8), and is recognised as an unsustainable practice that drives capitalist economies (Martin 2016). The hyper-consumption of objects and services (Ritzer 2005, 2011; Ritzer and Miles 2019) is exacerbated in mass tourism. Promoting volunteer tourism within hyper-consumption models underpins critiques of volunteer tourism as sharing the values and practices of neoliberalism (Lyons et al. 2012). If not carefully regulated, volunteer tourism perpetuates a consumer mindset, further perpetuating structural inequality.

However, framing volunteer tourism as development is unrealistic given the short-term nature of volunteering and is also problematic due to multiple underlying issues of neocolonialism in the development aid sector more broadly (Palacios 2010; Everingham 2015, 2016, 2018). In this context, volunteer tourism is too often marketed in ways that skim over the complexities of structural issues related to development by producing and perpetuating the idea of 'development consumers' (Kipp, Hawkins, and Gray 2021). Framing volunteer tourism in these ways perpetuates structurally violent geographies of colonialism. The marketing of volunteer tourism as 'development' also perpetuates neoliberal consumerism, depoliticising development through placing the responsibilities of development onto 'development consumers', such as volunteer tourists (Kipp, Hawkins, and Gray 2021). Critiques of volunteer tourism have focused on neoliberalism perpetuated through neocolonial development aid models, co-opting the emotional sentiments of volunteers through, for example, the notion of cosmopolitan citizenship and individualism (Crossley 2012; Lyons et al. 2012; Mostafanezhad 2013; 2014). Neoliberal ideologies promote the idea that people are consumers rather than co-producers (Comaroff and Comaroff 2001). This shift in the mode of contemporary citizenship, from production to consumption, has become the axis upon which identity and experience are constructed in free-market societies and does little to address underlying structural inequalities (Butcher and Smith 2015; Lyons et al. 2012). As a form of 'humanitarian consumption', volunteering becomes more about the performativity of 'self' rather than altruism (Sin 2009).

Locating peace: reading for difference in volunteer tourism

Condemnations of commodified volunteer tourism and the relationship between volunteer tourism and neocolonial forms of development are undoubtedly valid and necessary, particularly concerning how structural violence is perpetuated. However Wearing, Young, and Everingham (2017) note the overarching tendency for critical accounts of volunteer tourism to homogenise volunteer tourism and development. In so doing, there is a failure to recognise decommodified forms of volunteer tourism and the *heterogeneity* of volunteer tourism organisations (Wearing, Young, and Everingham 2017). When volunteer tourism and volunteer tourists present as singular and monolithic, little attention is given to their multiple and nuanced motivations. The actual lived experiences of both volunteers and host communities, and other forms of social development such as creative development and intercultural learning and exchange (Everingham 2016; 2017; 2018; Wearing, Stevenson, and Young 2010), provide an opportunity for new insights *with* the community rather than the development *of* the community.

Locating potentialities of peace and justice in volunteer tourism, we take a diverse economies perspective of volunteer tourism following feminist geographers Gibson-Graham (2006; 2008). This approach is a postcapitalist politics, an 'anti-essentialist approach' that theorises the "contingency of social outcomes rather than the unfolding of structural logics" (Gibson-Graham 2008, p. 615). Our approach to unpacking the role of volunteer tourism in promoting peace necessitates a critique of neoliberal capitalism that promotes hyper-consumption by unpacking the diverse economies that *also* exist within these models. Gibson-Graham (2008, p. 626) argue that privileging the dominance of capitalocentric analyses to the social world endows capitalism with an all-encompassing power, thereby "generalising its dynamics and organisations, and enlarging the spaces of its agency". Taking a postcapitalist approach requires not only a shift in hegemonic framings of the economy as 'capitalocentric' (Gibson-Graham 2006, p. 56) but, also, foregrounding invisibilised transactions and working towards more ethical forms of economy. Capitalocentrism refers to how 'the economy' is homogenised, where the complexities of economic practice (for example, the non-market transactions that exist within economic transactions) are, in turn, invisibilised (Wright 2010). Capitalocentrism can, thus, strengthen the dominance of capitalism. By understanding the economy as a site of struggle, we acknowledge it as a site of contestation, one that is "always open to redefinition" (Wright 2010, p. 300).

Gibson-Graham (2006) construct a language of economic diversity to reclaim and reimagine the economy: to 'read for difference' to dislodge the discursive dominance of capitalism as the necessary and naturally dominant form of the economy. In doing so, the economy is brought back into the discursive political sphere of contestation. The terms diverse economy and postcapitalism give us a language for (re)thinking the economic landscape towards a myriad of contingent forms and interactions. For example, Gyimóthy and Meged (2018) utilise a diverse economies perspective in their description of grassroots initiatives of Danish coastal communities, where tourism initiatives are leveraged on the Camøno walking trail. They describe how non-market resources are turned into commodities through processes of collaborative consumption. Their analysis reveals the importance of refocusing attention away from capitalocentric approaches (where the agency of various actors are confined to hegemonic market relations) towards negotiated boundaries of various actors within specific and diverse tourism organisations (Gyimóthy and Meged 2018). Highlighting such diverse practices can foreground ethical and inclusive ways of doing tourism.

Critiques of volunteer tourism that privilege analysis of dominant market transactions tend to invisibilise the non-market relationships that *also* occur within these forms of tourism (Griffiths 2014). Such capitalocentric analyses can reify the power of neoliberal capitalism. This is especially the case when research and practice reinforce narrow and homogenous models of a commodified 'volunteer tourism industry' (Everingham 2016, 2018). For these reasons, Wearing, Young, and Everingham (2017) and Grabowski et al. (forthcoming) argue the need for more nuanced analyses of volunteer tourism to capture the diversity of volunteers, local communities, and the tourism industry itself. In line with a performative approach to this research agenda, highlighting the possibilities for rethinking the economy away from macro-neoliberal capitalist doctrine and hyper-consumption models, the following sections present insights into volunteer tourism as an agent of change. *Existing* instances of diversity within volunteer tourism are highlighted as exemplars for a rethinking of volunteer tourism based on collaboration and sharing models of consumption.

Revealing diverse economies through collaborative consumption

Collaborative consumption includes various exchange forms with implications for tourism to move away from "traditional tourism industrial systems" (Dredge and Gyimóthy 2015, p. 287). In this context, collaborative consumption in tourism can be associated with the 'moral economy' and prioritising alternative forms of tourism that mutually benefit host communities (German Molz 2013). A collaborative consumption approach centres ethical exchanges and moves away from privileging hyper-consumption models of economic exchange. Dredge and Gyimóthy (2015) argue for a more comprehensive application of how tourism researchers discursively understand collaborative consumption. Centring collaborative consumption within a diverse economies perspective reveals hybrid models of economic exchange where analysis can foreground ethical consumption and the intangible aspects of the economic transaction – in the case of volunteer tourism, the intercultural exchanges.

Research of collaborative consumption as an example of diverse economies exists in case studies of diverse forms of tourism. Lai et al. (2020) explore collaborative consumption in their study of social exchange in World Wide Opportunities on Organic Farms (WWOOF). In positioning WWOOF as a 'sharing economy', Lai et al. (2020) raise awareness of particular tourism initiatives, specifically those that involve relational exchanges and mutually beneficial social exchanges, as working to decentre commodification and profit. These tourism transactions are non-monetary and non-binding, exemplifying a community-based economy where mutual intercultural exchange is at the centre of the transaction (Lai et al. 2020). This collaborative consumption perspective positions the community's needs as far more important than the monetary exchange itself.

Intercultural exchange through cultural knowledge sharing is also central to cultural tourism research. Brueckner et al. (2016) and McGinnis, Harvey, and Young (2020) demonstrate how cultural knowledge sharing of land and environment in tourism translates into benefits for local communities. The Dudungurr Nursery, a not-for-profit environmental organisation in Australia, blends semi-formal employment with local tradition as a two-way learning space enriched by Indigenous and non-Indigenous cultures (Brueckner et al. 2016). Yolgnu Aboriginal cultural knowledge of plants is a key resource in the business enterprise, ensuring that local knowledge is retained and shared and enriched through engagement with the formal market economy. McGinnis, Harvey, and Young (2020) demonstrate the significance of reciprocity between the Wagiman Aboriginal community and non-Indigenous residents in Northern Australia to deliver intercultural knowledge of the area for tourists. Through the collaborative process of sharing biocultural knowledge, the Wagiman language is preserved for future Wagiman generations and shared with tourists to the community (McGinnis, Harvey, and Young 2020). Wright, Suchet-Pearson, and Lloyd (2007, p. 155) define intercultural knowledge sharing between Aboriginal and non-Aboriginal groups as 'interwoven learning exchanges' characterised by the potential to transform and reconfigure power relationships in tourism.

These intercultural learning and collaboration instances break down barriers between cultures and need to be deliberately highlighted in research. A focus on facilitating two-way collaborative learning processes is vital for working towards peace and justice. However, organisations need to purposely foster mutually beneficial learning exchanges (Everingham 2015, 2016; Raymond and Hall 2008). Two-way collaboration in international volunteering can lead to mutual learning experiences for both volunteers and hosts, essential for building international solidarity (Devereux 2008). Cheung Judge (2017, p. 164) argues that the potential for volunteer tourism to foster global citizenship must be considered in light of the "particular configurations of power in diverse volunteer tourism initiatives". Indeed, this diversity and the possibilities for exchange (based on collaboration and mutuality) are key for centring volunteer tourism to work towards peace and justice.

Wearing, Young, and Everingham (2017) argue for research that incorporates flexible and in-situ models to acknowledge exchanges between the actors in volunteer tourism spaces. Their research highlights instances where beneficial mutual collaboration is *already happening* between volunteer tourists and local communities (Wearing, Young, and Everingham 2017). By highlighting examples of volunteer tourism where mutual spaces of learning and collaboration are provided, there is significant scope to pursue the decommodified agenda for volunteer tourism through promoting collaborative consumption. However, there remains a need for explicit avenues that advocate alternative ways to operate and practice volunteer tourism that centre on the original aims of alternative tourism: peace, equity and justice.

Recentring peace in volunteer tourism through a diverse economies lens

For volunteer tourism to promote peace and justice, sustainability must exist at its core. Higgins-Desbiolles (2008; 2009), in her call for a 'justice tourism agenda', seeks to address the underlying structural violence inherent in hyper-consumption models of tourism. Justice tourism, when positioned within grassroots community focused and community-centred frameworks (Higgins-Desbiolles et al. 2019, p. 1927), is key to "achieving [tourism] degrowth that is socially sustainable". Justice tourism has the potential to be impervious to co-option by mainstream tourism (Higgins-Desbiolles 2009, p. 333) when encompassing tourist activities that are: ethical and equitable, build solidarity, promote mutual understanding, and prioritise sharing and respect, *at the same time as* optimising the positive impacts of tourism for destination host communities (Higgins-Desbiolles 2009; Scheyvens 2002).

A diverse economies approach highlights the co-existence of capitalist exchanges and non-capitalist exchanges. Applying a diverse economies lens to volunteer tourism enables an analysis that critically considers the potentialities of encounters to work towards peace through, for example, promoting intercultural understanding. This approach also enables a nuanced analysis of diverse and heterogenous organisations within the volunteer tourism industry and diverse forms of development. The implications of this approach include the "mapping [of] both market and non-market transactions, labour and business forms simultaneously, the diverse economy mapping technique may identify opportunities for new social enterprises together with members of the local community" (Gyimóthy and Meged 2018, p. 502). The diverse economies approach is one way of performativity representing how we might reimagine how new economies can be 'brought into being' (Gibson-Graham 2008, p. 616). Table 1 is a visual framing of an alternative ethical diverse economies model.

While the diverse economy approach presented in this table contains binary understandings of the economy, these categories can also overlap and be deconstructed.

We now present a case study whereby 'reading for difference' illustrates how volunteer tourism projects can be peace-promoting through intercultural exchange between tourists and host communities. The diverse economies lens focuses on the social and cultural development that can promote creativity and mutual intercultural connections and is a valuable framework to reconceptualise volunteer tourism. The focus on diversity in the case study unpacks these interactions under the categories of diverse economies in transactions, labour and enterprise.

Case study: *Fundación Arte del Mundo* in Ecuador, South America

In this case study, we 'read for difference' to illustrate opportunities for volunteer organisations to be collaborative spaces, where mutual intercultural exchange is at the core of economic

Table 1. Gibson-Graham (2008, p. 4).

Transactions MARKET	Labor WAGE	Enterprise CAPITALIST
ALTERNATIVE MARKET	*ALTERNATIVE PAID*	*ALTERNATIVE CAPITALIST*
Sale of public goods	*Self-employed*	*State enterprise*
Ethical 'fair-trade' markets	*Cooperative*	*Green capitalist*
Local trading systems	*Indentured*	*Socially responsible firm*
Alternative currencies	*In kind*	*Non-profit*
Underground markets	*Work for welfare*	
Co-op exchange		
Barter		
Informal market		
NON-MARKET	*UNPAID*	*NON-CAPITALIST*
Household flows	*Housework*	*Communal*
Gift giving	*Family care*	*Independent*
Indigenous exchange	*Neighborhood work*	*Feudal*
State appropriations	*Volunteer*	*Slave*
Gleaning	*Self-provisioning labor*	
Hunting, fishing, gathering	*Slave labor*	
Theft, poaching		

transactions between volunteers and the volunteer tourism organisations. We draw on the case of *Fundación Arte del Mundo,* located in Baños de Agua Santa, Ecuador, South America, as an example of a volunteer program providing opportunities for local communities and visiting tourists to work together on projects designed to meet the needs of the local community.

Background to the case study

Fundación Arte del Mundo was established in 2008 as a small not-for-profit organisation registered with the Ecuadorian Ministry of Culture and Education. The mission of the *Fundación* involved encouraging "literacy, creativity and a love of the arts by providing a free after school program for children and theatrical entertainment for the greater Baños de Agua Santa community" (Centro Cultural Arte del Mundo, 2015, 'About Us'). To meet its goals, the *Fundación* was dependent on volunteers to cover the costs of running the literacy program. Initially, the emphasis and energy of the directors and volunteers was the creation of a library, *La Biblioteca Interactivo.* The library grew to house *Intercambio*, a language exchange program for adults, a community space for creative activities, and a small theatre that hosted entertainment for Baños residents. Volunteers at the *Fundación* participated in projects, volunteered and lived together on site. Thereby, international tourists volunteering at the *Fundación* developed friendships – between themselves and the local community encouraging intercultural exchange and understanding.

An autoethnographic approach

Our study of *Fundación Arte del Mundo* is based on a total of nine weeks of field work at the organisation. The research employed autoethnography, as well as participant observation and interviews with volunteers, the directors and local community members.[1] An autoethnographic approach involves researcher reflection on their own emotions in the exploration of society and culture (Pelias 2003). According to Ellingson and Ellis (2008), autoethnograhers aim to be inclusive (rather than exclusive), focus on commonalities rather than differences, and place importance on intersubjectivity in knowledge production. There is strong resistance to process/product dichotomies in autoethnographic methods. Instead, the process takes the form of revealing the complex roles of researchers and recognises "the messiness and mistakes that inevitably imbue the process of conducting such research" (Ellingson and Ellis 2008, p. 453). Everingham (2018, pp. 69-70) explains autoethnographic methodologies as providing "spaces for speaking about the ambiguities and contradictions, the messiness of lived experience". This statement reflects the embodied process that is "inextricably bound to the final products of our research" (Ellingson and Ellis 2008, p. 453). Autoethnography, thus, destabilises fixed interpretations, "opening up possibilities for multiple understandings" (Ellingson and Ellis 2008, p. 455). Everingham, for example, engages autoethnographic methodologies in her tourism research to be "aware and attentive to process, emotions, partiality, ambiguity and the possibilities for enriching multiple understandings of volunteer tourism" (Everingham, 2018, p. 69). Part of the autoethnographic process involves reflecting on, and learning from, our research experiences and constantly re-evaluating our research critically (England 1994).

Alongside autoethnographic methods, in-depth semi-structured interviews were conducted with volunteers and members of the local community involved in the volunteer projects. A total of 33 interviews were conducted around 1 hour long; 18 with volunteers, 3 with program directors, and 12 with local people studying the English language program run by the *Fundación*. A social constructionist method was employed in the interview context, with attention given to the relational aspects of the interviews whereby recounting experience is shaped by interactions between the interviewee and the researcher (Miller and Glassner 2011; Longhurst 2016; Valentine 2005). These relational considerations were particularly important as the researcher in the field was also a volunteer participant who had social relationships with the volunteers, the locals, and the directors (outside the volunteer and research context). Subjectivities were, therefore, taken seriously in making sense of these lived experiences. We now turn to the findings by examining the first *structure* of the *Fundación*, followed by the *activities* at the *Fundación* to highlight a decommodified agenda enabling collaborative consumption in volunteer tourism.

Findings: Arte del Mundo as 'alternative capitalist enterprise'

Drawing on Gibson-Graham (2008) diverse economies framework, presented above, we classify the *Fundación* as an alternative capitalist enterprise. A not-for-profit organisation is part of the Third Sector: 'the social economy' where social objectives are prioritised over business-related objectives. This Third Sector plays an important role in creating and maintaining social and cultural well-being (Gibson-Graham 2008, p. 5). To meet its goals, the *Fundación* was dependent on volunteers to cover the costs of running the programs. Volunteers paid around $5 (USD) a night to stay in rooms above the library. The volunteers paid no other administrative costs or fees. Thus, the wage-labour transaction included volunteer work, not only by the volunteer tourists but also the directors, two of whom operated a business in the town while volunteering ad-hoc at the *Fundación*. The other two directors, who lived on-site, can be classified as 'alternative paid', as they were self-employed. Their wages were derived from the small fee paid by the volunteers and fluctuated drastically depending on how many individuals were volunteering at any given time. When money was tight, the directors' living expenses were affected, and their roles in the organisation tended more towards 'volunteer'.

The interactions between the volunteers, directors and local community occurred through 'non-market transactions.' As Wright (2010, p. 315) notes, within diverse economies of development, "logics of sharing, kin, friendship, or redistribution are entangled. The precise distribution of these entanglements, however, is open to change". In other words, within any economic transaction, there is a variety of entanglements between non-market and market transactions. In the case of the *Fundación*, non-market transactions are centred at the core of the exchange. These transactions revolved around intangible processes of exchange, mediated by certain activities offered by the organisation that centre on social good for the community.

We now examine the different *activities* that occur at the *Fundación* and how these activities facilitated non-market transactions through cross-cultural exchanges. As non-market transactions, the focus is on mutual intercultural learning opportunities for both volunteers and local community members, with the *Fundación* facilitating alternative market exchanges between volunteers and host communities. These collaborative exchanges contain possibilities for disrupting neocolonial and neoliberal relations in volunteer tourism encounters. As noted above, the three key projects of *Fundación Arte del Mundo* were the library, the language exchange, and the theatre.

La Biblioteca Interactivo

The main project at the *Fundación* was *La Biblioteca Interactivo*, an interactive library where children came to read for pleasure and participate in games, creative activities, and the performing arts. *La Biblioteca* held around 1200 Spanish language children's books, all donated or purchased with donated money. The establishment of *La Biblioteca* highlights the gift-giving dimensions of diverse economic transactions. Gift-giving reinforces social ties between the donor and recipient, where the gift acts as a mediator of obligation and reciprocity (Mosedale 2012).

For the volunteers, the first part of their day involved reading Spanish language books with local children for both pleasure and learning. The second part of the volunteers' day saw the space devoted to creative activities and creative play, depending on what skills the individuals volunteering at any particular time had to offer. These activities included *papier mache* art, photography workshops, cooking classes, drama games, *capoeira* classes, and science experiment days.

Diverse communities use creative play globally to build community capacity and, significantly, for community organising concerning campaigns for peace and justice (Shepard 2005). Citing Johan Huizinga's account of the play element in cultural history, Shepard (2005, p. 47) explains "all play means something" and even the "mere act of imagining alternative visions of the world … offers important clues for social work organisation and practice". The important role of creativity and play was highlighted by one of the *Fundación* directors in her interview:

> I think the bottom line for me [is] offering something for children – a situation where children can be creative basically supports the development of self-esteem. There are too many situations where children again and again are stuck in only one answer – the school – the teachers – they are looking for one answer you know? In an art or creative situation there is not just one right answer, whatever you come up with – the creative process – and support and opening up to the creative process is just not available out there too much.

For the children participating in these programs, creative play allowed them freedom from caring for siblings while their caretakers worked. Relationships between peers also flourished, as well as intercultural exchange through the facilitation of play with volunteers from all over the world. In writing about geographies of friendship, Bunnell et al. (2011) urge researchers to take friendship seriously in fieldwork. Friendship is about maintaining relationships both proximate and distant and is an integral part of what makes us human (Bunnell et al. 2011). By providing a space for children in Baños to engage in creative play, the *Fundación*, allowed for cross-cultural friendships (even if they were fleeting) and language exchange through the medium of 'doing play' together. Play allowed these relationships to flourish in non-paternalistic ways, providing a sharp contrast to the neocolonial helping narrative so prominent in volunteer tourism.

Similarly, the volunteers and children reading books together subverted some of the paternalistic neocolonial stereotypes of 'who helps who' in volunteer tourism (Everingham 2015; Everingham and Motta 2020) because the children became Spanish teachers for the non-Spanish speaking volunteers, correcting their grammar and pronunciation. Our reading for difference here involves moving away from neocolonial helping discourses. When encounters in volunteer tourism are not framed in this way, mutual intercultural learning and exchange can occur, as evidenced at the *Fundación*. These kinds of interactions can lead to 'other-than-neoliberal worlds';

however, these moments and practices of mutual exchange can easily be missed and excluded from view within capitalocentric frameworks of understanding (Williams 2017).

Intercambio

Alternative market transactions also took place with *Intercambio*, the language exchange program hosting low-cost English classes for adults in the Baños community. The volunteers ran the English language classes three times a week. One of the Fundación directors explained the organisation of English class delivery for community members involved in the *Intercambio* in his interview:

> As Baños is located near the gateway to the Amazonian region some of the adults that take English classes are tour guides. They might do jungle treks which means they will be working in the jungle for two weeks and back in town for two weeks. The flexibility of the classes means these students are not excluded.

The actualisation of collaborative consumption in volunteer tourism is evident here. English classes were designed to allow flexibility, so local community members did not have to commit to classes with a beginning and end date, which better aligned with their paid work within market transactions. Such alternative market transactions were beneficial to locals within the community looking to develop their English language skills and, in turn, provide increased opportunities for locals working in the tourism sector. In addition to the example of tour guides mentioned above, the local residents interviewed in this study explained how their involvement in the *Intercambio* developed their English language skills and opportunities to enhance employment in the local tourism industry.

For example, Martha, a local woman working as a cleaner in one of the local hotels, hoped the English classes would enable her to branch out into other local tourism ventures. Alfonzo, a local taxi driver, discussed his involvement in the language classes as giving him the confidence to converse with tourists. Alfonzo discussed his interactions with tourists in his taxicab, saying that he enjoyed learning about the world outside Ecuador and giving tourists advice about what to see and do in Baños. Within the alternative market transactions through collaborative consumption, non-market transactions occurred through the intangible exchanges between locals working in the Baños tourism sector and the volunteers. As exemplified by Alfonzo, these intangible exchanges included volunteers who boosted the confidence of local tourism workers, while locals gave volunteers insights into life in Baños.

As well as the English language classes, free *Intercambio* nights also took place, where local residents would exchange language with the volunteers – mainly English and Spanish, and other languages of the volunteers, such as Italian and the local indigenous language Quechua. The *Intercambio* was mutually beneficial for both locals and volunteers, providing opportunities for learning language, achieved through the medium of conversation and games. Both the locals and the volunteers experienced unique opportunities to engage in conversations and learn about cultural differences and cultural traditions.

In light of framing peace and justice in volunteer tourism, this example presents the possibilities of intercultural encounters in building cross-cultural understanding. Key to the peace agenda in tourism is facilitating these kinds of meaningful encounters across cultural difference. Farmaki (2017, p. 538) recognises that "peace building is multifarious as it extends social milieus", in the context of *Intercambio*, possibilities for peacebuilding come from learning about 'other'. While it should not be assumed that intercultural contact through tourism can simply reduce cultural and psychological gaps across difference (Farmaki 2017), particular forms of tourism and particular types of intercultural connections can be fostered in ways that build intercultural understanding and awareness. It is these collaborative and mutual outcomes that are foundational for building more peaceful global cultures.

Promoting peace and reconciliation in tourism requires approaches that offer "meaningful tourist experiences that cultivate cultural and social understanding" (Farmaki 2017, p. 538). The *Intercambio* cultivated cross-cultural understanding through meaningful intercultural encounters. One of the Italian volunteers, Maria, explains *Intercambio* as:

> [A] really really an interesting thing… you have to speak in Spanish [and] they have to speak in English and you are both nervous about it – you have to push yourself at something that you are not really good at – so in that case – we were at the same level.
>
> For the very first time we weren't the ones teaching English – they were also teaching us Spanish – and that is extremely interesting – it was the first time it seems to be…
>
> yes, equal… They feel so proud to correct you – it's like finally I can teach something to you…

Intercultural encounters between volunteers and the local community in Baños were promoted by the directors, who encouraged all locals to participate in *Intercambio*. As well as language exchange, social and cultural exchange was facilitated through activities that encouraged friendships between volunteers, the local community, and throughout the community. Friendships were developed at regular *churrasco* parties, where volunteers and locals would come together to eat, drink, dance and sing. Locals taught volunteers about cooking Ecuadorian style and how to dance salsa. These instances of intercultural exchange continue to provide insights into volunteer tourism where neocolonial stereotypes of 'who helps who' are subverted because volunteer tourists were deeply engaged in cultural learning activities led by the local community.

The theatre

Another important part of the project for intercultural exchange, containing alternative market transactions, was the theatre. The directors always had a vision that *La Biblioteca* would house a 'black box' theatre because, just as there was no children's library in Baños, there was also no community theatre. Non-market transactions through gift-giving underpinned the development of the theatre. The directors raised $13,000 (USD) through fundraising, turning the backyard garage into a small theatre. As well as being a performing arts space, the theatre was also used to show films. Family-friendly films were presented at a low cost to the public every Friday night. Volunteers helped to set up the theatre and make popcorn and drinks to sell. All money raised from the theatre helped cover the organisational costs. The premises were also available as a public community space depending on circumstances and interest. Some of the activities that occurred within the community space included literacy workshops for teachers, and training for teachers in the art of dynamic storytelling. The community space was also used as a rehearsal space for a community choir.

While the theatre was still relatively new when this research was conducted, several children's performances had already been held, and local children used the space throughout the day to play and perform. Travelling artists performed shows and alternative transactions took place with locals paying a small fee to see the shows. One theatre show for children and families was a circus theatre performance by two young Argentinian actors. Another show was a political comedy performed by a Colombian actor. This particular performance, aimed at both children and adult audiences, explored issues around the Columbian dictatorship. Through storytelling, the actor kept these memories alive to warn the audience about the importance of not repeating history. Having a space for such creative development through performance and storytelling, particularly around important issues related to politics and war, presents a way to raise awareness of these issues, with an explicit objective to promote peace in the community.

Challenges to the sustainability of Fundación Arte del Mundo as a non-capitalist enterprise

A key priority of the *Fundación* was to find ways to involve the local community in the project. According to one of the directors interviewed, the relationship with the local community was the best it had ever been. The organisation was awarded a medal of recognition by the local Mayor in 2012. The fact that the organisation had been there for several years and was somewhat established meant that trust had developed, in the sense that the locals knew the organisers were not just passing through. The directors established a Board of Directors with six local community members who were consulted on how to keep the organisation sustainable and to ensure the goals of the *Fundación* aligned with local community needs.

However, despite these positive interactions with the community, it was difficult for the *Fundación* to remain financially sustainable, demonstrating the challenges for non-capitalist enterprises. The organisation has since closed down at the Baños location and moved to another small town called Pelileo. This decision was made based on the deterioration of the building and high rents, which were challenging to meet in months when there were few volunteers. The theatre was also poorly supported by the community. It remains to be seen if the organisation will continue in the new location. Being a not-for-profit that is dependent largely on the income from foreign tourists is not a sustainable model. The COVID-19 pandemic has further highlighted the vulnerability that comes from tourism-dependent economies. While the organisation did try to include the local community as much as possible, the project itself was not a grassroots community initiative. While the locals benefited from the project, they did not have ownership over it, and this also may have contributed to its closure in Baños.

Indeed, community-based development tends to be more successful when it is conceptualised and managed by the communities themselves. Adie, Amore, and Hall (2020) point out that process perspectives are meaningful when developing partnerships in community-based tourism development. The process perspective considers the range of stakeholder and interests, values and power relationships between the various actors before and during the project – where there is a clear common goal of 'success'. This involves a comprehensive and straightforward approach towards a common goal and productive partnerships involving "multi-stakeholder agency, collaborative planning, mobilization of resources, and knowledge sharing" (Adie, Amore, and Hall 2020, p. 2).

The ever-evolving nature of *Fundación Arte del Mundo* demonstrates the process perspective regarding the fluid nature of volunteer tourism organisations. They are "co-produced through networked connections to other spaces, bodies and practices" (Pallett and Chilvers 2015, p. 146). Directors changed hands, many volunteers came and went, all with different personalities and skills. Different locals were involved with the organisations at various points in time. All of these changes affected the subjective experiences as well as the research analysis. Pallett and Chilvers (2015, p. 150) discuss the "intersubjective spaces of affect within organisations" and the important role that emotions play in organisational processes. They further explain "identities, institutions, discourses and representations" are mutually constructed and, by attending to these processes, dualisms between "grand structures and individual agency, dynamism and stability, single events and long-term trends, and human and non-human" can be problematised (Pallett and Chilvers 2015, p. 150). Conceptualising volunteer tourism organisations as discursive spaces is key to moving away from binary and static analyses of volunteer tourism. Volunteer tourism occurs in non-fixed and fluid spaces, and, therefore, any analysis is partial and contextual. Therefore, researchers need to be careful not to make grand universal claims about 'volunteer tourism' and be more attentive to the situated embodied interactions and the relationality of people, places and organisations.

Discussion: diverse economies of development

The diverse economies approach to development and volunteer tourism, as conceptualised in this paper, is an approach that opens up how we understand peace in tourism. In our 'reading for difference' in the diverse economies of volunteer tourism, the potentialities for credible diverse practices of development – beyond the narratives of 'aid' through 'helping' that perpetuates the volunteer white saviour narrative (Bandyopadhyay and Patil 2017; Everingham and Motta 2020) are recognised and negotiated. Indeed, just as capitalism contains diversity (Ferguson 2004; Gibson-Graham 2006, 2008; Tham and Evers-Swindell 2018), so does development. As Sharpley and Telfer (2015, p. 5) suggest, the "concept of development has evolved from simply economic growth to a broader achievement of the 'good life' that encompasses social, cultural, political, environmental and economic aims and processes". In the same way that reading for difference in the realms of capitalism provides critical insights into building other possible worlds (Gibson-Graham 2008), reading for difference within development can also present possibilities for more peaceful worlds. Our case study, presented above, illustrates that 'reading for difference' in volunteer tourism brings attention to development practices that bring joy, creativity and mutual intercultural understanding. In this way, volunteer tourism can be valuable to providing mutually transformative experiences for both hosts and guests.

Galvagno and Giaccone (2019) discuss the importance of creative experiences co-created between tourists and local actors for the development of tourism destinations. They place importance on the framing of grassroots perspectives with participatory strategies for tourists and local communities (Galvagno and Giaccone 2019). Creative development, evident in our case study, presents an opportunity to address some of the structural violence that comes from the hyper-consumption models of tourism and neocolonial forms of development. In a tourism setting, we believe that local communities should benefit from tourism dollars and benefit from the human exchanges that tourism can facilitate. In this way, volunteer tourism can have enormous benefits for both local communities and the volunteers.

We have highlighted the diversity of development through creativity and language exchange provided at *Fundación Arte del Mundo*. These forms of development can lead to mutually beneficial intercultural encounters by facilitating community spaces where tourists can get to know locals and vice versa. This case illustrates that when communities are actively involved in tourism, the experience can subvert the stereotypes of 'otherness'. Mutual learning exchanges between the host community and tourists become more central. If volunteer tourism is to be framed and promoted as collaborative consumption where intercultural encounters can be facilitated in meaningful ways, then it is vital that the expertise and social value of host communities are upheld. At the *Fundación*, the local children and adults became expert language teachers, dance instructors and intercultural mediators for the volunteer tourists who stayed for more extended periods than other tourists who were passing through. Because profit was not centred on economic or social transactions, these matters demonstrate that diverse economies can exist within global capitalism itself, potentially facilitating peace within tourism. However, there is a focus on placing the benefits of these transactions into the hands of the local communities as well as the volunteers, both of whom are looking for interactions that are meaningful and transformative.

A decommodified volunteer tourism agenda that promotes peace is evident in our case study. Volunteers and locals engaged in encounters and tourism spaces that both resist and subvert hegemonic constructions of their 'othered' and 'exoticised' selves, and hegemonic cultural constructions that define them as passive (Wearing and Wearing 2001). Wearing and Wearing (2001, p. 146) insist that tourism scholars must resist imposing "a crude cultural inferiorisation thesis" that attributes all "relations of cultural definition and dominance to the hegemonic culture of capitalist markets". Tucker (2014, p. 206) also argues for the importance of our focus on the "potentialities" of tourism encounters as "moral spaces for interrupting, and for destabilising, normalised relationships and subject positions". For the creation of a shift in thinking, meaningful representations of cultures and alternative progressive practices are understood and promoted as possible. Places are not simply gazed upon in the context of volunteer tourism. Places are spaces of interaction, resulting in possibilities for tourists and hosts to cross over their cultural boundaries (Wearing, Stevenson, and Young 2010). Tourists need not be viewed as invaders but, instead, as imaginative co-creating travellers. Rather than positioning the hosts in theory and practice as stuck in an existing static culture, their agency becomes key within the tourist experience (Benali and Oris 2019; MacCannell 1992; Wearing, Stevenson, and Young 2010; Young et al. 2020).

Some critics of the diverse economies framework argue that equity issues related to gender, class and race, have been largely excluded (Fickey 2011). Therefore, as Wright (2010) notes, important questions must be asked concerning power and the possibilities that exist in relation to making equitable changes to broader political-economic landscapes. These questions are crucial to volunteer tourism because most volunteers come from the Global North to the Global South. For this reason, it is difficult to disentangle volunteer tourism from neocolonial and neoliberal geographies of mobility and inequity. This article, however, demonstrates the multiple possibilities inherent in the intangible affective and emotional aspects of the encounters. These encounters can challenge the inequitable types of exchanges that hyper-consumption models of capitalism often engender. The interpretive power given to encounters within neoliberalism is reduced to a "singular logic of rationality" (Ferguson 2004, p. 176) in approaches that ignore the inherent contradictions and subversions of neoliberalism within the on-the-ground practices.

While critiques of volunteer tourism as being connected to neoliberal macroeconomic doctrines are essential and necessary, ignoring heterogeneity in tourism results in a silencing of the multiplicities, relationalities and ambiguities of the tourism industry and the voices of extremely diverse destination communities (Young et al. 2020; Amoamo, Ruckstuhl, and Ruwhiu 2018). When the lived experiences of host communities and the volunteers are overlooked, the agency of host communities in mediating these relationships are silenced (Benali and Oris 2019). This silencing inhibits the promotion of doing volunteer tourism in ways that *promote* peace and justice between different cultures by limiting how we come to understand the respectful intercultural encounters that are constantly occurring in the everyday spaces of volunteer tourism (Wearing, Young, and Everingham 2017). One of these limitations is the broad-stroke approach taken to the application of neoliberalism in tourism analysis.

Reflecting on tourism economies in relation to the work of Gibson-Graham, Brouder (2018) urges scholars to open space to rethink tourism models that are creative and be a positive force for community development. For Brouder (2018) dominant discourses in tourism development and even tourism analysis deems hyper-consumption neoliberal capitalist models inevitable. Equitable and socially just tourism models can often go unnoticed within capitalocentric approaches, ignoring the non-market transactions that paradoxically *also* happen within these economic models. Attention to the emotional affective dimension of fieldwork, seen in the autoethnographic approach to our research, can uncover friendships across cultures, human-to-human connection, and moments of creativity, laughter, and joy (Everingham 2015, 2016; Everingham and Motta 2020; Griffiths 2014; Guiney 2018; Palacios 2010).

As Cave and Dredge (2020) note, in the context of resetting tourism post-COVID-19, a move away from 'business as usual' towards regenerative tourism that is sustainable and benefits local communities requires attention towards the already existing prototypes. In the case of volunteer tourism, this includes highlighting projects and organisations working with host communities in ways that centre peace and justice. This can be through ensuring money flows into local communities, that local communities benefit from the volunteer projects, and facilitating encounters across cultural differences in ways that centre mutual intercultural understanding. These models of volunteer tourism can go some way towards addressing the underlying structural violence of hyper-consumption models of tourism.

These views align with Gibson-Graham (2006) performative agenda that highlights diverse economies within capitalism. In taking a postcapitalist approach to the economy, Gibson-Graham (2006) challenge the 'naturalisation of the economy' and remind us that the economy is a social construct that follows "historical path-dependent patterns and cannot be separated from its specific historical, cultural and geographical contexts". Like Tham and Evers-Swindell (2018), we challenge the perceived binaries between capitalist economic activity and diverse economic/non-economic activity within tourism economies. Volunteer tourism is not contrary or outside of capitalist economic exchange; rather non-capitalist exchanges already happen *within* and capitalist market exchanges in volunteer tourism. Once the economy is discursively unpacked in this way, it becomes possible to reimagine the economy away from exploitative forms of the economy that are underpinned by and perpetuate structural violence.

Conclusion: peace and justice through intercultural exchange in volunteer tourism

Recentring the values that position volunteer tourism as an ethical alternative form of tourism capable of promoting peace and justice generally involves broadening our research philosophies in tourism generally, and volunteer tourism, specifically. This research approach argues for the decommodification of volunteer tourism through the prioritisation of social, cultural and ecological value on local communities and destination environments. Through informal and formal modes of sharing (specifically, collaborative consumption), a decommodified agenda for volunteer tourism is provided through alternative ways to operate and practice. By taking a diverse economies approach embedded in collaborative consumption for volunteer tourism, tourism research and practice enables a discursive space to centre peace and justice at the heart of volunteer tourism.

By making explicit 'other' stories of volunteer tourism, we enact performative research of 'reading for difference' (Gibson-Graham 2006; 2008) that is attentive to the everyday and the difference this emphasis can make in working towards peace and justice through intercultural learning and understanding. However, as Higgins-Desbiolles (2009) notes, volunteer tourism for justice only becomes possible when the dominant economical neoliberal paradigms of tourism research and practice are challenged. This is because these hyper-consumption models are underpinned by and perpetuate structural violence. Our research illustrates that while volunteer tourism has been co-opted into hyper-consumption models of tourism, furthering the exploitation of local communities and perpetuating neocolonial forms of development, the industry is *also* diverse and heterogeneous.

There are instances where some commodified volunteer tourism operators perpetuate othering practices and do more harm than good in local communities. However, we see possibilities for decommodified volunteer tourism to bring a different understanding and framework for host-guest interactions and exchange of cultural identity and symbols in local places if promoted within collaborative consumption models. These collaborative consumption models can be framed as alternative capitalist enterprises, containing alternative market and non-market transactions. In highlighting these diverse forms of transactions within *Fundación Arte del Mundo*, we have provided a framework for understanding how volunteer tourism can be peace-building. Critical to peace-building outcomes is a more equitable distribution of power and resources between travelling and travelled cultures (Wearing, Stevenson, and Young 2010), where decision making responsibility involves host communities who receive social returns based on exchanges that go beyond economic transactions.

Volunteer tourism has the potential to foreground shared values that can potentially lead to profound structural changes and address forms of structural violence that come from hyper-consumption models of tourism. However, there needs to be a shift in both theory and practice that promotes ethical, collaborative consumption and highlights possibilities for decommodified agendas. Volunteer tourism is unique in this respect because of the amount of time that volunteers spend living - and sharing - in local communities. This cultural immersion can mean that volunteers learn to appreciate and value the people and places in which they volunteer. In these collaborative models, tourism need not be exploitative of local populations. The benefits of tourism can flow to residents, as volunteers can enhance the communities where they volunteer by sharing their knowledge, skills, enthusiasm, energy, etc. Through mutual cultural respect, intercultural exchange is transformative. Collaborative consumption shifts the relationships of power between tourists and hosts by enabling both to interact and learn from each other, allowing for a (re)conceptualisation of self/other understandings. These are modes of tourist and host experience that allow for a fluid two-way process of co-presence and co-creation, with mutual benefits for both, crucial to achieving peace and justice in volunteer tourism.

Notes

1. The case study presented here draws from a broader volunteer tourism research project conducted by the lead author during field work in South America in 2012 and 2013. Our analysis of the structure and activities of *Fundación Arte Del Mundo* is based on data collected during the field work period, however, in this paper we also reflect on changes to the volunteer organisation that time.

Disclosure statement

No potential conflict of interest was reported by the authors.

ORCID

Tamara N. Young http://orcid.org/0000-0001-8132-4194

References

Adie, B., Amore, A., & Hall, C. M. (2020). Just because it seems impossible, doesn't mean we shouldn't at least try: the need for longitudinal perspectives on tourism partnerships and the SDGs. *Journal of Sustainable Tourism*, https://doi.org/10.1080/09669582.2020.1860071

Amoamo, M., Ruckstuhl, K., & Ruwhiu, D. (2018). Balancing indigenous values through diverse economies: a case study of Māori Ecotourism. *Tourism Planning & Development*, *15*(5), 478–495. https://doi.org/10.1080/21568316.2018.1481452

Bandyopadhyay, R., & Patil, V. (2017). The white woman's burden' – The racialized, gendered politics of volunteer tourism. *Tourism Geographies*, *19*(4), 644–657. https://doi.org/10.1080/14616688.2017. 1298150

Benali, A., & Oris, M. (2019). Consuming poverty: volunteer tourism in an orphanage in Nepal. In J. M. Cheer, L. Mathews, K. E. van Doore, & K. Flanagan (Eds.), *Modern day slavery and orphanage tourism* (pp. 110–123). CABI.

Brouder, P. (2018). The end of tourism? A Gibson-Graham inspired reflection on the tourism economy. *Tourism Geographies*, *20*(5), 916–918. https://doi.org/10.1080/14616688.2018.1519721

Brueckner, M., Spencer, R., Wise, G., & Banduk, M. (2016). A third space social enterprise: closing the gap through cross-cultural learning. *Australian Aboriginal Studies*, (2), 18–32.

Bunnell, T., Yea, S., Peake, L., Skelton, T., & Smith, M. (2011). Geographies of friendships. *Progress in Human Geography*, *35*(4), 490–507.

Butcher, J., & Smith, P. (2015). *Volunteer tourism: the lifestyle politics of international development*. Routledge.

Cave, J., & Dredge, D. (2020). Regenerative tourism needs diverse economic practices. *Tourism Geographies*, *22*(3), 503–513. https://doi.org/10.1080/14616688.2020.1768434

Cheung Judge, R. (2017). Class and global citizenship: perspectives from non-elite young people's participation in volunteer tourism. *Tourism Recreation Research*, *42*(2), 164–175. https://doi.org/10.1080/02508281.2017.1303894

Comaroff, J., & Comaroff, J. (Eds.). (2001). *Millenuial capitalism and the culture of neoliberalism*. Duke University Press.

Coren, N., & Gray, T. (2012). Commodification of volunteer tourism: a comparative study of volunteer tourists in Vietnam and in Thailand. *International Journal of Tourism Research*, *14*(3), 222–234. https://doi.org/10.1002/jtr.848

Crossley, É. (2012). Poor but happy: volunteer tourists' encounters with poverty. *Tourism Geographies*, *14*(2), 235–253. https://doi.org/10.1080/14616688.2011.611165

Devereux, P. (2008). International volunteering for development and sustainability: outdated paternalism or a radical response to globalisation? *Development in Practice*, *18*(3), 357–370. https://doi.org/10.1080/09614520802030409

Dredge, D., & Gyimóthy, S. (2015). The collaborative economy and tourism: Critical perspectives, questionable claims and silenced voices. *Tourism Recreation Research*, *40*(3), 286–302. https://doi.org/10.1080/02508281.2015.1086076

Ellingson, L. L., & Ellis, C. (2008). Autoethnography as constructionist project. In J. A. Holstein & J. F. Gubrium (Eds.), *Handbook of constructionist research* (pp. 446–467). The Guilford Press.

England, K. (1994). Getting personal: reflexivity, positionality, and feminist research. *The Professional Geographer*, *46*(1), 80–89. https://doi.org/10.1111/j.0033-0124.1994.00080.x

Everingham, P. (2015). Intercultural exchange and mutuality in volunteer tourism: The case of Intercambio in Ecuador. *Tourist Studies*, *15*(2), 175–190. https://doi.org/10.1177/1468797614563435

Everingham, P. (2016). Hopeful possibilities in spaces of 'the-not-yet-become': relational encounters in volunteer tourism. *Tourism Geographies*, *18*(5), 520–538. https://doi.org/10.1080/14616688.2016.1220974

Everingham, P. (2018). Speaking Spanglish: embodying linguistic (b)orderlands in volunteer tourism. *Emotion Space and Society*, *27*, 68–75. https://doi.org/10.1016/j.emospa.2018.04.001

Everingham, P., & Motta, S. C. (2020). Decolonising the 'autonomy of affect' in volunteer tourism encounters. *Tourism Geographies*. https://doi.org/10.1080/14616688.2020.1713879

Farmaki, A. (2017). The tourism and peace nexus. *Tourism Management*, *59*, 528–540. https://doi.org/10.1016/j.tourman.2016.09.012

Ferguson, J. (2004). Power topographies. In D. Nugent & J. Vincent (Eds.), *A companion to antropology politics* (pp. 383–399). Blackwell.

Fickey, A. (2011). The focus has to be on helping people make a living: exploring diverse economies and alternative economic spaces. *Geography Compass*, *5*(5), 237–248. https://doi.org/10.1111/j.1749-8198.2011.00418.x

Fletcher, R., Mas, I. V., Blanco-Romero, A., & Blázquez-Salom, M. (2019). Tourism and degrowth: an emerging agenda for research and praxis. *Journal of Sustainable Tourism*, *27*(12), 1745–1763. https://doi.org/10.1080/09669582.2019.1679822

Galtung, J. (1969). Violence, peace and peace research. *Journal of Peace Research*, *6*(3), 167–191. https://doi.org/10.1177/002234336900600301

Galvagno, M., & Giaccone, S. C. (2019). Mapping creative tourism research: reviewing the field and outlining future directions. *Journal of Hospitality & Tourism Research*, *43*(8), 1256–1280. https://doi.org/10.1177/1096348019862030

German Molz, J. (2013). Social networking technologies and the moral economy of alternative tourism: The case of Couchsurfing. *Annals of Tourism Research*, *43*(2), 210–230. https://doi.org/10.1016/j.annals.2013.08.001

Gibson-Graham, J. K. (2006). *A postcapitalist politics*. University of Minneapolis Press.

Gibson-Graham, J. K. (2008). Diverse economies: performative practices for "other worlds. *Progress in Human Geography*, *32*(5), 613–632. https://doi.org/10.1177/0309132508090821

Grabowski, S., Everingham, P., & Young, T. (forthcoming). Widening the scope of evaluating volunteer tourism: beyond impact measurement. In K. Holmes, L. Lockstone, K. Smith, & R. Shipway (Eds.), *Routledge handbook of volunteering in events, sport and tourism*. Routledge.

Gray, N. J., & Campbell, L. M. (2007). A decommodified experience? Exploring aesthetic, economic and ethical values for volunteer ecotourism in Costa Rica. *Journal of Sustainable Tourism*, *15*(5), 463–482. https://doi.org/10.2167/jost725.0

Griffiths, M. (2014). The affective spaces of global civil society and why they matter. *Emotion, Space and Society*, *11*, 89–95. https://doi.org/10.1016/j.emospa.2013.08.003

Guiney, T. (2018). Hug-an-orphan vacations": "love" and emotion in orphanage tourism. *The Geographical Journal*, *184*(2), 125–137. https://doi.org/10.1111/geoj.12218

Gyimóthy, S., & Meged, J. (2018). The Camøno: A communitarian walking trail in the sharing economy. *Tourism Planning & Development*, *15*(5), 496–515. https://doi.org/10.1080/21568316.2018.1504318

Higgins-Desbiolles, F. (2008). Justice tourism and alternative globalisation. *Journal of Sustainable Tourism*, *16*(3), 345–364. https://doi.org/10.1080/09669580802154132

Higgins-Desbiolles, F. (2009). International solidarity movement: A case study in volunteer tourism for justice. *Annals of Leisure Research*, *12*(3–4), 333–349. https://doi.org/10.1080/11745398.2009.9686828

Higgins-Desbiolles, F., Carnicelli, S., Krolikowski, C., Wijesinghe, G., & Boluk, K. (2019). Degrowing tourism; rethinking tourism. *Journal of Sustainable Tourism*, *27*(12), 1926–1944. https://doi.org/10.1080/09669582.2019.1601732

Jafari, J. (1989). Tourism and peace. *Annals of Tourism Research*, *16*(3), 439–443. https://doi.org/10.1016/0160-7383(89)90059-5

Kipp, A., Hawkins, R., & Gray, N. (2021). Gendered and racialized experiences and subjectivities in volunteer tourism. *Gender, Place and Culture*, *28*(1), 45–65. https://doi.org/10.1080/0966369X.2019.1708274

Lai, P. H., Chuang, S., Zhang, M., & Nepal, S. (2020). The non-profit sharing economy from a social exchange theory perspective: a case from worldwide opportunities on organic farms in Taiwan. *Journal of Sustainable Tourism*, *28*(12), 1970–1987. https://doi.org/10.1080/09669582.2020.1778709

Longhurst, R. (2016). Semi-structured interviews and focus groups. In N. Clifford, M. Cope, T. Gillespie, & S. French (Eds.), *Key methods in geography* (pp.143–157). Sage.

Lyons, K., Hanley, J., Wearing, S., & Neil, J. (2012). Gap year volunteer tourism: myths of global citizenship? *January Annals of Tourism Research*, *39*(1), 361–378. https://doi.org/10.1016/j.annals.2011.04.016

MacCannell, D. (1992). *Empty meeting grounds: The tourist papers*. Routledge.

Martin, C. (2016). The sharing economy: a pathway to sustainability of a nightmarish form of neoliberal capitalism. *Ecological Economics*, *121*, 149–159. https://doi.org/10.1016/j.ecolecon.2015.11.027

McGinnis, G., Harvey, M., & Young, T. (2020). Indigenous knowledge sharing in Northern Australia: Engaging digital technology for cultural interpretation. *Tourism Planning & Development*, *17*(1), 96–125. https://doi.org/10.1080/21568316.2019.1704855

McGregor, S. (2003). *Consumerism as a source of structural violence*. http://www.kon.org/hswp/archive/consumerism.html, accessed January 2021

Miller, J., & Glassner, B. (2011). The "inside" and the "outside": finding realities in interviews. In D. Silverman (Ed.), *Qualitative research* (pp. 131–149). Sage.

Mosedale, J. (2012). Diverse economies and alternative economic practices in tourism. In I. Atelvevic, N. Morgan, & A. Pritchard (Eds.), *The critical turn in tourism studies: Creating an academy of hope* (pp. 194–207). London: Routledge.

Mostafanezhad, M. (2013). The geography of compassion in volunteer tourism. *Tourism Geographies*, *15*(2), 318–337. https://doi.org/10.1080/14616688.2012.675579

Mostafanezhad, M. (2014). *Volunteer tourism: popular humanitarianism in neoliberal times*. Ashgate Publishing Limited.

Palacios, C. (2010). Volunteer tourism, development and education in a postcolonial world: conceiving global connections beyond aid. *Journal of Sustainable Tourism*, *18*(7), 861–878. https://doi.org/10.1080/09669581003782739

Pallett, H., & Chilvers, J. (2015). Organizations in the making: learning and intervening at the science-policy interface. *Progress in Human Geography*, *39* (2), 146–166. https://doi.org/10.1177/0309132513518831

Pelias, R. (2003). The academic tourist: an autoethnography. *Qualitative Inquiry*, *9*(3), 369–373. https://doi.org/10.1177/1077800403009003003

Pratt, S., & Liu, A. (2016). Does tourism really lead to peace? A global view. *International Journal of Tourism Research*, *18*(1), 82–90. https://doi.org/10.1002/jtr.2035

Raymond, E., & Hall, C. (2008). The development of cross-cultural (mis)understanding through volunteer tourism. *Journal of Sustainable Tourism*, *16*(5), 530–543. https://doi.org/10.1080/09669580802159610

Ritzer, G. (2005). *Enchanting a disenchanted world: Revolutionizing the means of consumption*. Pine Forge Press.

Ritzer, G. (2011). *The McDonaldization of society* (6th ed.). Sage.

Ritzer, G., & Miles, S. (2019). The changing nature of consumption and the intensification of McDonalidization in the digital age. *Journal of Consumer Culture*, *19*(1), 3–20. https://doi.org/10.1177/1469540518818628

Scheyvens, R. (2002). *Tourism for development: empowering communities*. Prentice Hall.

Scheyvens, R., Banks, G., & Hughes, E. (2016). The private sector and the SDGs: the need to move beyond 'business as usual. *Sustainable Development*, *24*(6), 371–382. https://doi.org/10.1002/sd.1623

Sharpley, R., & Telfer, D. (2015). *Tourism and development: Concepts and issues*. Channel View.

Shepard, B. (2005). Play, creativity, and the new community organizing. *Journal of Progressive Human Services*, *16*(2), 47–69.

Sin, H. L. (2009). Volunteer tourism: "Involve me and I will learn"? *Annals of Tourism Research*, *36*(3), 480–501. https://doi.org/10.1016/j.annals.2009.03.001

Smith, M., & Krannich, R. (1998). Tourism dependence and resident attitudes. *Annals of Tourism Research*, *25*(4), 783–802. https://doi.org/10.1016/S0160-7383(98)00040-1

Tham, A., & Evers-Swindell, B. (2018). Stand up and be counted – a diverse economy perspective of Air New Zealand. *Tourism Planning & Development*, *15*(5), 567–583. https://doi.org/10.1080/21568316.2018.1504816

Tucker, H. (2014). Mind the gap: opening up spaces of multiple moralities in tourism encounters. In M. Mostafanezhad & K. Hannam (Eds.), *Current developments in the geographies of leisure and tourism: Moral encounters in tourism* (pp.199–208). Routledge.

Valentine, G. (2005). Tell me about… ' using interviews as research methodology. In R. Flowerdew & D. Martin (Eds.), *Methods in human geography: a guide for students doing a research project* (pp. 110–127). Addison Wesley Longman.

Vrasti, W. (2013). *Volunteer tourism in the global south: giving back in neoliberal times*. Routledge.

Wearing, S. (2001). *Volunteer tourism: Experiences that make a difference*. CABI.

Wearing, S., & Grabowski, S. (2011). Volunteer tourism and intercultural exchange: exploring the 'Other' in this experience. In A. M. Benson (Ed.), *Volunteer tourism: Theoretical frameworks and practical application* (pp. 193–210). Routledge.

Wearing, S., & McGehee, N. G. (2013). Volunteer tourism: a review. *Tourism Management*, *38*, 120–130. https://doi.org/10.1016/j.tourman.2013.03.002

Wearing, S., Stevenson, D., & Young, T. (2010). *Tourist cultures: Identity. Place and the traveller*. SAGE.

Wearing, S., & Wearing, B. (2001). Conceptualizing the selves of tourism. *Leisure Studies*, *20*(2), 143–159. https://doi.org/10.1080/02614360110051631

Wearing, S., & Wearing, M. (1999). Decommodifying ecotourism: rethinking global-local interactions with host communities. *Society and Leisure (Losire et Societe)*, *22*(1), 39–70.

Wearing, S., Young, T., & Everingham, P. (2017). Evaluating volunteer tourism: has it made a difference? *Tourism Recreation Research*, *42*(4), 512–521. https://doi.org/10.1080/02508281.2017.1345470

Williams, M. (2017). Searching for actually existing justice in the city. *Urban Studies*, *54*(10), 2217–2231. https://doi.org/10.1177/0042098016647336

Wright, S. (2010). Cultivating beyond-capitalist economies. *Economic Geography*, *86*(3), 297–318. https://doi.org/10.1111/j.1944-8287.2010.01074.x

Wright, S., Suchet-Pearson, S., & Lloyd, K. A. T. E. (2007). An interwoven learning exchange: transforming research-teaching relationships in the Top End, Northern Australia. *Geographical Research*, *45*(2), 150–157. https://doi.org/10.1111/j.1745-5871.2007.00441.x

Young, T., Reindrawati, D., Lyons, K., & Johnson, P. (2020). Host gazes from an Islamic island: challenging homogenous resident perception orthodoxies. *Tourism Geographies*, 1–24. https://doi.org/10.1080/14616688.2020.1733067

OPEN ACCESS

Te Awa Tupua: peace, justice and sustainability through Indigenous tourism

Jason Paul Mika and Regina A. Scheyvens

ABSTRACT

Te Awa Tupua is an ancestor of the Māori people of Whanganui, and is also the Whanganui River, who in 2017 was formally recognised as a person. While legally conferring personhood upon an element of nature is relatively novel, it recognises a fundamental principle of indigeneity, that all things—human and nonhuman—are related. We explore intersections of peace, justice, and sustainability through Indigenous tourism in case studies of three Māori tourism enterprises on Te Awa Tupua (the Whanganui River). Our paper spotlights three findings. First, that treaty settlements elevate the status of Māori knowledge and contain elements of peace-making and economy-making as decolonising projects of self-determined development. Second, while indigeneity is foundational, we found that syncretism is evident in the sustainability of Māori tourism enterprises. Third, we uncovered a socioecological dissonance in attitudes towards commercial growth, with Māori tourism enterprises opting for slower and lower growth in favour of environmental and community wellbeing. We propose a model of Indigenous tourism called kaupapa tāpoi. We conclude by suggesting that reconciling differences in viewpoints on sustainability and growth between Māori and non-Māori tourism enterprises will require involvement of several institutional actors, starting with Te Awa Tupua.

Introduction

Tourism has some lofty ambitions: beyond being simply an enterprising activity, tourism strives to promote peace, justice, and sustainability (D'Amore, 1988; Hall, 2019; Higgins-Desbiolles & Whyte, 2013; Raymond & Hall, 2008). In relation to Indigenous peoples, we can add that tourism aims to contribute to Indigenous self-determination and development (Amoamo, Ruckstuhl, et al., 2018; Bunten & Graburn, 2018; Whitford & Ruhanen, 2016; Zapalska & Brozik, 2017). Expectations for peace, justice, and sustainability manifest as social, moral, legal, and institutional ethical codes because of universal intolerance for their opposites: war, injustice, inequity, unsustainability, and subjugation (Dutta & Elers, 2020; Pereiro, 2016; Verbos et al., 2017b). Thus, tourism may be viewed as a catalyst for harmony within humanity and with nature (D'Amore, 1988). The prospect of harmony through tourism seems vital in post-conflict situations. In the context of Aotearoa New Zealand, although the colonial government's conflict with Māori—the Indigenous people of Aotearoa New Zealand—may have subsided over 140 years ago, its postcolonial effects remain. Persistent socioeconomic disparities, the formation of a tribunal to enquire into historical injustices, and the reaffirmation of Māori knowledge across multiple domains allude to intersections of peace, justice, and sustainability within te ao Māori (Māori society) intended to create dialogic space for epistemological and ontological equity (Cole, 2017; Harris & Wasilewski, 2004; Smith, 1999).

The research question this paper addresses is how do peace, justice, and sustainability intersect in Indigenous tourism in Aotearoa New Zealand? This paper seeks to understand whether and how peace, justice, and sustainability feature within Indigenous tourism by exploring the experiences of three Māori tourism enterprises on the Whanganui River. We focus on the Whanganui River because it offers important insights into the indigeneity of Māori tourism in the context of a treaty settlement between Māori and the Crown pertaining to the river (Awatere et al., 2017; Mika et al., 2018; Reid et al., 2019). We proceed by contextualising the research within the structure of Māori tourism, before discussing tourism in relation to the Whanganui River. Second, we define four key concepts—peace, justice, sustainability, and indigeneity—which function as the theoretical framework for the analysis of Indigenous tourism. Third, we outline the methodology we employed for the research. Fourth, we present the findings as case studies of three Māori tourism enterprises on the Whanganui River. Fifth, we discuss findings on peace, justice, and sustainability in Indigenous tourism and propose a model of Indigenous tourism. We conclude by identifying principles that emerge and their implications for tourism research, policy, and practice.

Māori tourism and Te Awa Tupua

Māori have long been involved in tourism in Aotearoa New Zealand (Te Awekotuku, 1981), with a tradition of generosity in the hosting of people, characterised as manaakitanga (Barnett, 2001; Mika, 2014; Papakura, 1991). Although Māori initially controlled the development of tourism (Bremner, 2013), colonialism and its racist constructions (Meihana, 2015; Smith, 1999) reduced Māori participation to the roles of cultural performer, artisan, or guide (Amoamo & Thompson, 2010, cited in Amoamo, Ruckstuhl, et al., 2018). Colonial legacies continue to hamper Māori tourism (Horn & Tahi, 2009), as they do for other Indigenous groups (Bunten & Graburn, 2018). Furthermore, what constitutes success in tourism has tended to be moulded on normative economic conventions of self-interest and accumulation, devoid of political, identity and equity considerations (Amoamo, Ruckstuhl, et al., 2018). Peredo and McLean (2013) argue that the economic logic of market-based assistance for Indigenous entrepreneurs results in development mistakes that fail to account for the importance of social ties, social value and social exchange. In the vein of this plurality, Māori entrepreneurs increasingly see tourism as a means of securing self-determined economic development (Tretiakov et al., 2020), which requires novel forms of economy and enterprise to account for their indigeneity (Amoamo, Ruckstuhl, et al., 2018), and the "coexistence of traditional and nontraditional methods of organising economic activity" (Mika et al., 2019, p. 374).

The Whanganui Māori Regional Tourism Organisation (2020, p. 1) was established in 2003 by Māori tourism operators with the support of Whanganui iwi (tribes) to "lead and empower Māori tourism from the mountain to the sea." It does this by promoting the authenticity and cultural and spiritual significance of the Whanganui River as a destination, and by coordinating Māori tourism activities on the river. The organisation's chair, Hayden Potaka, initiated a process to update the organisation's strategy based on principles of sustainability, quality infrastructure, and ecocultural tourism (Whanganui Māori Regional Tourism Organisation, 2020). Hayden sees

E rere kau mai te Awa nui

Mai i te Kāhui Maunga ki Tangaroa

Ko au te Awa, ko te Awa ko au

The Great River flows

From the Mountains to the Sea

I am the River, and the River is me

Figure 1. Tribal narrative of the Whanganui River.
Source: Whanganui Iwi and The Crown (2014a, p. 1)

the organisation as leveraging off and contributing to existing strategies for Māori economic development in the region (Mika et al., 2016).

Te Awa Tupua—the Whanganui River—has significant spiritual and material meaning for the Māori people of Whanganui as the site of an 800-year long relationship that was severed through colonial settlement and is only now being restored through treaty settlements. The whakataukī (tribal proverb) in Figure 1 speaks to the indivisibility of the Whanganui awa (river) and the identity of local iwi (tribes) (Waitangi Tribunal, 1999). The Whanganui River runs for 290 kilometres from Mount Tongariro to the Tasman Sea, and passes through the three local government districts of Ruapehu, Rangitīkei, and Whanganui and the whenua (lands) of three related tribal groupings—Hinengākau of the upper river, Tama Ūpoko of the middle reaches, and Tūpoho of the lower Whanganui (Young, 2017). It flows through the Whanganui National Park, past several settlements of historical and cultural significance, and through the city of Whanganui. The Māori tourism enterprises we engaged with are situated along a depopulated stretch of the river.

On 14 October 1990, Hikaia Amohia and others of the Whanganui Māori Trust Board submitted a treaty claim (Wai 167) to the Waitangi Tribunal, arguing that the Crown had breached their unextinguished customary rights of tino rangatiratanga (self-determination) and kaitiakitanga (guardianship) over the Whanganui River (Waitangi Tribunal, 1999). The government's development of the Whanganui River as a tourist route in the 1890s, for example, effectively forced Te Ātihaunui-a-Pāpārangi (the principal tribe of the Whanganui River) from the river, their lands, and their fishing grounds (Parliamentary Commissioner for the Environment, 2019). On 8 June 1999, the tribunal established that Ātihaunui did indeed own the Whanganui River "in its entirety" and that the Crown had deprived Ātihaunui of "their possession and control of the Whanganui River" (Waitangi Tribunal, 1999, p. xi).

Of significance to Whanganui iwi is a treaty settlement signed at Ruakā Marae in Rānana on 5 August 2014 (Office of Treaty Settlements, 2014; Whanganui Iwi & The Crown, 2014a). The settlement—passed into law on 20 March 2017 via Te Awa Tupua (Whanganui River Claims Settlement) Act 2017—represents the culmination of over 100 years of struggle by the iwi to protect their special bond with the Whanganui River (Ngā Ngā Tāngata Tiaki o Whanganui, 2020). The deed of settlement—Ruruku Whakatupua—comprises two distinct documents. The first—Te mana o Te Awa Tupua—recognises the mana (status, power, and authority) of the Whanganui River and its tributaries as a single physical and spiritual entity (Whanganui Iwi & The Crown, 2014a). The second—Te mana o te iwi o Whanganui—is a deed recognising the mana of the tribes of the Whanganui River (Whanganui Iwi & The Crown, 2014b), affirming Ngā Tangata Tiaki o Whanganui as their post-settlement governance entity and outlining the agreed redress. Under the settlement, legal personhood is conferred on Te Awa Tupua—a novel legal principle (Morris

(1) Ko te awa te mātāpuna o te ora—The River is the source of spiritual and physical sustenance.

(2) E rere kau mai te Awa nui mai i te Kahui Maunga ki Tangaroa—The great River flows from the mountains to the sea.

(3) Ko au te Awa, ko te Awa ko au—I am the River and the River is me.

(4) Ngā manga iti, ngā manga nui e honohono kau ana, ka tupu hei Awa Tupua—The small and large streams that flow into one another and form one River.

Figure 2. Tupua te Kawa—Intrinsic values of Te Awa Tupua.
Source: Whanganui Iwi and The Crown (2014a, p. 7)

& Ruru, 2010) but one consistent with a Māori world view (Royal, 2003). Te Pou Tupua is the human face of the river, vested in two people appointed by Māori and the Crown who are to act and speak on behalf of Te Awa Tupua, and uphold the status of Te Awa Tupua and Tupua te Kawa, as the natural law and values which bind the people to the river and the river to the people (see Figure 2) (Whanganui Iwi & The Crown, 2014a).

Contextually, then, Māori tourism on the Whanganui River occurs amid the collective aspirations of Māori tourism operators for equitable, quality, sustainable, and culturally authentic tourism. This is a system supposedly predicated upon treaty-based relationships with Māori and a treaty settlement that privileges a Māori view of the river, and provides legal and financial means by which to restore the relationship of the people with the river, and with one another. Implicit within this context are the notions of indigeneity, peace, justice, and sustainability, which are foregrounded as theoretical anchors for analysis in the next section.

Theoretical positioning

Indigeneity and tourism

Indigenous tourism requires an understanding of indigeneity and how tourism is shaped by it. Although comprising more than 6 per cent of the world's population, Indigenous peoples make up 15 per cent of the extreme poor, face life expectancy that is up to 20 years lower than that of non-Indigenous people, and their status and rights may not be recognised by the states in which they live (World Bank, 2019). Yet, Indigenous peoples are increasingly engaging in economic development as emancipatory processes of Indigenous self-determination and sustainable development across multiple sectors (Henry et al., 2017; Mika et al., 2017; Peredo & Anderson, 2006), including in tourism (Amoamo, Ruckstuhl, et al., 2018; Jacobsen, 2017; Pereiro, 2016; Whitford & Ruhanen, 2016).

Indigeneity is evident in Indigenous people's world views, which Harris and Wasilewski (2004) characterise as relational, responsible, reciprocal, and redistributive. Venkateswar et al. (2011) see indigeneity as 'a way of being' expressed in relationships between people and nature,

Indigenous activism, and the assertion of Indigenous rights. Indigeneity tends to be viewed by non-Indigenous peoples as a threat to state sovereignty and the presumption of hegemonic postcolonial cultures and institutions (Durie, 2004; Dutta, 2015; Lightfoot, 2016). Bunten and Graburn (2018) suggest that indigeneity is thus contingent on permissible postcolonial categorisations, where states 'otherise,' objectify, and appropriate representations of indigeneity as semiotic projections for consumption of the touristic gaze.

For Indigenous tourism to develop equitably, the spirituality of Indigenous peoples must be respected (UNWTO., 2020). An appreciation of Indigenous spirituality (Venkateswar et al., 2011) requires personal experience of it with community consent (UNWTO., 2020), lest unfortunate cross-cultural encounters ensue (see for example, Venkateswar, 2018). Spirituality is embedded within Indigenous world views, which are "both spiritual and material … [expressions] of identity and culture" (Mika et al., 2020, p. 261). A fundamental principle of Indigenous world views is that all things—people, flora, fauna and planet—are interrelated, obliging people to care for the Earth as one would a mother, and for the Earth to care for her people as kin (Colbourne, 2017; Gladstone, 2018; Hēnare, 2001; Knudtson & Suzuki, 1997).

A Māori world view is broadly consistent with Harris and Wasilewski (2004) principles, but is grounded in tribally-specific knowledge about the origins of the universe (Mead, 2017; Royal, 2003). While Royal (2005) uses aronga (outlook) to focus on a Māori orientation to the world, Durie (2017) uses te ao Māori (the Māori world) to denote the totality of the Māori world, both its physical and metaphysical elements. In Aotearoa New Zealand, te ao Māori—a Māori world view—imbues Māori tourism with the interrelatedness and the intertemporality of all things as manifestations of localised indigeneity, but this is also inflected with the colonial Christian legacy with which it has interacted since early settlement.

Peace and tourism

Whether peace is a predicate or consequence of Indigenous tourism, some understanding of the term from Indigenous and non-Indigenous perspectives is necessary to move toward consensus about its relevance and meaning for Māori tourism on the Whanganui River. D'Amore (1988) contends that peace through tourism is possible and desirable as an unofficial way of generating goodwill, trust, and understanding between culturally diverse peoples, catalysing trade, investment, and exchanges of people and ideas. Moreover, D'Amore (1988) urges a move away from negative definitions of peace as the absence of war or violence toward positive conceptualisations of peace as unity, tranquillity, and freedom between humans and nonhumans. Litvin (1998) challenges D'Amore's (1988) view as conjectural, instead seeing tourism as benefiting from peace rather than inducing it because without peace tourism cannot proceed. D'Amore (1988) and Litvin (1998) leave unresolved a positive vision for peace; Indigenous perspectives offer possibilities for resolution. Verbos et al. (2017a, p. 1) refer to a "hopeful vision for Mother Earth if we begin to remember our interconnectedness," a view generated by Indigenous grandmothers. In recognising Indigenous rights and imploring educators and enterprises to appropriately include Indigenous peoples, the United Nations presents a vision for the peaceful coexistence of Indigenous and non-Indigenous peoples (Verbos et al., 2017a).

Indigenous forms of peace in Aotearoa New Zealand necessarily defer to Indigenous knowledge of the Māori god of peace—Rongo, also known as Rongomaraeroa and Rongomātāne—whose domain extends to peace-making, agriculture, and human generosity (Best, 1976). Karena (2020) characterises the epistemology of Rongo as relying less on state coercion and more on culturally-centred ontologies, such as hohou i te rongo (allowing peace to enter), as metaphors for peace-making and peace pacts.

As much as the Treaty of Waitangi of 1840 was a treaty of cession it was also a treaty of peace because it contained within it provision for power sharing between Māori and the British

Monarch (Orange, 1987). The treaty was not preceded by civil conflict, but such conflict soon followed through the colonial government dishonouring it and the courts setting it aside (Belich, 1998; Durie, 1998). The pathway to peace in Aotearoa New Zealand had been obvious to Māori from the signing of the treaty—simply, to honour it (Kelsey, 1990; Mutu, 2019; Orange, 1987). Honouring the treaty for Māori means enacting a decolonising agenda underpinned by constitutional reform that upholds mana motuhake (tribal sovereignty) (Burrows et al., 2013; Mutu, 2012, 2019).

Justice and tourism

Indigenous tourism must be just, but what constitutes justice is a complex matter. In the non-Indigenous world, justice has an illustrious Western tradition, helping decide what is right by reference to, for example, what is fair and lawful (Aristotle), what is due (Augustine), what is rational (Aquinas), and the protection of property (Hume), freedom (Kant), and liberty (Rawls) (Pomerleau, 2019). Using the Justinian (1913, p. 5) definition of justice as "the set and constant purpose which gives to every man his due," Miller (2017) explicates justice in ways that make sense and have credence in non-Indigenous settings. Watene (2016), however, finds that the omission of Indigenous peoples from justice theorising has not been sufficiently rectified by liberal theories, such as Kymlicka's (1995, 2002) multiculturalism and Anaya's (2004) self-determination. Instead, she argues that actual engagement with Indigenous peoples is necessary to adequately comprehend an Indigenous view of justice, which has potential to reframe how justice is understood and applied generally. Watene (2016) argues that Indigenous peoples see justice as healing and as relational, evident in processes that redress colonial trauma, legitimise indigeneity, and restore relationships. An Indigenous view, she argues, is consistent with the capabilities approach of Sen (1999), which holds that meaningful development is tied to the freedom to pursue and sustain lives that matter to those people. A precolonial Māori view of justice is to be found in the maintenance of balance achieved through concepts of utu (reciprocity) (Firth, 1929), which may manifest as gift-giving (Hēnare, 2018) on the one hand, or muru (ritual compensation) on the other (Walker, 2004).

Under the Treaty of Waitangi Act 1975 Māori have been able to make claims to the Waitangi Tribunal for breaches of the treaty by the Crown from 1840 and seek redress from government through treaty settlements (Kawharu, 1989; Tapsell et al., 2007). For the Crown, treaty settlements offer a middle-ground in which the government defines the process toward the just resolution of past wrongs within the confines of political settlements (Office of Treaty Settlements, 2015). Kawharu (2018) is critical of treaty settlements, however, because they rely on Crown-defined processes, which have been divisive, adversarial, exhausting, and protracted, leaving post-settled tribes with capacity challenges and a public unprepared for iwi whose rangatiratanga has been restored. Instead, Kawharu (2018) favours the recent reconciliation process of Parihaka with its attention on a future-focused relationship of peaceful coexistence with the Crown. Hēnare (2011) concurs with Kawharu (2018), arguing that tikanga Māori (Māori values), underscored by the assurances of the treaty (Kawharu, 1989; Waitangi Tribunal, 2016) provide principles for lasting peace, justice, prosperity, and guidance for contemporary enterprise among Māori. Thus, we contend that treaty settlements appear as devices for both peace-making and economy-making, with tourism a logical candidate for investment (Amoamo, Ruwhiu, et al., 2018; Barry et al., 2020).

Sustainability and tourism

When concerns are raised about the negative effects of tourism on natural ecosystems (Butler, 2018), the denigration of significant cultural sites (Harrison, 2001), and disparities in the spread of benefits (Scheyvens, 2011), sustainability has been proffered as the answer. The World Commission on Environment and Development defines sustainable development as "development that meets the needs of the present without compromising the ability of future generations to meet their own needs" (Brundtland, 1987, p. 41). With its focus on meeting human needs from within ecological and socio-technological limits, sustainable development appears to present an unrivalled opportunity for a better life (Brundtland, 1987). Yet, even at its inception, Brundtland (1987) recognised the enormity of the challenge for diverse societies to accept integrative and comprehensive approaches to development. Furthermore, the Sustainable Development Goals (SDGs) have been criticised for favouring growth over sufficiency (Hall, 2019), inconsistencies, measurement and evaluative difficulties (Swain, 2018), and insufficiently including Indigenous perspectives (Yap & Watene, 2019).

Virtanen et al. (2020) advocate for the inclusion of Indigenous views on sustainability because of the bio-cultural diversity within Indigenous communities. Sustainability from an Indigenous perspective, according to Virtanen et al. (2020), is contextual, emanating from particular landscapes, entailing human and nonhuman sociality and a capacity for the coexistence of all entities. Indigenous conceptualisations of sustainability challenge the universalism of the SDGs, which rely on individualistic, human-centred, and singular approaches rather than viewing sustainability as locally lived (Virtanen et al., 2020). While the SDGs point to the potential for sustainable development through tourism (Scheyvens, 2018), including empowering Indigenous participation in tourism (Mendoza-Ramos & Prideaux, 2018), reservations remain. Boluk et al. (2019), for example, argue that sustainable tourism requires greater consideration of Indigenous world views because the lingering effects of colonisation entrench inequity, but also because "Indigenous ontologies can offer… a transformed global order" (p. 853).

A Māori perspective on sustainability is evident in the epistemology of mātauranga Māori (Māori knowledge), the ontology of kaitiakitanga (stewardship, guardianship), and the axiology of whakapapa (genealogy), which are indicative of a socioecological framework that is instructive for sustainability in tourism (Harmsworth & Awatere, 2013; Reid & Rout, 2020; Rout et al., 2020). Kaitiakitanga is accepted within resource management law and policy as a fundamental principle guiding the relationship between tribes and their ancestral lands (Crengle, 1993; Kawharu, 2000). The effect of kaitiakitanga is to prioritise environmental considerations not as an arbitrary moral restraint on growth, but because the environment is kin, an ancestor who precedes humanity in the hierarchy of genealogical time and space (whakapapa) and upon whom humans depend for their wellbeing (Hēnare, 2015; Te Rito, 2007). Such a view may be problematic for those unused to evaluating the pragmatism of metaphysical Indigenous knowledge (Pratt, 2002), yet such knowledge illuminates a relational view of sustainability in tourism as an intrinsic ethical code rather than enforced compliance with an external legal code (Amoamo, Ruckstuhl, et al., 2018; Boluk et al., 2019).

Methodology

Indigenous methodologies

Our research is embedded in Māori cosmologies and foregrounds Indigenous methodologies (Martin & Hazel, 2020). Indigenous methodologies are culturally embedded in the diversity of Indigenous peoples' world views, institutions, circumstances and aspirations (Henry & Foley, 2018; Hudson et al., 2010; Kovach, 2010; Smith, 1999; Vivian et al., 2017). In Aotearoa New Zealand, Indigenous methodologies are those that develop Māori knowledge, improve Māori outcomes, involve Māori at all levels, and use Māori and non-Māori methods (Cunningham, 2000; Martin & Hazel, 2020). Māori-centred research and kaupapa Māori research are two such methodologies, distinguished by the degree to which Māori control and benefit from research. While Māori-centred research produces Māori knowledge, the knowledge is evaluated against dual standards—Māori and mainstream (Cunningham, 2000). Kaupapa Māori research is research by, for, and with Māori (Smith, 1999), which is culturally affirming and transformative (Smith, 1997). In kaupapa Māori research, non-Māori contributions are possible, provided the researcher's record of research "lends legitimacy to their work" (Smith et al., 2012, p. 20).

To explore our research question we used Māori-centred organisational ethnographic methods, including interviews, observation, and use of published information to produce three case studies of Māori tourism enterprises on the Whanganui River (Ybema et al., 2009). To facilitate our engagement with these enterprises, we contacted Hayden Potaka, a Māori tourism entrepreneur who agreed to assist the research. Hayden arranged visits to four Māori tourism enterprises along Te Awa Tupua over two days in April 2019, dinner with several other Maori tourism entrepreneurs at a Māori-owned restaurant, and opportunities to share preliminary findings with Māori and iwi tourism enterprises in the Whanganui-Manawatū region. Four interviews with Māori entrepreneurs were conducted ranging in duration from 60 minutes to over two hours, which were primarily in English. Participants were provided with information sheets and consent forms, and interviews were audio recorded (or notes were taken), consistent with the low risk ethics notification we received (reference number: 4000020654). Low risk means the researchers are responsible for the ethical conduct of the research.

Of the three case studies, two are private, family-owned Māori enterprises (Unique Whanganui River Experience and Whanganui River Adventure) and the third is a hapū-based enterprise (Te Ao Hou Marae). Interviews began with questions about where they were born and raised. Beginning interviews with 'origin stories' helps establish common ground through the sharing of mutually known people, places, and interests, and is an important method of Māori research centring on whakawhanaungatanga—reaffirming genealogical connections (Bishop, 1996). Questions then turned to how the enterprise came to be, its cultural values, approach to sustainability and entrepreneurial aspirations.

Positionality of the researchers

Positionality in Indigenous research involves giving an account of oneself to expose differences in perception, power, and privilege and establish grounds for commonality, relationality, and decoloniality between the researcher and the researched (Moffat, 2016). While collaboration between Indigenous and non-Indigenous researchers brings diverse positionalities, Jones and Jenkins (2014) suggest a more uneasy relationship is desirable based on learning from, rather than about, Indigenous people lest the coloniality that one seeks to displace is instead invigorated. The authors of this paper are of different ethnicities and genders. The first coauthor is male and identifies as Māori (New Zealander of Māori descent), but recognises his Scottish, French and Irish ancestry, whose research adheres to kaupapa Māori theory, but integrates within this elements of Western knowledge. The second coauthor is female and identifies as Pākehā (New Zealander of European descent) with Netherlands and Slovenian ancestry, whose upbringing imbued her with an affinity for the bio-cultural diversity of Aotearoa, a strong sense of social justice, and a deep respect for Māori and Pasifika peoples. Our respective positions have influenced the research by affirming a Māori world view; by the complementarity of using non-Māori knowledge for Māori-defined purposes; and the requisite acknowledgement of university standards for what counts as knowledge.

Findings

Unique Whanganui River experience

Unique Whanganui River Experience is a private Māori tourism enterprise owned and operated by Hayden Potaka, which he started in 2018. The enterprise specialises in providing guided canoe tours through the "heart of the Whanganui River, places inaccessible by land" (Unique Whanganui River Experience, 2020, p. 1). Visitors are hosted by a "team of cultural navigators" who have all completed the Tira Hoe Waka—an annual 12-day spiritual journey by canoe along the Whanganui River for iwi members who learn the culture, history, and language of their iwi (Stowell, 2020). A unique feature of some of the products offered is a gourmet dining experience incorporating produce of the bush (Unique Whanganui River Experience, 2020). Trips range from 3-5 days for groups of 4-10 guests.

On starting the business, Hayden says "I gave up my job and I saw this business on [an online auction site]—[a] canoeing business on the Whanganui River. My son is called Ahurei. Ahurei means unique. So I thought: why not? So, I bought the Unique Whanganui River Experience." Hayden grew up on the river and wanted to share his experience and his love of the river with others, while incorporating elements that he had developed in other ventures:

> *At the end of last year… I thought, 'Bloody hell, I'm going to get this sort of thing going on the river, to mix in sort of fine dining and then also canoeing'. Just really connecting with the river and really gaining that spiritual connection… get them trained for river crafting, and then back to the site to… have a cocktail before dinner and just chill out and mingle.*

A key motivation for Hayden in starting the venture was ensuring his children were well connected to their culture. He explains: "I think the biggest aspiration is for my kids to know and identify themselves with the river." Hayden found it was essential to have confidence in his ability to get the business going because "there was nobody out there supporting businesses … [so I] just went and did it."

On the spirituality of the river, Hayden says:

> *I developed [these tour packages] because I believed that there's a spiritual side to the river and people who want to disengage from city life and come out and just engage with nature is magic. When you're able to do that, in the solitude of the middle reaches of the river, and get the real understanding of the whakataukī [proverb]– ko au te awa, ko te awa ko au. So, I am the river and the river is me, and what that means.*

We elaborate further on the spiritual dimension to Māori tourism in the next case study.

Te Ao Hou Marae

A marae is a sacred courtyard in the front of a tribal meeting house surrounded by other buildings, which has a central role in Māori communities (Karetu, 1992; Tauroa & Tauroa, 1986). Te Ao Hou Marae was established in 1978. The marae is situated in Aramoho on the banks of the Whanganui River near the Whanganui township. Geoff Hipango, who was raised in a house adjacent to the marae and is the marae chairperson, provided an account of the marae and its engagement in tourism. Geoff has resided full time on the marae since 2017. His mission is to revive the marae. In the context of "the door being open for our community [and] our tribal gatherings, tourism came out as a possible strategy to look at some sustainable economic development." This "coincided with… the recognition of the Whanganui River and its status," which has received worldwide attention.

As more people arrived wanting to learn about the Whanganui River, the marae was offered as a space for Māori tourism operators to prepare visitors or debrief them. A collaboration emerged that worked—the operators organise the tours; the marae hosts the visitors and "make them feel welcome." This negated the need for the marae to engage in direct marketing of its offering. For Geoff, "rather than catering for the bus loads," the focus has been on hosting smaller groups with an environmental research interest. This has "allowed us to share stories… of why we keep saying, 'I am the river and the river is me,' [and] actually our iwi narrative is important for the world to know, especially… around global climate change." The intimacy of the river experience affects some visitors very deeply: "[When] they did come back here, some of them were breaking down crying; there was real transformation that occurred. Really, I think part of that is… a connection back to nature to see yourself through a different lens."

An experience of the river from the vantage point of the marae is something that not only appeals to international visitors, but also to domestic visitors. Geoff explains: "You know, there's this saying that we use, 'globally hot, locally not'. We've got a lot within our community that live here that have never been on a marae… [and] if we do need to shift in terms of relationships then there has to be more importance on the [domestic market, particularly, among non-Māori]." There is a view that sees the legal status of the Whanganui River as an unrealised economic gain, but there is an important distinction to be grasped, Geoff suggests, and that is, that the Whanganui River and Te Awa Tupua "are not necessarily the same thing." Te Awa Tupua is an "iwi narrative… that is best articulated in [the] authentic environment of marae… with our people." Yet, when looking through local tourism publicity, Geoff notes that "[the marae is] not visible," which he attributes in part to a slower pace of development. With the Māori tourism operators, though, "there's a trust, we've got decades… of knowing one another… we're on the same wavelength; it's storytelling in a combination."

Whanganui River Adventures

Whanganui River Adventures (2019) is a Māori tourism enterprise based at Pipiriki, a settlement along the middle reaches of the Whanganui River, owned and operated by Ken and Josephine Haworth as a family business. Josephine's husband Ken carries out the maintenance, and Ken, Josephine's brother, and their nephew take the tours. Both Ken and Josephine's families have been involved in jet boating for many years. In 1998, they purchased the family business, Pipiriki Jet Boat Tours, and in 2003 purchased Riverspirit Jet Boat Tours, combining it with their existing business to form Whanganui River Adventures. The offering includes guided jet boating, canoe and biking tours, ranging from a few hours to several days. In the busy season (October to April), the business employs 5-7 staff, almost all of whom are whānau (family), which is a deliberate strategy to keep locals employed. Josephine recalls that there was once 300 or more people living at Pipiriki, but now the population is around 30 people. Without tourism, Josephine says there would be even fewer.

The business is located in an old school that is leased from a Māori land incorporation: this "puts the pūtea [funds] back into our own people," says Josephine. Josephine says that "because [the business is on] Māori land it makes it hard for us as a Māori business—with banks, we can't get loans. [We] have to do things the hard way—saving. We built this business slowly to what we could afford."

Josephine says a distinctive feature of the tour is the history:

> *It's something we don't have to read about. Outsiders try to copy what we do, especially the history side of things, but it's a lived history of growing up on the river and on Māori land. We get asked all the time since the river gained legal status, what does the awa [river] mean to you? It's our food basket, our workplace, our place to play… It's always been that.*

On sustainability, Josephine says,

> *We do one tour per day, so there's less impact. With four boats, that's 60 people max a day. We won't do extra trips even if there's more demand—we'll pass those customers on to others; we'd prefer to share the load. [In Māori tourism fora] our discussions are about quality, not quantity, and culture. At general tourism forums it's all about the numbers.*

There is a strong ethic of reciprocity and wanting to support the community. The business invested in the school when it was operating, helped with the playground, and provided employment for locals. But, as Josephine adds, "it's hard in a little community, too. Some see us as 'Oh, they've got plenty of money'… we just focus on moving forward."

Discussion

Indigeneity

Indigenous tourism cannot exist without indigeneity. Indigeneity denotes a metaphysical world view whose central premise is the interrelatedness of all things. Indigeneity orients human thought toward holism and human action toward the wellbeing of all things. What is contentious is the extent to which the Indigenous people, whose indigeneity is on offer in tourism, have control of the journey through local narrative. Although coloniality has substantially displaced indigeneity, recognition of Indigenous rights, Indigenous self-determination, and Indigenous knowledge provide epistemic space for indigeneity to inform Indigenous enterprising ontologies (Awatere et al., 2017; Dana, 2015; Mika, 2020; Peredo & McLean, 2013).

The tribal narrative—ko au te awa, ko te awa ko (I am the river, the river is me)—evidences a socioecological unity between the Whanganui iwi (tribe) and the Whanganui awa (river) that has been promulgated in law, enabled by treaty settlement, and enacted by tribal institutions. Māori tourism operators act as authentic voices and stewards of tribal indigeneity because they are of the Whanganui River.

Three key findings on indigeneity in Indigenous tourism are evident in the case studies: (1) tribal indigeneity has meaning for Māori and non-Māori that is both spiritual and material, but it must be governed, preserved and shared by tribes; (2) sustaining tribal indigeneity is important for the identity and wellbeing of current and future generations of tribal members, which Māori tourism operators contribute to and draw from as motivation for initiating enterprise and the capability for delivering culturally authentic offerings; (3) while legal personality of the river via settlement has brought public attention to the innovation and substance of tribal indigeneity, it is susceptible to misrepresentation, appropriation, and indifference because the extent to which tribal values (Tupua te Kawa) are generally applicable on the river is still being established.

The research finds that the spiritual dimension of tribal indigeneity is consistent with syncretistic spirituality of the Rangiwai (2018) framework of atuatanga—a blending of Christian religiosity and Māori spirituality. This syncretism manifests as spiritual capital embodied within Māori entrepreneurs (Mika, 2018), with the potential for tourists to experience spiritual wellbeing in their encounters with the river. Further, the maintenance of tribal knowledge through the immersion of tribal members in their culture facilitates and supports the identity work of indigeneity, exemplified by tira hoe waka. The case studies reveal that the unique identity of the river is prone to being commodified by industry and government for the collective good of all without adequate consideration of Indigenous perspectives, rights, and interests. Yet, Māori institutional capacity at the iwi (Ngā Tangata Tiaki o Whanganui), industry (Whanganui Māori Regional Tourism Organisation) and firm-level (Māori tourism operators) exists to contribute to a more balanced view of the role of indigeneity in the wellbeing and prosperity of the community through tourism.

Peace

In the literature, tourism is viewed as either a catalyst or beneficiary of peace, where it is generally defined negatively as an absence of war and violence. Yet, idealisations of peace in transcendent terms as values of unity and harmony come close to the Indigenous predicate of the interrelatedness of all things, constituted in reciprocal relations among humans and between humans and nonhumans. Beyond the idealism of peace as harmonious relations, deference to the Māori epistemology of the god of peace—Rongomātāne—reveals cultural ontologies of peace-making, which have historically manifested as post-conflict peace pacts between tribes and nonviolent resistance to colonial conquest (Karena, 2020). The Treaty of Waitangi provided terms for peace conditional on preserving tribal self-determination and shared prosperity that were negated by colonisation and are now finding restitution via treaty settlements. For Māori, realising a positive vision of peace is contingent on appropriate and equitable power sharing through constitutional reform, which honours the treaty and enacts decoloniality (Boulton & Te Kawa, 2020; Durie, 1998).

The research suggests that peace in Indigenous tourism is about power sharing, access to fair and adequate means by which to live a good life, and acceptance of the legitimacy of Indigenous notions of peace and peace-making. The case studies show that peace has several dimensions: (1) an openess to encountering the river as a spiritual and material being; (2) engaging in tourism enterprise on the river to fulfil aspirations for cultural continuity and shared well-being; and (3) an evolving power sharing between Māori and the Crown.

In two of the cases, tourists found peace as personal tranquility and healing through curated encounters with the river. As these cases highlight, peace is conditional on the means to sustain physical, cultural, and spiritual needs. Māori tourism enterprises demonstrated this by participating at scales consistent with their identity and aspirations. For example, Te Ao Hou Marae saw its role in tourism as a medium for authentic sharing of the tribal narrative and sustaining itself as a marae through enterprise. Economic gain was a lower-order priority for Unique Whanganui River Experience than ensuring family and tribal members had the opportunity to maintain and share their culture. Peace for Whanganui River Adventures is characterised as village harmony. In Māori terms, this may be expressed as hau kāinga—the vitality of the village, where the people of the land on which the enterprise operates contribute to and benefit from the tourism offering. Fair and appropriate power sharing between Māori and non-Māori lies at the heart of peaceful coexistence and shared prosperity for Māori generally, and for Māori tourism in particular. The Whanganui River treaty settlement provides a basis for continuing peace between Māori and the Crown, local government, and mainstream industry, but the role of the settlement in tourism is still unfolding.

Justice

Our analysis of the literature suggests that justice in Indigenous tourism is about fair and equitable access to the means by which to live a good life, restoring relationships by healing past trauma and legitimising indigeneity (Hēnare, 2011; Mutu, 2019; Watene, 2016). Treaty settlements are the primary policy instrument through which justice for Māori in these terms is available. Settlements are intended to provide legal and financial means for rebuilding tribal nations. Tribal nation-building is, however, contingent on the sufficiency of the redress, the quality of post-settlement governance, and the value of properties available, among other considerations (Te Aho). Treaty settlements have been criticised as preserving postcolonial dominance and granting tribes limited statutory power and resources (Kawharu, 2018). Thus, it falls to tribal entrepreneurs to pursue Indigenous tourism rather than the tribe itself (Mika et al., 2019).

In the Whanganui River settlement, four provisions for justice are evident: (1) recognising Te Awa Tupua in law as a person; (2) the codification of Indigenous values for the river (Tupua te Kawa, see Figure 2); (3) the provision of a human voice to speak for the river (Te Pou Tupua); and (4) financial redress and a governance entity to fulfil post-settlement obligations and aspirations. The effect of these provisions is to oblige tourism institutions to engage in treaty-based relationships with the iwi, establish an ethical, moral and pragmatic framework for sector and firm-level engagement with the river, and to guide entrepreneurs on their obligation to uphold the indigeneity of the river. All three case studies recognise the significance of the Whanganui River settlement, but its implications for tourism on the river generally, and for their particular enterprises, is less clear. Unique Whanganui River Experience found enterprise assistance from the iwi was unavailable for investment in a new business. Te Ao Hou Marae spoke about the importance of sharing the tribal narrative with the world, and with local non-Māori who may not have had a marae experience. Whanganui River Adventures is concerned with providing livelihoods for community members to remain living on Māori land by the river. Justice in Indigenous tourism on the river arises in three ways: (1) what the tribe can do for the river and tribal members as a consequence of the treaty settlement despite the constraints; (2) what tribal entrepreneurs like those in our case studies can do to provide means by which to determine and lead a 'good life' (Hēnare, 2011); and (3) what the tourism system can do to uphold the mana (power and authority) of the river and of the tribe.

Sustainability

Sustainability in Indigenous tourism is about the extent to which a kinship-based approach, rather than simply a goal-oriented approach, can be operationalised to help the tourism system, sector, and operators meet the needs of current and future generations. When human and non-human elements are viewed as kin (Harmsworth & Awatere, 2013), a new socioecological framework for sustainability grounded in Indigenous knowledge emerges that requires careful articulation and evaluation. This framework fundamentally centres on the principle of kaitiakitanga (stewardship, guardianship), which obliges responsible use and protection of natural resources for the coexistence and shared wellbeing of people and nature (Kawharu, 2000). While kaitiakitanga is widely used in environmental policy and practice across mulitple sectors (Cherrington, 2019; Hutchings et al., 2020; Thompson, 2018), operationalising and evaluating its effect in business is relatively recent (Maxwell et al., 2020; Reid & Rout, 2020; Spiller et al., 2011).

In this research, we found that tourism is not all about the numbers—customers and revenue. Unbridled growth of both tourists and enterprises on the river falls outside the ambit of what constitutes sustainable tourism, according to the Māori tourism operators with whom we spoke. Rather than burden the river with the waste that additional visitors create, operators would forgo tours and pass them on to other operators who have a higher propensity and capacity for growth. It appears, then, that there may be a socioecological dissonance between Māori and non-Māori tourism operators regarding commercial growth. By this we mean that the social and ecological limits that Māori tourism operators work to are set at levels that differ from industry norms; however, we cannot argue this with certainty as the views of non-Māori operators were not canvassed in this research.

This presents a conundrum: how to reconcile different perspectives on the growth of the tourism sector with regard to planetary boundaries? This is a question of sustainability that is unlikely to be easily settled by tourism operators alone. Instead, institutional actors must be invited to help, starting with Te Awa Tupua itself and the voice of the awa (river), Te Pou Tupua. Ngā Tangata Tiaki o Whanganui, and the peoples of Hinengākau, Tama Ūpoko, and Tūpoho, must also be involved, followed by the principal regulators—the Department of Conservation, the Whanganui District Council, and tourism sector organisations, including the Whanganui Māori Regional Tourism Organisation. The perspectives of these actors remain questions for further research, policy, and practice. Next, we propose a model of Indigenous tourism to bring theory and findings together.

Foundations for Indigenous tourism—kaupapa tāpoi

In Figure 3, intersections of maungārongo (peace), manatika (justice), kaitiakitanga (sustainability), te ao Māori (indigeneity), and iwi (tribe) are depicted as kaupapa tāpoi—foundations for Indigenous tourism. The model privileges Māori language, Māori concepts and Māori symbolism because of the focus on Te Awa Tupua, the Whanganui River, but the principles may be generalisable as foundations for Indigenous tourism elsewhere.

Figure 3. Foundations for indigenous tourism—kaupapa tāpoi.

The foundations—kaupapa tāpoi—are enclosed in a circle, which signifies the unity, perpetuity, and holism of Papatūānuku, Earth Mother, represented in the model by te taiao—the natural environment—as the foundation from which other elements grow. While individual tribal entrepreneurs as Māori tourism operators were the focus of this research, they acknowledge the tribe as the custodian of tribal narrative. Thus, te ao Māori (a Māori world view) offers a view of indigeneity particular to Aotearoa New Zealand, but it more precisely manifests locally as the indigeneity of hapū and iwi knowledge around which Indigenous tourism *in situ* is configured. Intersecting principles of indigeneity appear in the three koru (spiral motifs) that represent new life, growth and development, and a strong connection to a place of origin (te taiao). These principles are maungārongo—peace, manatika—justice, and kaitiakitanga—sustainability. The English and Māori words of kaupapa tāpoi are neither equivalent nor substitutes, but represent different perspectives on related principles for Indigenous tourism. For instance, treaty settlements are intended as instruments of manatika (justice) by addressing past grievances, maungārongo (peace) by redefining Māori and Crown relationships, and kaitiakitanga (sustainability) by enabling shared responsibility for tribal wellbeing. The effect is self-determined tribal development, with Indigenous tourism an example of this.

A further example of the intersection of kaupapa tāpoi is the immersion of tribal members in tribal culture, which we see as the identity work of indigeneity. The effect is to strengthen cultural capacity for Indigenous tourism, producing, for example, cultural navigators as knowledgeable, authentic and engaged operators. Relatedly, the syncretism of atuatanga (blending of religiosity and spirituality) represents a distinct basis for Indigenous tourism, which operators report has had therapeutic effects for tourists. While inward (tribal) and outward (tourist) acculturations serve different ontologies, they depict reciprocal intersections of maungārongo, manatika, and kaitiakitanga for host and visitor. Finally, kaitiakitanga is indicative of an Indigenous socioecological framework that prioritises environmental considerations because of the interrelatedness of all things. Taken together, kaupapa tāpoi may provide foundations for Indigenous tourism research, policy and practice.

Conclusion

This paper set out to discuss peace, justice, and sustainability through Indigenous tourism using case study research of Māori tourism enterprises on Te Awa Tupua, the Whanganui River. The Whanganui iwi have, through the treaty settlement process, had their relationship with their ancestor—Te Awa Tupua—reaffirmed, and are now embarking on new era of development. While the legal status of Te Awa Tupua as a person is a modern-day marvel, it gives effect to an 800-year-old spiritual socioecological relationship that has sustained the people and the river.

Indigenous tourism—the tourism enterprises, assets and activity of Indigenous peoples—offers a unique perspective on notions of peace, justice, and sustainability. As descendents of the original peoples of colonised territories, a peaceful postcolonial existence is problematic as inequity, inequality, and precarity prevail for many Indigenous peoples. Yet, the promise of a self-determining and decolonising development agenda, while possibly unsettling for some, presents a potential path to peace, justice, and sustainability through Indigenous tourism. This paper identifies five main principles that support this effort. First, treaty settlements present a framework for self-determined tribal development, with Indigenous tourism an example of this. Second, the identity work of immersing tribal members in their tribal indigeneity strengthens individual and collective capacity for Indigenous tourism. Third, Indigenous spirituality, evident in the syncretism of atuatanga, represents a distinct basis for Indigenous tourism. Fourth, an Indigenous spiritual socioecological framework that prioritises environmental considerations because of the interrelatedness of all things differentiates Indigenous tourism. Fifth, a process for recalibrating the boundary of growth in tourism is intimated by the kaupapa tāpoi model, but requires further research. These principles are constitutive of an Indigenous approach to tourism that present development advantages for both Indigenous and non-Indigenous entrepreneurs.

Glossary of Māori words

ahurei	unique
ao Māori	Māori world view, Māori society
ao	world, light
Aotearoa	Land of the long white cloud, New Zealand
Ātihaunui	principal tribe of the Whanganui region
atua	god, deity
atuatanga	all things pertaining to gods and God
awa	river
hapū	subtribe
hau kāinga	vitality of the village, home people
hohou i te rongo	peace-making process
hou	new
iwi	tribe
kai	food, prefix added to verbs to make them nouns
kaitiakitanga	guardianship, stewardship, sustainability
kaupapa	purpose, principle, philosophy
koha	gift
koru	spiral motif, looped, coiled, fold
mana motuhake	separate identity, autonomy, self-determination
mana	power authority, prestige acquired by divinity and deed
manaakitanga	hospitality, kindness, generosity, care for people and land
manatika	justice
manuhiri	visitor
Māori	Indigenous people of New Zealand
marae	courtyard in front of a carved meeting house
mātauranga	knowledge
maungārongo	peace
muru	ritual compensation, forgive, pardon

Ngā Tangata Tiaki o Whanganui	Post-settlement governance entity of the Whanganui tribes
Pākehā	New Zealanders of European descent
Papatūānuku	Earth mother
Pipiriki	A settlement on the Whanganui River
pūtea	fund, finance, bank account, sum of money
Rānana	A settlement on the Whanganui River
rangatira	chief (male or female), leader, noble, high ranking person
Ranginui	Sky father
Rangitīkei	A river valley and district bordering Whanganui
Rongomaraeroa	god of peace
Rongomātāne	god of peace
Ruakā Marae	Tribal village courtyard and buildings on Whanganui River
Ruapehu	Ancestral mountain within the Tongariro National Park
Ruruku Whakatupua	Deed of Settlement of the Whanganui River
taiao	world, earth, natural environment
tangata whenua	people of the land, local people, Indigenous people
tangata	person, man, human
taonga	treasure, highly prized object, something of value
tāpoi	tourism
Te Ātihaunui-a-Pāpārangi	tribe of the Whanganui River
Te Awa Tupua	Ancestral Māori name for the Whanganui River
Te Pou Tupua	Human agents who act on behalf of Te Awa Tupua
tikanga	correct procedure, custom, rule, value and practice
tino rangatiratanga	self-determination, sovereignty, autonomy
Tira Hoe Waka	A canoe journey on the Whanganui River for iwi members
Tupua te Kawa	Customary laws and values for the Whanganui River
utu	reciprocity, payment
wānanga	meet, discuss, deliberate, consider
whakapapa	genealogy
whakataukī	proverb
whakawhanaungatanga	process of establishing relationships, relating well to others
whānau	family, extended family
Whanganui	A district in the lower North Island
whenua	land, placenta

Acknowledgements

We acknowledge te mana o Te Awa Tupua (the Whanganui River) and te mana o te iwi o Whanganui (the authority of the Whanganui tribes). We thank Associate Professor Hone Morris of Te Pūtahi-ā-Toi, Massey University, for sharing his mātauranga (knowledge), helping bring clarity and depth to Māori cultural concepts. We thank Hayden Potaka of the Whanganui Māori Regional Tourism Organisation for his support in facilitating our research on the Whanganui River. We thank the Māori tourism operators of Whanganui for their manaakitanga (generosity) in hosting us and sharing their knowledge. We thank June Lincoln and Fiona Brown of the Massey University Design Studio for turning a hand drawn depiction into a publishable image.

Disclosure statement

No potential conflict of interest was reported by the authors.

ORCID

Jason Paul Mika http://orcid.org/0000-0002-7614-8349
Regina A. Scheyvens http://orcid.org/0000-0002-4227-4910

References

Amoamo, M., Ruckstuhl, K., & Ruwhiu, D. (2018). Balancing indigenous values through diverse economies: A case study of Māori ecotourism. *Tourism Planning & Development*, *15*(5), 478–495. https://doi.org/10.1080/21568316.2018.1481452

Amoamo, M., Ruwhiu, D., & Carter, L. (2018). Framing the Māori economy: The complex business of Māori business. *MAI Journal: A New Zealand Journal of Indigenous Scholarship*, *7*(1), 66–78. https://doi.org/10.20507/MAIJournal.2018.7.1.6

Amoamo, M., & Thompson, A. (2010). (Re)imaging Māori tourism: Representation and cultural hybridity in postcolonial New Zealand. *Tourist Studies*, *10*(1), 35–55. https://doi.org/10.1177/1468797610390989

Anaya, J. S. (2004). *Indigenous peoples in international law*. Oxford University Press.

Awatere, S., Mika, J. P., Hudson, M., Pauling, C., Lambert, S., & Reid, J. (2017). Whakatipu rawa mā ngā uri whakatipu: Optimising the 'Māori' in Māori economic development. *Alternative: An International Journal of Indigenous Peoples*, *13*(2), 80–89. https://doi.org/10.1177/1177180117700816

Barnett, S. (2001). Manaakitanga: Maori hospitality–A case study of Maori accommodation providers. *Tourism Management*, *22*(1), 83–92. https://doi.org/10.1016/S0261-5177(00)00026-1

Barry, P., Graham, L., & Barry, M. (2020). *Iwi investment report 2019*. https://www.tdb.co.nz/wp-content/uploads/2020/02/Iwi-Investment-Report-2019.pdf

Belich, J. (1998). *The New Zealand wars and the Victorian interpretation of racial conflict*. Penguin.

Best, E. (1976). *Māori religion and mythology part 1: An account of the cosmogony, anthropogeny, religious beliefs and rites, magic and folk lore of the Māori folk of New Zealand* (2nd ed.). Government Printer.

Bishop, R. (1996). Collaborative research stories: Whakawhanaungatanga. Dunmore Press.

Boluk, K. A., Cavaliere, C. T., & Higgins-Desbiolles, F. (2019). A critical framework for interrogating the United Nations Sustainable Development Goals 2030 Agenda in tourism. *Journal of Sustainable Tourism*, *27*(7), 847–864. https://doi.org/10.1080/09669582.2019.1619748

Boulton, A., & Te Kawa, D. (2020). *Raising waka, and not just yachts*. https://www.psa.org.nz/assets/Uploads/ProgressiveThinking-Dr-Amohia-Deb-Te-Kawa-Raising-Waka-edited.pdf

Bremner, H. (2013). Tourism development in the Hot Lakes District, New Zealand c. 1900. *International Journal of Contemporary Hospitality Management*, *25*(2), 282–298. https://doi.org/10.1108/09596111311301649

Brundtland, G. H. (1987). *Report of the World Commission on Environment and Development: Our common future*. https://sustainabledevelopment.un.org/content/documents/5987our-common-future.pdf

Bunten, A. C., & Graburn, N. H. H. (Eds.). (2018). *Indigenous tourism movements*. University of Toronto Press.

Burrows, J., O'Regan, T., Chin, P., Pihama, L., Coddington, D., Poutu, H., Cullen, M., Smith, L. T., Luxton, J., Tennent, P., Mene, B., & Walker, R. (2013). *New Zealand's constitution: A report on a conversation: He kōtuinga kōrero mō te kaupapa ture o Aotearoa*. Constitutional Advisory Panel - Te Ranga Kaupapa Ture, Department of the Prime Minister and Cabinet and Office of the Minister of Māori Affairs.

Butler, R. (2018). Sustainable tourism in sensitive environments: A wolf in sheep's clothing? *Sustainability*, *10*(6), 1789–1711. https://doi.org/10.3390/su10061789

Cherrington, M. (2019, November). Environmental social and governance sustainability: Ka mua, ka muri. *Scope: Contemporary Research Topics (Learning & Teaching)*, *2019*(8), 51-56. https://doi.org/10.34074scop.4008019

Colbourne, R. (2017). An understanding of Native American entrepreneurship. *Small Enterprise Research*, *24*(1), 49–61. https://doi.org/10.1080/13215906.2017.1289856

Cole, A. (2017). Towards an indigenous transdisciplinarity. *Transdisciplinary Journal of Engineering & Science*, *8*(1), 127–150. https://doi.org/10.22545/2017/00091

Crengle, D. (1993). *Taking into account the principles of the Treaty of Waitangi: Ideas for the implementation of Section 8, Resource Management Act 1991*. M. f. t. Environment.

Cunningham, C. (2000). A framework for addressing Māori knowledge in research, science and technology. *Pacific Health Dialog*, *7*(1), 62–69.

D'Amore, L. (1988). Tourism: The world's peace industry. *Journal of Travel Research*, *27*(1), 35–40. https://doi.org/10.1177/004728758802700107

Dana, L.-P. (2015). Indigenous entrepreneurship: An emerging field of research. *International Journal of Business and Globalisation*, *14*(2), 158–169. https://doi.org/10.1504/IJBG.2015.067433

Durie, M. H. (1998). *Te mana, te kāwanatanga: The politics of Māori self-determination*. Oxford University Press.

Durie, M. H. (2004, February 17). *Public sector reform, indigeneity and the goals of Māori development*. In *Commonwealth Advanced Seminar*.

Durie, M. H. (2017). *Pūmau tonu te mauri: Living as Māori, now and in the future*. file:///C:/Users/jpmika/Downloads/Pumau%20Tonu%20te%20Wairua%20Mason%20Durie%20(1).pdf

Dutta, M. J. (2015). Decolonizing communication for social change: A culture-centered approach. *Communication Theory*, *25*(2), 123–143. https://doi.org/10.1111/comt.12067

Dutta, M. J., & Elers, S. (2020). Public relations, indigeneity and colonization: Indigenous resistance as dialogic anchor. *Public Relations Review*, *46*(1), 101852. https://doi.org/10.1016/j.pubrev.2019.101852

Firth, R. (1929). *Primitive economics of the New Zealand Māori* (2nd ed.). Routledge.

Gladstone, J. (2018). All my relations: An inquiry into a spirit of a Native American philosophy of business. *American Indian Quarterly*, *42*(2), 191–214. https://doi.org/10.5250/amerindiquar.42.2.0191

Hall, C. M. (2019). Constructing sustainable tourism development: The 2030 agenda and the managerial ecology of sustainable tourism. *Journal of Sustainable Tourism*, *27*(7), 1044–1017. https://doi.org/10.1080/09669582.2018.1560456

Harmsworth, G. R., & Awatere, S. (2013). Indigenous Māori knowledge and perspectives of ecosystems. In J. R. Dymond (Ed.), *Ecosystem services in New Zealand: Conditions and trends* (pp. 274–286). Manaaki Whenua Press.

Harris, L. D., & Wasilewski, J. (2004). Indigeneity, an alternative worldview: Four R's (relationship, responsibility, reciprocity, redistribution) vs. two P's (power and profit). Sharing the journey towards conscious evolution. *Systems Research and Behavioral Science*, *21*(5), 489–503. https://doi.org/10.1002/sres.631

Harrison, D. (Ed.). (2001). *Tourism and the less developed world: Issues and case studies*. CABI.

Hēnare, M. (2001). Tapu, mana, mauri, hau, wairua: A Māori philosophy of vitalism and cosmos. In J. A. Grim (Ed.), *Indigenous traditions and ecology* (pp. 197–221). Harvard University Press.

Hēnare, M. (2011). Lasting peace and the good life: Economic development and the 'āta noho' principle of Te Tiriti o Waitangi. In V. M. H. Tawhai & K. Gray-Sharp (Eds.), *Always speaking: The Treaty of Waitangi and public policy* (pp. 261–276). Huia.

Hēnare, M. (2015). Tapu, mana, mauri, hau, wairua: A Māori philosophy of vitalism and cosmos. In C. Spiller & R. Wolfgramm (Eds.), *Indigenous spiritualities at work: Transforming the spirit of enterprise* (pp. 77–98). Information Age Publishing.

Hēnare, M. (2018). "Ko te hau tēnā o tō taonga ... ": The words of Ranapiri on the spirit of gift exchange and economy. *Journal of the Polynesian Society*, *127*(4), 451–463. https://doi.org/10.15286/jps.127.4.451-463

Henry, E., & Foley, D. (2018). Indigenous research: Ontologies, axiologies, epistemologies and methodologies. In L. A. E. Booysen, R. Bendl, & J. K. Pringle (Eds.), *Handbook of research methods in diversity management, equality and inclusion at work* (pp. 212–227). Edward Elgar.

Henry, E., Newth, J., & Spiller, C. (2017). Emancipatory Indigenous social innovation: Shifting power through culture and technology. *Journal of Management & Organization*, *23*(6), 786–802. https://doi.org/10.1017/jmo.2017.64

Higgins-Desbiolles, F., & Whyte, K. P. (2013). No high hopes for hopeful tourism: A critical comment. *Annals of Tourism Research*, *40*, 428–433. https://doi.org/10.1016/j.annals.2012.07.005

Horn, C., & Tahi, B. (2009). Some cultural and historical factors influencing rural Māori tourism development in New Zealand. *Journal of Rural and Community Development*, *4*(1), 84–101.

Hudson, M., Milne, M., Reynolds, P., Russell, K., & Smith, B. (2010). *Te ara tika: Guidelines for Māori research ethics: A framework for researchers and ethics committee members*. https://www.hrc.govt.nz/sites/default/files/2019-06/Resource%20Library%20PDF%20-%20Te%20Ara%20Tika%20Guidelines%20for%20Maori%20Research%20Ethics_0.pdf

Hutchings, J., Smith, J., Taura, Y., Harmsworth, G., & Awatere, S. (2020). Storying kaitiakitanga: Exploring Kaupapa Māori land and water food stories. *MAI Journal: A New Zealand Journal of Indigenous Scholarship*, *9*(3), 183–194. https://doi.org/10.20507/MAIJournal.2020.9.3.1

Jacobsen, D. (2017). Tourism enterprises beyond the margins: The relational practices of Aboriginal and Torres Strait Islander SMEs in remote Australia. *Tourism Planning & Development*, *14*(1), 31–49. https://doi.org/10.1080/21568316.2016.1152290

Jones, A., & Jenkins, K. (2014). Rethinking collaboration: Working the indigene-colonizer hyphen. In N. K. Denzin, Y. S. Lincoln, & L. T. Smith (Eds.), *Handbook of critical and indigenous methodologies* (Online ed., pp. 471–486). Sage.

Justinian, F. (1913). The Institutes of Justinian: Translated into English by J. B. Moyle. *Project Gutenberg EBook of The Institutes of Justinian, by Caesar Flavius Justinian*. http://www.gutenberg.org/files/5983/5983-h/5983-h.htm

Karena, T. (2020). Reclaiming the role of Rongo: A revolutionary and radical form of nonviolent politics. In R. Jackson, J. Llewellyn, G. Leonard, A. Gnoth, & T. Karena (Eds.), *Revolutionary nonviolence: Concepts, cases and controversies* (1st ed., pp. 200–223). Zed Books.

Karetu, T. (1992). Language and protocol of the marae. In M. King (Ed.), *Te ao hurihuri: Aspects of Māoritanga* (pp. 28–41). Reed.

Kawharu, I. H. (Ed.). (1989). *Waitangi: Maori and Pakeha perspectives of the Treaty of Waitangi*. Oxford University Press.

Kawharu, M. (2000). Kaitiakitanga: A Maori anthropological perspective of the Maori socio-environmental ethic of resource management. *The Journal of the Polynesian Society, 109*(4), 349–370.
Kawharu, M. (2018). The 'unsettledness' of treaty claim settlements. *The Round Table, 107*(4), 483–492. https://doi.org/10.1080/00358533.2018.1494681
Kelsey, J. (1990). *A question of honour? Labour the Treaty.* Allen & Unwin.
Knudtson, P., & Suzuki, D. (1997). *Wisdom of the elders* (2nd ed.). Allen & Unwin.
Kovach, M. (2010). *Indigenous methodologies.* Toronto University Press.
Kymlicka, W. (1995). *Multicultural citizenship: A liberal theory of minority rights.* Clarendon Press.
Kymlicka, W. (2002). *Politics in the vernacular: Nationalism, multiculturalism and citizenship.* Oxford University Press.
Lightfoot, S. R. (2016). *Global indigenous politics: A subtle revolution.* Routledge.
Litvin, S. W. (1998). Tourism: The world's peace industry? *Journal of Travel Research, 37*(1), 63–66. https://doi.org/10.1177/004728759803700108
Martin, W.-J., & Hazel, J.-A. (2020). *A guide to vision mātauranga: Lessons from Māori voices in the New Zealand science sector.* https://mcusercontent.com/afe655700044a9ad6a278bac0/files/5e6b7d88-a39e-4de2-817f-23b6748c84a2/Rauika_Ma_ngai_A_Guide_to_Vision_Ma_tauranga_FINAL.pdf
Maxwell, K. H., Ratana, K., Davies, K. K., Taiapa, C., & Awatere, S. (2020). Navigating towards marine co-management with Indigenous communities on-board the waka-taurua. *Marine Policy, 111*(2020), 103722. https://doi.org/10.1016/j.marpol.2019.103722
Mead, H. M. (2017, August 23). Mātauranga Māori: te mahi, te mohio, te māramatanga – Mātauranga Māori: doing, knowing and understanding. Te Whare Wānanga o Awanuiārangi: 25 Year Anniversary 1992–2017 Lecture Series. Te Whare Wānanga o Awanuiārangi.
Meihana, P. N. (2015). The paradox of Māori privilege: Historical constructions of Māori privilege circa 1769-1940 [Unpublished PhD thesis]. Massey University.
Mendoza-Ramos, A., & Prideaux, B. (2018). Assessing ecotourism in an Indigenous community: Using, testing and proving the wheel of empowerment framework as a measurement tool. *Journal of Sustainable Tourism, 26*(2), 277–291. https://doi.org/10.1080/09669582.2017.1347176
Mika, J. P. (2014, February 4–7). Manaakitanga: Is generosity killing Māori enterprises? In P. Davidsson (Ed.), *Proceedings of the Australian Centre for Entrepreneurship Research Exchange Conference* (pp. 815–829). Queensland University of Technology.
Mika, J. P. (2018). The role of the United Nations Declaration of the Rights of Indigenous Peoples in building indigenous enterprises and economies. In S. Katene (Ed.), *Conversations about indigenous rights: The UN Declaration of the Rights of Indigenous People in Aotearoa New Zealand* (pp. 156–175). Massey University Press.
Mika, J. P. (2020). Indigenous entrepreneurship: How Indigenous knowing, being and doing shapes entrepreneurial practice. In D. Deakins & J. M. Scott (Eds.), *Entrepreneurship: A contemporary & global approach* (pp. 1–32). Sage. https://study.sagepub.com/deakins/student-resources/e-chapter-on-indigenous-entrepreneurship
Mika, J. P., Bishara, S., Kiwi-Scally, T., Taurau, M., & Dickson, I. (2016). *Te pae tawhiti: Manawatū-Whanganui Māori economic development strategy 2016 - 2040.* Te Rūnga o Ngā Wairiki-Ngāti Apa.
Mika, J. P., Colbourne, R., & Almeida, S. (2020). Responsible management: An Indigenous perspective. In O. Laasch, R. Suddaby, E. Freeman, & D. Jamali (Eds.), *Research handbook of responsible management* (pp. 260–276). Edward Elgar.
Mika, J. P., Fahey, N., & Bensemann, J. (2019). What counts as an Indigenous enterprise? Evidence from Aotearoa New Zealand. *Journal of Enterprising Communities: People and Places in the Global Economy, 13*(3), 372–390. https://doi.org/10.1108/JEC-12-2018-0102
Mika, J. P., Smith, G. H., Gillies, A., & Wiremu, F. (2019). Unfolding tensions within post-settlement governance and tribal economies in Aotearoa New Zealand. *Journal of Enterprising Communities: People and Places in the Global Economy, 13*(3), 296–318. https://doi.org/10.1108/JEC-12-2018-0104
Mika, J. P., Warren, L., Foley, D., & Palmer, F. R. (2017). Perspectives on indigenous entrepreneurship, innovation and enterprise. *Journal of Management & Organization, 23*(6), 767–773. https://doi.org/10.1017/jmo.2018.4
Mika, J. P., Warren, L., Palmer, F. R., & Jacob, N. (2018, November 13–16). *Entrepreneurial ecosystem efficacy for indigenous entrepreneurs* [Paper presentation]. Indigenous Futures: 8th Biennial International Indigenous Research Conference, Auckland University.
Miller, D. (2017). Justice. *Stanford Encyclopedia of Philosophy.* https://plato.stanford.edu/entries/justice/
Moffat, M. (2016). Exploring positionality in an Aboriginal research paradigm: A unique perspective. *International Journal of Technology and Inclusive Education, 5*(1), 750–755. https://doi.org/10.20533/ijtie.2047.0533.2016.0096
Morris, J. D. K., & Ruru, J. (2010). Giving voice to rivers: Legal personality as a vehicle for recognising Indigenous peoples' relationships to water? *Australian Indigenous Law Review, 14*(2), 49–62. http://www.jstor.org/stable/26423181
Mutu, M. (2012). Māori issues. *The Contemporary Pacific, 24*(1), 184–190. https://doi.org/10.1353/cp.2012.0020
Mutu, M. (2019). To honour the treaty, we must first settle colonisation' (Moana Jackson 2015): the long road from colonial devastation to balance, peace and harmony. *Journal of the Royal Society of New Zealand, 49*(sup1), 4–18. https://doi.org/10.1080/03036758.2019.1669670

Ngā Tāngata Tiaki o Whanganui. (2020). *Our story: Ngā tāngata tiaki.* https://www.ngatangatatiaki.co.nz/our-story/
Office of Treaty Settlements. (2014). *Ruruku whakatupua—Whanganui River deed of settlement between the Crown and Whanganui Iwi: Summary of the historical background to the Whanganui River claims of Whanganui Iwi.* https://www.govt.nz/assets/Documents/OTS/Whanganui-Iwi/Whanganui-Iwi-Whanganui-River-Deed-of-Settlement-Summary-5-Aug-2014.pdf
Office of Treaty Settlements. (2015). *Healing the past, building a future: A guide to Treaty of Waitangi claims and negotiations with the Crown.* https://www.govt.nz/assets/Documents/Red-Book-Healing-the-past-building-a-future.pdf
Orange, C. (1987). *The treaty of Waitangi.* Allen & Unwin.
Papakura, M. (1991). *Makareti: The old-time Maori.* New Women's Press.
Parliamentary Commissioner for the Environment. (2019). *Pristine, popular… imperilled? The environmental consequences of projected tourism growth.* https://www.pce.parliament.nz/media/196983/report-pristine-popular-imperilled.pdf
Peredo, A. M., & Anderson, R. B. (2006). Indigenous entrepreneurship research: Themes and variations. *International Research in the Business Disciplines, 5*, 253–273.
Peredo, A. M., & McLean, M. (2013). Indigenous development and the cultural captivity of entrepreneurship. *Business & Society, 52*(4), 592–620. https://doi.org/10.1177/0007650309356201
Pereiro, X. (2016). A review of Indigenous tourism in Latin America: Reflections on an anthropological study of Guna tourism (Panama). *Journal of Sustainable Tourism, 24*(8–9), 1121–1138. https://doi.org/10.1080/09669582.2016.1189924
Pomerleau, W. P. (2019). Western theories of justice. *Internet Encyclopedia of Philosophy.* https://iep.utm.edu/justwest/
Pratt, S. L. (2002). *Native pragmatism: Rethinking the roots of American philosophy.* Indiana University Press.
Rangiwai, B. (2018). Atuatanga and syncretism: A view of Māori theology. *Indigenous Pacific Issues, 11*(1), 653–661.
Raymond, E. M., & Hall, C. M. (2008). The development of cross-cultural (mis)understanding through volunteer tourism. *Journal of Sustainable Tourism, 16*(5), 530–543. https://doi.org/10.2167/jost796.0
Reid, J., & Rout, M. (2020). The implementation of ecosystem-based management in New Zealand – A Māori perspective. *Marine Policy, 117*, 103889–103886. https://doi.org/10.1016/j.marpol.2020.103889
Reid, J., Rout, M., & Mika, J. P. (2019). *Mapping the Māori marine economy.* https://www.sustainableseaschallenge.co.nz/tools-and-resources/mapping-the-maori-marine-economy/
Rout, M., Reid, J., & Mika, J. (2020). Māori agribusinesses: The whakapapa network for success. *AlterNative: An International Journal of Indigenous Peoples, 16*(3), 193–201. https://doi.org/10.1177/1177180120947822
Royal, T. A. C. (Ed.). (2003). *The woven universe: Selected writings of Rev Māori Marsden.* Estate of Rev Māori Marsden.
Royal, T. A. C. (2005, June 10–12). *An organic arising: An interpretation of tikanga based upon the Māori creation traditions* [Paper presentation]. Tikanga Rangahau Mātauranga Tuku Iho: Traditional Knowledge and Research Ethics Conference, Te Papa Tongarewa.
Scheyvens, R. (2011). *Tourism and poverty.* Routledge.
Scheyvens, R. (2018). Linking tourism to the sustainable development goals: A geographical perspective. *Tourism Geographies, 20*(2), 341–342. https://doi.org/10.1080/14616688.2018.1434818
Sen, A. K. (1999). *Development as freedom.* Anchor Books.
Smith, L. T. (1999). *Decolonizing methodologies: Research and indigenous peoples* (1st ed.). Zed Books.
Smith, G. H., Hoskins, T. K., & Jones, A. (2012). Kaupapa Māori: The dangers of domestication. *New Zealand Journal of Educational Studies, 47*(2), 10–20.
Smith, G. H. (1997). *The development of kaupapa Māori: Theory and praxis* [Doctor of Philosophy]. University of Auckland.
Spiller, C., Pio, E., Erakovic, L., & Henare, M. (2011). Wise up: Creating organization wisdom through an ethic of kaitiakitanga. *Journal of Business Ethics, 104*(2), 223–235. https://doi.org/10.1007/s10551-011-0905-y
Stowell, L. (2020, January 22). Big crowd of newbies for Whanganui River journey Tira Hoe Waka. *Whanganui Chronicle.* https://www.nzherald.co.nz/wanganui-chronicle/news/article.cfm?c_id=1503426&objectid=12302140
Swain, R. B. (2018). A critical analysis of the Sustainable Development Goals. In W. L. Filho (Ed.), *Handbook of sustainability science and research* (1st ed., pp. 341–355). Springer. https://doi.org/10.1007/978-3-319-63007-6
Tapsell, D., Sole, T., Armstrong, D., Mulligan, W., Skelton, K., & Bettjeman, W. (2007). *Aratohu mō ngā rōpū kaitono: Guide for claimants negotiating treaty settlements.* Crown Forestry Rental Trust.
Tauroa, H., & Tauroa, P. (1986). *Te marae: A guide to customs & protocol.* Reed Methuen.
Te Aho, L. (2017). The 'false generosity' of treaty settlements: innovation and contortion. In A. Erueti (Ed.), International Indigenous rights in Aotearoa New Zealand (pp. 99–117). Victoria University Press.
Te Awekotuku, N. (1981). *The sociocultural impact of tourism on the Te Arawa people of Rotorua, New Zealand* [Doctor of Philosophy]. University of Waikato.
Te Rito, J. S. (2007). Whakapapa: A framework for understanding identity. *MAI Review LW, 1*(3), 10.
Thompson, A. (2018). *Kaitiakitanga and genomic data: Health Research Council report.* University of Waikato.

Tretiakov, A., Felzensztein, C., Zwerg Anne, M., Mika, J. P., & Macpherson Wayne, G. (2020). Family, community, and globalization: Wayuu indigenous entrepreneurs as n-Culturals. *Cross Cultural & Strategic Management, 27*(2), 189–211. https://doi.org/10.1108/CCSM-01-2019-0025

Unique Whanganui River Experience. (2020). *Who we are.* https://uniquewhanganuiriver.co.nz/about/

UNWTO. (2020). *Recommendations on sustainable development of Indigenous tourism.* https://www.e-unwto.org/doi/epdf/10.18111/9789284421299

Venkateswar, S., Hughes, E., Kidd, C., Kendrick, J., Glauser, B., Waitere, H., McKinnon, K., & Singh, S. J. (2011). Introduction. In S. Venkateswar & E. Hughes (Eds.), *The politics of indigeneity: Dialogues and reflections on Indigenous activism* (pp. 1–15). Zed Books.

Venkateswar, S. (2018, December 5). How should we think of the Sentinel islanders? *Fieldsights, Member Voices.* https://culanth.org/fieldsights/how-should-we-think-of-the-sentinel-islanders

Verbos, A. K., Henry, E., & Peredo, A. M. (2017a). Business's effects on Indigenous aspirations and rights: An introduction. In A. K. Verbos, E. Henry, & A. M. Peredo (Eds.), *Indigenous aspirations and rights: The case for responsible business and management* (1st ed., pp. 1–9). Routledge.

Verbos, A. K., Henry, E., & Peredo, A. M. (2017b). *Indigenous aspirations and rights: The case for responsible business and management* (1st ed.). Routledge.

Virtanen, P. K., Siragusa, L., & Guttorm, H. (2020, April). Introduction: Toward more inclusive definitions of sustainability. *Current Opinion in Environmental Sustainability, 43*, 77–82. https://doi.org/10.1016/j.cosust.2020.04.003

Vivian, A., Jorgensen, M., Bell, D., Rigney, D., Cornell, S., & Hemming, S. (2017). Implementing a project within the Indigenous research paradigm: Yhe example of nation building research. In A. Porter (Ed.), *Ngiya: Talk the Law - Volume 5: Indigenous methodologies* (Vol. 5, pp. 47–74). UTS.

Waitangi Tribunal. (1999). *The Whanganui River report.* https://forms.justice.govt.nz/search/Documents/WT/wt_DOC_68450539/Whanganui%20River%20Report%201999.pdf

Waitangi Tribunal. (2016, September 19). *Translation of the te reo Māori text.* https://www.waitangitribunal.govt.nz/treaty-of-waitangi/translation-of-te-reo-maori-text/

Walker, R. (2004). *Ka whawhai tonu matou: Struggle without end* (2nd ed.). Penguin.

Watene, K. (2016). Indigenous peoples and justice. In K. Watene & J. Drydyk (Eds.), *Theorizing justice: Critical insights and future directions* (pp. 133–151). Rowman & Littlefield International.

Whanganui Iwi, & The Crown. (2014a). *Ruruku Whakatupua—Te mana o te Awa Tupua: Whanganui River deed of settlement.* https://www.govt.nz/assets/Documents/OTS/Whanganui-Iwi/Whanganui-River-Deed-of-Settlement-Ruruku-Whakatupua-Te-Mana-o-Te-Awa-Tupua-5-Aug-2014.pdf

Whanganui Iwi, & The Crown. (2014b). *Ruruku Whakatupua—Te mana o te Iwi o Whanganui: Whanganui River deed of settlement.* https://www.govt.nz/assets/Documents/OTS/Whanganui-Iwi/Whanganui-River-Deed-of-Settlement-Ruruku-Whakatupua-Te-Mana-o-Te-Iwi-o-Whanganui-5-Aug-2014.pdf

Whanganui Māori Regional Tourism Organisation. (2020). *Journeys on the Whanganui, Aotearoa New Zealand: Whanganui MRTO.* http://www.whanganuiriver.co.nz/mrto

Whanganui River Adventures (2019). *About Whanganui River Adventures.* https://www.whanganuiriveradventures.co.nz/about-whanganui-river-adventures/

Whitford, M., & Ruhanen, L. (2016). Indigenous tourism research, past and present: where to from here? *Journal of Sustainable Tourism, 24*(8–9), 1080–1099. https://doi.org/10.1080/09669582.2016.1189925

World Bank. (2019). *Indigenous peoples.* https://www.worldbank.org/en/topic/indigenouspeoples

Yap, M. L.-M., & Watene, K. (2019). The Sustainable Development Goals (SDGs) and Indigenous peoples: Another missed opportunity? *Journal of Human Development and Capabilities, 20*(4), 451–467. https://doi.org/10.1080/19452829.2019.1574725

Ybema, S., Yanow, D., Wels, H., & Kamsteeg, F. (Eds.). (2009). *Organizational ethnography: Studying the complexities of everyday life* (1st ed.). Sage.

Young, D. (2017). Whanganui tribes. *Te Ara - the Encyclopedia of New Zealand.* https://teara.govt.nz/en/whanganui-tribes/print

Zapalska, A., & Brozik, D. (2017). Māori female entrepreneurship in tourism industry. *Tourism: An International Interdisciplinary Journal, 65*(2), 156–172.

Part IV

Postscript

WWOOFing in Australia: ideas and lessons for a de-commodified sustainability tourism

Adrian Deville[a]*, Stephen Wearing[b] and Matthew McDonald[c]

[a]*UTS Business School, Management Discipline Group, University of Technology, Sydney, Australia, Kuringai Campus, Eton Rd Lindfield NSW 2070, PO Box 123, Broadway, NSW 2007, Australia;* [b]*UTS Business School, Management Discipline Group, University of Technology, Sydney, Australia, Kuringai Campus, Eton Rd Lindfield NSW 2070, PO Box 123, Broadway, NSW 2007, Australia;* [c]*Department of Management, RMIT University Vietnam, 702 Nguyen Van Linh, District 7, Ho Chi Minh City, Vietnam*

(*Received 18 August 2013; accepted 3 May 2015*)

This paper considers Willing Workers on Organic Farms (WWOOFing) as a form of sustainable tourism with particular focus on its social and cultural dimensions and the means by which deep engagement in these appear to lead participants to a better awareness or understanding of ecological sustainability issues. It draws upon a grounded theory-based exploration of the perspectives and interactions of WWOOFers and WWOOF hosts in Australia, using 323 formal written surveys of hosts and 188 surveys of WWOOFers, together with 16 in-depth unstructured WWOOFer interviews, which collectively enhances understanding of WWOOFing as an emerging, unique and valuable form of sustainable tourism. By virtue of the highly engaged and symbiotic basis of the exchange involved, WWOOFing is commonly perceived to facilitate a transcendence of the role of tourist. The research indicates this is the product of a unique relationship forged in the WWOOFing context, which differs markedly to relationships forged in more typical fee-for-service tourism contexts in which there is a different relationship at play between power, authenticity and sustainability. This relationship is outlined in order to articulate the notion that WWOOFing represents a type of "sustainability tourism" that is unexplored in the sustainable tourism literature.

Keywords: WWOOF; guest-host relationships; authenticity; sincerity; sustainable tourism; volunteer tourism

Introduction

Since emerging in the UK in the early 1970s as a work exchange network to support the organic farming movement and foster knowledge about its practices, the Willing Workers on Organic Farms (WWOOF) network has grown "organically" (Coppard, 2006; Green, 1980; Pollard, n.d.b; Vansittart, 2002; Ward, 1995). More recently, it has become closely linked to contemporary "tourism" practices. The activity of WWOOFing, involving living and labouring on host member properties, has been significantly expanding both in terms of the number of participants and the number of sites in the world where it occurs. Membership and activity in Australia in particular has expanded exponentially since the mid-1990s and today there are over 16,000 new annual memberships, obtained largely by international travellers to Australia. While membership relative to overall inbound tourism numbers, now typically over 6 million annually, is still small, its growth has significantly outstripped that of tourism and "backpackers" since the mid-1990s (Deville,

*Corresponding author. Email: adrigen123@gmail.com

2011). However, comparatively little research has been undertaken on WWOOFing by scholars in Australia.

The idea of WWOOFing has spawned, and/or now sits alongside a variety of other exchange-based travel types such as Help–X, Couch Surfing and house swapping. These travel-related types have in common with WWOOFing the characteristic of operating beyond mainstream commodity consumption culture, offering non-commercial alternatives to the way people can travel, interact and create experiences. If considered to be a tourism form, these all offer scope for a broadening of conceptualisations of tourism that are anchored in a generalised adherence to a neoliberalist, free market, fee-for-service based perspective (Hughes & Stitt, 2008).

Tourism research seems to have also largely overlooked these "collaborative consumption" (Botsman & Rogers, 2010) forms of travel, which provide opportunities within cashless "alternative economies" (Gibson-Graham, 2008). It is argued here that such de-commodified mechanisms should, however, be seen as significant for sustainable tourism in particular. They offer opportunities to radically reduce the various costs or impacts typically associated with tourism infrastructure. They also decentralise and democratise tourism through maximising the possibility of generating alternative types of valuable benefits for a greater range of participants – both tourists and hosts. As discussed below, consideration of WWOOFing is relatively new territory, but WWOOFing perhaps has the greatest significance for this field given also the nature and specific sustainability oriented aims of the WWOOF organisation.

This paper has two main aims. The first is to offer an update on the current state of knowledge of the WWOOF phenomenon. While aspects of this have been explored from the perspective of other countries, little empirical data has been collected in the Australian context. This is an important outcome given the phenomenal growth of WWOOF in Australia over the last two decades.

Second, the paper offers an analysis of WWOOFing guest–host relationships, which illustrates that some unique elements are at play there that have significance for how we might conceptualise contemporary "sustainable tourism". In particular, the intense interpersonal, non-monetary or de-commodified aspects of the WWOOFing exchange appear to enable WWOOFers to transcend "tourism" experiences to the degree that they are rich in experiential or existential authenticity, with exchanges frequently approximating a symbiosis – a unique form of engagement with the "other" that is more sincere and authentic, compared with other types of tourism. In turn this appears to facilitate an opening and orienting of these tourists to key messages of the sustainability practitioners found in the host community, despite initially having relatively self-interested orientations about experiencing Australian people, culture and landscapes rather than towards "sustainability" per se. This paper argues that tourism that orientates, educates and even transforms tourists in relation to theories and practices of sustainable food production and living, and which facilitates active contributions by tourists to practical sustainability orientated efforts of hosts, could be articulated as "sustainability tourism". This concept, which differs in emphasis from sustainable tourism, has not been previously identified, although some of its conceptual forebears are discussed.

The paper begins with an introduction to WWOOFing and its basic tenets and how the WWOOF phenomenon has been depicted previously, followed by an outline of the study's research methods and its findings. It goes on to discuss the unique nature of the host–guest relationship in the WWOOF context and the ways in which this seems to lead many participants to become more sustainability oriented.

WWOOF and WWOOFing

Willing Workers on Organic Farms is a now global network that offers individual "WWOOFers" the opportunity to exchange approximately half a day of labour on an organically oriented farm, for a day's food and accommodation provided by a registered WWOOF "host". Host properties conform to a range of criteria established by a national (or regional) WWOOF organisation, generally through involvement in organic farming in some way, such as by "growing or producing organic products" (WWOOF Australia, 2008). The term "organic" indicates chemical-free or "natural" forms of agriculture and WWOOF's core aims and values derive from involvement in and support of the broader and now fast growing movement concerned with organic food or fibre production (Biological Farmers of Australia, 2006, 2010; IFOAM, 2008; Lockie, Lyons, Lawrence, & Mummery, 2002; Organic Farm Association, 2006; Organic Trade Association, 2011). Underpinning that aim is the assumption that organic techniques are more sustainable for people and nature than those centred around or reliant upon synthetic chemicals for fertilizing and pest control, given their well-understood ecosystem impacts.

In addition, WWOOF hosts are frequently involved in local "earth repair" or "landcare" projects that seek to repair damaged natural landscapes for the benefit of local biodiversity and reversing the impacts of unsustainable practices.

WWOOF Australia is run as a not-for-profit organisation, seeking to return surplus funds to hosts in the form of grants to enable them to carry out conservation and reforestation projects, which WWOOF emphasises are often on lands that would otherwise be ineligible for government grants (Cosgrove, 2000). WWOOF also points out that using WWOOFers to assist also has no cost to domestic taxpayers.

WWOOF *a priori* accepts and promotes the view that it is, and increasingly will be through the significant inputs of human labour, good design and the application of sound practical knowledge, that sustainable agricultural production and consumption are to be achieved. This is because a triple-bottom-line understanding is required in evaluating any consideration of sustainability per se, while "conventional", reductionist single (i.e. economic) bottom line approaches to food production involving the application of synthetic chemicals (and increasingly genetically engineered crops) which, by definition, derive from a paradigm promoted for and maintained by economic self-interest, rather than a deep evaluation of sustainability focused upon human and non-human well-being.

Whether or not the WWOOF paradigm (or an "ecocentric" paradigm generally) is accepted or embraced, the increasingly significant numbers of tourists that WWOOF and who thus encounter and are sometimes personally transformed by the aims and practices of their hosts (Deville & Wearing, 2013) should make a close focus upon WWOOFing an area ripe for research in the inchoate field of sustainable tourism. Indeed, given its educational orientation (English, 2007; Maxey, 2006; Maycock, 2008; McIntosh & Bonnemann, 2006; Stehlik, 2002), its role in assisting people, and its frugality in terms of tourism infrastructure, WWOOFing has been described as the quintessential form of low impact, ethical/altruistic, sustainable and/or responsible tourism by some observers (Clarke, 2004; Doherty, 1997; Fenton Huie, n.d.; Hughes & Stitt, 2008; Idelbrook, 2007; Maycock, 2008; Pollard, n.d.a; Trainor, 2008). Significantly, it has also been described in the *Scientific American (Earth) Magazine* as an exemplary success in fostering practical sustainability (Chinn, 2008). Again however, tourism scholars have yet to analyse WWOOFing specifically with regard to these issues.

Depictions of WWOOFing

There has been much attention to WWOOFing in the popular media, particularly in alternative lifestyle and rural media and via websites, social media, blogs and internet video posts. While offering a wide range of interpretations of WWOOFing, available data driven research remains limited.

One New Zealand study largely portrayed WWOOFing in terms of a neglected aspect of farm tourism (McIntosh & Campbell, 2001), or as an "alternative farm stay experience". Several important elements are shown to present in contrast with regular, commercially oriented farm-based tourism (McIntosh & Bonnemann, 2006), including *rurality* (as a source of "otherness" for urban travellers), *opportunities to learn about organic farming, personal meaningfulness* and the element of *sincerity* in the experience. By analysing interviews with WWOOFers, McIntosh and Bonneman (2006) developed some important insights about the nature of the WWOOFing experience. Along with numerous accounts in the broader media, they brought attention to the interactive nature of the experience between WWOOFers and hosts. However, to date, a coherent understanding of this key aspect of "engagement" among or between participants, built from sound empirical data, has not been offered. Additionally, these micro-social elements have not coherently been put in to a macro-social or sociological context.

Ateljevic and Doorne (2001, p. 169) note that two ends of a continuum exist among backpackers: "traditional long-term travellers" who "live with the citizens of the country" for long periods and "mainstream" backpackers (c.f. Riley, 1988) who travel to "holiday spots" (quoting a respondent, p. 175), echoing Cohen's (1979) recreational tourist. They describe overlaps between WWOOFers and "long-term budget travellers", who generally are portrayed as seeking "escape from mainstream 'institutionalized' tourism flows". WWOOFers are painted as travelling lightly and cheaply, often staying in specific places for long periods as part of an effort made to remedy Western anomie/alienation from constant exposure to the "pervasive values of an overarching capitalist system" (Ateljevic & Doorne, 2001, p. 185) and homogenised global culture.

While the character of "the backpacker" will perhaps continue to evolve (Ateljevic & Doorne, 2001, 2004; Bennett, 2007; Cohen, 2004; Curtis, 2005; Huxley, 2004; Krippendorf, 1987; Loker-Murphy & Pearce, 1995; Pearce, 1990; Riley, 1988), Ateljevic and Doorne (2001) suggest WWOOFing will play a key role at the surviving "traditional" end of the spectrum, where the "richness of experience (cultural immersion, social relations, 'back to nature')" will continue to be of "primary importance". They observe that WWOOF is a program "increasingly oriented towards a growing interest from these travellers" as, combined with travel, WWOOFing assists them in their goal of avoiding the "beaten track" of institutionalized tourism (in the language of Cohen 1972, 1973). Importantly, schemes such as WWOOF "facilitate the spontaneity they seek" as "freedom and flexibility are regarded as crucial for the whole experience" (Ateljevic & Doorne, 2001, p. 175).

MacIntosh and Campbell (2001) and more recently Yamamoto and Engelsted (2014) have focused on the motivations of hosts in utilising the WWOOFing system in New Zealand and the USA, respectively, with some critical consideration of the degree of match in the expectations of both parties. In contrast, Stehlik (2002, p. 221) in considering the basis upon which the WWOOFing exchange operates, concluded positively that "the system runs entirely on goodwill and non-monetary exchange without anyone creaming off a profit or being exploited". While his primary focus is upon the range of learning opportunities for adult participants in unfamiliar situations, he also dwells upon the need for a "willingness to engage and adapt to local customs and environments", which can bring about material and personal enrichment, including the building of social capital for both WWOOFer and host (Stehlik, 2002, p. 224). Crucially, Stehlik points out the importance of human engagement within a de-commodified context, while Rother (2009) offered views on the techniques required of WWOOFers to ensure that both parties are able to maximise the available positives of the experience. When this takes place in practice, Hughes and Stitt (2008) suggest WWOOFing deserves attention as an exemplary "energy exchange" in which hosts "communicate their needs to their visitors, who in turn provide what they can, resulting in a wonderful give-and take dynamic" (p. 41). When this kind of "symbiosis" occurs, it situates the WWOOF exchange rather differently from economically driven, commodified host–guest relations found elsewhere in tourism, with hosts retaining power in respect of meeting their needs directly through hosting.

Maycock (2008) casts the growth and spread of WWOOF in the context of a general "greening" arising from growing awareness of both dependence on and disconnection from nature. The growing (re)popularisation of farmers markets, community supported agriculture schemes and local food co-ops are seen as part of a movement for the re-localisation of food and efforts to reconnect with the sources of food eaten (see Lockie et al., 2002). Maycock (2008) asserts that many look to make local efforts to address global problems through modified consumption choices, there is also increased interest in attempting backyard food production, sometimes through involvement in local community gardens. The desire to return to and connect with the land as various crises associated with modern living and/or economic circumstances become more apparent is not new, but most people "have forgotten (or, rather, never learned) the basic skills of agricultural living" and it is in this context along with others, that one might see a growing interest in WWOOF, as it connects "people wanting to try their hand at farming/growing with farms" (Maycock, 2008, p. 284).

Moscardo (2008) has argued that "sustainable tourism development" is a term still primarily limited to a concern about "the continuity of tourism" than "the contribution of tourism to sustainable outcomes". Citing Wall (1997, p. 33), Moscardo argues that the real question to be asked about tourism and sustainability is "whether and in what form might tourism contribute to sustainable development?" The challenge is to rework our assumptions about the existence and nature of sustainable tourism, beginning with the idea that tourism might be properly viewed as potentially able to contribute to sustainable development locally. Tourism should be seen as a "potential resource for communities seeking sustainable development options" through "wider and more innovative types" of tourism development. Thus tourists themselves "can be seen as more than just customers, they can also be seen as human resources for regional development" (p. 8). Moscardo (2008) then identifies several instances in which innovative "alternative" forms of tourism are already working in this way, specifically describing WWOOF hosting as "an example of the use of tourist volunteers to support local organic agriculture" (see also McIntosh and Campbell (2001), Maxey (2006) and Strong (2008)). Observing that the "initial establishment of organic agriculture can be challenging", often associated with limited income during the transition period and thus low levels of productivity, Moscardo (2008) observes that "the ability to access low-cost labour can be an important benefit". In terms of framing the contribution of the present research, she noted that the potential use of 'volunteer tourists' assisting with "traditional economic activities has yet to be fully explored" (Moscardo, 2008, p. 9).

Others have also sought to challenge narrow economic development driven perspectives underpinning some versions of the sustainable tourism paradigm. Potts and Harrill (2002, p. 55) argued that the tourism industry has "the potential to improve the world, not just sustain itself" and that their suggested proactive "travel ecology" approach could provide "a form of tourism development that would encourage a more holistic form of community development", one in which the needs of community are clearly articulated and placed before that of development proponents. Though not articulated by them, their position has some overlap with the concept of WWOOF in that it advocates "backyard activism" or the "geography of everywhere", through the recognition "that all landscapes, no matter how mundane, contribute to the community tourism product" (Potts & Harrill, 2002, p. 51).

Stanford (2008) argues that "responsible tourism" is not the exclusive domain of eco, green or other "niche" *types* of tourism, but should be applicable to all, including conventional mass tourism. Responsible tourism must be considered through quadruple bottom-line thinking – economy, society, environment and culture – and Stanford therefore suggests that responsible tourism aims for *respect and awareness*, *reciprocity*, local *economic contributions* and *engagement* with landscapes and people. Stanford argues that WWOOFers are "very responsible" tourists with "a deep engagement with local people", but that few tourists are likely to achieve large degrees of responsibility in each of the above dimensions without locally suitable guidance (Stanford, 2008, p. 270). The WWOOF model does directly offer this form of local guidance via individual hosts (Deville, 2011; McIntosh & Bonnemann, 2006) and arguably beyond Stanford's focus on *engagement* built into the core exchange between hosts and WWOOFers are also *reciprocity* and *respect/awareness*, and of course labour, which can be readily seen as being a substitute form of *economic benefit* for participating hosts.

This array of insights in conjunction with a significant body of popular media raises questions about the nature of the relationship between WWOOFing and tourism (particularly 'alternative' and sustainable tourism) and to wider concepts of sustainability. As there has not yet been a systematic consideration of this relationship based upon empirical research, this paper addresses this knowledge gap, drawing from a body of demographic and psychographic research data on WWOOFing in Australia.

Research methods

The research presented here was undertaken as part of a broader examination of the WWOOFing phenomenon, conducted by Deville (2011) between 2006 and 2009. It presents the first in-depth empirical study of WWOOFing in Australia, which was initially motivated by recognition of a marked increase of WWOOF membership in that country. Given the unique research field and the obvious importance of social relationships (Crooks, 2001) among the WWOOFing participants, the approach taken was based in grounded theory, which is itself influenced by interactionist theory (Strauss & Corbin, 1990; Glaser, 1992). Given also the prior involvement of the researcher in WWOOFing as a traveller, there was recognised value in working with a post-positivist or "constructivist" form of grounded theory building (Charmaz, 2003), such that the researcher is acknowledged to be represented to some degree as an active participant and even collaborator with other participants in the co-production of shared knowledge.

Mixed methods were used involving the collection of quantitative and qualitative data gathered through direct participation, participant observation, surveys and recorded interviews with hosts and WWOOFers (refer to Deville, 2011, for detailed discussion of its rationale).

Written surveys were trialled on a small number of known hosts and WWOOFers, using a range of open and closed questions about demographics, motivations, experiences and outcomes relevant to hosts and to WWOOFers. The questionnaires were then refined based on feedback. Hard copy and emailed links to an online version of host surveys were distributed to a random sample of 50% of all Australian WWOOF hosts with the assistance of WWOOF Australia. In this process, hosts were asked to also administer WWOOFer Surveys to any WWOOFers they had staying with them during a nominated "census" month. This resulted in the retrieval of 323 surveys of hosts and 188 surveys of WWOOFers, producing a range of useful demographic and psychographic data which was analysed using SPSS software. Copies of the questionnaires can be provided by the author upon request.

In parallel, in-depth unstructured interviews (Miniciello, Aroni, Timewell, & Alexander, 1995) were conducted and recorded with 16 WWOOFers and hosts along the Eastern coast of Australia between 2005 and 2008. Data from each interviewee was transcribed, thematically coded in detail and analysed using NVIVO software.

Finally, participant observation (Veal, 1997) also provided a means of achieving researcher immersion which not only created conditions for empathic, "insider" understanding of others' perspectives, but served to sharpen "theoretical sensitivity" during the research process and effectively assisted in creating a "triangulated" effort to interpret the nature and meaning of the Australian WWOOFing phenomenon. It also positioned the researcher as an agent in the constructed grounded theory that was ultimately generated (Deville, 2011). While all efforts were made to work towards a sound and well-formed knowledge construction, the findings presented are naturally limited by the degree to which data is "incomplete", or interpretation is contestable.

Research findings

The findings presented here were selected from a larger study (Deville, 2011), including basic demographic information about WWOOF in the Australian context, an examination of the motives of hosts and WWOOFers and some key insights into the WWOOFing experience. This selection of results provides a basis for consideration of the WWOOF host–guest relationship and how this intersects with issues of sustainability.

Key characteristics and motives of participants

A brief characterisation of participants to contextualize discussion of findings is provided in Tables 1 and 2 below.

Hosts were distributed across Australia, largely, but not only, living in regional and rural areas. A wide range of types of productive activities were reported among hosts, with an average age being in the early 50s. The host population are relatively well educated with about 57% obtaining tertiary qualifications compared to 21% among the wider Australian labour force.

Less than 15% are certified "organic producers", with the majority found not to be strongly commercially oriented at the time of research. Indeed, much of the focus among hosts was found to be related to the establishment and especially the maintenance of kitchen gardens and home orchards. However, many hosts are nevertheless involved in commercial farming on a variety of scales, and in general, help in getting work done was found to be a principle motive for hosting WWOOFers.

WWOOFers range in age and nationality, but were most likely to be international, long-term travellers from urban centres in their early 20s. Females were more prevalent than males (2:1) and WWOOFers were generally a highly qualified group, with 72% having tertiary qualifications or higher.

Table 1. Demographic profile of WWOOF hosts ($n = 323$).

Age	
Age range	24–85 years
Sex of person most responsible for WWOOFers*	
Female: Male ratio	1.5: 1.0
Host family situation	
Nil children	19.2%
Dependent children	36.8%
Independent and dependent	2.8%
Independent	41.2%
People in host households	
Host's partner	72.8% of cases
Host's children	53.7%
Other family members	30.0%
No one else (i.e. a 'solo' host)	14.7%
Employees	13.7%
Property/community shareholders	10.9%
Formal education attainment (compared with national average)**	
Postgraduate degrees	20% (3.2%)
Graduate diplomas and certificates	16.8% (2.4%)
Bachelor degree	20.3% (15%)
Secondary school qualifications	13.9% (47.6%)

*Refers to sex of survey respondent nominated as the person in household with overall responsibility for WWOOFers.
** Australian Bureau of Statistics, various years.

Key stated reasons for WWOOFing were *meeting, living with* and *interacting with local people; seeking experiences in natural, outback and rural areas* and *having encounters with wildlife* (Deville, 2011). Developing social relations, meeting with and learning about other people, as well as experiencing non-urban places were important to WWOOFers. It is notable that WWOOF Australia strongly promotes itself in these terms, emphasising the prospects of encountering and experiencing "the real Australia" through WWOOFing (Statham, 2003, www.wwoof.com.au), while sustainability related issues, though central to WWOOFs charter and purpose, appear to be promoted as a secondary consideration to prospective new WWOOFers. This corresponds with Deville's (2011) finding that much of their motivation for joining can be reasonably seen in the context of the most common motivations for "journeying" among long-term budget travellers, particularly "social interaction with local people" (Obenour, 2004, p. 3). WWOOFers are also keen to *save money* while travelling, while also having valuable *life experiences. Learning about organics* was ranked fifth by WWOOFers, important in only about 11% of cases (see Table 3).

Hosts also perceived that WWOOFers are primarily keen to *save money*, ahead of *experiencing cultural exchange* or *rural and/or natural environments*, with the opportunity to *meet people/enjoy social interaction* being the next most important driver from their perspective. Various other self-interested motivations such as *improving language*

Table 2. Demographic profile of WWOOFers ($n = 188$).

Age and sex	
Age range	17–65 years
Female:male ratio	2:1
Relationship status of WWOOFers	
Single: partnered ratio	3:1
Percentage WWOOFers with children	5%
Country of origin	
Percentage of domestic WWOOFers	8.1%
Germany	Rank 1
South Korea	Rank 2
Britain/UK	Rank 3
Japan	Rank 4
Australia	Rank 5
Dwelling characteristics	
Large or small cities	43%
Large or small towns	31%
Small village and other communities	27%
Formal education attainment	
Postgraduate degrees	5%
Graduate diplomas and certificates	12%
Bachelor degree	26%
Other tertiary/trade	35%
Secondary school qualifications (only)	17%
Other	4%

Source: Deville, 2011.

Table 3. WWOOFers' top 10 reasons for joining WWOOF.

Reasons for joining WWOOF	Freq. n = 188	%	% of cases	Cumulative %
(1) Meet local people	60	15.2	34.1	15.2
(2) Experience Australian life/culture	54	13.6	30.7	28.8
(3) Save money/cheap travel	43	10.9	24.4	39.6
(4) Life experience	20	5.1	11.4	44.7
(5) Learn about organic growing	19	4.8	10.8	49.5
(6) Improve English	18	4.5	10.2	54.0
(7) Alternative way to travel	16	4.0	9.1	58.1
(8) Experience farm life/work	16	4.0	9.1	62.1
(9) Live with people/family	16	4.0	9.1	66.2
(10) Experience Australian landscape/places	13	3.3	7.4	69.4

skills and *enjoying home comforts* were considered by hosts to be more important to WWOOFers than *learning about and contributing to environmental repair* and *learning about and experiencing organic farming* and related skills, which they ranked last.

The WWOOFing experience

Personally meaningful, authentic and non-touristic experiences "off the beaten track" are key motivations among long-term budget travellers generally (Cohen, 1988; Conran, 2006; Kim & Jamal, 2007; McCabe, 2002; McIntosh & Bonnemann, 2006; Obenour, 2004; Pearce, 2005; Uriely, 2005; Steiner & Reisinger, 2006; Taylor, 2001; Wearing & Wearing, 2001; Welk, 2004), while increasingly more mainstream tourists also seek to avoid feeling "like a tourist" (Tourism Australia, 2006). However, close, personal, sincere encounters with others in the context of their daily lives can be frustratingly elusive in a touristic world where relationships are underpinned and influenced by commercial imperatives (see Minca & Oakes, 2006). Where tourism is constructed as a leisure form growing out of economic production and commodity consumption (Meethan, 2001) - usually removed from proximity to the daily chores and routines of those outside the tourism service industry – a type of "tourist angst" is commonly experienced around identity and the authenticity of tourists' experiences (Richards & Wilson, 2004, discuss this term).

WWOOFing however, provides a useful, affordable mechanism for undertaking immersed (Nimmo, 2001) forms of experience with difference or "otherness", which is an increasingly important aim for many tourists (Franklin, 2003; Gillespie et al., 2008; Matos, 2004; Woehler, 2004). Significantly, host members of the WWOOF network *invite* people to meet, stay and work with them in their places on projects they deem important, and it is typical that WWOOFers:

> just fit in with what I'm doing at the particular time. I don't change anything to have the WWOOFers with me. They just come in … and be part of my everyday life… (Host M) (Deville 2011, p. 192).

Thus hosts retain power, and compared to commercial home based hosting ventures, can "be themselves" to whatever degree they feel inclined:

> … you look after them well, but you don't run around after them. I don't do B & B anymore because I really hated having everything spotless and perfect. (host quoted in McIntosh & Bonnemann, 2006, p. 96)

For hosts – who demonstrated no interest in providing tourism-related services or being the object of tourist curiosity – this exchange accomplishes getting work done for which there are otherwise no time or resources available to achieve this. For tourists, the exchange can help achieve some refuge from the touristic (Franklin, 2003). For both parties, social interaction, cultural exchange, learning and increased self-understanding are also important outcomes. But importantly, host and WWOOFer participants in this study rated their experiences within this exchange network very highly. Deville (2011) found that:

- 91% of surveyed hosts considered it to be at least a *mostly positive* experience; 92% would, or already had recommended hosting to others; and 95% would, or had recommended WWOOFing to travellers.

- 52% of surveyed WWOOFers found it to be *all positive* and 40% regarded it as *mostly positive;* 94% would, or had already recommended WWOOFing to other travellers.

This does not imply that problematic and non-reciprocal WWOOFing experiences do not occur. Indeed they do, particularly where there is a mismatch of expectations or ineffective communication among parties (refer to the WWOOF website and Rother 2009). But the positive assessment by the majority of participants indicates there are qualities inherent in the experience that reverberate with the broader ideals and goals of many participants, and that these produce a range of positive benefits or outcomes for them (both predicted and unexpected).

To understand these qualities and therefore grasp the phenomenal popularity of WWOOFing among contemporary tourists in Australia, the following sections explore this unique host-guest relationship.

The host–guest relationship

Exploration of the host–guest relationship in WWOOFing is framed by two key issues, namely:

- a disparity between hosts and WWOOFers motivational orientations toward sustainability issues and
- the significance for WWOOFers of having experiences with "real people not tourism operators" (host in McIntosh & Bonnemann, 2006, p. 96).

The first is important because it has been observed that many WWOOFers have been transformed in regard to their orientations to sustainability-related issues as a result of WWOOFing (Deville & Wearing 2013). As suggested above, the processes of engagement at the heart of the host–guest encounter appear to be instrumental in achieving such transformation, which may have some significance to the study of sustainable tourism.

The second is important because a number of WWOOFers and hosts interviewed for this study reported that the act and experience of WWOOFing requires a "tourist" to shed that role or identity and become more like a *person*. This highlights the quality of "sincerity" that some have described as critically characteristic of WWOOFing experiences (McIntosh & Bonnemann, 2006). It also suggests that the relationship of WWOOFing to tourism is ambiguous, with "authenticity" presenting a liminal threshold for the tourist-WWOOFer to negotiate through sincere engagement with the host other.

The following discussion attempts to illuminate some of the dimensions of this relationship and thereby provide an updated understanding of the place and role of WWOOFing in contemporary sustainable tourism discourse.

WWOOFers and sustainability

While WWOOFers had a quite limited initial interest in sustainability-related issues such as organic production techniques or earth repair projects, this was nevertheless not perceived by hosts to be an intractable barrier to a successful or valuable exchange, provided WWOOFers had the appropriate general approach and attitude to the experience (Deville, 2011). Conversely, this has been expressed by some hosts in regard to shorter term WWOOFers, that they sometimes do try to treat hosts' places "like a backpackers" hostel and therefore do not "have the right spirit" (New Zealand host in McIntosh & Bonnemann, 2006, pp. 95–96).

This reveals a broader concern that the character of WWOOF has changed over a period of time, manifested by new "types" of WWOOFers with limited interest or knowledge about core WWOOF goals, although there is no clear agreement among today's hosts on this matter. There was a deliberate expansion in the mid-1990s into the backpacker market by WWOOF Australia, explained by its founder (Pollard cited in Doherty, 1997, p. 17) as part of a desire to expose a wider audience chiefly interested in getting close to Australian families to the overall sustainability-related "good" that WWOOF was considered to bring about through exposure of people to organic approaches and alternative thinking (Pollard, 1998), with "at least some seeds of change in their minds" (Pollard, 1993, pp. 81–82). Today, a minority participate *because* they have a direct interest in learning about or contributing to efforts towards organic production and earth repair (the original goals of WWOOF); others combine an interest in travel with independent concern for a sustainable future. Surveys demonstrated that such sustainability oriented tourists are significantly fewer than the majority of general "experience seeking" tourists who now WWOOF.

Undoubtedly, as WWOOF accepted a broader clientele (hosts and WWOOFers), WWOOFing then grew in popularity as "an alternative way of travelling" with the "added benefit of the chance to live as part of the host family and experience local customs and lifestyle", in contrast to the options available to most tourists (Stehlik, 2002, p. 221). Tourism and the organic movement began to become strongly and irreversibly entwined in this moment, which was also seen by many hosts to have irrevocably changed the character of WWOOFing in Australia. As it lends itself so well to the achievement of important pursuits and goals of long-term, low budget travellers, WWOOF began to be used increasingly for purposes quite specific to that group. Indeed, since 2005, undertaking three months of WWOOFing became a legitimate means to extend a working tourist visa in Australia. This step towards embracing a more formal realm of tourism and labour market management has led to the perception of further change impacting on the hosting experience (perhaps the subject of future research).

The relationship between WWOOFers and sustainability is, therefore, not well understood, but, the success of WWOOF (and perhaps other energy exchange based enterprises) among tourists testifies that new modes of travel exist which should be further empirically analysed at micro-sociological and macro-social levels to understand the basis of that success.

Commodity, engagement and symbiosis

Through WWOOFing, tourists are offered a flexible experiential pathway into the sociocultural realm of Australian hosts which is interactive and economical, but perhaps crucially for some, is removed from the usual techniques of "touring" rooted in and determined by commodified or market based frames of reference. In conventional tourism forms, the tourist is often engaged in a diversionary mode seeking distraction and escape from their boredom and routinised daily lives and various degrees of alienation.

In contrast, the growth of WWOOFing suggests the existence of a demand for decommodified experiences as part of "a distinctive form of escape from mainstream 'institutionalized' tourism flows" (Ateljevic & Doorne, 2001, p. 169). WWOOFing allows its participants the opportunity to "avoid the beaten track" of institutionalised tourism (in the language of Cohen, 1972, 1973), which can be otherwise difficult to achieve

(MacCannell, 2001; Richards & Wilson, 2004), while allowing them to extend their time and resources allocated to regular aspects of tourism.

Opening up WWOOF, combined with growing tourist demand for existentially authentic experiences is certain to have underpinned the phenomenal growth in WWOOFing. The emergence of alternative, "new", post-modern and post-tourism forms in response to conventional tourism, hints at broader forces at play in the social and psychological horizon of those who come to WWOOF. These, combined with pre-packaged arrangements by travel agents that include WWOOF membership; the expense of travel in Australia; and the emergence of the "experience economy" do lend weight to some hosts' view that WWOOFing is increasingly "part of the tourist industry" (host participant). Yet, it is simultaneously difficult to describe the exact relationship between tourism and WWOOFing (Deville, 2011). Indeed, this ambiguity might be part of its appeal among those tourists with a penchant for "rejection of the market driven paradigm" (Ateljevic & Doorne, 2001, p. 175) seeking to transcend perceived negative aspects of tourism, and/or to assert their "distinction" (c.f. Bourdieu, 1984) from other tourists.

Existing literature and recent data indicate that what makes the experiences of WWOOFing tourists most distinct from even the most interactive of tourism forms is the degree and character of human *engagement* involved, the achievement of which is effectively a mutual "obligation" in WWOOFing, albeit experienced in many variations in practice (Deville, 2011). The centrality of interpersonal engagement gives rise to "personal meaningfulness", "intimacy", "sincerity" and "authenticity" in WWOOFing encounters. Framing this study within interactionist theory, operationalised through the employment of a grounded theory method, made it possible to capture these unique interactive elements. Understanding the interpersonal, micro-social processes that occur and how they produce unique effects upon participants is important to grasping WWOOFing as the phenomenon it has become, within the macro-sociological context of tourism encounters. It also highlights the need to take a wide view of WWOOFers, without presuming too much about their identities as tourists, as much pre-existing literature seems to have done.

As suggested by McIntosh and Bonneman (2006), the sincerity of both parties in the encounter (Taylor, 2001) is of particular significance here. The sincerity of either party can be called into doubt at times, but, the WWOOFing arrangement mitigates against one-sided benefits working in favour of one party for very long, as both hosting and WWOOFing take significant amounts of physical and emotional energy. Ultimately, either party can terminate unsuccessful experiences. Upon engagement, work itself often facilitates micro-social dynamic interactions which govern the progress, nature and outcomes of the experience (Deville, 2011). WWOOFing requires and ultimately forces frequent engagement between strangers, brought together in an effort to satisfy individual motives, without the assistance of intermediary "culture brokers" (Smith, 2001) or other commercial agents. From an individual perspective, being in contexts of otherness removed from the influence of traditional reference groups may encourage the WWOOFer to open their mind to different ways of thinking and behaving (Wearing, Deville, & Lyons, 2008). It facilitates transcendence of otherwise alienating barriers found in tourism forms built upon cash-based exchange (Oakes, 2006), despite limited shared interests or initial motivations, and frequently generates mutually beneficial outcomes. Experiences are created with others and in the process, when it works well it appears to approximate *symbiosis*. Of course some might consider symbiosis is at work in the exchange of cash for a desired service within commercial tourism. But the qualities of sincerity and thus authenticity in fully engaged WWOOFing encounters undoubtedly more fully addresses the motivations of those tourists who, as MacCannell (1976/1999, p. 197) suggested, want to "abjure commercialised entertainments" and to set their own "touristic itineraries".

This naturally best suits the interests of those who truly value experiential engagement and thus seek escape from mainstream or institutionalized travel (Ateljevic & Doorne, 2001; Cohen, 1979; Obenour, 2004; Wilson & Richards, 2004), with opportunities for journeys of self-discovery, self-awareness and self-development through the search for, and experience of, life alternatives. It is the adoption of different contexts of 'other' places and people and the relative freedom offered to individuals in WWOOFing that makes possible changes in the self, of beliefs and values. WWOOFing it is argued, is more likely to foster significant relationships that lead to change because of the inherent mutuality of effort required and the generally more informal and open environment than might be typically found in any equivalent commodified tourism setting (Conway, 1999).

The sometimes powerful and transformative effect of participation for some participants was explained in terms of the removal of the human exchange process from the effects of monetary interchange, with discoveries of the 'commonalities of humanity' and the richness of life "when no money stuff is involved" (participant host). Frow (1997) sees this is as key to the authentic tourist experience of late modernity.

Compared with even the most interactive of known tourism forms, WWOOFing is an experience built upon a unique basis of exchange with a uniquely structured reciprocity (Molm, 2010), with characteristic results for those that try it, even as a minor component of their travels (Deville & Wearing, 2013). The structure not only ensures a customised experience, but modulates the extent to which the interaction becomes one occurring between people, rather than between service providers and consumers. Research illustrates that one cannot effectively gaze upon WWOOF hosts, or use other "common-sense understandings of how to be a tourist" (McRae 2003, quoting Edensor, p. 241) when WWOOFing.

The circumstances of WWOOFing effectively force a move along the ontological spectrum, experienced in terms of increasing degrees of human connection, ranging from alienation, through immersion, towards engagement at the other end. Traversing this spectrum, the epistemological foundation of the experience also shifts from positivist/objectivist towards constructivist. That is, the experience needs to be assessed less in relation to an objective reality, and more as one which is created in and relative to co-constructed moments and the efforts that underpin them. Again, increasing degrees of engagement of tourist and host selves shifts the relevance and viability of various modes of authenticity, away from objective, towards existential authenticity (Pearce, 2005).

Not all WWOOFers embrace changed ontological and epistemological standpoints. But, for the time they are in a WWOOFing encounter, such shifts are necessary to some degree, while the extent to which they are embraced and occur helps in producing a better fit and better outcomes in that circumstance. It seems that achieving host–guest symbiosis is partly an issue of "flow" (Csikszentmihalyi, 1988), and it is perhaps conversely true that resistance to these shifts in the WWOOFing situation is an important element in many a failed experience (Rother, 2009).

Authenticity and sustainability in tourism

WWOOFing generates a range of bottom-up benefits for Australian hosts who retain power and autonomy over their situation in respect of tourism in their midst. Hosts' involvement is essentially reversible: no new infrastructure is required; they are able to refuse visitors or regulate their flow and behaviour as needed, or entirely withdraw. The relatively even distribution of power between hosts and guests appears to underpin an important and uniquely configured relationship between authenticity and sustainability, further elaboration of which has value in conceptualisation of sustainable tourism.

Authenticity: objective and existential

The degree to which tourism itself begins to diminish the possibility for objectively authentic tourist experiences has long been a concern for tourists and scholars (Cohen, 1988, 1995, 2004; Conran, 2006; Huxley, 2004; Kim & Jamal, 2007; McCabe, 2002; Minca & Oakes, 2006; Oakes, 2006; Parsons, 1997; Rojek, 1993; Shepherd, 2002; Steiner & Reisinger, 2006; Taylor, 2001). Because successful tourism products carry with them the dilemma of managing increasing demand, carrying capacity must be addressed for the benefit of future tourists and dependent host communities (Butler, 1992; Wahab & Pigram, 1997; Wall, 1997). If unaddressed, destinations may undergo ultimately destructive life cycle processes at the whim of market forces. Interventions to guard the ecological and cultural base for tourism to be able to continue are key concerns; distinctive qualities of the tourism product are preserved (restrictions on visitation) and promoted (i.e. a "pristine rainforest" or a "typical street scene"); fortifications and concentrations of the essential objectively authentic qualities (local distinctiveness) may also be sought in the effort to *sustain* such a tourism "product".

One choice for tourists and the industry is to move on to new "unspoiled" frontiers in search of the still authentic (Bennett, 2008; Cohen, 2004; Coleman & Crang, 2002) and alternative tourism techniques, destinations and stances have been pioneered that hope to achieve authentic experiences to some extent. Eventually, however, there is an inevitable need to accommodate the tourist and tourism infrastructure in the frame, ultimately or eventually undermining the degree of objective authenticity within the tourist's grasp (Minca & Oakes, 2006; Oakes, 2006).

Tourists may also accept the "inauthentic" or "incorporate" it and therefore not be disappointed by it. Celebration of the "touristic" is a reasonable stance to the world more generally for some (Franklin, 2003), accommodating tourism within the frame of reference, with multiple subjective relationships to the authenticity of individual situations.

Yet, these these possibilities do not seem to alter the belief, despite many careful analyses that suggest its futility, that somewhere out there it is still possible to experience true authentic otherness, fuelling further touristic endeavours by "pioneering" individuals (Pryor, 1997), while a responsive industry creates new authentic tourism frontiers. Therefore, the sustainable tourism discourse is commonly focused on the notion of sustaining tourism products and the experiences built around them as they approach 'carrying capacity' that would otherwise threaten their authenticity (Saarinen, 2006), but this understandable focus may ignore important questions as to *what* is to be sustained? W*ho* is to sustain it? And *for whom* it is to be sustained?

The above consideration of WWOOFing tourists experiencing, learning about and contributing to more sustainable production and consumption efforts in relatively prosaic environments and existentially authentic contexts of "collaborative consumption" prompts consideration of sustainability in tourism with a different emphasis to that rooted in sustainable tourism development.

Sustainable tourism and sustainability tourism

Because the WWOOF network largely avoids the construction of specialized accommodation and other infrastructure, it is likely to be significantly more sustainable than other tourism forms in terms of (biophysical) resource use and waste management issues.

The behaviour of tourists with respect to the socio-cultural sustainability of the host community is also critically different in the case of WWOOFing (Stanford, 2008). Research suggests that WWOOFing tourists become more like people who must co-create an experience, without applying touristic evaluative criteria to the description of its success (Deville, 2011). Transcendence of this liminal threshold can be personally significant to tourists, but in the case of WWOOFing, it also seems to effectively facilitate the transmission of an increased awareness about local scale efforts towards sustainability, about responsibilities in respect of global sustainability issues, and a range of practical techniques for attempting to contribute to its achievement in the process. WWOOF uniquely offers exposure to ideas and practices that orient them to more sustainable forms of producing, consuming and living (Jamieson 2007), while gaining the opportunity to travel in a more sustainable and intimate fashion. Naturally, there is some important contestation regarding definitions of, let alone the means for achieving a "sustainable future" (see Beder, 1993; Robinson, 2004; Saarinen, 2006), and there is certainly variability of local practices among hosts within the global host community. It is, however, the important task of WWOOF organisations and WWOOFers together, to help to ensure that their host members stay true to this overriding goal.

The data from this study shows that the majority of WWOOFers are international tourists, travellers or backpackers seeking to meet locals, try out a novel experience and/or seek to travel cheaply, away from the beaten track (horizontal axis, Figure 1). However,

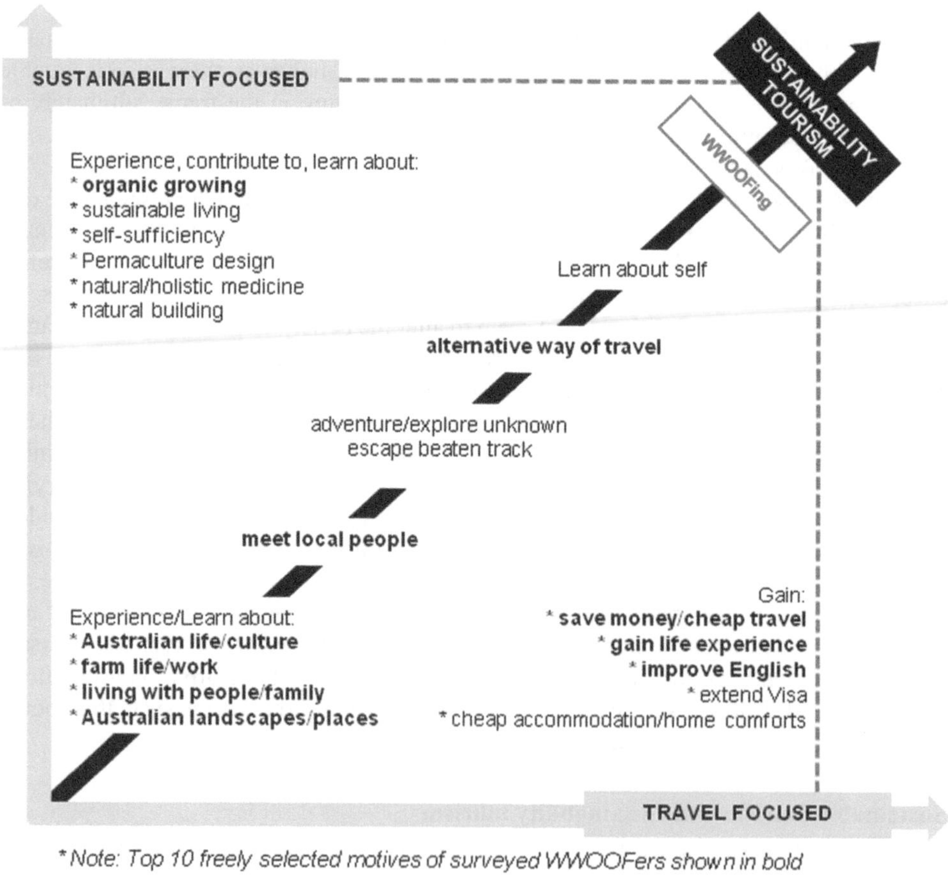

Figure 1. WWOOFing as sustainability tourism*.

despite these primary motivations, changes in attitudes towards nature or the environment more generally were common following WWOOFing. These changes included increased appreciation of organic produce/practices and the nexus between organic production and wider sustainability goals. Many WWOOFers reported experiencing significant inspiration, with related resolutions made to alter their lives in terms of how they travelled, engaged with community, or in regard to future involvement in environmental protection activities and/or living sustainably (Deville, 2011; Deville & Wearing, 2013). These findings are in line with those from other studies conducted on WWOOFers and of volunteer and ecotourists more broadly in which important transformations in the self have occurred (Wearing et al., 2008).

A number of WWOOFers were also found to be specifically focused in the manner of special interest tourists, on *learning about* and/or learning *for sustainability* (vertical axis, Figure 1). Indeed, most of the interviewed WWOOFers had particular, purposeful and sometimes unique intentions regarding WWOOFing, either deliberately learning (and offering) specific skills and methods for sustainable living, or fulfilling career or research oriented objectives relating to the achievement of a more sustainable governing paradigm (Oskamp, 2000).

Maycock (2008) described WWOOFers as people seeking to return to and connect with the land in response to various crises associated with modern living, using WWOOFing to gather "the basic skills of agricultural living" (p. 284). While very much in the minority, this characterisation resonates with Cohen's (1979) phenomenological typology characterising some tourists as experiential or existential pilgrims, in this instance moving towards a sustainability-oriented spiritual centre.

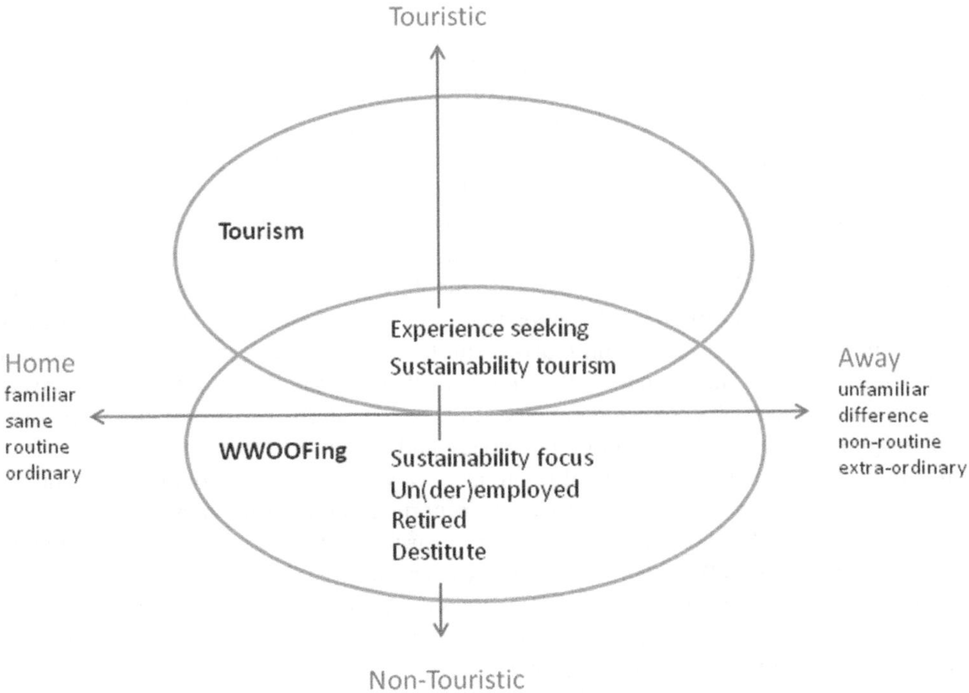

Figure 2. WWOOFing and tourism.

When an interest in sustainability itself is *combined* with tourism, tangible learning outcomes can be produced and valuable contributions to sustainability-oriented hosts result. This might be articulated as *sustainability tourism*, with WWOOFing representing its initial, pioneering expression (diagonal, Figure 1). Such WWOOFers doubtless overlap with those described by Nimmo (2001) as those having a "critical political awareness" of global issues of sustainability (Type 1 WWOOFers). They may also be connected with the so-called "new" or "other postmodern" tourists that (rightly or naively) want to travel in line with this awareness and their ideals (Fennell, 2004; Holden, 2000; Munt, 1994; Swarbrooke & Horner, 1999; Uriely, Reichel, & Ron, 2003; Weaver, 2002).

In terms of the contemporary WWOOFing phenomenon, there are also other known WWOOF user "types" to be explored and described, including travelling retirees, the unemployed and the destitute (see Figure 2). Further targeted research might identify the extent to which these "types" are distinct or blend together, or what the effects might be as both tourism and interest in organics and consciousness about global sustainability expand and blended.

Conclusion

WWOOFing has experienced phenomenal growth in Australia and its spread encourages a new look at some old issues in tourism studies. WWOOFing shares many similarities with tourism in broad terms, but has paradoxically always operated largely outside of it, being attractive to those tourists seeking "refuge from a touristic world" (Franklin, 2003). The majority of WWOOFers seek to meet and have experiences with local people, which indicates dissatisfaction with tourism forms that maintain a distance between tourist and host (Frow 1997).

WWOOFing focuses our attention on the primary importance of host–guest relations and the significance of de-commodified forms of engaged and reciprocal interactivity that are rich in existential authenticity. Although this raises the issue of liminal thresholds between tourism and non-tourism, existential authenticity in WWOOFing has been explored to highlight also how it might expand conceptualisations of "sustainable tourism" focused on carrying capacity, which appear to arise primarily as managerial questions about maintenance of the *objective* authenticity attributes (Pearce, 2005; Minca & Oakes 2006; Saarinen, 2006; Moscardo 2008) of developed tourism places/products/essences.

WWOOFing and its phenomenal expansion, however, cannot be properly grasped in relation to sustainable tourism through that lens. Indeed, WWOOFing perhaps offers an example of Conran's (2006, p. 275) hope that the humanist desire for reciprocal interactions will ultimately prevail within tourism and that the "intimate experience supersedes the desire for object authenticity". If it has been demonstrated that WWOOFing breaks from a concern with objective authenticity, it is hoped that this paper provides some useful insights into emerging experiential modes and hands on practices oriented to the macro-social concept of sustainability in new ways, including through new tourism related mechanisms that are closely based on important micro-social interactions.

To the extent that hosts are pioneers and guardians of techniques for a sustainable future and teaching WWOOFers, some of these end up transforming their entire worldviews in relation to sustainability issues, WWOOFing expands the way we need to think about the interactions possible between guests and hosts within the realm of sustainable tourism and about the effects of these interactions.

As well as WWOOFers being altered by their experiences, Deville (2011) determined that hosts benefit in multiple ways from the interactions and inputs of WWOOFers who contribute to the sustainability of their enterprises in a variety of ways as well (Maxey 2006, Strong 2008). In this regard, Moscardo (2008) suggested that sustainable tourism should be conceived in terms of the extent to which it is an activity that contributes towards achieving sustainable development, specifically identifying WWOOFing in this light. Potts and Harill (1998, 2002) argued for a "travel ecology" that contributes to backyard restoration efforts everywhere, which being reflected in the concept and practice of WWOOFing, offers another useful framing that captures the sometimes prosaic nature of this form of "sustainable tourism". When we acknowledge that there are numerous, complex and difficult ecological restoration problems to overcome in Australia, such as climate change, salinity and water quality issues (Australian Department of Climate Change and Energy Efficiency, 2007; Australian Government Natural Resource Management Team, 2000), as well as the need to supply more and more food and fibre through sustainable farming techniques that ultimately relies on physical labour, it is hard not to take a keen interest in the role that WWOOFers might already be playing, or could be more effectively playing towards meeting these ends.

Maycock (2008) optimistically characterised WWOOFers as individuals increasingly aware of organic farming and the importance of localising of food sources, with WWOOFing being part of the expression of their search for "opportunities to become more ethical consumers by forming meaningful connections with their food sources" (p. 83). Though a relatively small minority, sustainability focused WWOOFers do prompt consideration of the possibility that WWOOFing is at the heart of an emerging movement not only away from alienated consumptive modes of travel, but towards a more purposeful, educative travel form that has been expressed tentatively here as *sustainability tourism*. A start has been made in establishing an understanding of this phenomenon in Australia, but what this means for tourism in the future and elsewhere is unclear. It seems likely though that if there is a growing desire for contact with nature (among tourists), an increasing scrutiny of the sustainability of our actions (including as tourists), an interest in practical solutions to address sustainability issues, and particularly an expansion of interest in travelling in ways that maximise engagement, WWOOFing will continue to play an important role in the future of tourism and sustainable tourism research.

Acknowledgements

The authors thank WWOOF Australia for their excellent support and contributions to the research underpinning this paper.

Disclosure statement

No potential conflict of interest was reported by the authors.

References

Ateljevic, I., & Doorne, S. (2001). Nowhere left to run: A study of value boundaries and segmentation within the backpacker market of New Zealand. In J.A. Mazanec, A.G. Woodside, & G.I. Crouch (Eds.), *Consumer psychology of tourism, hospitality and leisure (vol 2)* (pp. 169–186). Oxfordshire: CAB International.

Ateljevic, I., & Doorne, S. (2004). Theoretical encounters: A review of backpacker literature. In J. Wilson & G. Richards (Eds.), *The global nomad: Backpacker travel in theory and practice* (pp. 60–76). Clevedon: Channel View Publications.

Australian Department of Climate Change and Energy Efficiency. (2007). Climate change in Australia: Technical report 2007. Retrieved 13 June 2014 from http://www.climatechangeinaustralia.gov.au/documents/resources/TR_Web_FrontmatterExecSumm.pdf

Australian Government Natural Resource Management Team. (2000). *A national action plan for salinity and water quality*. Canberra, ACT: Department of Agriculture, Fisheries and Forestry. Retrieved 13 June 2014 from http://nrmonline.nrm.gov.au/catalog/mql:2573

Beder, S. (1993). *The nature of sustainable development*. Newham: Scribe Publications.

Bennett, R.J. (2007). *Moving off the beaten track: Developing a critical literacy in backpacker discourse* (Unpublished doctoral dissertation). School of Media, Communications and Culture, Murdoch University, Perth.

Bennett, R.J. (2008). Entering the global margin: Setting the 'Other' scene in independent travel. In P.M. Burns & M. Novelli (Eds.), *Tourism and mobilities: Local-global connections* (pp. 133–145). Oxfordshire: CAB International.

Biological Farmers of Australia. (2006). *Organic annual 2006 - edition 3*. Brisbane: Author.

Biological Farmers of Australia. (2010). *Australian organic market report 2010*. Brisbane: Author.

Botsman, R., & Rogers, R. (2010). *What's mine is yours: How collaborative consumption is changing the way we live*. London: HarperCollins.

Bourdieu, P. (1984). *Distinction: A social critique of the judgement of taste*. London: Routledge.

Butler, R. (1992). Alternative tourism: The thin end of the wedge. In V.L. Smith & R.W. Eadington (Eds.), *Tourism alternatives: Potentials and problems in the development of tourism* (pp. 31–46). West Sussex: Wiley and Sons Ltd.

Charmaz, K. (2003). Grounded theory: Objectivist and constructivist methods. In N.K. Denzin & Y. Lincoln (Eds.), *Strategies of qualitative inquiry* (2nd ed., pp. 249–291). Thousand Oaks, CA: Sage.

Chinn, A. (2008). Organic farms say "WWOOF". *Scientific American, Earth 3.0 (Special Edition), 18*(5), 12.

Clarke, P. (2004). Global spread for organic skills. *Farmers Weekly, 141*, 89–92.

Cohen, E. (1972). Towards a sociology of international tourism. *Social Research, 39*(1), 164–182.

Cohen, E. (1973). Nomads from affluence: Notes on the phenomenon of drifter tourism. *International Journal of Comparative Sociology, 14*, 89–103.

Cohen, E. (1979). A phenomenology of tourist experiences. *Sociology, 13*, 179–201.

Cohen, E. (1988). Authenticity and commoditization in tourism. *Annals of Tourism Research, 15*(3), 371–386.

Cohen, E. (1995). Contemporary tourism – trends and challenges: Sustainable authenticity or contrived post-modernity? In R. Butler & D. Pearce (Eds.), *Change in tourism: People, places, processes* (pp. 12–29). London: Routledge.

Cohen, E. (2004). *Contemporary tourism: Trends and challenges*. Oxford: Elsevier.

Coleman, S., & Crang, M. (2002). Grounded tourists, travelling theory. In S. Coleman & M. Crang (Eds.), *Tourism: Between place and performance* (pp. 1–19). New York, NY: Berghahn Books.

Conran, M. (2006). Commentary: Beyond authenticity: Exploring intimacy in the touristic encounter in Thailand. *Tourism Geographies, 8*(3), 274–285.

Conway, H. (1999). Working holidays: Its better to unwind doing something active. *Resurgence, 194*, 30–31.

Coppard, S. (2006). How it all began. Retrieved 16 December 2007 from http://www.wwoofinternational.org/home/index.php?option=com_content&task=view&id=16&Itemid=34
Cosgrove, L. (2000). WWOOFing in Tasmaina. *Issues, 53*, 38–41.
Crooks, D.L. (2001). The importance of symbolic interaction in grounded theory research on women's health. *Health Care for Women International, 22*, 11–27.
Csikszentmihalyi, M. (1988). *The flow experience and its significance for human psychology*: Cambridge: Cambridge University Press.
Curtis, R. (2005). *Backpacker's field manual: A comprehensive guide to mastering backcountry skills*. Westminster, MD: Crown Publishing Group.
Deville, A. (2011). *Alice in WWOOFerLand: Exploring symbiotic worlds beyond tourism* (Unpublished doctoral dissertation). University of Technology, Sydney.
Deville, A., & Wearing, S. (2013). WWOOFing tourists, beaten tracks and transformational paths. In Y. Reisinger (Ed.), *Transformational tourism: Tourist perspectives* (pp. 151–168). Oxfordshire: CAB International.
Doherty, M. (1997, November 23). The visitors who like hard work: WWOOFers know holidays weren't meant to be easy, *Sunday Examiner*, pp. 16–17.
English, J. (2007). Cultivating new farmers: new organic programs are teaching young people the skills they need to make a living on the farm. *Mother Earth News, 220*, 80–85.
Fennell, D.A. (2004). Deep ecotourism: Seeking theoretical and practical reverence. In T.V. Singh (Ed.), *New horizons in tourism: Strange experiences and stranger practices* (pp. 109–120). Cambridge, MA: CAB International.
Fenton Huie, S. (n.d.). Way to go on a shoestring. (Pamphlet). Retrieved from WWOOF archive in Victoria, Australia.
Franklin, A. (2003). *Tourism: An introduction*. London: Sage.
Frow, J. (1997). *Time and commodity culture: Essays in cultural theory and postmodernity*. Oxford: Clarendon Press.
Gibson-Graham, J.K. (2008). Diverse economies: Performative practices for 'Other worlds'. *Progress in Human Geography, 32*(5), 613–632.
Glaser, B.G. (1992). *Basics of grounded theory analysis*. Mill Valley: Sociology Press.
Green, D. (1980). Working weekends on organic farms. *Organic Growing*, Spring.
Holden, A. (2000). *Environment and tourism*. New York, NY: Routledge.
Hughes, E., & Stitt, L. (2008). As is: Secrets to having enough. *Communities, 141*, 38–75.
Huxley, L. (2004). Western backpackers and the global experience: An exploration of young people's interaction with local cultures. *Tourism Culture and Communication, 5*(1), 37–44.
Idelbrook, C. (2007). Live on less and love it! Try these 75 inspiring ideas and enjoy life more while spending and consuming less. *Mother Earth News, 224*, 34–38.
IFOAM (International Federation of Organic Agriculture Movements). (2008). Uniting the organic world. Retrieved 23 February 2008 from http://www.ifoam.bio/
Jamieson, P. (2007). *The wild green yonder: Ten seasons volunteering on New Zealand's organic farms*. Auckland: New Holland.
Kim, H., & Jamal, T. (2007). Touristic quest for existential authenticity. *Annals of Tourism Research, 34*(1), 181–201.
Krippendorf, J. (1987). *The holiday makers: Understanding the impact of leisure and travel.* Oxford: Heinemann Publishing.
Lockie, S., Lyons, K., Lawrence, G., & Mummery, K. (2002). Eating 'Green': Motivations behind organic food consumption in Australia. *Sociolgia Ruralis, 42*(1), 23–40.
Loker-Murphy, L., & Pearce, P.L. (1995). Young budget travelers – backpackers in Australia. *Annals of Tourism Research, 22*(4), 819–843.
MacCannell, D. (1976/1999). *The tourist: A new theory of the leisure class* (revised edition). Berkeley: University of California Press.
MacCannell, D. (2001). Remarks on the commodification of culture. In V. Smith & M. Brent (Eds.), *Hosts and guests revisited: Tourism issues of the 21st century* (pp. 380–390). New York, NY: Cognizant Communications Offices.
Matos, R. (2004). Can slow tourism bring new life to alpine regions? In K. Weiermair & C. Mathies (Eds.), *The tourism and leisure industry: Shaping the future* (pp. 93–105). New York, NY: Haworth Hospitality Press.
Maxey, L. (2006). Can we sustain sustainable agriculture? Learning from small-scale producer-suppliers in Canada and the UK. *The Geographical Journal, 172*(3), 230–244.

Maycock, A. (2008). World wide opportunities on organic farms (WWOOF). *Journal of Agricultural & Food Information, 9*(4), 282–288.

McCabe, S. (2002). The tourist experience and everyday life. In G.M. Dann (Ed.), *The tourist a metaphor of the social world* (pp. 61–75). Wallingford: CAB International.

McIntosh, A., & Bonnemann, S. (2006). Willing workers on organic farms (WWOOF): The alternative farm stay experience? *Journal of Sustainable Tourism, 14*(1), 82.

McIntosh, A., & Campbell, T. (2001). Willing workers on organic farms (WWOOF): A neglected aspect of farm tourism in New Zealand. *Journal of Sustainable Tourism, 9*(2), 111–127.

McRae, L. (2003). Edward said and a politics of meeting and movement. *Tourist Studies, 3*(3), 235–251.

Meethan, K. (2001). *Tourism in global society: Place, culture, consumption*. New York, NY: Palgrave.

Minca, C., & Oakes, T. (2006). Introduction: Travelling paradoxes. In C. Minca & T. Oakes (Eds.), *Travels in paradox: Remapping tourism* (pp. 1–21). Oxford: Rowman and Littlefield.

Miniciello, V., Aroni, R., Timewell, E., & Alexander, L. (1995). *In-Depth interviewing: Principles, techniques, analysis* (2nd ed.). Sydney: Longman.

Molm, L. (2010). The Structure of Reciprocity. *Social Psychology Quarter, 73*(2), 119–131.

Moscardo, G. (2008). Sustainable tourism innovation: Challenging basic assumptions. *Tourism and Hospitality Research, 8*, 4–13.

Munt, I. (1994). The "Other" postmodern tourism: Culture, travel and the new middle classes. *Theory, Culture and Society, 11*, 101–123.

Nimmo, K. (2001). *Willing workers on organic farms: A case study* (Unpublished master's thesis), Victoria University of Wellington, Wellington.

Oakes, T. (2006). Get Real! On being yourself and being a tourist. In C. Minca & T. Oakes (Eds.), *Travels in paradox: Remapping tourism* (pp. 229–253). Oxford: Rowman and Littlefield.

Obenour, W. (2004). Understanding the meaning of the 'Journey' to budget travellers. *International Journal of Tourism Research, 6*, 1–15.

Organic Federation of Australia. (2006). Organic industry booming (media release June 23 2006): Organic Federation of Australia. Retrieved 15 April 2007 from http://www.ofa.org.au/

Organic Trade Association. (2011). Industry statistics and project growth. Retrieved 7 April 2011 from http://www.ota.com/organic/mt/business.html

Oskamp, S. (2000). Psychological contributions to achieving an ecologically sustainable future for humanity. *Journal of Social Issues, 56*(3), 373.

Parsons, M.G. (1997). *Encounters in touriculture: Indigenous cultural tourism in contested domains* (Unpublished doctorate dissertation). School of Tourism and Hospitality, Southern Cross University, Lismore.

Pearce, P. (1990). *The backpacker phenomenon: Preliminary answers to basic questions*. Townsville: James Cook University.

Pearce, P. (2005). *Tourist behaviour: Themes and conceptual schemes*. Clevedon: Chanel View Publications.

Pollard, L. (1993). WWOOFing about. *Earth Garden, 86*, 80–82.

Pollard, L. (1998). WWOOFing and other cultures. *Earth Garden*, 72–73.

Pollard, L. (n.d.a). WWOOFing around the world. *New Traveller* (Pamphlet), pp. 20–21. Retrieved from WWOOF archive in Victoria, Australia.

Pollard, L. (n.d.b). The WWOOF story. Unpublished manuscript (retrieved from WWOOF archive in Victoria, Australia).

Potts, T.D., & Harrill, R. (1998). Enhancing communities for sustainability: A travel ecology approach. *Tourism Analysis, 3*(3,4), 133–142.

Potts, T.D., & Harrill, R. (2002). Travel ecology and developing naturally: Making theory - practice connections. In R. Harris, T. Griffin & P. Williams (Eds.), *Sustainable tourism: A global perspective* (pp. 45–57). Sydney: Butterworth Heinemann.

Pryor, M. (1997). The traveller as a destination pioneer. *Progress in Tourism and Hospitality Research, 3*, 225–237.

Richards, G., & Wilson, J. (2004). Backpacker icons: Influential literary 'nomads' in the formation of backpacker identities. In G. Richards & J. Wilson (Eds.), *The global nomad: Backpacker travel in theory and practice* (pp. 123–145). Buffalo, NY: Channel View Publications.

Riley, P.J. (1988). Road culture of international long-term budget travellers. *Annals of Tourism Research, 15*, 313–328.

Robinson, J. (2004). Squaring the circle? Some thoughts on the idea of sustainable development. *Ecological Economics, 48*, 369–384.

Rojek, C. (1993). *Ways of escape: Modern transformations in leisure and travel.* London: Macmillan.
Rother, F. (2009). Adventures and misadventures with the WWOOF volunteer program. *Transitions Abroad.* Retrieved 2 October 2012 from http://www.transitionsabroad.com/
Saarinen, J. (2006). Traditions of sustainability in tourism studies. *Annals of Tourism Research, 33*(4), 1121–1140.
Shepherd, R. (2002). Commodification, culture and tourism. *Tourist Studies, 2*(2), 183–201.
Smith, V. (2001). The culture brokers. In V. Smith & M. Brent (Eds.), *Hosts and guests revisited: Tourism issues of the 21st century* (pp. 275–282). New York, NY: Cognizant Communications Offices.
Stanford, D. (2008). 'Exceptional visitors': Dimensions of tourist responsibility in the context of New Zealand. *Journal of Sustainable Tourism, 16*(3), 258–275.
Statham, C. (2003). WWOOF to it. *Get up and Go* (online). Retrieved 2 October 2012 from http://getupandgo.com.au/
Stehlik, T. (2002). Willing workers on organic farms: Cultural exchange and informal adult learning in an organisation that is voluntary, non-profit and environmentally friendly! *Australian Journal of Adult Learning, 42*(2), 220–226.
Steiner, C., & Reisinger, Y. (2006). Understanding existential authenticity. *Annals of Tourism Research, 33*(2), 299–318.
Strauss, A., & Corbin, J. (1990). *Basics of qualitative research: Techniques and procedures for developing grounded theory.* Thousand Oaks, CA: Sage.
Strong, G. (2008). *Regenerative agriculture: The case for dialogue with nature.* In T. Lefroy, K. Bailey, G. Unwin & T. Norton (Eds.), *Biodiversity: Integrating Conservation and Production* (pp. 75–87). Collingwood: CSIRO.
Swarbrooke, J., & Horner, S. (1999). *Consumer behaviour in tourism.* Oxford: Butterworth-Heinemann.
Taylor, J. (2001). Authenticity and sincerity in tourism. *Annals of Tourism Research, 28*(1), 7–26.
Tourism Australia. (2006). A uniquely Australian invitation: The experience seeker (Brochure). Canberra: Commonwealth Government of Australia.
Trainor, P. (2008, Jan/Feb). See the country, save the planet. *This Magazine,* pp. 32–33.
Uriely, N., Reichel, A., & Ron, A. (2003). Volunteering in tourism: Additional thinking. *Tourism Recreation Research, 28*(3), 57–62.
Vansittart, K. (2002). Helping hands. *Canadian Geographic, 122,* 38.
Veal, A. (1997). *Research methods for leisure and tourism: A practical guide.* Great Britain: Pearson Profession Ltd.
Wahab, S., & Pigram, J. (1997). The challenge of sustainable tourism growth. In S. Wahab & J. Pigram (Eds.), *Tourism, development and growth: The challenge of sustainability* (pp. 3–13). London: Routledge.
Wall, G. (1997). Sustainable tourism – unsustainable development. In S. Wahab & J. Pigram (Eds.), *Tourism, development and growth* (pp. 33–49). London: Routledge.
Ward, J. (1995). *A brief history of WWOOF.* East Sussex: WWOOF UK.
Wearing, S., Deville, A., & Lyons, K. (2008). The volunteer's journey through leisure into the self. In K. Lyons & S. Wearing (Eds.), *Journeys of discovery in volunteer tourism: International case study perspectives* (pp. 63–71). Oxfordshire: CAB International.
Wearing, S., & Wearing, B. (2001). Conceptualizing the selves of tourism. *Leisure Studies, 20*(2), 143–159.
Weaver, D.B. (2002). The evolving concept of ecotourism and its potential impacts. *International Journal of Sustainable Development, 5*(3), 251–264.
Wilson, J., & Richards, G. (2004). Backpacker icons: Influential literary "nomads" in the formation of backpacker identities. In G. Richards & J. Wilson (Eds.), *The global nomad: Backpacker travel in theory and practice* (pp. 123–145). Clevedon: Channel View Publications.
Woehler, K. (2004). The rediscovery of slowness, or leisure time as one's own and as self-aggrandizement? In K. Weiermair & C. Mathies (Eds.), *The tourism and leisure industry: shaping the future* (pp. 83–92). New York, NY: Haworth Hospitality Press.
WWOOF Australia. (2008). Guidelines for hosts. Retrieved 23 February 2008 from http://www.wwoof.com.au/WWOOF_host_info.html#Host%20Guidelines
Yamamoto, D., & Engelsted, K.A. (2014). World wide opportunities on organic farms (WWOOF) in the United States: Locations and motivations of volunteer tourism host farms. *Journal of Sustainable Tourism, 22*(6), 964–982.

Gender and sustainability – exploring ways of knowing – an ecohumanities perspective

Kumi Kato

ABSTRACT
This article takes up the challenge to apply critical enquiry to the interface between tourism and the United Nations' Sustainable Development Goals (SDGs). Applying a political ecology perspective, it examines the intersectionality of gender and (ocean) sustainability through a study of traditional women divers in Japan. Recognizing the SDGs as an agenda setting platform, this work engages with Goal 5: *Gender Equality,* and proposes that engaging with multiple and diverse ways of knowing is critical to promoting a sustainability agenda with gender perspectives an essential component. Employing ecohumanities as a methodological foundation, a qualitative study of women divers in Japan (*ama*) is reported with a focus on their particular relationship with the ocean. The study identifies the power of women's knowledge in its inclusiveness, reciprocity and intuitive way of knowing. The example also shows that while tourism can be an important social and economic force, it can also devalue these core qualities as a result of the negative impacts caused by gender stereotypes. These findings indicate that the sustainability agenda can be advanced by challenging hierarchical systems of knowledge and valuing alternative ways of knowing, in this case, women's knowledges.

Introduction

This article examines the intersectionality of gender and (ocean) sustainability from a political ecology perspective (Mostafanezhad, Norum, Shelton, & Thompson-Carr, 2016; Mosedale, 2015; Peterson, 2005). The Sustainable Development Goals (SDGs, UN, 2017) is used as an agenda setting platform for critical thinking and inquiry in its implementation in tourism systems. With particular reference to Goal 5: *Gender Equality: Achieve gender equality and empower all women and girls,* the article proposes that acknowledging the characteristic of multiplicity in knowledge and ways of knowing is critical in promoting sustainability agenda, for which gender equality is one fundamental cornerstone.

Taking ecohumanities (Rose & Robin, 2004; Weir, 2008) as a methodological foundation, a qualitative inquiry is conducted into the example of women divers in Japan (*ama*), focusing on their particular relationship with the ocean. Ama free-dive in coastal seas to harvest shellfish, seaweed, and other food resources. This traditional practice has existed since the country's early history, and is even depicted in some of the early literature, including the oldest chronicles, *Kojiki* (711–712) and *Nihonshoki* (720). The practice is believed to have originated in the Ise-Shima region, the central east coast of Japan, and has a mythological connection with the Sun Goddess Amaterasu, who is regarded as Japan's origin. In recent years, this traditional practice has become a popular tourism resource through regional promotion and attention from various media sources. While tourism can be an important social and economic force, as recognized in academia (Higgins-Desbiolles, 2006, 2009; Jamal & Camargo, 2014; McCabe & Johnson, 2013), it can also generate gender stereotypes, which devalue the core qualities of the practice. The paper proposes that the women's power lies in their knowledge and ways of knowing, and acknowledgement of this would allow a true advancement of the sustainability agenda, and tourism as a "system of knowledge production" (Figueroa-Domecq, Pritchard, Segovia-Pérez, Morgan, & Villacé-Molinero, 2015, p. 89). This approach would also initiate an alternative discourse on gender and sustainability, and present an opportunity for a critical and creative research direction.

To begin this exploration into women, knowledge, and sustainability, the SDGs are used as an agenda-setting platform, which is then discussed from a political ecology perspective to locate power (or the lack thereof) within the knowledge hierarchy. This topic is then further explored through the example of women working with the ocean, interpreting the stories regarding their particular relationship with it. This is a qualitative and interpretive inquiry grounded by ecohumanities as its methodological foundation, which also directed the style of writing employed in this paper.

Sustainable development goals – *knowledge* as a missing agenda

Among the 17 interrelated SDGs (UN, 2015), Goal 5: *Gender Equality: Achieve gender equality and empower all women and girls* promises to put an end to barriers that prevent women and girls from realizing their full potential. This has been an ongoing focus following the UN Millennium Development Goals (MDGs) 3: *to promote gender equality and empower women)*. Promoting gender equality is, in fact, critical to the achievement of all 17 SDGs, as asserted in the UN Women report, *Turning promises into Action: gender equality in the 2030 agenda for sustainable development* (UN Women, 2018, p. 4).

Clearly, sustainable development is not possible if the female population, which comprises half of humanity "continues to be denied its full human rights and opportunities, including equal access to quality education, economic resources and political participation, employment, leadership, and decision-making at all levels" (UN, 2015, para 20). Empowering all women and girls is therefore crucial for "improving all parts of society"(Jahnsen, 2018, p. 22). Tackling gender issues would also improve other policy-related issues included in the 2030 Agenda's focus on sustainability, equality, peace, and human progress, which aims to provide "a powerful counter-narrative to current practices of extraction, exclusion and division" (UN Women, 2018, p. 2). This implies *inclusiveness*, as stated in Goal 16, which calls for efforts to "promote peaceful and inclusive societies for sustainable development, provide access to justice for all and build effective, accountable and inclusive institutions at all levels" (UN, 2015). However, as Goetz and Jenkins (2016) assert in their critique, the concept of inclusiveness is not clearly defined. This also raises a question regarding the true meaning of the phrase *"Leaving no one behind"* (Stuart & Woodroffe, 2016).

Inclusiveness has been discussed as a means of removing social barriers and obstacles from diverse perspectives, including those of multiculturalism (Fleras, 2009; Rivera, 2017), disability, and other social equity viewpoints (Ravneberg & Soderstrom, 2017; Zhao & Zhang, 2018) as well as development (Agola & Hunter, 2016), environmental vulnerability (Leal Filho, 2018; Fankhauser & McDermott, 2014) and resilience (Ruiz-Ballesteros, 2011; Shaw, Rahman, Surjan, & Parvin, 2016; West & Haug, 2017). In a social-environmental context, inclusiveness is related to sustainability, as indicated by the following statement: "If our ecological relationships are unsustainable, so are the social relations that produce them" (Norum, Mostafanezhad, Shelton, & Thompson-Carr, 2016, p. 305); conversely, "to remedy this lack of a suitable ethics, we must seek out sustainability in relationships, not just between all humans, nonhuman species, and our material setting, but also relations among various humans based on respect and tolerance, freedom of thought, equality, and equity" (Norum et al., 2016, p. 305). This understanding of inclusiveness connects sustainability and gender equality, for which it is critical to value different qualities of knowledge and ways of knowing.

While gender equality is a major global challenge, for tourism, it presents an opportunity to advance the sector as a "knowledge-generating system" (Figueroa-Domecq et al., 2015, p. 89), especially in the context of sustainability (Ferguson & Moreno Alarcón, 2015). The research opportunity for tourism is clear, as suggested by Ferguson's call for more research on "grassroots feminist tourism projects across the world" to "offer alternative ways of understanding the relationship between gender and tourism development, and provide inspiration for creative and progressive ways of harnessing tourism to meet this goal" (Ferguson, 2011, p. 246).

This implies that the gender equality discourse has the power to open up possibilities for non-divisive thinking; tourism has a role to fulfil in the advancement of knowledge production. One such possibility is explored in the discussion on political ecology in the next section, which attempts to specifically connect women, knowledge, and sustainability.

Women's knowledge and sustainability: a political ecology perspective

Political ecology of knowing

Political ecology is defined as "a branch that considers the environment as critical to our understanding of the relationship between the social and environmental disparities that are experienced unequally". It aims "to understand the complex relations between nature and society through a careful analysis of access and control over resources and their implications for the environment" (Cole, 2016, p. 31). Among the range of political ecology debates that have been conducted (Mosedale, 2015; Norum et al, 2016; Peterson, 2005), gender perspective is one critical variable "in shaping resource access and control, interacting with class, caste, race, culture, and ethnicity to shape processes of ecological change" (Rocheleau et al., in Cole, 2016, p. 33).

Women as a community can be "a group of individuals who share characteristics such as ethnicity, heritage, religious, spiritual, cultural beliefs and values as well as geographical spaces" (Thompson-Carr, 2016, p. 25). They can be seen as a community that holds a specific power of knowledge – ecology of the environment, and resource management (Cole, 2016, Cole & Ferguson, 2015; Hanson & Buechler, 2015; Williams & Golovnev, 2015). Goebel suggests that "Given this shared position of oppression between women and the earth, and women's association with life-giving and conserving work, women are conceived as having a "privileged epistemological approach to nature, a better understanding of both what is wrong with human/ environment relations and how to promote positive change"(Goebel, 2004, p. 3). In the context of ecotourism, Scheyvens (2000) also argues that women who tend to be placed in close connection with the natural environment develop specialist knowledge of that environment, and therefore also have an interest in protecting it. Thus, an effective tourism knowledge production system should "encourage the active involvement of women, even if their primary concern is not gender equity" (Scheyvens, 2000, p. 235).

Women's ecological knowledge as power

In the promotion of sustainability, women's ecological knowledge is referenced by Goebel (2004), who claims that the role of women in subsistence activities makes them "crucial actors in contexts of environmental degradation, reclamation and sustainable use" (Goebel, 2004, p. 3). At the same time, systems of gender inequality prevent women's attempts to "meet subsistence needs or manage resources sustainably" (Goebel, 2004, p. 4). In reference to women's knowledge about the ocean environment and its changes, Hanson states that women's stories and narrative oral histories are "critical forms of knowledge for confronting garbage-induced water, health, and development challenges in tropical coastal areas .... Women's oral histories of socio environmental change express their embodied knowledge and the lived experiences navigating and confronting global patterns of consumption and conservation" (Hanson, 2015, p. 166).

Understanding environmental degradation and resource access requires an understanding of "social relations (such as class, gender, and culture) to see the multiple roles they play. Doing so allows for changes that are not just technical in nature, but social as well" (Norum et al., 2016, p. 304). Here, the relationship between people and the land becomes "the template for society and social relations" (Graham, 2008, p. 182) and shifts "our stories to the earth [or sea] others on the moral principle that we live by enabling others to live" (Matthews, 2015, p. 73). Such views pay particular attention to "the mutual embeddedness of hierarchies of power based on gender, class, ethnicity, and race in the politics of control and access to environmental resources" (Hanson, 2015, p. 167). The articulation of women's knowledge is a form of empowerment that establishes their position in the society.

As an overarching goal, empowerment will help individuals to "develop their capabilities, have a sense of control over their wellbeing and lead dignified lives" (Knight & Cottrell, 2016, p. 34); it is also connected to people's sense of cultural and social integrity (Scheyvens & Russell, 2012). Achieving the goal of empowerment will unlock stereotypes and shift the focus to more fundamental values, in addition to counteracting perpetuated inequalities (Wearing, 2002). In this sense, empowerment not only entails addressing economic status; it also requires the assertion and assignment of the fundamental power that exists in knowledge and knowledge production. This fact is asserted by Tucker and Boonabaana (2012), who challenge an instrumentalist approach and a Eurocentric ideal by suggesting that we need "to move beyond simplistic and fixed ideas of women's economic empowerment, particularly in non-western contexts, and consider more fully the cultural complexity and the shifting dynamics of how gender norms, roles and inequalities affect, and are affected by, development and poverty reduction outcomes" (Tucker & Boonabaana, 2012, p. 438). Similarly, Swain and Wallentin argue that in tourism, women's empowerment can only take place when "the existing norms and cultures of a society" are challenged (Swain & Wallentin, 2008, p. 24). Higgins-Desbiolles also suggests that tourism is a powerful social force that can "achieve many important ends when its capacities are unfettered from the market fundamentalism of neoliberalism and instead are harnessed to meet human development imperatives and the wider public good" (Higgins-Desbiolles, 2006, p. 1192). Tourism's achievement may include "fostering cross-cultural understanding, facilitating learning, contributing to cultural protection supplementing development, fostering environmental protection, promoting peace" (Higgins-Desbiolles, 2006, p. 1197) towards a globally conscious society. Harris also suggests that alternative and pluriversal knowledge is necessary to foster alternatives to the "hegemonic ways of doing things" (Harris, 2015, p. xxii). In the following section, such multiplicity of knowledge and knowing is explored through the specific example of women divers in Japan, drawing on the ecohumanities perspective as a methodology for this qualitative inquiry, which is briefly described below.

Methodological foundation – ecohumanities of knowing

The qualitative approach, Ecological humanities or ecohumanities (Rose & Robin, 2004; Weir, 2008), is defined by its founding scholar, Deborah Bird Rose, as "a new inter-discipline that has emerged specifically to address the fact that current ecological problems, including extinctions, climate change, toxic death zones, water degradation, and many others, are anthropogenic events originating from the nature/culture divide, reflected also in the academic division between arts and sciences" (Rose, 2015a, p. 1). Another ecohumanities thinker, the late environmental philosopher, Val Plumwood, also wrote, "The real threat is not so much global warming itself… as our own inability to see past the post-enlightenment energy, control and consumption extravaganza we so naively identify with the good, civilized life to a sustainable form of human culture. The time of *Homo Reflectus*, the self-critical and self-revising one, has surely come.… We will go onwards in a different mode of humanity, or not at all" (Plumwood, 2007, p. 1).

The ecohumanities approach is positive, striving to "engage in life and the living world in an unconstrained and expansive way" (Gibson, Rose, & Finch, 2015, p. ii), which is further detailed in the manifesto issued by the group formed by Gibson, Rose and other scholars.

> Our thinking needs to be in the service of life—and so does our language. This means giving up preconceptions, and instead listening to the world. This means giving up delusions of mastery and control, and instead, to see the world as uncertain and yet unfolding. So our thinking needs to be - *Curious; Experimental; Open; Adaptive; Imaginative; Responsive; and Responsible. We are committed to* think with the community of life and contributing to healing. (Gibson et al., 2015, p. ii)

The approach includes feminist political ecology perspectives, where methodological and theoretical perspectives from the feminist discipline are linked with an analysis comprising multiple layers of ecological, economic, and political power relations (Ateljevic, Pritchard, & Morgan, 2012; Carney, 1993; Cole, 2016; Cole & Ferguson, 2015; Jarosz, 2011; Truelove, 2011).

The primary guiding theme of ecohumanities is: *connectivity, connectedness* or *being connected* (Weir, 2008). Rose and Robin (2004) argue that "the imperative of learning to think about and with connectivity can be operationalized as an imperative to enlarging the boundaries of thought and thinking itself – to enhance our ability to think in dialogue and, perhaps, in empathy with others". The Ecohumanities perspective also challenges the boundaries and divides that are prevalent in academia, and the ranking of knowledge systems that places western science at the top of an epistemological ladder. This impedes "our capacity for knowledge sharing within fields of plural and diverse knowledges" (Rose and Robin, 2004). Such ways of knowing relate to Indigenous knowledge, which is a "complex and intimate knowledge of how to live sustainably in some very harsh and forbidding environments" (Higgins-Desbiolles, 2009, p. 147).

In the context of sustainability, the concept of connectivity may be expressed as reciprocity, which Abram defines as "the ceaseless give and take, the flow that moves in two directions" and is "the foundation of any real ethics" (Abram, 2004, p. 81). A reciprocal human-nature relationship is characterized by dialogical, sensory/experiential, and place-specific qualities. Such qualities signify "a capacity that recognizes the elements that supports our lives… in the essentially narrative terms of naming and interpreting the land, of telling its story in ways that show a deep and loving acquaintance with it and history of dialogical interaction" (Plumwood, 2002, pp. 229–230). This is rephrased by Rose (2018) as life that is flowing out of the earth, and contains not only biodiversity and ecosystems but also "the languages, senses, emotions, and timeless connections with the earth other". The place-specific quality of reciprocity is an orientation towards *genius loci* (spirit of place)—the authenticity of place and integrity sustained over time (Norberg-Schulz, 1980), which may be expressed by care, sentiment, concern, warmth, love, and sacredness. Reciprocity may be expressed as a dialogue (Rose, 2015b) or an offering (care, gratitude, prayer) that is physical, conceptual, or spiritual in nature. Rose (2018) defines such exchange in her lecture, "Gift of life in the shadow of death", where she asserts the importance of celebrating the power of life – not only human life but all kinds of life, and its will to give.

Inclusiveness, connectivity, and reciprocity can be situated as fundamental qualities for sustainability, a positioning which challenges the hierarchical division of knowledge. This also applies to women's knowledge and the ways of knowing presented in the next section.

Qualitative inquiry of the women and the ocean

Ama divers

Ama (海女, literally *sea women*) dive primarily for shellfish such as abalone (*awabi*), sea snails (*sazae*), seaweeds, and other small shore creatures. The term ama generally refers to female divers. In certain regions, men (ama, 海士) also take part in diving, which is a more recent phenomenon since the introduction of wetsuits and mechanized boats from the 1950s onwards. A similar diving practice exists in Jeju province, South Korea, where the divers are called *Haenyo* or sea women. According to a national survey, 2,174 ama existed in Japan in 2011, nearly one-sixth of the figure established in 1931: 12,426 (Tōyō suisan kagaku kyōkai, 2011). The average age of the ama continues to rise, putting pressure on the local areas to recruit from younger generations to continue the tradition.

Previous research on ama focused on anthropological, historic, or religious aspects (Martinez, 2004; Maraini, 1962, Tanabe, 1993; Kawaguchi, 2013). Certain works gave specific attention to the women's traditional ecological knowledge through their observation of the environmental changes and ethical orientation towards the ocean environment (Kato, 2007a, 2017). This was developed into a radio documentary, *Waiting for the tide* (Kato, 2007b), using environmental sound to articulate this practice beyond words. It has been shown that this simple harvesting practice is equipped with traditional ecological knowledge, being sensitive to environmental changes, safety and wellbeing of the divers themselves, and the ocean environment (Kato, 2007a; Martinez, 2004). Their embodied knowledge and life experiences, often expressed in stories, rituals, and cultural practices as well as various kinds of self-imposed regulations, form an important foundation for living as part of the environment, similar to Indigenous perspectives discussed in tourism, environmental management, and ecosophy literatures (Booth & Harvey, 2001; Higgins-Desbiollos, 2009; Kirksey & Helmreich, 2010; Liao, Relle, & Kassam, 2016; Low, 1999; Rose, 1996).

Qualitative inquiry of ama practice

Building on these works, the current study is a qualitative and interpretive inquiry (Atkinson & Delamont, 2011; Willis, 2007), specifically attempting to draw out the multiplicity and complexity of the voices on the ground (Fine, 2011; Jackson & Mazzei, 2009). Ethnographic fieldworks, using methods such as direct participant observation and semi-structured interviews, were conducted between 2015 and 2017 at various coastal regions in Japan to elicit oral histories of the women. Regions visited include Toba in Mie Prefecture, Minami-chō (Abu, Izari) in Tokushima Prefecture, Iki Island in Nagasaki, Kuji in Iwate, and Chikura in Chiba Prefecture. At these locations, visits were made to the local fishery unions, fish markets, community halls, museums, history rooms, local restaurants, shops, guesthouses, as well as shrines, temples, and monuments. To gain an insider perspective, on one occasion, the researcher participated in a harvest, swimming with the women off the boat.

In each site, the local fishery union office was the first contact. This enabled introduction to one or two women leaders in the area, who then contacted other women willing to participate in the study. Seniority and cultural protocol were important considerations in selecting participants. A consent form, clarifying the ethical condition of the study, was signed by each participant, and after the analysis, a summary of the comments was returned to the fishery union and each participant for verification and approval. All communication took place in Japanese, and excerpts of the women's comments, presented in this paper, were translated into English by the author.

Twenty women, who were currently working or had worked as ama divers, aged 30 to 86, participated in this study. A certain number of women were involved in tourism to various degrees, formally or informally. Each interview lasted about 60 minutes, with questions focused on such issues as: what are some of the environmental changes observed; what are the rules followed; and what are the fears and concerns confronted. Although a set of questions were used as guidelines for the interview, it is more appropriate to define these sessions as story-telling. Storytelling is an established form of qualitative enquiry (Charmaz, 2005; Ozyıldırım, 2009; Rooney, Lawlor, & Rohan, 2016; Stapleton & Wilson, 2017; Willis, 2007), which is "a social process that constructs meaning through interaction". The approach does not attempt to present "the truth of the matter as objective description does, but instead to represent, in detail, the perspectives of participants in the process or setting being studied" (Willis, 2007, p. 10). Considering Charmaz's view, a story, as one narrative type, in this case, is much more than a fable, imagination or metaphor and is "concrete reality, rather than a construction that we place on these data" (Charmaz, 2005, p. 526). Accordingly, its interpretation is hermeneutic, holistic, and exploratory, seeking to clarify the collective reality of the world the women share.

Ama and tourism development

An earlier study on ama included visits to a few remaining ama huts, where the value of women's stories was acknowledged (Kato, 2007a). Subsequently, one of these ama women developed an ama hut tourism experience called *Hachiman Kamado* (open fire kitchen). This became a prototype of tourism activity, in which guests sample ama cuisine and participate in cultural experiences. This was adopted in many other regions.

> I never thought I'd start a tourism business myself, but it's a good way to use my knowledge and stories from my experience – the 70 year worth of stories from the sea (R, 86 years old)

> We now all have jobs other than diving, and don't spend time anymore hanging around the ama hut. So this tourism kamado is a place we meet. It's our new ama-hut (K, 72 years old)

The ama hut, housing a *kamado,* was traditionally the place where ama practiced and related culture evolved. It is a place to rest between dives, warming up the body, eating, chatting, and sharing stories. Recently, however, majority of ama dive between other jobs, coming and going by motorbike or car. As a result, the original cultural traditions associated with the ama hut are disappearing. Instead, tourism experiences centring around the *kamado* and hut have proved to be popular among families, groups of women, and student groups, and even visitors coming from overseas. A popular tour activity is story telling by an ama while cooking the ama's harvest, dressed in traditional ama costume and taking photos with the tourists. Certain huts provide *ama* merchandise, including talismans (*o-mamori*) from the shrine worshiped by ama, and talismans featuring their lucky charms. In this region, a pentagram-and-lattice motif (*sei-man dō-man*), drawn with purple shell dye, was traditionally used. They are believed to bring good luck and the strength of the ama women, who work in the ocean, to the holder. These souvenirs are popular, particularly with young women seeking to "gain power from the strong ama women" (Pers. Com., 14 August 2017).

Recently, collecting *shuin* (red seal stamp) has become popular among visitors of Shinto shrines and Buddhist temples. Shuin is stamped alongside a calligraphy denoting the main deity at the shrine and temple, also marking the date of the visit. Visitors take their own books (shuin-cho) to receive shuin on one of the pages, for which a small fee is charged. The shuin at one ama shrine (Ishigami in Shinmei Shrine) is popular among the women, as its female deity is believed to allow one wish to come true for all women. The shrine's shuin features the *sei-man dō-man* symbols (Figure 1).

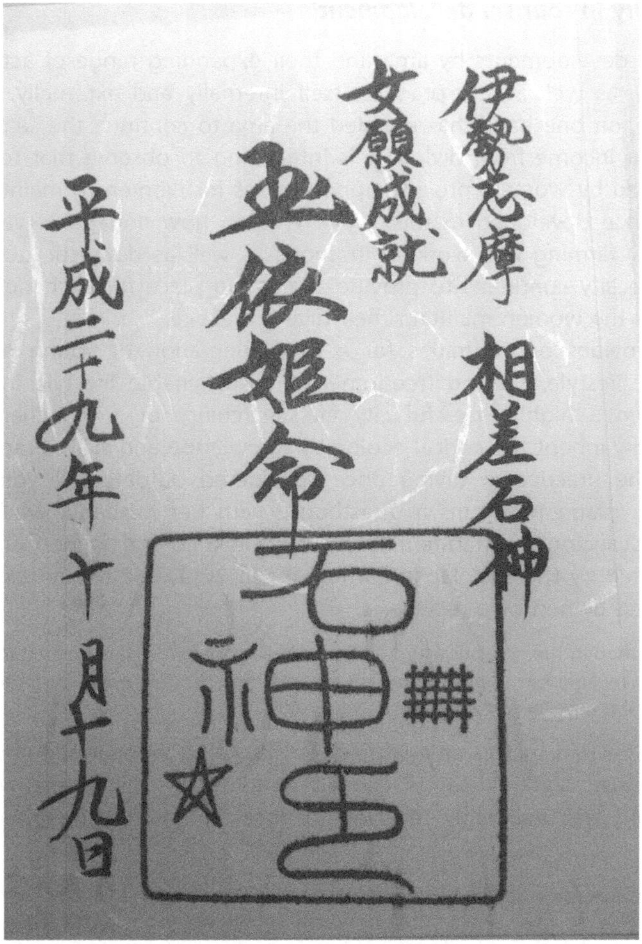

Figure 1. Shuin from Ishigami shrine with sei-man dō-man symbols (Author's photograph).

In another region, Izari community of Minami-chō, Tokushima Prefecture, the ama *hut* is now a restaurant, Izari Café, with an accommodation facility. The café is open for lunch, and two or three ama women take turns to manage the café.

> At the beginning we all thought, what is a café? No one drinks coffee in this remote village. But our lunch special (tiger prawn donburi) became famous through a magazine article, and people come from quite far. It's good to know we (ama) can do work other than diving, and it's also fun to work with other ama women (M, 65 years old).

> I started diving quite late so I'm still not really good at it. Working here, as we talk while we cook in the kitchen, I get to hear the stories of other women who are much more experienced than me. I like that aspect. (S, 50 years old).

Alongside the café, the community also runs an ama experience course. It is a one-day program introducing ama work, including diving and harvesting shells, targeting women aged 20 to 40. Three courses were held in 2016, and two in 2017, each offering ten places, which filled up quickly. The two courses in 2017 were held on the third and fourth Saturdays of August, and a total of 20 women from across Japan participated. The women running the course told me that part of their motivation is to recruit new ama women who may settle in the area. So far, they have managed to recruit one such person, who "brought new vibrance to the village" (Pers. Comm, 20 June 2017).

Media and publicity in tourism development

The recent tourism developments by ama and their expanding range of activities are reshaping the image, identity, as well as the practice itself internally and externally, both positively and negatively. Tourism, on one hand, has enabled the ama to continue the seasonal practice, while supplementing their income from diving. It is interesting to observe that tourism ventures, collaboratively managed by women, provide opportunities for women to maintain their inner community network, once developed in their huts. Women now engage in various kinds of work other than diving – farming and working in shops as well as domestic duties and community work. The tourist facility continues to provide a meeting place for exchanging information and stories, which helps the women maintain their diving practice.

Tourism also provides opportunities for a new generation of young men and women to choose a different lifestyle, seeking freedom and a sustainable lifestyle in rural communities, often escaping from a highly stressful city life, searching for a sea change. These "lifestyle migrants" unwittingly inherit the critical ecological knowledge and skills, empowering themselves and maintaining the practice of diving once considered outdated. A young novice ama (O, 27 years old) on Iki Island now runs a guesthouse with her husband, while being one of the young members advancing the tradition. O grew up in a coastal section of the Tohoku region, which was devastated by the East Japan Earthquake in 2011. The family relocated to the region to work in a business owned by a relative.

> I was working in a shop in the city, but as I grew up with the ocean, I always wanted to do a job related to the ocean. When I saw an advertisement about an ama position to help revive the local economy, I jumped at the opportunity. I've never regretted it. (O, 27 years old)

The guesthouse, owned by this ama woman, is becoming a community hub, popular for both local youth and tourists. Various events, concerts, and small festivals are regularly organized at the guesthouse, and are favourably received by local community members (Pers. Comm., 17 Aug. 2017).

> Young people bring new ideas to the island. We have to value them, otherwise we won't survive ourselves. We older ama can only do what we can do, but young people have to live differently, and they can bring in new ideas to ama work too. (T, 69 years old)

Inevitably, tourism can assist in overcoming inherent gender division and inequality and provide leadership opportunities for women. When women especially utilize their ecological knowledge, senses, and holistic communication approach, new insights and values can be brought into traditional primary industries of fishing, farming, and forestry, as creative forces. Tourism can provide these women opportunities to articulate their connection, power and knowledge. At the same time, however, tourism promotion often reinforces stereotypical images of women, misrepresenting the reality, as seen in the example below. This works to dismiss and devalue the essence of the women's power, ecological knowledge and connection with the ocean.

Stereotypical images in tourism

Recent depictions of ama, including Toba City tourism promotion (Toba City, 2018, Figure 2) and one in All Nippon Airlines magazine *Innovative Voice* (ANA, 2018), feature women divers dressed in traditional white robes, which they no longer wear. Nor is the sea they dive in so blue or clean. Women are often depicted as mystic "mermaids," usually highly sexualized. Traditionally, the women dived semi-naked. This complete natural exposure with the elements, placed the women in a position to sense everything from tidal movements to currents or temperature changes. Despite this, headlines persist as demonstrated in this web-based article, "Japan's tough, topless free-diving mermaid" (Shubach, 2015). Another example is an anime character,

Aoshima Megu (Maribon, 2015) (Figure 3), who is a young girl aspiring to become a skillful ama. The long-haired girl is a "moe" character, which implies a somewhat erotic attractiveness in anime, manga, and game sub-culture. Another popular image of ama originated from a TV

Figure 2. Toba City Tourism Promotion Brochure (Reproduced with permission).

Figure 3. Aoshima Megu, MARIBON (Reproduced with permission).

drama series, *Ama-chan* or "Little Darling Ama" broadcasted from 1st April to 28th September 2013 (NHK, 2013). The 15-minute morning series, popularly referred to as Asa-dora (morning drama), is shown daily for six months. Each series generally features the life of a woman, her family and community, her personal and career development, and is mostly based on a true story. Among the 98 series aired so far since 1961, *Oshin* (1983) gained international fame. The series Ama-chan was based on a radio drama "Hokugen-no Ama" (northernmost ama, written by Mizuki Yoko, aired in 1959) (Hokugen-no Ama, 2013). The series told the story of a young city girl going to live in her mother's home village, where her grandmother was one of the last remaining ama. There, *chan* is a diminutive suffix attached to personal names or family roles, and the word Ama also reads as a verb "amaeru," which implies the sweet, dependent, pampered or spoiled nature of a young girl who grows into an independent woman.

The practicing ama women laugh at the images of young ama.

> You have to be so careful so that you don't get tangled in ropes and other things. No one has long hair for that reason. (T, 69 year old)

> You get sweaty, dirty and sun, wind and salt burnt – you cannot be pretty doing this job. We are nothing like them! (S, 50 years old)

In Japan, the ama women are idealized and celebrated as they are believed to be connected to Japan's origin, the Sun Goddess Amaterasu (Kawaguchi, 2013; Tanabe, 1993). Such idealized depiction was included in the G7 Summit report, which referred to ama as the "heart of Japan" when the spouses of the heads of states met 85 ama women (Mainichi Shimbun, 2016; Sankei West, 2016). Interestingly, little was mentioned about the nine *okami*, or female managing directors of traditional Japanese hotels (ryokan) who were also present, nor the first female Executive Chef at the Summit venue, Shima Kanko Hotel, who hosted the Summit dinner (Shima Kanko Hotel, 2014).

While ama women are understood to be free, strong, and capable, stereotypical images place them in a fixed tourist scape of romanticized heritage. This was observed in the previous example of a new young ama in Oki island, where the focus of attention was only given to her appearance and novelty, not the importance of her being a new bearer of ecological knowledge. In this context, tourism does not "improve well-being, but in fact reinforces traditional gender roles and unequal divisions of labour" (Jiménez-Esquinas, 2017, p. 311), making the people "national objects of attention" (Venkatesan, 2009, p. 92) and transforming people into "picturesque bearers of an obsolescent tradition" (Herzfeld in Jiménez-Esquinas, 2017, p. 316). This trivializes the women's power and identity and denies the essence of their practice and knowledge, which are observational and experiential, articulating their intuitive understanding of sustainability.

Changes in the ocean and women's ecological knowledge

Since the ama practice free-diving without any special equipment, the harvest is controlled by the physical limits of each breath. There is little risk of overfishing, nonetheless, all ama communities visited developed their own regulations to prevent such an eventuality. Apart from harvest size (no less than 10.5 cm for abalone), many of the rules are unwritten, existing in their beliefs, taboos, and mythologies, which practically serve as self-regulation.

> We want to come back every year ... if the shell is too small, we'll put them back in the ocean ... so that we can come back next year. (T, 69 years old).

> We don't go into the ocean if there are bad things in family (eg death, illness) so that we don't bring bad luck into the ocean. (R, 86 years old).

At the same time, the inevitable changes in the environment over the past decades have seriously affected the ocean and its resources. Women describe the dramatic change in their catch over the last 30 years both in number and size, despite the fact that their practice remains mostly unchanged. Women believe, based on their observation, that these changes are the result of infrastructural developments (sand erosion, building of walls, barriers, roads, bridges, and riverside developments), contamination (sewage, chemicals such as detergents), and the (global) warming as well as unregulated overfishing by others.

> We used to get the shells this big (30 cm), but they have become so much smaller now. We used to get them this big, almost half a kilo, but now, they are so small. (H, 89 year old)

> Since the new road was built, the ocean has changed – lots more rubbish, and dirty. So less fish, seaweed and everything. The sea used to be so rich. (M, 70 years old)

> The sea is now dirty – rubbish at the bottom, less seaweed, and different kinds of fish. (T, 69 years old)

> We were told to leave one if there were three, when we were young, but now we may not find any. (E, 55 years old)

Maintaining ama practice and knowledge

Each ama community has debated on whether to introduce goggles (introduced from around 1890), wetsuits (from the 1960s), and other clothing. Although none of the communities restrict the use of goggles, wetsuits were introduced along with certain control measures, such as a time restriction, allowing only one or two hours per day to dive, and a catch rotation system, or a one-wetsuit per family policy. Various other restrictions apply, such as allowing only the upper or lower part of the suit or short pants. One group who did not introduce wetsuits is the Yahata community, Ashibe-chō in Iki Island, where only the wetsuit hoods are allowed. Instead, women wear dance "leotards" on top of layers of clothes – tracksuits, cotton shirts or woollen tops and bottoms. The colourful dance costumes are, according to women, for safety as well as expressing their fashion identity (Figure 4). Here, the author's observation is included, as the women invited her to join them on the boat trip on a calm sunny day. The women explained how they decided against the use of wetsuits (Pers. Comm., 12 Aug, 2017).

> If we wear them, we have to limit diving time. We didn't want to be controlled by the Union. We want to be free. (T, 69 years old)

This desire for freedom is framed by a self-confidence in their own form of intuitive self-regulation.

> Not wearing wetsuits, it is actually easier to swim. Wetsuits make you tired, as you have to swim harder. (N, 57 years old)

> Not wearing wetsuits… It's good so we won't be greedy, although we cannot get that many now. (T, 69 years old)

There were five boats on the bay, including one occupied by sisters, aged 80 and 82, the oldest in the community.

> It is something we look forward to. Feeling the ocean is good for you. (F, 80 years old)

> We never wear any wetsuits – we didn't fit into them. They are so tight. (laughing) (Y, 82 years old)

As the boat came back into the port, all women stood on the boat, put their hands together and bowed towards the shrine gate facing the sea. The gate leading to the local shrine was built on the edge of the shore by the group of ama women, so that they could pray when they left and came back into the port.

> That is the local sea gods we worship every day, before we go, and as we come home. We thank them that we came home safe. (N, 57 years old)

Prayer is an essential part of ama culture, praying for and being grateful for safety and a good harvest. Praying, a woman explained, is also a way to remind them that the "Sea can be dangerous. It can change anytime. So we pray, always pray." One such reminder is a group of six small Jizo (earth deities *Ksitigarbha*) statues standing on the Yahata shore, Ashibe-town, Iki Island (Figure 5).

Figure 4. Ama diver (Author's photograph).

Figure 5. Six Jizo Statues in the sea (Author's photograph).

At low tide, the tide washes gently at their feet, and on high tide, the water rises up to their chest. These Jizo are called *Harahoge (chest hole)-jizo*, as these stone statues have a small dip in the chest level. The dip was apparently made so that offerings given to these statues, such as sweets, fruits, or coins, would not get washed away in high tides. There is no record of who made them or why; however, the local women believe that they were made to commemorate the lives lost in the sea, including those of women divers in the region. Consequently, they "always give offerings to these statues, so that they look after the ocean, and us of course," a woman said as she placed a small rice cake on each of the six statues. For these women, prayer is an articulation of knowledge, and is a way of knowing the world.

Conclusion

The women's stories, observations, and senses illustrate the nature of knowledge and knowing based on their particular relationship with the ocean environment. Their power lies in their connectivity with their ocean environment, which guides them in maintaining their strict self-regulation despite their clear understanding that the ocean is changing because of many external forces. Their ways of knowing rely on their senses of connectivity, inclusiveness, and reciprocity, which are also expressed through their stories as care, gratitude, fear, ethics, offerings, and prayer. The recognition of these qualities as one socio-environmental foundation allows a move towards sustainability, in which gender equality presents a strong narrative to counter discourse that is divisive, exclusive, or hierarchical. As the cornerstone of sustainability, gender equality, and thus the promotion of all SDGs, encourages inclusiveness at all levels, propelling the oft-discussed aspirations, *"leave no-one behind"* and *"turning promises into action"* (United Nations, 2015).

The year 2017 was designated as The United Nations *International Year of Sustainable Tourism for Development* based on the Sustainable Development Goals towards 2030 (UN, 2017). The colourful banner comprising the 17 goals now appears in a wide range of activities, such as peace education, welfare, and conservation, as well as business marketing strategies. While the popularization of the term *sustainability* is a positive development, its fundamental values should be constantly reiterated, which will, according to Plumwood (2008), "restore place honesty". She suggested that we follow the path that allows us "to recognize the reality of multiple relationships to place but insist that they be reshaped as meaningful and responsible". Such multiplicity of relationships characterized by sensitivity and respect comprise women's contribution to the system of knowledge generation.

Regarding knowledge systems, Rose states that we are "called to build dialogical bridges between knowledge systems: between ecological sciences and the humanities, between Western and other knowledge systems" (Rose, 2015a, p. 1). The basic axioms that underlay these socio-environmental systems are: *land is the law*; and *you are not alone in the world*, which she refers to as a "practice of connectivity" (Rose, 2015a, p. 4). This implies that all living things, including humanity as a participant in a larger living system, must recognize and submit to the laws of the living world. Here, "respect is a matter of knowledge—of knowing the connections so that one knows the many contexts in which respect is due, and knowing how to look after things so that one can fulfil one's role in life … Web of life is a web of mutual inter-dependencies, which is in essence the earth" (Rose, 2015a, p. 5). The ama women's perspectives demonstrate an inclusive way of knowing through their sense of connectivity and reciprocity, which is critical to our understanding of sustainability. Maintaining such inter-dependencies through stories, prayers, senses, gratitude, and care may be what sustainability looks like. Women's capacities to foster such uniting practices supports weaving together the diverse elements addressed through SDGs, ensuring that we *"leave no-one behind."*

As suggested earlier by Ferguson (2011), grassroots feminism has the ability to inject a creative and progressive energy into tourism. Higgins-Desbiolles articulates that tourism can be a social force through its "ability to foster contact between peoples who increasingly need to understand each other and cooperate harmoniously in a world where space, resources and options are shrinking quickly" (Higgins-Desbiolles, 2006, p. 1205). Women, such as the ama in this article, clearly play an important role as holders of knowledge whose ability to live with both the land and the seas is critical today. At the same time, the imposition of stereotypes, quite separate from the reality, devalues the women's knowledge and their ways of knowing.

Finally, it is suggested that tourism's potential is only limited by "our imaginations to harness its powers for the public good" (Higgins-Desbiolles, 2006, p. 1205). From the ecohumanities perspective, it is imperative for academia to promote such imagination by incorporating more creativity into thinking and writing, and being unafraid to use the words of spirituality, hope, and love. The example of the ama of Japan reveals another system of knowledge concerning living sustainably. These findings have shown that the sustainability agenda can be advanced by challenging hierarchical systems of knowledge and valuing alternative ways of knowing. The ama women's knowledge presented here reveals a way of knowing ocean ecology and a way of being with nature and others that is valuable in our quest to build a more sustainable, inclusive and reciprocal future.

Acknowledgement

This article is dedicated to late Deborah Bird Rose, who led us to the new world of ecohumanities, and whose words of beauty and love keep us inspired.

References

Abram, D. (2004). Reciprocity. In B. Foltz & R. Frodeman (Eds.), *Rethinking nature: Essays in environmental philosophy*. Bloomington: Indiana University Press.

Agola, N., & Hunter, A. (Eds.). (2016). *Inclusive innovation for sustainable development theory and practice*. Palgrave Macmillan.

ANA (2018). Ama culture, *Innovative Voices No. 78*, https://www.ana.co.jp/dom/promotion/iv/1802/.

Ateljevic, I., Pritchard, A., & Morgan, N. (2012). *The critical turn in tourism studies: Promoting an academy of hope*. Oxford: Elsevier.

Atkinson, P., & Delamont, S. (Eds). (2011). *Sage qualitative research methods*. Thousand Oaks.

Booth, A., & Harvey, J. (2001). Ties that Bind: Native American Beliefs as a Foundation for Environmental Consciousness. In M. Boylan (Ed.). *Environmental ethics*. New Jersey: Prentice Hall.

Carney, J. (1993). Converting the wetlands, engendering the environment – the intersection of gender with agrarian change in the Gambia. *Economic Geography*, *69*(4), 329–348. doi:10.2307/143593

Charmaz, K. (2005). Grounded theory in the 21st century. In Denzin, Y., & Lincoln, Y. *The SAGE handbook of qualitative enquiry* (3rd ed.). Thousand Oaks: SAGE.

Cole, S. (2016). A gendered political ecology of tourism and water. In M. Mostafanezhad, R. Norum, E. Shelton & A. Thompson-Carr (Eds.), *Political ecology of tourism. Community, power and the environment*. New York: Routledge, Chapter 1, 31–49.

Cole, S., & Ferguson, L. (2015). Towards a gendered political economy of water and tourism. *Tourism Geographies*, *17*(4), 511–528. doi:10.1080/14616688.2015.1065509

Fankhauser, S., & McDermott, T. (2014). Understanding the adaptation deficit: Why are poor countries more vulnerable to climate events than rich countries? *Global Environmental Change*, *27*, 9–18. doi:10.1016/j.gloenvcha.2014.04.014

Ferguson, L. (2011). Promoting gender equality and empowering women? Tourism and the third millennium development goal. *Current Issues in Tourism*, *14*(3), 235–249. doi:10.1080/13683500.2011.555522

Ferguson, L., & Moreno Alarcón, D. (2015). Gender and sustainable tourism: Reflections on theory and practice. *Journal of Sustainable Tourism*, *23*(3), 401–416. doi:10.1080/09669582.2014.957208

Figueroa-Domecq, C., Pritchard, A., Segovia-Pérez, M., Morgan, N., & Villacé-Molinero, T. (2015). Tourism gender research: A critical accounting. *Annals of Tourism Research*, *52*, 87–103.

Fleras, A. (2009). *The politics of multiculturalism. Multicultural governance in comparative perspective*. New York: Palgrave Macmillan.

Fine, G. (2011). Towards a peopled ethnography. Developing theory from group life. In Atkinson, P., & Delamont, S. (Eds). *Sage qualitative research methods*. Thousand Oaks: Sage, 42–60.

Gibson, K., Rose, D.B., & Finch, R. (Eds.). (2015). *Manifesto for living in the Anthropocene*. Sydney: Puncton Books.

Goebel, A. (2004). Women and sustainability: What kind of theory do we need? *Canadian Woman Studies*, 23(1), 77–84.

Goetz, A., & Jenkins, R. (2016). Gender, security, and governance: The case of Sustainable development goal 16. *Gender & Development*, *24*(1), 127–137. doi:10.1080/13552074.2016.1144412

Graham, M. (2008). Some thoughts on the philosophical underpinnings of aboriginal worldviews. *Australian Humanities Review*, *45*, 181–194.

Hanson, A. (2015). Shoes in the seaweed and bottles on the beach Global garbage and women's oral histories of socio-environmental change in coastal Yucatán. In S. Buechler & A. Hanson (Eds.), *A political ecology of women, water and global environmental change*. New York: Routledge, Chapter 9, 165–184.

Hanson, A., & Buechler, S. (2015). Introduction. Towards a feminist political ecology of women, global change, and vulnerable waterscapes. In S. Buechler & A. Hanson (Eds.), *A political ecology of women, water and global environmental change*. New York: Routledge, 1–16.

Harris, L. (2015). A quarter century of knowledge and change: pushing feminism, politics, and ecology in new directions with feminist political ecology. In S. Buechler & A. Hanson (Eds.), *A political ecology of women, water and global environmental change*. New York: Routledge, xix–xxiii.

Higgins-Desbiolles, F. (2006). More than an industry: The forgotten power of tourism as a social force. *Tourism Management*, *27*(6), 1192–1208. doi:10.1016/j.tourman.2005.05.020

Higgins-Desbiolles, F. (2009). Indigenous ecotourism's role in transforming ecological consciousness. *Journal of Ecotourism*, *8* (2), 144–160. doi:10.1080/14724040802696031

Hokugen-no-ama Promotional Committee (2013). *Hokugen-no-ama (The northernmost ama)*. Morioka: Hokugen-no-ama Promotional Committee.

Jackson, A., & Mazzei, L. (Eds.). (2009). *Voice in qualitative inquiry: Challenging conventional, interpretive, and critical conceptions in qualitative research*. New York: Routledge.

Jahnsen, C. (2018). Women are key to achieving the sustainable development goals. *International Trade Forum*, *1*, 22–23.

Jamal, T., & Camargo, B. A. (2014). Sustainable tourism, justice and an ethic of care: Toward the Just Destination. *Journal of Sustainable Tourism*, *22*(1), 11–30. doi:10.1080/09669582.2013.786084

Jarosz, L. (2011). Nourishing women: Toward a feminist political ecology of community supported agriculture in the United States. *Gender, Place & Culture*, *18*(3), 307–326. doi:10.1080/0966369X.2011.565871

Jiménez-Esquinas, G. (2017). "This is not only about culture": On tourism, gender stereotypes and other affective fluxes. *Journal of Sustainable Tourism*, *25*(3), 311–326. doi:10.1080/09669582.2016.1206109

Kato, K. (2007a). Waiting for the tide, tuning in the world. Traditional knowledge, environmental ethics and community. In R. Bandt, M. Duffy & D. MacKinnon (Eds.), *Hearing places: Sound place time culture*. Newcastle: Cambridge Scholars Publishing.

Kato, K. (2007b). Waiting for the tide. ABC Radio National Radio Eye (now Earshot, first broadcasted on 21 July, 2007, http://www.abc.net.au/radionational/programs/earshot/waiting-for-the-tide/6660984

Kato, K. (2017). Traditional industry and sustainable tourism. Traditional ecological knowledge of ama women divers. *Tourism Culture*, 23–24.

Kawaguchi, Y. (2013). *Ama*. Tokyo: Hokuto Books.

Kirksey, S. E., & Helmreich, S. (2010). The emergency of multispecies ethnography. *Cultural Anthropology*, *25*(4), 545–576. VolIssuedoi:10.1111/j.1548-1360.2010.01069.x

Knight, D., & Cottrell, S. (2016). Evaluating tourism-linked empowerment in Cuzco, Peru. *Annals of Tourism Research*, *56*, 32–47. doi:10.1016/j.annals.2015.11.007

Leal Filho, W. (Ed.). (2018). *Climate change impacts and adaptation strategies for coastal communities*. Cham: Springer.

Liao, C., Ruelle, M. L., & Kassam, K.-A. S. (2016). Indigenous ecological knowledge as the basis for adaptive environmental management: Evidence from pastoralist communities in the Horn of Africa. *Journal of Environmental Management*, *182*, 70–79. doi:10.1016/j.jenvman.2016.07.032

(1999). Low, N. (Ed.). *Global Ethics and Environment*. London: Routledge.

McCabe, S., & Johnson, S. (2013). The happiness factor in tourism: Subjective wellbeing and social tourism. *Annals of Tourism Research*, *41*, 42–65. Vol. doi:10.1016/j.annals.2012.12.001

Mainichi Shimbun (2016). Spouses meet Ama. Encountering the heart of Japan (26 May, 2018. https://mainichi.jp/articles/20160527/k00/00m/040/084000c.

Maribon (2015). Aoshima Megu. http://ama-megu.com

Maraini, F. (1962). *The island of the fisherwomen*. San Diego: Harcourt, Brace & World.

Martinez, D. P. (2004). *Identity and Ritual in a Japanese Diving Village. The Making and Becoming of Person and Place*. Honolulu: University of Hawai'i Press.

Matthews, F. (2015). Earth as ethic. In K. Gibson, D. B. Rose & R. Fincher (Eds.), *Manifesto for living in the Anthropocene*. New York: Punctum books, 72–74
Mosedale, J. (2015). Critical engagements with nature: Tourism, political economy of nature and political ecology. *Tourism Geographies, 17*(4), 505–510. doi:10.1080/14616688.2015.1074270
Mostafanezhad, M., Norum, R., Shelton, E., & Thompson-Carr, A. (2016). Introduction, In M. Mostafanezhad, R. Norum, E. Shelton & A. Thompson-Carr (Eds.), *Political ecology of tourism. Community, power and the environment*. New York: Routledge, 1–21.
NHK (2013). *Amachan*. http://www6.nhk.or.jp/drama/pastprog/detail.html?i=asadora88.
Norberg-Schulz, C. (1980). *Genius loci: Towards a phenomenology of architecture*. New York: Rizzoli.
Norum, R., Mostafanezhad, M., Shelton, E., & Thompson-Carr, A. (2016). Conclusion. Towards future intersections of tourism studies and political ecology. In M. Mostafanezhad, R. Norum, E. Shelton & A. Thompson-Carr (Eds.), *Political ecology of tourism: Community, power and the environment*. New York; Routledge, 302–308.
Ozyıldırım, I. (2009). Narrative analysis: An analysis of oral and written strategies in personal experience narratives. *Journal of Pragmatics, 41*, 1209–1222.
Peterson, V. (2005). How (the meaning of) gender matters in political economy. *New Political Economy, 10*(4), 499–521. doi:10.1080/13563460500344468
Plumwood, V. (2002). *Environmental culture. The ecological crisis of reason*. London: Routledge.
Plumwood, V. (2007). A review of Deborah Bird Rose's 'Reports from a Wild Country: Ethics for Decolonisation'. Australian Humanities Review (22). http://australianhumanitiesreview.org/2007/08/01/a-review-of-deborah-bird-roses-reports-from-a-wild-country-ethics-for-decolonisation/
Plumwood, V. (2008). Shadow places and politics of dwelling. *Australian Humanities Review, 44* http://australianhumanitiesreview.org/2008/03/01/shadow-places-and-the-politics-of-dwelling/ accessed 16 April, 2018.
Ravneberg, B., & Soderstrom, S. (2017). *Disability, society and assistive technology*. New York: Routledge.
Rivera, V. (2017). A path to social inclusion in multicultural Australia. https://www.internationalaffairs.org.au/news-item/a-path-to-social-inclusion-in-a-multicultural-australia/.
Rooney, T., Lawlor, K., & Rohan, E. (2016). Telling Tales: Storytelling as a methodological approach in research. *The Electronic Journal of Business Research Methods, 14*(2), 147–156.
Rose, D. (1996). *Nourishing Terrains*. Canberra: Commonwealth of Australia.
Rose, D. B. (2015a). The ecological humanities. In K. Gibson, D. B. Deborah & R. Fincher (Eds.), *Manifesto for living in the Anthropocene*. New York: Punctum books, Chapter 1, 1–6.
Rose, D. B. (2015b). Dialogue. In K. Gibson, D. B. Deborah & R. Fincher (Eds.), *Manifesto for living in the Anthropocene*. New York: Punctum books, Chapter 20, 127–132.
Rose, D. B. (2018). *Gift of life in the shadow of death*. Human Nature Lecture Series, Australian Museum (8 March, 2018). https://soundcloud.com/australianmuseum/live-at-the-am-humannature-lecture-series-deborah-bird-rose.
Rose, D., & Robin, L. (2004). The ecological humanities in action: An invitation. *Ecological Humanities* (31–32). http://australianhumanitiesreview.org/category/issue/issue-31-32-april-2004/.
Ruiz-Ballesteros, E. (2011). Social-ecological resilience and community-based tourism: An approach from Agua Blanca, Ecuador. *Tourism Management, 32*(3), 655–666. doi:10.1016/j.tourman.2010.05.021
Sankei West. (2016). Foreign media impressed with Ama and Hospitality, *Sankei West* (May 28, 2018. https://www.sankei.com/west/news/160528/wst1605280045-n1.html.
Scheyvens, R. (2000). Promoting Women's Empowerment Through Involvement in Ecotourism: Experiences from the Third World. *Journal of Sustainable Tourism, 8*(3), 232–249. doi:10.1080/09669580008667360
Scheyvens, R., & Russell, M. (2012). Tourism, land tenure and poverty alleviation in Fiji. *Tourism Geographies, 14*(1), 1–25. doi:10.1080/14616688.2011.593188
Shaw, R., Rahman, A., Surjan, A., & Parvin, G. (Eds.). (2016). *Urban disasters and resilience in Asia*. Amsterdam: Butterworth-Heinemann.
Shima Kanko Hotel (2014). *Press release: The first female executive chef*. https://www.miyakohotels.ne.jp/file.jsp?id=97854.
Shubach, A. (2015). *The* disappearing Ama: Japan's tough, topless free-diving mermaids. https://jezebel.com/the-disappearing-ama-japans-tough-topless-free-divin-1679290183
Stapleton, K., & Wilson, J. (2017). Telling the story: Meaning making in a community narrative. *Journal of Pragmatics, 108*, 60–80. doi:10.1016/j.pragma.2016.11.003
Stuart, E., & Woodroffe, J. (2016). Leaving no-one behind: Can the Sustainable Development Goals succeed where the Millennium Development Goals lacked? *Gender & Development, 24*(1), 69–81. doi:10.1080/13552074.2016.1142206
Swain, R. B., & Wallentin, F. Y. (2008). Economic or non-economic factors: What empowers women? (No. 2008: 11). *Working Paper*. Sweden: Department of Economics, Uppsala University.
Tanabe, S. (1993). *Ama*. Tokyo: Hosei University Publishing.
Thompson-Carr, A. (2016). Introduction to community and power. In M. Mostafanezhad, R. Norum, E. Shelton & A. Thompson-Carr (Eds.), *Political ecology of tourism: Community, Power and the environment*. New York: Routledge, 25–30.

Toba City (Tourism Promotion) (2018). *Toba walking handbook (Toba Osanpo techo)*, http://tobakanko.jp/modules/pamphlet/detail.php?dtid=30.

Tōyō suisan kagaku kyōkai (2011). *Report on the current state of ama in Japan*. Toba Sea Folk Museum.

Truelove, Y. (2011). (Re-)conceptualizing water inequality in Delhi, India through a feminist political ecology framework. *Geoforum*, *42*(2), 143–152. doi:10.1016/j.geoforum.2011.01.004

Tucker, H., & Boonabaana, B. (2012). A critical analysis of tourism, gender and poverty reduction. *Journal of Sustainable Tourism*, *20*(3), 437–455. doi:10.1080/09669582.2011.622769

United Nations (2015). *Transforming our world: The 2030 agenda for sustainable development*. New York: UN, https://sustainabledevelopment.un.org/content/documents/21252030%20Agenda%20for%20Sustainable%20Development%20web.pdf.

United Nations (2017). UN *Sustainable development goals 17 goals to transform our world*. http://www.un.org/sustainabledevelopment/sustainable-development-goals/.

UN Women (2018). *Turning promises into action: Gender equality in 2030 agenda for sustainable development*. http://www.unwomen.org/en/digital-library/publications/2018/2/gender-equality-in-the-2030-agenda-for-sustainable-development-2018#view

Venkatesan, S. (2009). Rethinking agency: Persons and things in the heterotopia of "traditional Indian craft". *Journal of the Royal Anthropological Institute*, *15*(1), 78–95. doi:10.1111/j.1467-9655.2008.01531.x

Wearing, S. (2002). Re-centering the self in volunteer tourism. In G. Dann (Ed.), *The tourist as a metaphor of the social world*. New York: CABI, 237–262.

Weir, J. (2008). Connectivity. *Australian Humanities Review*, *45*, 153–164.

West, J., & Haug, R. (2017). The vulnerability and resilience of smallholder-inclusive agricultural investments in Tanzania. *Journal of Eastern African Studies*, *11*(4), 670–691. doi:10.1080/17531055.2017.1367994

Williams, C., & Golovnev, I. (2015). Pamiri women and the melting glaciers of Tajikistan. A visual knowledge exchange for improved environmental governance. In Buechler, S., & Hanson, A. (Eds.). *A political ecology of women, water and global environmental change*. New York: Routledge, Chapter 11, 207–225.

Willis, J. (2007). *Foundations of qualitative research: Interpretive and critical approaches*. Thousand Oaks, CA: Sage Publications.

Zhao, W., & Zhang, C. (2018). From isolated fence to inclusive society: The transformational disability policy in China. *Disability & Society*, *33*(1), 132–137. Vol. doi:10.1080/09687599.2017.1375246

"This is a holy place of Ama Jomo": *buen vivir*, indigenous voices and ecotourism development in a protected area of Bhutan

Heidi Karst

ABSTRACT

Common definitions of ecotourism address the need to benefit the well-being of local people; in reality, ecotourism is often supported or dominated by external agents. Local ideological, social and cultural values may be overlooked, thereby disenfranchising local and indigenous stakeholders. This paper examines indigenous perceptions of ecotourism and well-being, and how they relate to ecotourism practices in three Brokpa communities of the first protected area in Bhutan to become a special tourism destination. A range of methods (e.g. semi-structured interviews, participatory mapping) was used and data analysis was couched in the context of *buen vivir* (living well), an indigenous concept of well-being. Findings indicate that: (1) perceptions of ecotourism vary between community members and contrast with the official park definition; (2) well-being is conceived in diverse ways but corresponds to key features of *buen vivir*; (3) indigenous worldviews are pervasive, and shape social values and spiritual beliefs in connection to nature; and (4) indigenous cosmologies and traditional values may be strengthened through their integration into modern policies and institutions. This research contributes more non-Western voices to the literature, a concept and mode of analysis for well-being that embraces an indigenous ontology, and empirical insights on indigenous communities in protected areas.

Introduction

Ecotourism is widely considered a tool for increasing the economic well-being of local communities while conserving natural resources (Fennell, 2001; Harris, 2009). One of the most prevalent definitions of ecotourism is "responsible travel to natural areas that conserves the environment, sustains the well-being of the local people, and involves interpretation and education" (The International Ecotourism Society [TIES], 2015). The all-encompassing nature of this definition and its earlier adaptation (i.e. TIES, 1990) reflect the values of its origins in the mid-twentieth-century environmental movement (Higgins-Desbiolles, 2009). Such views have been critiqued for compartmentalising humans and nature, and serving as Western constructs that promote the interests of the Global North (Cater, 2006; Mowforth & Munt, 2009). Accordingly, there have been calls for more different, non-Western and distant voices in the literature (Cater, 2006; Prakash, 1994; Wearing & McDonald, 2002). These critiques challenge the discourse on what ecotourism entails to local people because ecotourism is often introduced, supported, or dominated by external actors or agencies (Blackstock, 2005; Kontogeorgopoulos, Churyen, & Duangsaeng, 2014). Local belief systems can have strategic implications for conservation-related policy and programmes, as academics and practitioners have argued that indigenous knowledge and beliefs in sacred places may encourage ecological integrity and promote sustainable development (Higgins-Desbiolles, 2009; Verschuuren, Wild, McNeeley, & Oviedo, 2010). This paper examines how local, indigenous definitions and understandings of ecotourism and well-being connect to ecotourism development in a Bhutanese wildlife sanctuary.

The link between the culture and spiritual beliefs of indigenous peoples and their natural environment has been explored extensively in various bodies of literature, from ecotourism (Fennell, 2008; Higgins-Desbiolles, 2011; Zeppel, 2006) and pilgrimage tourism (Andriotis, 2009; Timothy & Olsen, 2006) to traditional ecological knowledge (Armitage, 2003; Berkes, 2012; Colding & Folke, 2001) and sacred natural sites and landscapes (Bernbaum, 2006; Gadgil & Vartak, 1976; Ormsby & Bhagwat, 2010; Verschuuren et al., 2010). Ecotourism is part of the wider concept of sustainable tourism: conflicts and opportunities between sustainable tourism and indigenous peoples remain issues for research, discussion and contestation (Carr, Ruhanen, & Whitford, 2016). Sacred natural sites, which vary in scale and can be found around the world, may honour a deity, provide sanctuary for spirits, be a living expression of ancestors, or protect a holy, historic place (Dudley, Higgins-Zogib, & Mansourian, 2009; Rutte, 2011). These sites may be situated in ecologically sensitive and strategic locations, and often supply a range of ecosystem services, such as providing medicinal plants, protecting water and soil, and sustaining culture as sites for important socio-cultural events and rituals (Colding & Folke, 2001; MEA, 2005; Rutte, 2011).

The indigenous culture–spiritual–nature connection has also been realised as a philosophy across South America known as *buen vivir*, which complements the literature on sacred sites in nature. Roughly translated as "living well" or "collective well-being", *buen vivir* is closely related to expressions in many Andean and neighbouring cultures: *sumac kawsay* of the Kitchwa, Ecuador; *suma qamaña* of the Aymara, Bolivia; *küme mongen* of the Mapuche, Chile and Argentina; *ñande reko* of the Guarani, Paraguay and *shiir waras* of the Ashuar, Ecuador and Peru (Jiménez, 2011). These terms reflect fullness of one's life within a society, and unity with other people and nature (Gudynas, 2011; Jiménez, 2011; Walsh, 2010). Aspects of this concept are reflected in other global philosophical traditions and worldviews, such as Aristotelian values of *eudaimonia* (human flourishing) (Ryan & Deci, 2001; Sirgy et al., 2006); the indigenous Fijian concept of *vanua*, which interrelates social, ecological and spiritual elements of life (Farrelly, 2011); and Buddhist principles of compassion and interconnectedness (Brooks, 2013; Wangmo & Valk, 2012).

Several key features of the *buen vivir* perspective encompass aspects of relationships , and culture and spirituality. *Buen vivir* recognises nature as a subject and well-functioning ecosystems are paramount to economic and social objectives (Acosta, 2010). Relationships and reciprocity also matter in *buen vivir*, which addresses harmonious co-existence between human beings, and human service and care toward the environment (Deneulin, 2012). In this way, it is concerned with the human right to live in dignity (e.g. to have a good, meaningful livelihood) as well as the rights of nature. Furthermore, the material is not separate from cultural and spiritual dimensions (Villalba, 2013). The unseen realm sustains the material world, and a balanced sense of sacredness is required in daily life. *Buen vivir* is contextual and non-homogenous (Deneulin, 2012; Gudynas, 2011). It provides space for different articulations of the meaning "to live well" but specific ideas (e.g. *sumac kawsay*) cannot be transplanted to other locations. Finally, *buen vivir* is infused with idealism. Deneulin (2012) notes that its utopic dimension means that contradictions and tensions will always be present. These central characteristics can be distilled into three broad and measurable features: (1) human relationships; (2) human–nature relationships; and (3) culture and spirituality. This paper does not seek to systematically apply the entire *buen vivir* approach. Rather, it draws upon seminal features that have been simplified into "relationships" and "nature-spirituality"-related categories.

Well-being concepts, including *buen vivir*, have gained momentum in social and political arenas at the global and grassroots levels. Countries such as Bhutan, Canada, France and Thailand are developing and improving measurement tools and indices of well-being to track social progress (Brooks, 2013; Theerapappisit, 2003). As a practice, *buen vivir* has been fruitful as a social movement for indigenous and non-indigenous people. For instance, mutual respect for *Pachamama* (Mother Earth) and the sacredness of water has brought together indigenous communities, environmental groups, unions and the Catholic Church in a successful coalition to keep Chilean Patagonia free of dams (Latta, 2014). As an indigenous ontology, *buen vivir* is compatible with Buddhist emphasis on social and ecological integrity, spiritual over material growth, and moderation. Research on how indigenous perceptions of ecotourism relate to notions of well-being from a *buen vivir* perspective would contribute to the growing literature on local narratives and indigenous perceptions on ecotourism that includes cases from the Global South (cf., Farrelly, 2011; Higgins-Desbiolles, 2009; Hutchins, 2007; Lynch, Duinker, Sheehan, & Chute, 2010; Manyara & Jones, 2007).

The concept of reciprocity is elaborated through *tha damtshig*, one of the most fundamental social values that pervades contemporary Bhutan (Kinga, 2001). Based on Buddhist teachings and referenced in most Bhutanese folktales, *tha damtshig* revolves around a sense of social responsibility and reciprocity as it seeks to establish and maintain social harmony and justice through love, honour and loyalty in relationships with all people (Allison, 2004; Kinga, 2001; Whitecross, 2010). *Tha damtshig* contains a vast range of applications and meanings (e.g. generosity, honesty, filial piety), depending on the context for which it is used. More recently, it has been tied to notions of state loyalty and political allegiance (Phuntsho, 2004). Allison (2004) notes that *tha damtshig* may lead to potential abuses of power (e.g. nepotism, corruption) if people are motivated by socio-political obligations and considerations over personal integrity and virtue. Discussion and awareness of this complex concept are critical to the success and sustainability of any development project in Bhutan.

This paper begins with an outline of the research context, focusing on three indigenous *Brokpa* (nomad, highlander) communities currently engaged in ecotourism development in a protected area (PA) in eastern Bhutan. The methods used are briefly described, including semi-structured interviews, focus groups and participant observation. Study results examine local perceptions of ecotourism and well-being in relation to ecotourism development among indigenous people, couched in terms of the three main features of *buen vivir*, namely human relationships, human–nature relationships, culture and spirituality. The discussion offers four areas of insight for developing ecotourism with indigenous stakeholders, highlighting how challenges in practice present opportunities for future policy and planning. In closure, this paper reflects on lessons learned for PAs and conservation and development projects, and contributes new directions for future research in well-being.

Study location

Brokpa society

The Brokpa are a Himalayan tribe and ethnic minority group in Bhutan. They inhabit remote mountain villages and surrounding hamlets in Sakteng Wildlife Sanctuary (SWS), one of the newest and smallest Bhutanese PAs. According to the religious biography *Jomo's Namthar*, Brokpas are descendants of Tibetan immigrants who fled a tyrannical king and slave labour between 627 and 649 A.D. (Wangmo, 1990). Ama (or Aum) Jomo[1], the young woman who led the exodus to Bhutan, is considered the local protector goddess whose spiritual abode is a mountain towering over the valley of Merak and Sakteng (Pommaret, 2004; Wangmo, 1990). The total human population of Sakteng and Merak *geogs* (administrative blocks) is 4557 residents (WWF Bhutan & SWS, 2011), consisting of 702 households. This study focuses on the largest settlements in Merak, Sakteng and Joenkhar (Figure 1).

Brokpa culture and tradition has remained relatively intact as a result of limited external influence over the years. Brokpake is spoken locally and traditional clothing is still commonly worn in the villages. Female attire includes a long white or red tunic-style dress (*shingkha*), red jacket (*todung*) with colourful woven animal or geometric designs, and a thick black woollen apron (*otey*). Men don animal hide vests (*paksa*) over long-sleeved red or brown woollen tunics (*chupa*) and the progressively less common *kanggo* (deerskin half-pant) and leather chaps (*phishoop*). Both genders commonly wear turquoise stone earrings fastened by thread through the earlobes in addition to a black felted yak wool hat (*shamo*), which has five long, twisted tufts that radiate out like spider legs and function as rainspouts.

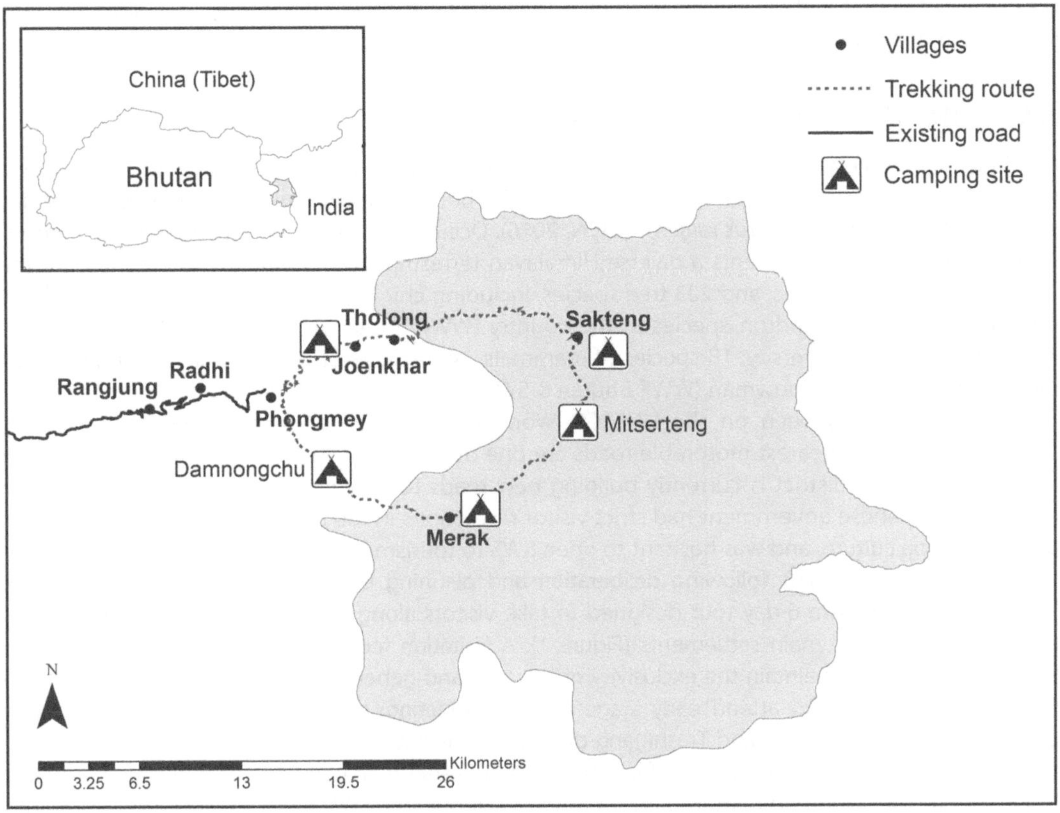

Figure 1. Map of Bhutan showing villages along the Merak-Sakteng Trek in Sakteng Wildlife Sanctuary. Source: Author.

Lifestyle and customs of the Brokpa differ greatly from mainstream Bhutanese society. Brokpas traditionally practice transhumance, a tradition that carries on in Merak and Sakteng but not Joenkhar, where inhabitants are sedentary agriculturalists. Maize and potato are their staple crops and some have small herds of cattle. For highlanders, the movement of livestock such as yaks and cattle from high altitude to lower in winter and return in spring for suitable grazing pastures is attributed to temporal factors and spatial distribution of different livestock breeds (Dorjee, 2012). Houses in all villages are customarily made of stone and wood. Sakteng is the seat of the sub-district administration and other government offices, including a police outpost. There is a basic health unit in Joenkhar and outreach clinics in the other two villages. Mobile communication service has existed in the area since 2010 and electricity since 2011.

Marriage customs include the increasingly rare practice of fraternal polyandry and sororal polygyny;[2] preference is given to whichever type of marriage will produce greater economic gain (Wangmo, 1990). This custom prevents division of family property and ensures that farm work is shared. For example, one husband takes care of the herds while the other trades butter, cheese and meat for grains and vegetables with neighbours in southern villages and across the Indian border. These traditional bartering practices are ceding to the modern money economy as a result of increased road access to villages and towns outside of Merak–Sakteng.

Religion plays a prominent role in the daily life of all Bhutanese citizens. Buddhism is practiced in all Brokpa communities with elements and rituals related to Bön, an ancient animist-shamanist religion (Brooks, 2011; Pelgen, 2007). Picturesque temples, monasteries and historic sites grace SWS, and locals hold various festivals with traditional song and *chams* (masked dances) throughout the year.

Brokpas are famed for their customary rites of hospitality, such as *tshogchang* (group drinks), and offering *ara* (local wine/spirits) or *suja* (butter tea) to visitors (Dorji, 2012).

Sakteng Wildlife Sanctuary and ecotourism development

The Sanctuary was established in 2003 following a revision of the national PA system in 1993. It protects the easternmost temperate ecosystems, which harbour some endemic and endangered species, including the red panda (*Ailurus fulgens*) (IUCN, 2016). Occupying 740.6 km^2 with an altitude range of 1600–4500 masl, SWS represents a diverse Himalayan terrestrial ecosystem of mixed coniferous forest, alpine scrub and screes, and 203 tree species, including chir pine (*Pinus roxburghii*) and the highest diversity of rhododendron species in the country (WWF Bhutan & SWS, 2011). It also houses a wide range of faunal diversity: 18 species of mammals, 147 species of birds, and according to locals, the yeti or Abominable Snowman (WWF Bhutan & SWS, 2011).

The Sanctuary has been on the UNESCO World Heritage List as a tentative site since 2012 (UNESCO, 2016). The nearest motorable roads are one day's walk to Merak or Joenkar and two days to Sakteng, but the district is currently building new roads to connect Merak and Sakteng to main roads. The Bhutanese government had strict visitor restrictions in the past to protect biological diversity and Brokpa culture, and was hesitant to open SWS to tourism.

Tourism began in 2010, following deliberation and planning by several government bodies. The Merak–Sakteng Trek is a 6-day tour designed to take visitors along an ancient and operative walking trail that connects the main settlements (Figure 1). A visitation fee, the first of its kind for any PA in Bhutan, is charged to maintain the exclusivity of the area and generate funds for the park. Community members were invited to attend early stage planning meetings with representatives from Tourism Council of Bhutan (TCB), SWS and Trashigang district administration. Funding for construction of campsites and other project amenities came from TCB, and SWS staff built garbage pits and created signage.

In tandem with community members and other stakeholders, SWS staff co-facilitated the creation of by-laws that outlined the rules and responsibilities of key stakeholders and management committees, collection for and usage of the Community Development Fund (CDF), rates and responsibilities for local products and services (e.g. local guides), coordination between local service providers, and a list of key contacts for services (Karst & Gyeltshen, 2016). The CDF allows a certain percentage of the profits from campsite or other tourism activities stipulated in the by-laws to be collected and used for facility maintenance and community development. Joenkhar is the only village to have a system of "tourist *tshogpas* (coordinators)", elected community members who make decisions and coordinate tourism services.

The government invested in capacity-building and other resources to develop ecotourism in SWS. The TCB subsidised a select number of households in Merak and Sakteng to build toilet facilities in their homes in order to operate as homestays, and held a training programme for a set of youth and school dropouts to gain skills as local guides and cooks who would then provide mandatory services to tour groups. A few homestays in Merak were in operation as of Spring 2014. The Nature Recreation and Ecotourism Division (NRED) of the Department of Forest and Park Services (DOFPS) published guidelines for ecotourism planning and management for the Bhutanese PA network, and copies of the guidelines were circulated to all park range offices from 2012 through 2013. Partially due to its remote location, the number of foreign visitors has remained relatively low: 147 visitors in 2014 and 85 in 2015 (pers. comm. Sonam Penjor, May 11, 2016).

Methods

This study adopted a case study approach and involved close collaboration with stakeholders. A literature review and document analysis were conducted prior to data collection in Merak–Sakteng to provide insights into where and how the term "ecotourism" was used and interpreted in academic texts, reports, government policies and documents (Corbin & Strauss, 2008). During Fall 2013 and

Spring 2014, 63 semi-structured household interviews and six focus groups consisting of 15 participants total between the ages of 23 and 67 were held, in addition to participant observation and informal discussions with community leaders and members. All community participants identified as ethnic Brokpa and Buddhist.

In addition to attending several household *pujas* (rituals), the author participated in three religious community events: a *Jomo sekha* (autumn) ceremony in Merak, which had been witnessed by only one other foreign visitor in the past; the annual school *rindu* (ritual for good health in the community); and a spring *chokhor* (procession) in Sakteng. Data from field notes were used to elicit key themes and codes (e.g. definitions of well-being), compare different perceptions and scenarios between communities, and capture the complexity of local perspectives (Creswell, 2009; Newing, 2011; Yin, 2009).

To prevent power imbalances, elected government officials only participated in interviews, with the exception of local *tshogpas*. These elected delegates also joined focus groups because community leaders and members felt they would represent the views and values of the community well. The author kept a research log to track her conceptual journey through the research process and to practice critical reflexivity as a non-indigenous researcher (Dowling, 2010; England, 1994).

A qualitative case study design that would provide deeper understanding of stakeholder perceptions and influences was employed to complement existing ecotourism research in Bhutan, which frequently concentrates on quantitative approaches and analysis (cf., Gurung & Seeland, 2008; Rinzin, Vermeulen, & Glasbergen, 2007). Participant selection was based on a mix of purposive, snowball, and convenience sampling strategies (Newing, 2011) to include local stakeholders located near the main trekking route between non-migratory periods and to counterbalance transitory lifestyles.

Household interviews and focus groups were conducted with field assistants in one or more language (Brokpake, Dzongkha, Sharchopkha, English) and notes were transcribed into English. Each interview lasted c. 60–90 minutes, and focus group sessions ran two to three hours. All semi-structured questions were designed with local stakeholder input to procure insights on themes corresponding to key features of *buen vivir* (Table 1). Open-ended questions applied brainstorming,

Table 1. Seminal features of *buen vivir* and key questions posed to study participants.

Features	Key questions
Human relations	• What does the word 'ecotourism' mean to you? • What types of ecotourism activities take place locally? • What are the 3 most important relationship levels (e.g., household, institutions) that impact your life and work? Of these 3, select which ones most influence your livelihood decisions, with 1 being the most important. Please explain. • In this village, how are decisions made about who participates in ecotourism activities? • What is important for living well together in this community? • Which rules (formal/informal) help create/reduce wellbeing in this community, and how? For each rule listed, what is the general degree of compliance? • What social changes have occurred since ecotourism initiatives first began? • Who benefits most/least from ecotourism activities? Why? • Does ecotourism contribute to your wellbeing?
Human-nature relations	• What does it mean for you to live well and be happy? • What natural resources do you need to live well and be happy? Why? • Identify and describe any natural resources that have spiritual or historical value for you/your community. • What are some things that constrain your access to ecological resources? • What ecological changes have occurred since ecotourism initiatives first began?
Culture and spirituality	• What is your religious affiliation? • Are/were there any religious beliefs, rituals and/or cultural traditions concerning the environment at present/in the past? If so, describe. • Identify and describe any man-made resources that have spiritual or historical value for you/your community.

pile sorting and ranking to create and organise ideas and suggestions (Puri, 2011). Participatory mapping was used to stimulate discussion, verify household interview responses, and assess tangible natural and built resources (Newing, 2011; Puri, 2011). Focus group participants itemised and decided upon main themes which were compared with household interview data and analysed to identify similar and differing perspectives related to *buen vivir* features.

Results

Following the main research objective to examine local perceptions of ecotourism and well-being, the results are presented in three parts: (1) local perceptions of ecotourism; (2) local perceptions of well-being; and (3) broader socio-cultural influences, which participants identified in relation to ecotourism and well-being. Each subsection emphasises the relevance of the results to one or more of the key features of *buen vivir* (human relationships, human–nature relationships, and culture and spirituality) apropos the questions in Table 1.

Local perceptions of ecotourism

Overall, ecotourism held different meanings for different people and reflected the importance of inter-human relations and some human–nature interactions, but some participants were unable to explain what ecotourism meant to them. For example, a participant from Joenkhar stated that ecotourism was "to see our culture and tradition, and to take photos of animals and plants here". Another from Sakteng said, "Tourists come to see the culture of Merak-Sakteng, they see sacred places, how community is staying good and our rituals". This particular comment draws attention to the perceived importance of culture and spiritual realms for ecotourism activities. However, almost a quarter of all interviewees in Sakteng and Merak said they were "unsure" or had "no idea" of what ecotourism meant, and nearly half of all Sakteng interviewees could not distinguish between ecotourism and general tourism. Even those who were able to articulate the concept in their own words had doubts. One Sakteng interviewee said, "I think ecotourism is the interchange of nature and people from other countries, foreigners visiting particular places and nature, sharing information about nature, their place…I'm not entirely sure". All Joenkhar participants were able to confidently define the term.

There was consensus on types of local ecotourism activities offered, predominantly around culture. Locals would provide "cultural programmes", consisting of traditional folk songs and dances, *tshogchang*, and ritual masked dances like the *yak cham* (yak dance) or the ancient dance-drama *Achey Lham* (Sister/Lady Goddess) performed during the annual folk festival. Selling handmade items such as *chupa*, *kanggo*, or *fatchung* and *raka* (types of bags) were listed as common activities in addition to camping or staying in guesthouses and portering. Some Joenkhar participants referenced ecotourism activities in terms of "community conserving the environment" or "not cutting down trees as much anymore", and a few Sakteng participants believed ecotourism meant "coming to visit the beauty of nature" and to "show types of trees". Several Sakteng participants likened ecotourism to "pilgrimages", as one stated, "Tourists gain more knowledge from coming here, they do pilgrimage to places – holy places". In this view, local traditions and religious practices provided a strong foundation for ecotourism activities and what they constitute, enhancing human (locals and tourists)–nature bonds.

The most common understanding of ecotourism reported by more than half of all Merak and Joenkhar households was that tourists come to visit a place, which would bring economic benefit to the community. Tourists buy handmade items and sometimes make donations at local schools and temples. In the words of one Merak man, "Tourists are rich men, they come and spend lots of money". A Joenkhar farmer stated, "Through ecotourism, we can conserve the environment, preserve our culture and tradition and *lhakhangs* [temples] – also get money through *tshogchang*". Most respondents repeatedly emphasised that economic gain "should be evenly distributed". Participants from all three communities mentioned that TCB officials explained ecotourism to villagers during the early planning stages in terms of economic benefits to community, yet many TCB-trained cooks and guides had difficulty finding employment. One Merak participant claimed, "[Ecotourism] means nothing for me. When it started, they promised benefits. But where are the benefits?" Another person from Merak said he thought ecotourism was "helpful and harmful – harmful because I received training but there's no work now". Ecotourism fostered discord in human relations while nature was seen as a commodity through which tourism could be used to harness its economic value.

The park guidelines and official NRED (DOFPS) definition of ecotourism specifically refer to human and human–nature relationships and the importance of preserving cultural heritage, but do not fully address spirituality or religion. The NRED guidelines for ecotourism were designed around Bhutan's vision for tourism and development strategy of Gross National Happiness (GNH), which is framed through the four pillars of sustainable and equitable socio-economic development, environmental conservation, preservation and promotion of culture, and good governance (NRED, 2012). Further informed by the "official" definition from TIES (1990, 2015), the guidelines define ecotourism in the Bhutanese context as:

> High value low impact travel that supports the protection of cultural and natural heritage; provides positive and enriching experiences for visitors and hosts; assures tangible benefits to local people; and contributes to the pillars of Gross National Happiness. (NRED, 2012, p. 15)

For DOFPS, key elements to consider for successful ecotourism in Bhutan encompass high revenue generation, quality and authenticity of experience and services (high value), and less negative socio-cultural and environmental impacts (low impact); protection for cultural and natural heritage; tangible socio-economic benefits to local people, community participation and empowerment through capacity building; educational and valuable experiences for visitors and hosts; and contribution to GNH goals.

Local perceptions of well-being

When asked what they needed to "live well and be happy", participants across all villages highly valued good relationships and peace, money and income, and good health. Merak and Sakteng participants most frequently mentioned the need to cultivate close relationships and spend time with friends, community and particularly family members. A Joenkhar man claimed, "If I can't keep good relationships with my family, how can I get food? We all have to work and cooperate, work and share". Another Merak participant stated, "I need my family to be happy, then local community. If I'm good to them, then automatically they will treat me very well and I will be happy".

Several participants from Sakteng and Merak acknowledged that peaceful relations and collaboration were not constant in their communities. According to one Sakteng participant:

> [People in Merak-Sakteng] sometimes quarrel and have disagreements and don't get along together much. Most of the public are peaceful and get along, but it's usually one or two individuals who instigate problems. They can talk well and will gossip to others, then fighting will start.

This thought was reinforced by a respondent in Merak, who bemoaned that her fellow citizens "don't show respect to elders and [know] how to keep *tha damtshig* with government. They quarrel with staff". These responses point to the importance of maintaining positive human relations in order for local people to have peace and access the resources they need to thrive.

Corresponding to the need of good relations for a good life was the desire for having "no sickness or misery". When asked which levels of relationships were necessary to achieve well-being, participants from all villages discussed the importance having religious leaders present. A Joenkhar resident said, "If someone in the family is sick, I go directly to religious leaders for help so the *lama* (priest) does *pujas* for [him/her]". People donated regularly for *pujas* and *tsechus* (religious festivals) to receive blessings to help with their work, family, community and daily life. Every family hires *gomchens* (lay priests) and monks to perform *lo chu*, an annual puja for overall household blessings.

Efforts to worship and celebrate together were considered extremely important, as one Merak man explained, "Water that comes from one source will come down to others through the same passage. For community to live well together the main thing is religion. When we perform a *puja* we all sit together and eat together". Individual well-being (e.g. health) is reinforced by local culture and the spiritual, which can enhance community (human) relationships.

Material resources, such as money, income and built resources, were also high priority needs for achieving well-being in all villages, particularly Joenkhar. One farmer stated, "If there is money, I can manage everything". Good facilities and services, specifically a road and electricity, were considered critical for communal well-being. One woman clarified why she wanted a road for Joenkhar, "It's not easy to sell vegetables to tourists because road is not nearby. A road is also better for getting things we need. By selling vegetables we'll make money and live peacefully". A weaver from Sakteng commented:

> In the past, the community had a hard life. We had to carry things on our backs from Trashigang and India. We earned less money then, but things are better now. I am praying things will continue to get better and be peaceful.

Therefore, having more material resources can positively impact human relations and generate peace in families and the community.

Human relationships were strained when ecotourism development did not result in widespread financial gains. Common occupations in all three villages were pastoralism or farming (92%) but most residents held two or more jobs concurrently, such as construction or performing cultural dances. Other primary occupations involved a trade (e.g. commerce) or skilled labour (e.g. weaving, hat making), and 10% of participants held elected positions in local government. When asked if ecotourism contributed to their well-being, nearly all participants reiterated that economic benefits were not shared equitably. Horse contractors, porters with yaks and tour operators from big cities were regarded as the greatest beneficiaries of ecotourism. Respondents noted that those in the community who had the most horses were the "same people who sell things to tourists". On the question of who benefits most from ecotourism and how decisions are made about participation and management, one retired man from Merak said, "Porters make all the money, but others in the community don't make any money or benefit. Porters make all the decisions". At the time of research, the village CDF accounts did not have enough savings to pay for campsite maintenance or charitable works.

Many participants in every community mentioned reliance on place and ecological resources for their well-being and ecotourism development. Fodder for animals, edible plants, and trees, plants, and flowers for incense and medicinal purposes could be found in the surrounding forests. Some animals were illegally hunted to make folk remedies or Brokpa clothing and bags (e.g. *paksa*, *raka*), including the near threatened Himalayan goral (*Naemorhedus goral)* (IUCN, 2016). A few participants in Merak and Joenkhar acknowledged that having a good environment was important for attracting tourists, as one Merak participant noted, "Ecotourism requires good improvement for environment – people come to see nature and need to have good scenery". Indigenous lifestyles that require dependence on their environment to subsist represent strong human–nature relationships for humans, but over-reliance or negligence could weaken ecosystem health.

Broader socio-cultural influences

In general, sacred sites in nature were symbolic of the direct connection between all three main features of *buen vivir*. All focus group participants described the presence of *neys*, holy places in nature that hold spiritual significance as the dwelling places of local deities (Allison, 2004). *Neys* can take many forms, ranging from mountains, lakes or trees to cliffs, rocks and bushes, and were easily identified and known to most villagers. Sacred sites are occupied by different types of deities and immortal beings. A *nepo* (protector/owner of the land) or *yulha* (village deity) may be considered a guardian or caretaker, whereas a "harmful" deity could be a *dud* (demon), an unappeased *yulha* or a disrupted *nepo*. *Phodrang* (deity citadels or palaces), such as mountain peaks or sacred groves, are areas of restricted access and activity because deities desire solitude (Allison, 2004).

In describing which formal or informal rules created or reduced community well-being, all focus group participants noted geographical constraints around *neys* and *phodrang*, and consequences for transgressions. Prohibitions at the sites include cooking garlic, pork or eggs, producing human waste, and bathing in sacred waters. Once disturbed, the angered deity will transmit disease or illness to the offender, other people or animals in the community, or send severe weather that can damage livestock and crops. For people of Merak, corpses cannot be cremated because the polluted air would offend Ama Jomo, and they were not permitted to cut down trees or "do dirty things" near Jomo Khungkhar, her mountain abode, otherwise there will be "storms and heavy rains". Likewise, Sakteng locals are not allowed to cut trees in the mountains surrounding Lake Tsorong Gomba out of respect for *yulha* Ama Jomo. In Joenkhar, locals reported that they do not cut trees or extract stone or sand from the peak of Yumzang Mountain, where the land god Yumzang Mo lives.

Strong bonds between people, their ecosystems and religion were evident in discussions of the transhumant lifestyle of Brokpas. Personal and cultural identity was closely tied to sense of place, as one Merak interviewee described seasonal migration:

> Older men and women will stay in Merak because they can't walk very well. Younger people will migrate down to Khaling, to Shetimey with their yaks. Some make cheese and milk, some care for food, while others transport food...people work together and cooperate.

Another Merak participant declared, "We are all people of Merak. This village was given to us by Aum Jomo, we have to stay here. This is a holy place of Aum Jomo. We are strong, holy people".

Community *pujas* were very important for mediating relations between humans and their environment. Villagers frequently show reverence for Ama Jomo and seek her blessings through numerous oblations, propitiation ceremonies and ritual invocations, some on a daily basis, others less often. Various rituals are performed at *rindus* while the *rinpoche* (reincarnated high priest) and monks chant blessings to protect the community from illness. The Jomo ceremony, where community members burn juniper branches and offer food and *ara* to Ama Jomo every spring and autumn, must be conducted at a precise time and location to receive her protection.

Consequences for not abiding by socio-religious norms are severe and will likely affect other people. If some residents do not attend or perform the ritual as expected, harm or disease may befall some individuals or animals. In response to the question of specific consequences, one Sakteng participant said he witnessed a boy who once became "full of rashes, like fire burns" and described how one man's skin "turned black" one year. Cases of a village youth or adult unexpectedly "running away", getting lost in the mountains and permanently disappearing from family and friends were mentioned by several participants in Merak and Sakteng.

The influence of culture and spirituality in the local context appeared to be changing for some residents while access to modern commodities have increased. The environment and sacred natural sites, despite their importance to communal well-being given their strong connection to religion and society, were not exempt from desecration. The influx of packaged goods and non-biodegradable waste in recent years has resulted in a growing waste disposal problem in remote villages. In Merak, IFAD funded the establishment of a stone-lined garbage dump at the edge of the village, which was almost completed in Spring 2014. However, smaller garbage disposal pits that had been created earlier by SWS staff along the trail and in the villages were not used as frequently as dumping waste over hillsides or creating informal, locally made pits, some of which were located near sacred (human made) structures or riverbanks. In addition, it appeared that some local villagers were burning garbage around *neys*.

Another indication of the changing significance of local culture and spirituality was the increased pilferage of sacred structures associated with sacred natural sites, which was partially accredited to modernisation. Some Merak participants reported that most *chortens* and *mani* walls (Buddhist monuments) in the villages were considered "fake" and "no longer authentic" because they have been looted for the religious relics and treasures they once stored. Robbing holy sites that were built to bless the community was considered "very sinful" by all participants since "the whole community will suffer". These subversive actions were attributed to recent shifts in mentality and behaviour, as one respondent stated, "Modernisation has brought increased competition. People are money-minded now. My Merak is now a modern Merak. I'm very sad to be here now".

Discussion

This section offers four areas of insights for developing ecotourism with indigenous stakeholders. Challenges in current practice are highlighted in light of the opportunities they present for future policy and planning of PA development projects.

Moving beyond the people–park divide: the need to find common ground

Ecotourism is not a term that is well and widely understood, nor confidently described at the community level in Merak–Sakteng but the outcomes of its activities greatly impact social and ecological dimensions of well-being. Several participants were not able to comment on the term itself; rather, it was easier to clearly articulate what ecotourism activities should resemble and whom they should benefit. Difficulty in defining ecotourism could infer lack of formal education or confidence, or limited participation in ecotourism development. For most participants, ecotourism should translate into economic gain for the community. Unequal distribution of benefits and lack of employment opportunities have fuelled resentment in communities where vast income disparity is not customary. The widening gulf of household incomes has weakened human relationships in the community toward TCB and government agencies, who were blamed for raising expectations on the economic benefits of ecotourism.

A disconnect exists between DOFPS guiding principles for ecotourism and real-world ecotourism practice. The NRED ecotourism guidelines are uniquely Bhutanese, involving class-based "high value, low impact" tourism that contributes to GNH goals and corresponds to the Buddhist philosophy of achieving a "Middle Path" of development and sufficiency (Brooks, 2013; NRED, 2012; Theerapappisit, 2003; Wangmo & Valk, 2012). The guidelines also embrace a pro-poor tourism approach by including tangible benefits to communities and bottom-up participation (Novelli & Hellwig, 2011; Scheyvens, 2007). However, providing socio-economic benefits and empowering communities have yet to be realised in a satisfactory way. In Merak–Sakteng, material well-being increased for few community members, echoing earlier research on unequal stakeholder access to economic and political resources among Mayan peoples in the Palenque rainforest, Mexico (Ramos & Prideaux, 2014). Communities in SWS were not aware of the NRED definition of ecotourism, nor were some park staff, even though NRED staff distributed copies of the guidelines to all PA ranger offices. The Sanctuary would benefit greatly from a park-wide concept to help bridge the divide between official and local understandings and expectations of ecotourism, because meanings build up to definitions, which are the basis of policy-making (Fennell, 2001).

Meanings of well-being: beyond the buen vivir perspective?

Study results show that local people conceived of well-being in diverse ways, but all views clearly align with the main features of *buen vivir*. Through its indigenous ontology, the impacts of ecotourism are considered through human relationships, human–nature relationships, and culture and spirituality. It highlights where greater attention needs to be placed in order to achieve more holistic outcomes. *Buen vivir* draws attention to reciprocity and responsibility in relationships, which can be embraced in ecotourism practice with indigenous communities. In SWS, many tour operators did not employ local guides who can share their knowledge of indigenous cosmology and customs. In Russia, indigenous Altai guides ask tourists to join them in observing traditional practices and dispelling negative, "polluting" inward thoughts and emotions before entering the sacred valley in Uch Enmek indigenous Nature Park (Dobson & Mamyev, 2010). In a study of sacred forests in India and Ghana, Ormsby (2012) found that local guides should be hired to preserve sacred sites, minimise negative impacts and increase local income. This practice may also reinforce traditional values among indigenous people, and impart traditional knowledge and consciousness of the environment to tourists.

Although *buen vivir* is a useful perspective for unpacking well-being in the indigenous worldview context, it de-emphasises personal aspects of the human experience, including thoughts and emotions. *Buen vivir* focuses on the inseparability of human–nature relations because nature encompasses human life. Therefore, humans must relate to nature as a subject, and all socio-economic goals are secondary to the ability of ecosystems to operate well (Acosta, 2010; Deneulin, 2012). This view follows ecocentric thinking, which some scholars note is intrinsic to the cosmovision of most indigenous and traditional peoples (Harmon & Putney, 2003).

The *buen vivir* analysis that uses the three main features in this paper provides insights on socio-relational and ecological dimensions. Complementary research in the region shows the growing friction between community and park stakeholders over inequitable distribution of ecological and built resources, and local people who feel powerless against wealthier horse contractors and tour operators who tend to reap the most benefits from ecotourism (Karst & Gyeltshen, 2016). The complexities of resource management and ecotourism development require analysis on additional fronts, suggesting the need for an approach that captures material, subjective, socio-relational and ecological dimensions.

Cosmological relationship to place: implications for ecotourism

The evidence presented here demonstrates the power of indigenous cosmology, which should be considered when implementing ecotourism with indigenous communities. Brokpas feel a strong connection with the land precisely because it was handed down by a guardian deity and ancestors. People rely heavily on local animals and nature for their livelihoods, and the tradition of seasonal migration forms the basis for all aspects of semi-nomadic Brokpa life (Wangmo, 1990). Maintaining cultural traditions is crucial to the attractiveness of indigenous ecotourism destinations. However, traditions must be advertised and interpreted to visitors in a sensitive and balanced manner to prevent representing indigenous groups without context or as being frozen in time, as in the case of the Ovahimbas of Namibia, where primarily women were depicted as "exotic" and "primitive" in tourism brochures (Saarinen, 2011).

Natural sacred sites and indigenous worldviews are critical factors to incorporate in the planning and development processes to support greater sustainability in ecotourism practice. Even as attitudes appear to be changing, devotion to deities and fear of ritual pollution are still manifest through a calendar of invocations and propitiation ceremonies across Bhutan because the actions of an individual can impact others, and deities are seen as meditating relations between humans and natural resources (Allison, 2004; Pommaret, 2004; Ura, 2001). This approach has been successful in the sacred grove at Tafi Atome Monkey Sanctuary, an example of shifting customary beliefs and how ecotourism helped reclaim traditional conservation practices in Ghana (Ormsby & Edelman, 2010).

In this vein, governance and management approaches that embody an indigenous cosmology may help address environmental issues arising from tourism or other development activities (e.g. garbage disposal) more effectively. Brooks (2011) argues that the top-down process of instilling environmentalism in Bhutan includes Buddhist principles and may contribute to sustainable development. This process can be extended to the tourism sector and expanded to include tourism development policy and programmes that embrace indigenous worldviews and values.

Times of change: challenges and opportunities

Research findings suggest that the increasing shift away from traditional norms in light of growing exposure to external sources are having some adverse and potentially long-term implications on individual and communal well-being in ecotourism development. Ecotourism development in Merak–Sakteng has not been entirely negative, and development projects in SWS have rendered positive impacts on humans and environment, including the introduction of metal roofing materials that replaced less durable bamboo tiles (Karst & Gyeltshen, 2016). Yet indigenous traditions and values are dissipating. The increasing desire for and importance of material needs to achieve well-being among Brokpas may be partially attributed to greater exposure to Western society and international demand for antiquities (Whitecross, 2000).

Some studies have found that urbanisation of rural societies (Pretty et al., 2009) and Westernisation of public services and belief systems (Ura, 2001) have eroded biological and cultural diversity. Burning garbage around *neys* and robbing monuments in SWS not only mark the decline of *tha damtshig* but can degrade the environment and related sacred sites to the point where they are unattractive to visitors. In addition, the weakened traditional system of internalised moral and social values appears to be further undermined, replaced by relatively externalised legislative and administrative controls, such as fines and imprisonment (Ura, 2001; Whitecross, 2000).

How can indigenous cosmology and values be preserved or revived to increase well-being and ensure sustainable ecotourism development? At national and regional levels, one approach would be to formally integrate traditional moral and social values into policies and institutions, which are modern systems and strategies for raising and strengthening consciousness and outcomes. This approach has been adopted in Ecuador and Bolivia, where local concepts of *buen vivir* have been translated into normative principles and incorporated into their constitutions (Acosta, 2010; Vanhulst & Beling, 2013; Walsh, 2010). By codifying the rights of nature, Ecuador granted *buen vivir* and nature legal gravitas, which has subsequently been applied in struggles to protect natural resources and indigenous rights as in the once lauded but now abandoned proposal against oil drilling in Yasuní National Park (Deneulin, 2012; Vallejo, Burbano, Falconí, & Larrea, 2015). Similarly, ecotourism policy and programmes could employ guidelines that are explicitly sensitive to indigenous worldviews and oriented toward sustainable projects.

At the local level, participation in ecotourism could be fostered through flexible governance arrangements (e.g. formal and informal rules, institutions) to combat the erosion of indigenous cosmology and values. The potential of sustainable tourism activities involving, if not initiated by, local community to provide economic opportunities and teach indigenous culture and values can be found in recent cases from South Australia (Higgins-Desbiolles, 2009) and eastern Canada (Lynch et al., 2010). In the Canadian sub-arctic, Holmes, Grimwood, King, and the Lutsel K'e Dene First Nation (2016) describe how indigenous people created a code of visitor conduct. In Merak–Sakteng, community members intend to create additional specialised walking tours focusing on sacred natural sites and semi-nomadic lifestyle. New tours may comprise visits to special *neys* beyond village boundaries, and travel with a Brokpa family during the seasonal migration period to partake in traditional activities such as livestock herding and collecting firewood.

Community-driven initiatives may not only teach tourists about indigenous culture and principles but foster pride and reinstate the value of historical and religious treasures to community members, particularly younger generations. But efficacy relies on communities having a certain degree of decision-making power through institutional support and capacity building (Kiss, 2004; Okazaki, 2008). Culturally appropriate, community-based methods of communication are equally compelling measures to be taken to avert negative outcomes, as in the case of Boumā National Heritage Park, where community members combined Western entrepreneurship with traditional Fijian values and processes to reduce conflict in ecotourism initiatives that were initially based on "European" democratic systems (Farrelly, 2011), or commodification of indigenous culture and re-interpretation of local meanings to meet consumer desires, as experienced by Kitchwa communities in the Ecuadorian Amazon (Hutchins, 2007).

Conclusions

This paper draws attention to the value and implications of definitions, their meanings and the broader influence of socio-cultural factors on well-being and ecotourism development involving indigenous people in Bhutan. It presents empirical insights on ecotourism and well-being from marginalised people in a marginalised place to contribute additional disparate and non-Western voices to ecotourism scholarship (Cater, 2006; Prakash, 1994; Wearing & McDonald, 2002). This study also adds to the growing literature on local narratives and indigenous perceptions on ecotourism.

Framed in the context of the *buen vivir* discourse, this study points to the need for local people and PA authorities to find a common, park-wide understanding of ecotourism to more effectively balance expectations and manage outcomes. This might be accomplished in SWS by openly examining the currently underutilised park guidelines on ecotourism with local communities, since the guidelines clearly reflect Bhutanese values and vision for conservation and development in PAs. Findings also show that well-being is understood in disparate ways but correspond to key *buen vivir* features. Depending on the context, this perspective can offer a culturally appropriate mode of analysis for development projects, implying the need for more empirical investigations in other countries and communities that embrace well-being from the indigenous *buen vivir* discourse. However, the *buen vivir* perspective minimises attention to personal components of the human experience, including thoughts, feelings and motivations. Further research to this analysis could explore an approach that incorporates subjective and material dimensions of well-being in addition to social, relational and ecological dimensions.

Remote communities living in PAs face complex challenges with the onset of development projects such as ecotourism. A decrease in the transfer of eco-cultural knowledge and shift in local knowledge bases, including the loss of a deep connection to ecology and sense of inviolability, can have profound impacts on natural and human habitat (Ura, 2001; Verschuuren et al., 2010). Despite the mounting challenges of ecotourism development in Bhutan and its impact on local communities, the landscape of Merak–Sakteng is still regarded as a holy place and requires careful attention. As in the case of the Brokpa, indigenous cosmologies may play an influential role in shaping social values and spiritual beliefs of a given society. Therefore, it is prudent to include natural sacred sites and indigenous worldviews during ecotourism planning and implementation to ensure more just and sustainable well-being outcomes for humans and ecosystems. Likewise, indigenous worldviews and traditional values may be strengthened through integration into modern policies and institutions, and institutional support for local participation and empowerment in ecotourism activities. Incorporating measures and observances that are sensitive to indigenous cosmovisions and values have the potential to enhance successful ecotourism development outcomes in PAs and elsewhere.

Notes

1. *Ama* (mother) or *Aum* (lady) are honorific titles the Brokpa people have bestowed upon Jomo.
2. Fraternal polyandry refers to the marriage of one woman to brothers from the same family, whereas sororal polygyny occurs when several sisters share one husband.

Acknowledgments

I wish to thank the communities of Merak, Sakteng and Joenkhar, particularly those who participated in this study. I am grateful to my partner organisation, Nature Recreation and Ecotourism Division (Department of Forest and Park Services), and to Sakteng Wildlife Sanctuary and Tourism Council of Bhutan for their support. Thanks to field assistants Jambay Dhendrup, Mindu Dema, Tenzin Cheda, Dorji Phuntsho, Dorji Chezom, Kelzang Namgay, and Sangay. Further thanks to Sanjay Nepal, Derek Armitage, Bob Sharpe and Suzan Ilcan for their insightful comments on earlier drafts, and to three anonymous reviewers for their helpful feedback. I gratefully acknowledge the support of the University of Waterloo.

Disclosure statement

No potential conflict of interest was reported by the author.

References

Acosta, A. (2010). *El buen vivir en el camino del post-desarrollo* [Buen vivir on the post-development path] (Policy Paper 9). Quito: Ebert Stiftung Foundation.

Allison, E. (2004). Spiritually motivated natural resource protection in eastern Bhutan. In K. Ura & S. Kinga (Eds.), *The spider and the piglet* (pp. 528–561). Thimphu: Centre for Bhutan Studies.

Andriotis, K. (2009). Sacred site experience: A phenomenological study. *Annals of Tourism Research, 36*(1), 64–84.

Armitage, D.R. (2003). Traditional agroecological knowledge, adaptive management and the socio-politics of conservation in Central Sulawesi, Indonesia. *Environmental Conservation, 30*(01), 79–90.

Berkes, F. (2012). *Sacred ecology* (3rd ed.). New York, NY: Routledge.

Bernbaum, E. (2006). Sacred mountains: Themes and teachings. *Mountain Research and Development, 26*(4), 304–309.

Blackstock, K. (2005). A critical look at community based tourism. *Community Development Journal, 40*(1), 39–49.

Brooks, J.S. (2011). Economic and social dimensions of environmental behavior: Balancing conservation and development in Bhutan. *Conservation Biology, 24*(6), 1499–1509.

Brooks, J.S. (2013). Avoiding the limits to growth: Gross National Happiness in Bhutan as a model for sustainable development. *Sustainability, 5*(9), 3640–3664.

Carr, A., Ruhanen, L., & Whitford, M. (2016). Indigenous peoples and tourism: The challenges and opportunities for sustainable tourism. *Journal of Sustainable Tourism, 24*(8–9), 1067–1079.

Cater, E. (2006). Ecotourism as a Western construct. *Journal of Ecotourism, 5*(1–2), 23–39.

Colding, J., & Folke, C. (2001). Social taboos: "Invisible" systems of local resource management and biological conservation. *Ecological Applications, 11*(2), 584–600.

Corbin, J., & Strauss, A. (2008). *Basics of qualitative research: Techniques and procedures for developing grounded theory* (3rd ed.). Thousand Oaks, CA: Sage.

Creswell, J.W. (2009). *Research design: Qualitative, quantitative and mixed methods approaches* (3rd ed.). Los Angeles, CA: Sage.

Deneulin, S. (2012). *Justice and deliberation about the good life: The contribution of Latin American buen vivir social movements to the idea of justice*. Claverton Down: The Centre for Development Studies, University of Bath.

Dobson, J., & Mamyev, D. (2010). Sacred valley, conservation management and indigenous survival: Uch Enmek indigenous Nature Park, Altai Republic, Russia. In B. Verschuuren, R. Wild, J. McNeeley, & G. Oviedo (Eds.), *Sacred natural sites: Conserving nature and culture* (pp. 244–262). London: Earthscan.

Dorjee, K. (2012). *Transhumant practices, its impact and threat to red panda habitat conservation in Sakteng Wildlife Sanctuary* (Unpublished undergraduate thesis). Royal University of Bhutan, College of Natural Resources, Lobesa, Bhutan.

Dorji, L. (2012). *Alcohol use and abuse in Bhutan*. Thimphu: National Statistics Bureau.

Dowling, R. (2010). Power, subjectivity and ethics in qualitative research. In I. Hay (Ed.), *Qualitative research methods in human geography* (3rd ed., pp. 40–68). Toronto: Oxford University Press.

Dudley, N., Higgins-Zogib, L., & Mansourian, S. (2009). The links between protected areas, faiths, and sacred natural sites. *Conservation Biology, 23*(3), 568–577.

England, K.V.L. (1994). Getting personal: Reflexivity, positionality and feminist research. *Professional Geographer, 46*: 80–89.

Farrelly, T.A. (2011). Indigenous and democratic decision-making: Issues from community-based ecotourism in the Boumā National Heritage Park, Fiji. *Journal of Sustainable Tourism, 19*(7), 817–835.

Fennell, D.A. (2001). A content analysis of ecotourism definitions. *Current Issues in Tourism, 4*(5), 403–421.

Fennell, D.A. (2008). Ecotourism and the myth of indigenous stewardship. *Journal of Sustainable Tourism, 16*(2), 129–149.

Gadgil, M., & Vartak, V.D. (1976). Sacred groves of Western Ghats in India. *Economic Botany, 30*(2), 152–160.

Gudynas, E. (2011). Buen vivir: Today's tomorrow. *Development, 54*(4), 441–447.

Gurung, D.B., & Seeland, K. (2008). Ecotourism in Bhutan: Extending its benefits to rural communities. *Annals of Tourism Research, 35*(2), 489–508.

Harmon, D., & Putney, A.D. (Eds.). (2003). *The full value of parks: From economics to the intangible*. New York, NY: Rowman and Littlefield.
Harris, R.W. (2009). Tourism in Bario, Sarawak, Malaysia: A case study of pro-poor community-based tourism integrated into community development. *Asia Pacific Journal of Tourism Research, 14*(2), 125–135.
Higgins-Desbiolles, F. (2009). Indigenous ecotourism's role in transforming ecological consciousness. *Journal of Ecotourism, 8*(2), 144–160.
Higgins-Desbiolles, F. (2011). Death by a thousand cuts: Governance and environmental trade-offs in ecotourism development at Kangaroo Island, South Australia. *Journal of Sustainable Tourism, 19*(4–5), 553–570.
Holmes, A.P., Grimwood, B.S.R., King, L.J. & the Lutsel K'e Dene First Nation. (2016). Creating an Indigenized visitor code of conduct: The development of Denesoline self-determination for sustainable tourism. *Journal of Sustainable Tourism, 24*(8–9), 1177–1193.
Hutchins, F. (2007). Footprints in the forest: Ecotourism and altered meanings in Ecuador's upper Amazon. *The Journal of Latin American and Caribbean Anthropology, 12*(1), 75–103.
IUCN (2016). The IUCN red list of threatened species. Version 2016-1. Retrieved May 1, 2016, from www.iucnredlist.org
Jiménez, R. (2011). *Recovering and valuing other ethical pillars: Buen vivir*. Paris: Forum for a new World Governance.
Karst, H., & Gyeltshen, N. (2016). The politics of community-based ecotourism in Sakteng Wildlife Sanctuary, Bhutan. In S. Nepal & J. Saarinen (Eds.), *Political ecology and tourism* (pp. 68–81). London: Routledge.
Kinga, S. (2001). The attributes and values of folk and popular songs. *Journal of Bhutan Studies, 3*(1), 134–175.
Kiss, A. (2004). Is community-based ecotourism a good use of biodiversity conservation funds? *Trends in Ecology and Evolution, 14*, 579–600.
Kontogeorgopoulos, N., Churyen, A., & Duangsaeng, V. (2014). Success factors in community-based tourism in Thailand: The role of luck, external support, and local leadership. *Tourism Planning & Development, 11*(1), 106–124.
Latta, A. (2014). Matter, politics and the sacred: Insurgent ecologies of citizenship. *Cultural Geographies, 21*(3), 323–341.
Lynch, M.F., Duinker, P., Sheehan, L., & Chute, J. (2010). Sustainable Mi'kmaw cultural tourism development in Nova Scotia, Canada: Examining cultural tourist and Mi'kmaw perspectives. *Journal of Sustainable Tourism, 18*(4), 539–556.
Manyara, G., & Jones, E. (2007). Community-based tourism enterprises development in Kenya: An exploration of their potential as avenues of poverty reduction. *Journal of Sustainable Tourism, 15*(6), 628–644.
MEA (2005). *Ecosystems and human well-being: Synthesis*. Washington, DC: Island Press.
Mowforth, M., & Munt, I. (2009). *Tourism and sustainability: Development, globalisation and new tourism in the Third World* (3rd ed.). London: Routledge.
Newing, H. (Ed.). (2011). *Conducting research in conservation*. New York, NY: Routledge.
Novelli, M., & Hellwig, A. (2011). The UN Millennium Development Goals, tourism and development: The tour operators' perspective. *Current Issues in Tourism, 14*(3), 205–220.
NRED (2012). *Ecotourism development in the protected areas network of Bhutan: Guidelines for planning and management*. Thimphu: Ministry of Agriculture and Forests.
Okazaki, E. (2008). A community-based tourism model: Its conception and use. *Journal of Sustainable Tourism, 16*(5), 511–529.
Ormsby, A. (2012). Perceptions of tourism at sacred groves in Ghana and India. *RASAALA: Recreation and Society in Africa, Asia and Latin America, 3*(1), 1–18.
Ormsby, A.A., & Bhagwat, S.A. (2010). Sacred forests of India: A strong tradition of community-based natural resource management. *Environmental Conservation, 37*(3), 320–326.
Ormsby, A., & Edelman, C. (2010). Community-based ecotourism at Tafi Atome Monkey Sanctuary, a sacred natural site in Ghana. In B. Verschuuren, R. Wild, J. McNeeley, & G. Oviedo (Eds.), *Sacred natural sites: Conserving nature and culture* (pp. 233–243). London: Earthscan.
Pelgen, U. (2007). Rituals and pilgrimage devoted to Aum Jo Mo Re Ma Ti by the'Brog Pas of Me Rag of eastern Bhutan. In J.A. Ardussi & F. Pommaret (Eds.), *Bhutan: Traditions and changes* (pp. 121–134). Leiden: Brill.
Phuntsho, K. (2004). Echoes of ancient ethos. In K. Ura & S. Kinga (Eds.), *The spider and the piglet* (pp. 466–481). Thimphu: Centre for Bhutan Studies.
Pommaret, F. (2004). Yul and yul lha: The territory and its deity in Bhutan. *Bulletin of Tibetology, 40*(1), 39–67.
Prakash, G. (1994). Postcolonial criticism and Indian historiography. In L. Nicholson & S. Seidman (Eds.), *Social postmodernism: Beyond identity politics* (pp. 87–100). Cambridge: Cambridge University Press.
Pretty, J., Adams, B., Berkes, F., de Athayde, S., Dudley, N., Hunn, E., … Pilgrim, S. (2009). The intersections of biological diversity and cultural diversity: Towards integration. *Conservation and Society, 7*(2), 100–112.
Puri, R.K. (2011). Participant observation. In H. Newing (Ed.), *Conducting research in conservation* (pp. 126–152). New York, NY: Routledge.
Ramos, A.M., & Prideaux, B. (2014). Indigenous ecotourism in the Mayan rainforest of Palenque: Empowerment issues in sustainable development. *Journal of Sustainable Tourism, 22*(3), 461–479.
Rinzin, C., Vermeulen, W., & Glasbergen, P. (2007). Ecotourism as a mechanism for sustainable development: The case of Bhutan. *Environmental Sciences, 4*(2), 109–125.

Rutte, C. (2011). The sacred commons: Conflicts and solutions of resource management in sacred natural sites. *Biological Conservation, 144*(10), 2387–2394.

Ryan, R., & Deci, E., (2001). On happiness and human potentials: A review of research on hedonic and eudaimonic well-being. *Annual Review of Psychology, 52*, 141–166.

Saarinen, J. (2011). Tourism, indigenous people, and the challenge of development: The representations of Ovahimbas in tourism promotion and community perceptions toward tourism. *Tourism Analysis, 16*(1), 31–42.

Scheyvens, R. (2007). Exploring the tourism–poverty nexus. *Current Issues in Tourism, 10*(2–3), 231–254.

Sirgy, M., Michalos, A., Ferriss, A., Easterlin, R., Patrick, D., & Pavot, W. (2006). The quality-of-life research movement: Past, present, and future. *Social Indicators Research, 76*(3), 343–466.

TIES (1990). Retrieved November 12, 2012, from http://www.ecotourism.org/what-is-ecotourism

TIES (2015). Retrieved August 10, 2015, from https://www.ecotourism.org/what-is-ecotourism

Theerapappisit, P. (2003). Mekong tourism development: Capital or social mobilization? *Tourism Recreation Research, 28*(1), 47–56.

Timothy, D., & Olsen, D. (Eds.). (2006). *Tourism, religion and spiritual journeys* (Vol. 4). London: Routledge.

UNESCO (2016). Sakteng Wildlife Sanctuary. Retrieved March 21, 2015, from http://whc.unesco.org/en/tentativelists/5701/

Ura, K. (2001). Deities and environment. Thimphu: Centre for Bhutan Studies.

Vallejo, M.C., Burbano, R., Falconí, F., & Larrea, C. (2015). Leaving oil underground in Ecuador: The Yasuní-ITT initiative from a multi-criteria perspective. *Ecological Economics, 109*, 175–185.

Vanhulst, J., & Beling, A.E. (2013). Buen vivir: La irrupción de América Latina en el campo gravitacional del desarrollo sostenible [Buen vivir: The emergence of Latin America in the gravitational field of sustainable development]. *Revista Iberoamericana de Economía Ecológica, 21*, 15–28.

Verschuuren, B., Wild, R., McNeeley, J., & Oviedo, G. (Eds.). (2010). *Sacred natural sites: Conserving nature and culture*. London: Earthscan.

Villalba, U. (2013). Buen vivir vs development: A paradigm shift in the Andes? *Third World Quarterly, 34*(8), 1427–1442.

Walsh, C. (2010). Development as buen vivir: Institutional arrangements and (de)colonial entanglements. *Development, 53*(1), 15–21.

Wangmo, S. (1990). The Brokpas: A semi-nomadic people in eastern Bhutan. In N.K. Rustomji & C. Ramble (Eds.), *Himalayan environment and culture* (pp. 141–158). Shimla: Indian Institute of Advanced Study.

Wangmo, T., & Valk, J. (2012). Under the influence of Buddhism: The psychological wellbeing indicators of GNH. *Journal of Bhutan Studies, 26*, 53–81.

Wearing, S., & McDonald, M. (2002). The development of community-based tourism: Re-thinking the relationship between tour operators and development agents as intermediaries in rural and isolated area communities. *Journal of Sustainable Tourism, 10*(3), 191–206.

Whitecross, R.W. (2000). Signs of the degenerate age: The desecration of chorten and lhakhang in Bhutan. *Journal of Bhutan Studies, 2*(2), 182–199.

Whitecross, R.W. (2010). "Virtuous Beings": The concept of tha damtshig and being a moral person in contemporary Bhutanese society. *Himalaya, 28*(1–2), 71–82.

WWF Bhutan & SWS (2011). *Participatory zoning for Sakteng Wildlife Sanctuary: Balancing conservation and development goals*. Thimphu: World Wildlife Fund and Sakteng Wildlife Sanctuary, Department of Forest and Park Services, Ministry of Agriculture and Forests.

Yin, R.K. (2009). *Case study research: Design and methods* (4th ed.). Thousand Oakes, CA: Sage.

Zeppel, H. (2006). *Indigenous ecotourism: Sustainable development and management* (Vol. 3). Wallingford (CT): CABI.

The land has voice: understanding the land tenure – sustainable tourism development nexus in Micronesia

T. S. Stumpf and C. L. Cheshire

ABSTRACT

Externally-generated prescriptions for development continue to be imposed on Pacific Island societies. Frequently lost in this process is provision for the idiosyncratic localized realities of the people actually living in these places. Using an interpretivist critical discourse analysis, this study examines Indigenous perspectives on land, and how these perspectives interface with such standardized development ideals. More specifically, this study explores how issues related to Indigenous land holding rights in Micronesia are at the core of the 17 Sustainable Development Goals (SDGs) set forth by the United Nations in the 2030 Agenda for Sustainable Development. The results elucidate how the current SDG targets lack sufficient provision for how Micronesian societies function, what is valued by the people that comprise these societies, and how Indigenous land holding rights form the bedrock of the sustainability of Micronesian livelihoods.

Introduction

In the vast swath of earth dominated by water, there has been nothing of more importance and consequence to its human civilizations than land. For the people and islands of Oceania, issues of land holding have defined their histories, current circumstances, and future trajectories in ways that are difficult to overstate. Indeed, those islands with governing bodies which have retained Indigenous land holding rights stand in stark contrast to those characterized by liberalized land laws which have opened the floodgate of foreign ownership, and the subsequent sociocultural and economic implications that inevitably follow. (Banner, 2007) Today, the remaining Indigenous land holding systems in Pacific Island countries (PICs) that prohibit foreign land ownership by law have been cited by external financial institutions, multinational organizations, and governments of aid-providing nations as a roadblock to development. (Asian Development Bank, 2015; Cheshire, 2010; Connell, 2007; Duncan, Codippily, Duituturaga, & Bulatale, 2014) According to the purview of some of the more influential of these entities operating in the Asia-Pacific region (see Abbott & Pollard, 2004; Asian Development Bank, 2015; Holden, Bale, & Holden, 2004), such PICs cannot attract the resources (i.e. foreign investment) required for development (and all the promised aspects of improved health, prosperity and livelihoods supposedly therein) via industries like tourism without substantial institutional reforms, including liberalized land holding laws.

However, it is imperative to recognize Indigenous perspectives on the role and value of land in Pacific Island societies, and how they interface with standardized development prescriptions imposed externally. To what extent the idiosyncrasies of localized realities are reflected in blanket development criteria is an issue that has been raised previously and referred to as "representativeness." (Bäckstrand, 2006) Reed, Fraser, and Dougill (2006, p. 407) state that such approaches "may miss critical sustainable development issues at the local level and may fail to measure what is important to local communities." For example, Taylor (2008) raises questions about the relevance of the United Nation's Millennium Development Goals to Indigenous people in Australia. While Rametsteiner, Pülz, Alkan-Olsson, and Frederiksen (2011) suggest the need for provisions of local norms in sustainable development indicators, there is a dearth of critical analyses on such issues in Pacific Islands.

The aim of this paper is to elucidate how issues related to Indigenous land holding rights in PICs are at the core of the 17 Sustainable Development Goals (SDGs) set forth by the United Nations in the 2030 Agenda for Sustainable Development. To this end, an interpretivist methodological approach is used to critically analyze discourse from 19 Indigenous residents regarding land holding and sustainable development via tourism across four island states in the Federated States of Micronesia (FSM). The findings suggest a need for more contextualized understandings of the SDGs and the land holding – sustainable tourism development nexus in the FSM and PICs like it. More specifically, the analysis highlights the disconnect between the content of the SDGs and how land holding rights are central to localized interpretations of the issues embedded within them.

This paper proceeds as follows. First, a historical and conceptual foundation of land and land holding systems in PICs is presented, with particular focus on Micronesia. This includes a discussion of the decolonization process and subsequent evolution of tourism as a driver of economic development in the islands. Next, the methodology for the data collection and analysis of the present research is detailed. This is followed by a critical analysis of how the findings elucidate a need for localized conceptualizations of how an illustrative sample of the 17 SDGs relate to Indigenous land holding systems and sustainable development via tourism in the FSM. Finally, some implications of the findings and analysis of this research are offered.

Literature review

Land and land holding in pacific islands

Various arrangements can exist between land and humans. The parameters of such arrangements can range from ownership to stewardship, individual to communal, private to public, fee hold to free hold, and traditional to modern. The present study focuses on Indigenous land holding systems in Micronesia. In this paper, "Micronesia" is the term applied to the regional sea of islands, communities, and social networks found in western Oceania (Hanlon, 1994; Hau'ofa, 2008), and in addition to the FSM includes the Republic of Palau, the Republic of the Marshall Islands, the Republic of Kiribati, the Republic of Nauru, the Commonwealth of the Northern Marianas Islands. (Ballendorf, 1988) It is recognized that the term "Micronesia" is itself a qualitative symbol designating a history of geopolitical imperialism characterized by artificial demarcation of islands into groups of states and nations by hegemonic Western powers. (Hanlon, 1994) Next, the term "land holding systems" (as opposed to alternatives such as "land tenure" or "ownership") is used in this paper to acknowledge the idea that people do not technically own the land. Rather, they own holding rights to a specified piece of land, and such holding rights occur within formal and informal systems developed by respective island societies. (Crocombe, 1971)

Finally, the term "Indigenous" in regards to land holding is used in lieu of alternatives such as "traditional" or "customary". There have been various invented traditions across Oceania in regards to the relationship between past and present day land holding systems. (Ward & Kingdon, 1995, p. 14) The current Indigenous land holding systems in Micronesia (and across

Oceania) often bear little resemblance to the traditional systems that were used prior to European contact. For instance, traditional land systems across Micronesia were often characterized by an autocratic ruling body which held all the land, and dispersed land use/cultivation rights and/or obligations among the people at its discretion. (Banner, 2007; Hanlon, 1994) This power could be consolidated in a single chief as in Kosrae, to an alliance of chiefs as in Palau, or to a network of decentralized chiefs as in Pohnpei. (Haglegam, 1998) In response to the socio-economic and geopolitical changes brought by various external catalysts of change (i.e. colonization and de-colonization processes, missionary activity, introduction of capitalist economic systems) (Farrell, 1972), these traditional systems were replaced by individual and familial holding systems backed by local government laws. As such, "Indigenous land holding" refers here to specified parcels of land in which Indigenous (non-foreign) individuals retain holding rights which are permanent, inheritable, and transferable to other Indigenous individuals only. (Ward & Kingdon, 1995)

For Micronesians, and other Indigenous Pacific Islanders, attachment to land and place is a central component of personal/familial identity and genealogy. (Diaz & Haulani Kauanui, 2001; Hanlon, 2004) As Diaz and Haulani Kauanui (2001, p. 318) state the sense of place rooted in Indigenous land holding rights is nothing less than "paramount to the continued existence and viability of Pacific Islanders." While the connection between Pacific Islanders and their land was evident in pre-colonial times through the Indigenous names for people and places that illustrate the depth of land-human identity connections, the focus on such connections has subsequently become even more heightened in colonial and post-colonial contexts. (Diaz & Haulani Kauanui, 2001) Specifically, in those Pacific Islands where Indigenous land holding systems were modified (involuntarily or voluntarily) to permit ownership by foreign parties, the sociocultural upheaval that followed was inevitable and substantial. (Banner, 2007) As all the associated adverse effects on islanders' identities, sociocultural institutions, and socioeconomic plights played out over time, issues relating to the imposition of foreign land holding rights became the basis for Indigenous sovereignty movements in places like Hawaii, New Zealand, and Guam. (Diaz & Haulani Kauanui, 2001) Today, in those PICs where Indigenous land holding rights were retained and foreign ownership prohibited by local law, the retention of such rights have been subjected to reform pressures from some influential external entities and the standardized blueprints for development they advocate. This is discussed in greater detail next.

Development ideology in Micronesia

Oceania is a critical region for examining (sustainable) development on account of its unique geographical, historical, sociocultural, and environmental characteristics. (Overton, 1999) In the eighteenth-century, the economic ventures of the Western world began to encroach upon Pacific Island societies in earnest. (Kiste, 1994) From beche-de-mer to whale oil to copra to sugar, these islands served as a breeding ground for centuries of international trade and colonial imposition. (Cheshire, 2010; Kiste, 1994) In the post-World War II period, the colonial superpowers of the Pacific began to instill the development ideologies considered essential to the future of the newly formed PICs, including an emphasis on productive economic activity. (Hanlon, 1998) In Micronesia, this included the use of a "very institutionalized, highly standardized view of what economic development should be" (Hanlon 1998, p. 104) by organizations like the UN Development Program in conjunction with the United States government. As Purdie (1999, p. 64) points out, there has been a "tendency to employ development blueprints constructed in and for contexts differing markedly from those in the Pacific Islands."

These colonial histories combined with the current paradigm of external development assistance from entities that seek to "enforce similar policies across diverse countries" (Dolowitz & Marsh, 2000, p.7) continue to create both implicit and explicit pressure on recently established governments to pursue standardized development ideals as a panacea for improved livelihoods. As Stumpf and Reynolds (2016) show such conformity pressures are also present in development planning around tourism. While thought by their purveyors to "occupy the moral high ground", a critical interpretation reveals "somewhat ill-defined objectives, clouded in good intentions and selectively or inadequately applied." (Connell, 2007, p. 121) As McMillan (2002, p. 225) suggests, notions of development "that are based more on preconceptions than on the specifics of the situation are still regrettably common."

Institutional reform, tourism development, and land

One result of this external influence on local governance across Micronesia is an imbalanced, if not totally misguided, focus on a top-down approach to realizing the tenets of development. (Cheshire, 2010; Stumpf & Swanger, 2017) This top-down approach emphasizes formal institutional reform, including liberalized land holding laws, to pave the way for increased levels of foreign investment in potential drivers of development such as tourism. The bottom-up, informal institutions stemming from sociocultural norms and values are not only ignored, but actually regarded as an "irrational influence" (Connell, 2007, p. 116) that stands in direct contrast to external notions of what constitutes good governance toward development. In the 1960s, a report commissioned by the UN Trusteeship Council referred to traditional land tenure systems in Micronesia as a "lingering anachronism" that must be reformed to achieve economic development, while "ignoring any consideration of what land tenure systems might mean to the different peoples of Micronesia." (Hanlon, 1998, p. 97)

The Asian Development Bank (ADB) continues to be active in advocating for the top-down approach to development across Oceania. (Cheshire, 2010) Headquartered in the Philippines, the ADB espouses the alleviation of poverty through economic development, and provides loans, grants, strategic development advice, and technical assistance to this end across Oceania. (Asian Development Bank, 2016) Since 1990, the FSM has received over $112 million in assistance from the ADB, much of which has centered on governance around private-sector business development through "strengthening of institutional capacity for land administration and management." (Asian Development Bank, 2017) The ADB has a number of formalized relations with the UN, including working arrangements on the UN Economic and Social Commission for Asia and the Pacific (ESCAP), UN Human Settlements Programme (HABITAT), UN International Labour Organization (ILO), UN Development Programme (UNDP), and UN Industrial Development Organization (UNIDO). (Asian Development Bank, 2017a) The ADB is listed as a partner organization for the UN's SDGs on various projects in Oceania including the Poverty-Environment Partnership, the Pacific Region Infrastructure Facility, the Global Platform for Sustainable Cities, and the Coral Triangle Initiative. (United Nations, 2016)

As Scheyvens and Momsen (2008, p. 495) point out, such external entities have discussed Indigenous land holding rights backed by local law as a "curse rather than a blessing", and thus posit the reform and clarification of laws which stymie foreign access to land as requisite to development. While land may be regarded as a relatively less important factor of production in the development of product-based ventures (Parker, 2006), quite the opposite is true for tourism due to the premium placed on ideal location. However, land in Pacific places like Micronesia is "managed under a complex mix of modern and traditional systems" (Federated States of Micronesia Economic Summit, 2004, p. 35), which are "not easily amenable to translation into western codes and conventions." (Connell, 2007a, p. 126) The complexities of Indigenous land holding systems has been documented in case studies across Micronesia and other PICs. (Bascom, 1965; Labby, 1976; Lingenfelter, 1975; Peoples, 1985) Consider the following passage which explains one aspect of the intricacies of land ownership in the FSM state of Yap (Labby, 1976, p. 35–36):

> The estate itself was seen to consist not only of the domestic group living on specific land, but also an entire extended network of relations of people and land ... Since the man's claim on his land began with his mother, who had worked to earn it for her children, the land was not simply his alone. It also belonged to his siblings ... Brothers, of course, would have been given pieces of their father's land, but sisters had to marry out ... Although the [sister] could not use the land she left at her natal estate, she and the children she bore stood as guardians to it, watching over those who had taken their place, the woman's brother's wife and her children ... It was only after four generations had passed that the claim of the people established on the land by an in-marrying woman finally expired, that the clan group was fully repaid for its investment in the land.

External organizations pushing for Indigenous land holding reform to facilitate development via tourism often cite such nuanced systems (seemingly inscrutable to Western sensibilities) as evidence of the difficulties outside investors face when trying to acquire land for developing commercial tourism ventures. Such organizations thus promote a form of economic dualism, whereby the ideals from their economic system are superimposed upon what they perceive to be less advanced systems. (McKee & Tisdell, 1990) New depths of understanding regarding the Indigenous land holding – tourism development nexus are needed to develop more holistic, if not accurate, understandings of the development ideals promoted by the UN and its partner entities. Paramount to this is juxtaposing the localized realities of the Indigenous people which actually comprise these societies with such externally-generated ideals.

Methodology

Indigenous voices and perspectives are an under-represented, yet essential component of scholarship on topics related to hegemony, history, identity, culture, economic and geopolitical issues in Pacific Islands. (Diaz & Haulani Kauanui, 2001; Stumpf, 2016) As Hanlon (2004, p. 200) states, "even the most liberal, empathetic, and avowedly radical studies can remain colonialist in their methodologies and colonizing in their inadvertent silencing or displacement of more local voices and histories." To avoid the problem of privileging non-Indigenous thought regarding localized Indigenous issues (Hanlon, 2004), this study uses qualitative interview data combined with an interpretivist critical discourse analysis (CDA) to elucidate a contextualized understanding of Indigenous perspectives regarding the intersection of land, tourism, and the 17 SDGs set forth by the United Nations across four island states in the Federated States of Micronesia (FSM).

The Federated States of Micronesia: an overview

Situated slightly above the equator, the FSM is a sovereign nation of roughly 708 islands (65 inhabited) in the Micronesia sub-region of western Oceania. (see Figure 1) The FSM is a fairly recent geopolitical construction that has been grouped into four states (west to east): Yap, Chuuk, Pohnpei, and Kosrae. Each state is comprised of many islands, with the exception of Kosrae, which is a single island state. The total combined land area of the four states is approximately 271 square miles. (Federated States of Micronesia, 2008) Previously part of the Trust Territory of the Pacific Islands formed by the UN and administered by the U.S. after World War II, the FSM formally gained its political independence in 1986. (Hanlon, 1998) Today, the FSM remains a self-governing nation in free association with the U.S. under the Compact of Free Association arrangement between the countries. The most recent Compact stipulates the U.S. provide roughly $3 billion in development aid to the FSM through 2023 in exchange for certain national defense and security-related provisions. (Federated States of Micronesia Development Partners Forum, 2012) For the FSM, the Compact funds have ostensibly been designed to support increased levels of development and economic self-reliance.

In its national strategic development plan created with technical assistance from external organizations (i.e. the ADB and U.S. Department of Interior) which consider themselves "conduits

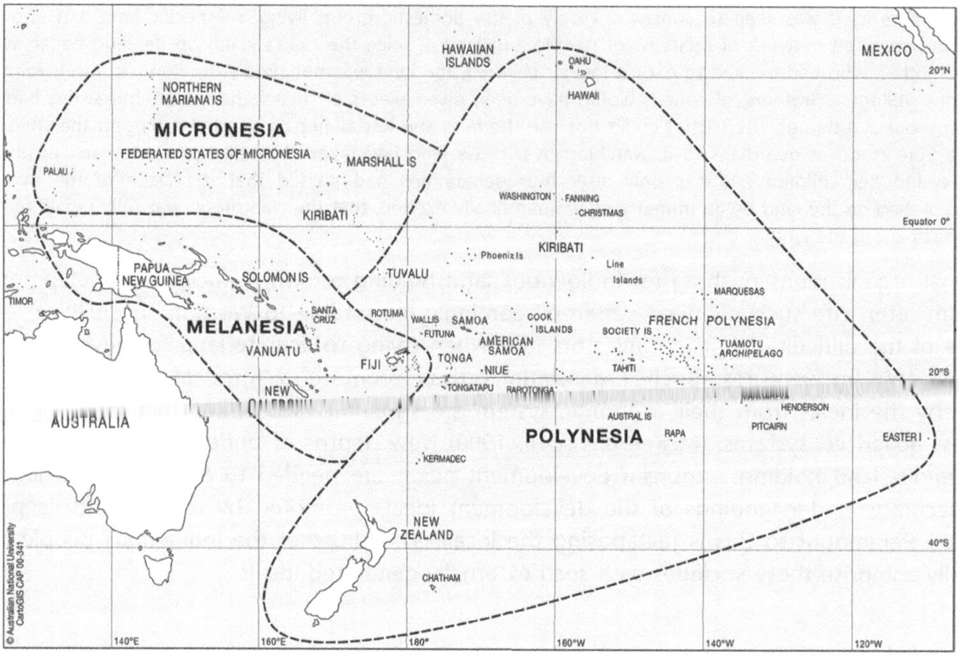

Figure 1. Map of Oceania (courtesy of CartoGIS Services, College of Asia and the Pacific, and The Australian National University).

for policies of good governance" (Larmour, 2005, p. 117) across Oceania, the FSM has singled-out tourism as the industry with the greatest potential to drive such self-reliance. (Federated States of Micronesia Economic Summit, 2004; Federated States of Micronesia Development Partners Forum, 2012) While reliable tourism statistics for the FSM are elusive, annual arrivals between the years 2000 to 2016 ranged from 15,300 to 47,500. (UNWTO, 2018) However, available data do not differentiate between international tourists and FSM citizens visiting friends and relatives, so the actual number of bona fide tourists is unclear.

The FSM faces a number of barriers to developing tourism into a viable industry. For instance, the lack of well-coordinated policies linking air transportation and tourism have been a roadblock to tourism development for island nations around the world (Abeyratne, 1999), and the FSM is no exception. In the 1960s, Continental Airlines led the formation of the United Micronesia Development Association (UMDA). The goals of the UMDA (in which Continental took a 32% interest) were centered on using tourism to spur economic development in the region. Continental concurrently launched Air Micronesia, an air service of which UMDA (49%) and Continental (31%) became the largest shareholders. As part of the proposal, Continental pledged that Air Micronesia would devote itself to the economic development of the islands by creating a tourism industry anchored by resort hotels built and managed by Continental. (Davies, 1984) Continental gained exclusive and permanent rights to service Micronesia in 1971. The company subsequently formed a wholly-owned subsidiary called Continental Hotels, Inc. to build a chain hotels across Micronesia (including the FSM states of Chuuk and Pohnpei) to integrate its tourism operations. (Carlile, 2000; Kelley & Penseyres, 1976) The Continental Hotel in Chuuk was built in 1970 at a size almost twice the original plan and $250,000 over budget. (Kelley & Penseyres, 1976) The proposed hotel in Pohnpei was blocked by the local government and never built. (Hanlon, 1998)

Thus, despite having a monopoly on the region, Continental's efforts to develop tourism into a viable industry in the FSM through an integrated supply chain did not prove successful. In 2011, Continental Airlines was bought by United Airlines, which at the time of this study remains the sole provider of regular air transportation service to the FSM. The FSM states of Kosrae,

Pohnpei, and Chuuk are serviced by United Airlines' "island hopper" route originating out of both Hawaii and Guam, and Yap is serviced separately by United Airlines out of Guam. Given the access challenges for potential tourist markets in the U.S., Japan, Australia, China, Russia, and Korea, the FSM is "unlikely ever to become a high volume destination for leisure tourists." (Federated States of Micronesia Economic Summit, 2004, p. 228)

Tourism in the Federated States of Micronesia: Yap, Chuuk, Pohnpei, and Kosrae

Each FSM state offers a distinct tourism product, and each has its own tourism bureau. The FSM Visitors Board is the national tourism body located in the capital of Palikir on Pohnpei, and ostensibly coordinates tourism activities between the states. Yap State is comprised of 139 islets which amount to about 46 total square miles of land. Approximately 98% of all dry land in Yap is privately owned, with the remaining 2% being public and commercial land. (Federated States of Micronesia, 2008) Yap is also generally thought to have the most intact traditional culture of all four FSM states, and Yapese society is still characterized by a complex sociopolitical organization of castes, councils, and chiefs. Some of the main attractions in Yap are its ancient stone money, cultural tours, deep-sea fishing, and manta ray diving. Some 900 miles to the east of Yap, Chuuk State is comprised of 542 islets amounting to 49 total square miles of land. Similar to Yap, approximately 98% of all dry land in Chuuk is privately owned, with the remaining 2% split evenly between public and commercial land. (Federated States of Micronesia, 2008) Tourism in Chuuk is almost entirely centered on historical resources from World War II. The centerpiece for this is Chuuk Lagoon, which is home to the largest fleet of downed World War II Japanese air and water craft in the world, and is thus considered the world's premiere site for wreck diving.

Four hundred fifty miles to the east of Chuuk is Pohnpei State. Pohnpei is comprised of 26 islets, which together amount to 133 total square miles of land. As opposed to Yap and Chuuk, only about 63% of all dry land in Pohnpei is privately owned, with 36% classified as public land and the remaining 1% as commercial land. (Federated States of Micronesia, 2008) Pohnpei typically ranks the highest among the FSM states in terms of total arrivals due to it being both home to the nation's capital and the nation's largest single island of Pohnpei proper (129 square miles). Pohnpei also offers the most diverse tourism experience of the FSM states, which includes deep-sea fishing, surfing, diving, hiking, World War II historical sites, and Nan Madol, the world's only ancient city built atop a coral reef.

Finally, Kosrae State is the easternmost FSM state, situated 340 miles from Pohnpei. Kosrae is the only FSM state comprised of a single island (42 square miles). Kosrae is unique among the FSM states in that only about 35% of all dry land is privately owned, with 64% classified as public land, and the remaining 1% as commercial land. (Federated States of Micronesia, 2008) Kosrae attracts mainly nature and adventure tourists looking for uncrowded diving, jungle hiking, and mangrove kayaking. Kosrae is also home to the Lelu Ruins, an ancient compound which once served as the ruling center for the high chiefs of the island.

Data collection and analysis

This study uses interpretivist CDA to account for Indigenous voices and perspectives on land, tourism, and the 17 SDGs set forth by the United Nations. CDA, sometimes referred to as "critical language awareness" or "critical language studies" (Billig, 2003), connects human discourse with varying underlying sociological structures and institutions. In particular, CDA examines the discursive linkages between qualitative text, influencing power structures, and localized social situations and cultural identities. (Weiss & Wodak, 2003) In the present study, semi-structured interviews were used to generate qualitative data from Indigenous residents of the FSM associated with the tourism industry. These individuals included hotel owners ($n = 7$), tourism bureau staff ($n = 1$), government officials ($n = 2$), college faculty of hotel and tourism management ($n = 1$), local land owners ($n = 6$), and staff of small business development agencies ($n = 2$). In these small islands, most local residents have at least some tangential association with tourism. While the interview participants in this study had a relatively higher direct association with tourism than most, all interview participants were first and foremost Indigenous Pacific Islanders and members of their respective communities.

These perspectives were then analyzed and interpreted from an ontology and epistemology of cultural (or anthropological) relativism which assumes that an individual's sociocultural context plays a major role in inscribing the contours of their reality, and that a trained researcher can elucidate these localized realities by interpreting the discursive perspectives within the sociocultural context in which they were stated. (Spiro, 1986; Van Dijk, 2003) The authors of this research had a total of approximately 30 years of combined experience living and/or working in the business and economic development, education, community development, and media sectors in the FSM prior to the data collection and analysis.

A total of 22 interviews with 19 different individuals (5 females, 14 males) were conducted for this research, and all interviews were completed face-to-face on site across all four FSM states. Three participants were interviewed twice to follow-up on information provided in their previous comments. The researchers leveraged their personal and professional networks in the FSM to identify potential interview participants, and these individuals were contacted via e-mail roughly 6 months prior to the data collection. Ten individuals provided initial consent to participate in the research as a result, and the other nine participants were located in-country during the field work through reference from other interview participants. Three of the 19 individuals were from Yap State, two from Chuuk State, nine from Pohnpei State, and five from Kosrae State. The interviews totaled 1605 min in length (26.75 h), and the average interview length was 72.95 min. A digital-voice recorder was used to capture the interview data with consent. Each interview was first transcribed word-for-word by the lead author immediately subsequent to completion. The text was then broken down into data incidents and coded.

When coding in CDA, the researcher can adopt various analytical approaches ranging from coding an entire qualitative text corpus in broad terms, coding topical areas of perceived importance, and/or more micro-analytical coding. (Fairclough, 1992) The present study adopted the latter approach whereby groups of 13 sentences (data incidents) were analyzed and assigned a code. Following Fairclough (1992), a three dimensional framework was used to move from text to analysis to code to interpretation. First, the *text analysis* dimension was used in the language analysis of the interview text, with focus on wording, metaphor, and grammar modality. Next, the *discursive practice dimension* was used to begin contextualizing the production and consumption of the discourse of the interview text with its sociocultural context.

Finally, the *discourse as social practice* dimension was used to situate the discourse from the interview text into "the institutional and organizational circumstances of the discursive event and how that shapes the nature of the discursive practice." (Fairclough, 1992, p. 4) The circumstances here were defined by the researchers as the juxtaposition of the individual interview participants in his/her sociocultural context, his/her involvement in tourism, and the 17 SDGs set forth by the United Nations in the 2030 Agenda for Sustainable Development. As part of the critical interpretive process, special attention was given to the interactions between Indigenous ideology and practice regarding land, and the contrasts between the external ideology of the SDGs and the UN's goal of tourism as a vehicle for achieving them. In most cases, the interview participants' knowledge of the SDGs was more a general sense of externally-imposed development criteria as opposed to intimate knowledge of each SDG. The job of the researchers then was to interpret, analyze, and situate the ideological contrasts elucidated via the context of the discourse. The resulting findings of this analytical process is discussed next.

Findings and discussion

Indigenous land holding systems have been targeted as a constraint to development in the FSM and the other island nations in Micronesia. The ADB, a UN partner organization, has played a particularly prominent role in advocating for land holding system reforms to local governments using the rhetoric and promise of development. These reforms are posited as an essential component for realizing the myriad advances (i.e. social, physical and economic well-being) that will supposedly result via increased foreign investments in the keystone industry of tourism. In the tradition of the nineteenth-century colonists and traders which actively sought to influence islands' ruling bodies into alienating Indigenous land rights through land holding reforms (Banner, 2007), such organizations view land through a prism of Western interests, norms, and values. However, conceptualizations of land are highly localized and both culturally and historically determined. As such, it is imperative to understand local perspectives on land from Indigenous individuals who are also involved in the tourism industry, and how they interface with the UN's SDGs.

Indigenous perspectives: land, tourism, and the sustainable development goals

In juxtaposing Indigenous and external ideologies through the CDA, three main thematic pathways emerged that illustrated salient contrasts between specific SDGs and the localized understandings of issues embedded within them. Thus, three of the UN's 17 SDGs are focused on here to illustrate the importance of understanding Indigenous perspectives on the land holding – sustainable tourism nexus in the FSM, and the implications for externally-generated development goals. These include *SDG 1: No poverty, SDG 3: Good health and well-being,* and *SDG 16: Peace, justice, and strong institutions.* The thematic insights from this select group of SDGs can be used for further critical analyses of those not included here at readers' discretion.

Sustainable development goal 1: No poverty

A critical interpretation of the discourse from the interview participants points toward the need for a more nuanced understanding of poverty, land, and tourism in the FSM. For instance, while securing land control and ownership rights are listed as targets of *SDG 1* (United Nations, 2017), the present findings indicate that much of the conceptual underpinnings of the *SDG 1: No poverty* goal are nonetheless couched in Western notions of the economics of wealth. Through his discussion of land, the following statement by an interview participant illustrates a variation on how wealth and poverty are conceptualized in the FSM:

> *Many outsiders come [to the FSM], see a piece of land that would work for their business, and attach only money value to it. But land here has not only financial value, land here has voice. It is land that determines your role in society, not your financial wealth. So locals would rather be land rich and cash poor than vice versa.*

As this statement illuminates, a substantial component of what constitutes wealth for Indigenous Micronesians is rooted in the sociocultural capital accrued through individual and familial land holding rights. Familial pedigree is tracked closely in FSM societies, and plays an important role in forging social identity and status, as well as regulating interpersonal relationships and communication practices. Much of this pedigree and identity is still connected to the ties to land established by one's bloodline. (Hezel, 2012) The sociocultural capital that can be accumulated via bloodline and land is often used in lieu of the financial capital and material symbols that constitute wealth (and therefore economic and social livelihood) in the West.

Land is also parlayed into wealth for Micronesians via its utility in facilitating communal activities that sustain sociocultural norms and values. For instance, in traditional Micronesian society "accumulated wealth meant nothing" and conversely "it was the surrender of this wealth to build up or strengthen interpersonal relationships that counted for everything." (Hezel, 2012, p. 52) While financial wealth has an increased role in FSM society today, engaging in activities like land cultivation in order to participate in ceremonial gift giving and community activities is still a primary means for maintaining and/or accumulating sociocultural capital as a localized form of wealth. (Hezel, 2012) Whereas Western metrics of poverty and livelihood are centered on economic indicators, issues of quality of life, identity, and personhood are hinged upon the social relationships fostered through gift exchange in many Indigenous societies. (Curry, 2003) As such, it is possible for a Micronesian to have almost no accumulated financial wealth, but have considerable sociocultural wealth accumulated through land and its use to maintain a desired community status. As one interview participant stated:

> *[In the FSM] it's not money that motivates everybody, because they know they can survive without money … It is relationships. That's a really strong thing. Relationships and collective interests are often more important than $1 million dollars based on individual interests.*

As the FSM has progressed through various planning stages of using tourism as the baseline for sustainable development, such sociocultural norms have been generally regarded by external organizations as the "foremost impediment to general commercial development." (Hanlon, 1998, p. 120) For those entities that desire homogeneity and thus seek to superimpose preordained standards regarding development and improved livelihoods, islands offer "distinct identities and spaces, an apparent combination of isolation and community. (Connell, 2003, p. 573) This combination has created an environment viewed as antithetical to development organizations and the standards they use which assume "that the capitalist business model works everywhere regardless of the cultural environment." (Cheshire, 2001, p. 3) As Hezel (2012, p. 23) states:

> Ironically, the very quality that makes Micronesians so charming in the eyes of the western world also hinders them from achieving the development and governance goals that are being held up for them. This personalization may go a long way in melting the hearts of tourists, but it is perceived as a stumbling block for good government and earns the new island societies a poor rating on the international index.

Thus, provision for localized understandings of what does and does not constitute wealth and poverty in the FSM is important. In a report for the ADB, Abbott and Pollard (2004, p. 90) cite the importance of "enabling land's dead capital value" as part of the poverty reduction plan for Pacific Islands. Contrary to the ADB, the results of this study highlight the reality that Indigenous land holding is paramount to poverty alleviation in the FSM and other places where sociocultural wealth is of equal, or greater, importance than financial wealth. Land in Oceania is far from a detached, perfunctory arena for commodity exchange, but rather deeply embedded in important sociocultural systems. (Swedberg, 1994) For many Pacific Island communities, the "maintenance of social and cultural practices is central to participatory development and just as important as income gains and poverty reduction." (Connell, 2007, p. 130) Land becomes the stage for sociocultural practices, and when islanders come together to work the land, "they are forming social relations of identity, communal labor practices, individual and group exchanges" that are all about community building. (Curry, 2003, p. 417) As one interview participant stated:

> *You are who you are because of your land and where you come from. That's what makes it hard for both investors and locals. Because if I give you a land with voice, and afterwards will I still be the same person with the same role and respect? Land here determines who you are and what you are.*

Sustainable development goal 3: Good health and well-being

Similar to the one-dimensional perspective on what constitutes wealth and poverty adopted by the UN in *SDG 1: No poverty*, the analysis also elucidated a relatedly narrow view on other issues relating to livelihood promoted by the UN in *SDG 3: Good health and well-being*. Specifically, *SDG*

3 is entirely focused on targeting aspects of physical wellness such as disease eradication and reduction, reproductive health, and child mortality. (United Nations, 2017) While unquestionably important, this is merely one piece of the multidimensional puzzle of what comprises good health and well-being in PICs like the FSM. As the results of this study underscore, a substantial emphasis is placed on community and social relationships as aspects of mental, emotional, and spiritual health and well-being in Pacific societies, for which land plays "a fundamental role in determining livelihood strategies." (Purdie, 1999, p. 77) As opposed to Western notions of land as arenas for various forms of economic exchange (Swedberg, 1994), land becomes the stage for aspects of health and well-being ranging from identity affirmation, communal activity, social interactions, and maintenance of the sociocultural order in Pacific Island societies. As such, land is regarded as that which feeds health and well-being both through agricultural cultivation, as well as through the cultivation of Indigenous personhood. (Diaz & Haulani Kauanui, 2001)

External organizations which advocate for tourism as the vehicle for sustainable development in places like the FSM must understand the impact on islanders' health and well-being when land is objectively treated like an emotionally-neutral commodity that should be optimized for productive capacity through liberalized land laws. An illustrative example of this revealed through the interviews of this study was the recent failed attempt to develop a resort hotel complex in the FSM state of Yap by a large, foreign investment company. According to interview participants with direct knowledge of this project, the foreign investment group vastly underestimated the central role of land and land ownership in the sociocultural landscape of Yap. For instance, the foreign investment group presented Yapese officials with detailed resort development plans on land currently owned by local individuals and families. However, the investors neglected to consult with these parties beforehand, instead choosing to view the land through a lens which assumed that it would undoubtedly be available for a long-term lease at a fair market rate. As one interview participant stated:

> *[The foreign investors] took a map of Yap and made a proposal for where the hotel would be and so forth without considering that was actually people's land! Then the local landowners saw the proposal and said, "Wait is that my banana farm? Is that my beach? Is that my taro patch? No way!"*

As is often the case with foreign entities operating in PICs, land was improperly treated like an arena for commodity exchange rather than a stage for myriad sources of health and well-being for islanders, and the project unsurprisingly fell apart as a result. Another illustrative example of the complications of developing land for commercial use can be found in the so-called "Battle of Map." This case involved a proposed resort complex called Yap Nature Life Garden, Inc. to be built in the district of Map in Yap State, FSM, and was to be funded by a Japanese investment company. (Hanlon, 1998) In response to the plans for the proposed resort, the district chiefs of Map successfully circulated a petition against what they felt was unlawful use of land inherited in trust to the villagers of Map. Their efforts to protect against what they considered to be foreign exploitation of Yapese land culminated in a government directive requiring Yap Nature Life Garden, Inc. to discontinue the project and withdraw from Yap entirely. (Hanlon, 1998) As one interview participant of this research summarized:

> *When outsiders come in here to make deals about land for a hotel, they don't appreciate the value of land to the people that own it. They might see a small parcel of land and attach money value to it, but to those people it means more than that.*

As external organizations advocate for sustainable tourism as a means for achieving SDGs, provision for how local conceptualizations of issues like health and well-being are tied to land is of primary importance. This is especially true given the UN's partnerships with the ADB, which has a history of advocating for Indigenous land holding system reform to stimulate foreign investment activity in tourism. Pacific places with less understood sociocultural norms like the FSM differ substantially from the context on which conventional business and economic knowledge is based, and thus work in these areas "must go beyond general schemata, and develop more detailed models of the specific situations and problems." (Dixit, 2004, p. 21) This study suggests that a more detailed model of health and well-being in PICs like the FSM must go beyond mere physical wellness, and include aspects of the mental, emotional, and spiritual health that is tied directly to the retention of Indigenous land holding systems.

Sustainable development goal 16: Peace, justice, and strong institutions

As with *SDG 1: No poverty* and *SDG 3: Good health and well-being*, the results of this research also illuminate how *SDG 16: Peace, justice, and strong institutions* (United Nations, 2017) is rooted in the standardized ideologies which both overlook and undervalue localized realities in the FSM. Specifically, the UN is promoting a one-dimensional view of institutions in *SDG 16* that seemingly does not recognize, understand, and/or value both sides of the institutional coin. This view of institutions places the emphasis solely on the formal institutions (i.e. banks, courts, and government), while ignoring the informal institutions supported by sociocultural norms and values which take precedence in many Pacific Island contexts. (Cheshire, 2010; Stumpf & Swanger, 2017) In the FSM, it is these all-important informal institutions which have a disproportionate influence on myriad aspects of social and economic life. One interview participant described Indigenous views on this relative importance of formal versus informal institutions using the following analogy:

> *You know laws don't mean much to us, people here don't really pay a lot of attention to legal rules. So doing business here for locals is like fishing for sea turtles. There are laws about it, but nobody pays attention. If we see someone take a turtle out of season, we pay more attention to who the person is and their standing in the community to determine the consequences much more than anything related to the actual law.*

Despite the prominence of informal institutions in the FSM, many external organizations only account for formal institutions, which they argue must be reformed for effectiveness in order to facilitate development. This is particularly true regarding reform around Indigenous land holding rights in order to support foreign investments in tourism as a driver of development. For instance, an ADB report on the FSM entitled "Understanding Land Issues and their Impact on Tourism Development" (2015, p. 5–6) laments that the FSM "has not fully liberalized foreign investment, including investment in tourism" which can deter "the kind of investment that is interested in establishing a long-term presence, expanding that presence, employing local people, contracting local services, and adding local economic value." The report goes on to state that while the current land holding system "remains in favor of domestic business interests … - land and institutional reforms may be considered to advance the productive use of land and to address the remaining weaknesses of land use and management system." (Asian Development Bank, 2015, p. 6) Another ADB report states that in order to develop the kinds of land markets that can effectively parlay into economic development, PICs need "mortgages, property deeds, land surveys, property valuation, and other institutional services that form part of a land management system." (Holden et al., 2004, p. 89)

The assumption is that without clear land holding laws which cater to foreign investment and are backed by Western-standard formal institutions, places like the FSM will not be able to meet preordained development standards. (Cheshire, 2010; Purdie, 1999) However, it is important to point out that successful investments in tourism and associated land use arrangements are being made by leveraging the *informal institutions* that have historically worked in PICs like the FSM. (Cheshire, 2010, Stumpf & Swanger, 2017) The rules supported by sociocultural norms and values may be in direct contrast to those supported by the courts. (Ostrom, 2005) This suggests a need to look beyond a standardized institutional ways and means considered requisite to sustainable development, and instead focus on leveraging the kinds of deeply-embedded informal institutions which actually produce effective outcomes around land use. (see Stumpf & Swanger, 2017)

Further, the one-dimensional emphasis placed on strengthening formal institutions around land holding systems ignores the substantial resistance and barriers to actually achieving this. As Duncan et al. (2014, p. 16) state, external organizations providing development assistance to PICs "have not understood the lack of interest in open markets, secure property rights, and impartial enforcement of contracts and have not given sufficient attention to how to progress economic reform in such circumstances." In the FSM, sociocultural norms and values often act as barriers to the kinds of formal institutional reforms advocated for by organizations which assume that externally-legitimated prescriptions for development should be used irrespective of the sociocultural reality. As Cheshire (2001, p. 3) states, this failure to account for the role of informal institutions of culture in FSM economic life (including land use and holding) is akin to attempting to "run through a river as if it were dry land." As one interview participant commented:

> *When you talk about foreign investment in tourism and institutional reform, it's a sensitive issue because most people think this means you want to start looking at changing land laws. That's something people don't want to get into here. [Local] people don't like things that have a great impact on the culture. That's what happened with previous hotel projects that didn't work. Culture is the way to keep the peace here on the island. Too much Western way and that peace won't be there.*

This interview participant went on to state that:

> *In general, many [outsiders] come in with an "I know better than you" frame of mind, and use an "I'm smarter than you" mentality. It's pretty demeaning and it creates resistance right away. One has to listen to what locals want and need. When outsiders come in and say "Here's what you need", it won't work. It's a difference between what you think local people need and what they actually want.*

An uncritical view of the UN's focus on strengthening institutions to promote a more peaceful and just society in Pacific Islands assumes that formal institutions are of primary importance in these contexts, that the barriers to institutional reform are surmountable, and that local institutions should adapt to external standards rather than vice versa. (Stumpf & Swanger, 2017) However, a critical discourse interpretation of Indigenous perspectives reveals the narrow view of institutions these ideas are predicated on. While the formal institutions may be assessed by external metrics as weak (World Bank, 2016), the informal institutions in the FSM are strong. It is these informal institutions that have been used successfully to develop a variety of economic ventures within Indigenous land holding systems across industries, including tourism. (Cheshire, 2010; Stumpf & Swanger, 2017)

While many of these successful ventures and their associated land dealings take substantial time, focus, and effort by Western standards, they nonetheless adhere to the rules-in-use (Ostrom, 2005) utilized by islanders and island societies, and should be recognized and valued as such. A recent study by Milne, Deuchar, Berno, Taumoepeau, Pusinelli, and Raymond (2017, p. 119) of the New Zealand Institute for Pacific Research on how private-sector investment can help PICs achieve sustainable development goals states that "there are no easy solutions to the challenges that surround land and investment." The report also suggests the complexities around land and investment be accepted while ensuring that Indigenous land holding rights are not compromised in the development process. (Milne et al., 2017) Insomuch as land has been, and continues to be tied to personal livelihood and identity in Oceania, diverting focus from strengthening institutions to valuing and building upon existing institutions may indeed represent a more conscientious and productive approach for external organizations to consider.

Conclusion: life on land

This study has used a critical discourse analysis and interpretive approach to juxtapose Indigenous perspectives of land with the UN's SDGs and the use of tourism as a principle driver of development in the FSM. Specifically, *SDG 1: No poverty, SDG 3: Good health and well-being, SDG 16: Peace, justice, and strong institutions* were used to illustrate practical and conceptual disconnects between such externally-validated notions of sustainable development, and how retention of Indigenous land holding systems underscores localized realities of these issues in the FSM and PICs like it.

Organizations which often assume it is their duty to impose development blueprints on the developing world often couch them in frameworks which do not align with local realities. Another, less euphemistic way of stating this is that "the local perspective is frequently seen as subordinate to the global." (Overton, 1999, p. 1) In particular, "rational development experts" often hold views and thus develop plans which treat land as commodified physical spaces detached from "cultural idiosyncracies." (Connell, 2007, p. 125) These individuals and their associated organizations view land as sites for latent development activity which could potentially lead to improved livelihoods if only local people and governments would get their acts together and treat it as such.

However, Micronesian societies are characterized by localized ideologies that do not always sync with Western conventions. (Connell, 2007; Hanlon, 1998) In these societies where land is much more than a latent economic asset and where society is "written in the ground" (Connell, 2007, p. 126), standardized development prescriptions miss the mark at best, and indeed may threaten the potential for prosperous Indigenous livelihoods at worst. As the results of this study help illustrate, *Life on Land* (*SDG 15*) is predicated not on liberalized regulations that may facilitate sustainable development via tourism in Pacific Islands, but on retention of Indigenous land holding systems as the lifeblood of the Indigenous livelihoods tied directly to it. For islanders, what is "paramount to the continued existence and viability" of society is the sense of place rooted in land ownership. (Diaz & Haulani Kauanui, 2001)

As such, PICs represent special cases which require alternative theoretical and practical development frameworks. This begins with seeking to understand what is actually valued in local societies, and the functional rules of the game which infuse life and meaning into Indigenous realities. This cannot be accomplished through detached analyses and interpretations of such realities by external consultants and their laptop ruminations from the hotel lobby in places like the FSM. Rather, this can only be accomplished through an engaged, bottom-up approach which understands land for what it is, what it does, and what it says in Pacific societies. Land does not have suppressed economic productive value that must be translated into financial capital as an economic development imperative; Land has culturally and historically created value which is translated into social (and sometimes financial) capital as a sociocultural imperative. Land does not represent an administrative stumbling block and bottleneck to development; Land enables Indigenous personhood and livelihood. Land is not a passive commodity; The land has voice.

What does this bottom-up approach look like in which local norms and values regarding Indigenous land holding rights retain priority? In the FSM, this has required formal institutions which reflect and maintain the respective, and often very different, informal institutions of the various islands. Stated differently, this has required officials and decision-makers in local courts and governments who selectively accept external development aid, technical assistance, and governance prescriptions as members of the global community, while simultaneously prioritizing and adhering to core the sociocultural norms and values that govern everyday life in the islands. Hezel (2012, p. vii) uses the term "dilemmas of development" to refer to this administrative high wire act required to balance "the points of conflict between traditional island culture and the demands of the modern world" in the FSM. There are inevitably give-and-takes throughout these processes of reconciling such dilemmas. However, the past three centuries of Pacific history have illustrated that giving up land holding rights is perhaps above all others, the one concession that is most detrimental to Indigenous livelihoods, and that is least likely to lay a foundation for the sustainable development of Indigenous communities.

Disclosure statement

No potential conflict of interests was reported by the author(s).

References

Abbott, D., & Pollard, S. (2004). *Hardship and poverty in the Pacific*. Manila, Philippines: Asian Development Bank.
Abeyratne, R. (1999). Management of the environmental impact of tourism and air transport on small island developing states. *Journal of Air Transport Management*, *5*(1), 1–37.
Asian Development Bank. (2015). *Understanding land issues and their impact on tourism development*. Mandaluyong City: Asian Development Bank.
Asian Development Bank. (2016). *Key indicators for Asia and the Pacific 2016*. Mandaluyong City: Asian Development Bank.
Asian Development Bank. (2017). *Fact sheet*. Retrieved from <https://www.adb.org/publications/federated-states-micronesia-fact-sheet>.
Asian Development Bank. (2017a). Retrieved from <https://www.adb.org/about/united-nations>.
Bäckstrand, K. (2006). Multi-stakeholder partnerships for sustainable development: Rethinking legitimacy, accountability and effectiveness. *European Environment*, *16*(5), 290–306.
Ballendorf, D. A. (1988). The new freely-associated states of Micronesia: Their natural and social environmental challenges. *GeoJournal*, *16*(2), 137–142.
Banner, S. (2007). *Possessing the Pacific: Land, settlers, and indigenous people from Australia to Alaska*. Cambridge, MA: Harvard University Press.
Bascom, W. (1965). Ponape: A Pacific economy in transition. *Anthropological Records*, *22*, 1–156.
Billig, M. (2003). Critical discourse analysis and the rhetoric of critique. In G. Weiss & R. Wodak (Eds.), *Critical discourse analysis: Theory and interdisciplinarity* (pp. 35–46). New York, NY: Palgrave Macmillan.
Carlile, L. (2000). Niche or mass market? The regional context of tourism in Palau. *The Contemporary Pacific*, *12*(2), 415–436.
Cheshire, C. L. (2001). Family and business in Micronesia. *Micronesian Counselor*, *37*, 1–19.
Cheshire, C. L. (2010). Swimming with the tide. *Micronesian Counselor*, *81*, 1–10.
Connell, J. (2003). Island dreaming: The contemplation of Polynesian paradise. *Journal of Historical Geography*, *29*(4), 554–581.
Connell, J. (2007). Islands, idylls and the detours of development. *Singapore Journal of Tropical Geography*, *28*(2), 116–135.
Crocombe, R. G. (1971). *Land tenure in the South Pacific*. Oxford: Oxford University Press.
Curry, G. N. (2003). Moving beyond postdevelopment: Facilitating indigenous alternative for "development". *Economic Geography*, *79*(4), 405–423.
Davies, R. E. (1984). *Continental airlines: The first fifty years, 1934–1984*. Woodlands, TX: Pioneer.
Diaz, V. M., & Haulani Kauanui, J. K. (2001). Native Pacific cultural studies on the edge. *The Contemporary Pacific*, *13*(2), 315–341.
Dolowitz, D., & Marsh, D. (2000). Learning from abroad: The role of policy transfer in contemporary policy-making. *Governance*, *13*(1), 5–24.
Dixit, A. K. (2004). *Lawlessness and economics*. Princeton, NJ: Princeton University Press.
Duncan, R., Codippily, H., Duituturaga, E., & Bulatale, R. (2014). *Identifying binding constraints in Pacific Island economies*. Honolulu, Hawaii: East-West Center.
Fairclough, N. (1992). *Discourse and social change*. Cambridge: Polity Press.
Farrell, B. H. (1972). The alien and the land of Oceania. In R. G. Ward (Ed.), *Man in the Pacific Islands: Essays on geographical change in the Pacific Islands* (pp. 34–73). Oxford: Oxford University Press.

Federated States of Micronesia. (2008). *Federated States of Micronesia Statistical Yearbook*. Palikir, Micronesia: FSM National Government.

Federated States of Micronesia Development Partners Forum. (2012). *Federated States of Micronesia development framework: Looking to the future*. Palikir, Micronesia: FSM Department of Economic Affairs.

Federated States of Micronesia Economic Summit. (2004). *Federated States of Micronesia's strategic development plan (2004-2023): The next 20 years: Achieving economic growth and self-reliance*. Palikir, Micronesia: FSM Department of Economic Affairs.

Haglegam, J. (1998). *Traditional leaders and governance in Micronesia. State, society and governance in Melanesia (Discussion Paper 98/1)*. Canberra: Australia National University.

Hanlon, D. (1994). Patterns of colonial rule in Micronesia. In K. R. Howe, R. C. Kiste, & B. V. Lal (Eds.), *Tides of history: The Pacific Islands in the twentieth century* (pp. 93–118). Honolulu: University of Hawaii Press.

Hanlon, D. (1998). *Remaking Micronesia: Discourses over development in a Pacific territory, 1944 –1982*. Honolulu: University of Hawaii Press.

Hanlon, D. (2004). Wone sohte lohdi; History and place on Pohnpei. In B. V. Lal (Ed.), *Pacific places, Pacific histories* (pp. 195–215). Honolulu: University of Hawaii Press.

Hau'ofa, E. (2008). *We are the ocean: Selected works*. Honolulu: University of Hawaii Press.

Hezel, F. (2012). *Making sense of Micronesia: The logic of Pacific Island culture*. Honolulu: University of Hawaii Press.

Holden, P., Bale, M., & Holden, S. (2004). *Swimming against the tide? An assessment of the private sector in the Pacific*. Manila, Philippines: Asian Development Bank.

Kelley, L., & Penseyres, M. (1976). *Continental Airlines and the development of tourism in the Marianas: TTPI*. Manoa: University of Hawaii.

Kiste, R. C. (1994). Pre-colonial times. In K. R. Howe, R. C. Kiste, & B. V. Lal (Eds.), *Tides of history: The Pacific Islands in the twentieth century* (pp. 3–28). Honolulu: University of Hawaii Press.

Labby, D. (1976). *The demystification of Yap*. Chicago, IL: University of Chicago Press.

Larmour, P. (2005). *Foreign flowers: Institutional transfer and good governance in the Pacific Islands*. Honolulu: University of Hawai'i Press.

Lingenfelter, S. G. (1975). *Yap: Political leadership and culture change in an island society*. Honolulu: University of Hawaii Press.

McKee, D. L., & Tisdell, C. A. (1990). *Developmental issues in small island economies*. New York, NY: Praeger.

McMillan, J. (2002). *Reinventing the bazaar: A natural history of markets*. New York, NY: W.W. Norton & Company.

Milne, S., Deuchar, C., Berno, T., Taumoepeau, S., Pusinelli, M., & Raymond, J. (2017). *Private sector investment in the Pacific*. Auckland: New Zealand Institute for Pacific Research.

Ostrom, E. (2005). Doing institutional analysis: Digging deeper than markets and hierarchies, In C. Menard & M. M. Shirley (Eds.), *Handbook of new institutional economics* (pp. 819–848). Dordrecht, Netherlands: Springer.

Overton, J. (1999). Sustainable development and the Pacific Islands. In J. Overton & R. Scheyvens (Eds.), *Strategies for sustainable development: Experiences from the Pacific* (pp. 1–18). London: Zed Books.

Parker, S. C. (2006). Entrepreneurs as producers. In S. C. Parker (Ed.), *The life cycle of entrepreneurial ventures* (pp. 337–360). New York, NY: Springer Science.

Peoples, J. G. (1985). *Island in trust: Culture change and dependence in a Micronesia economy*. London: Westview Press.

Purdie, N. (1999). Pacific Islands livelihoods. In J. Overton & R. Scheyvens (Eds.), *Strategies for sustainable development: Experiences from the Pacific* (pp. 64–79). London: Zed Books.

Rametsteiner, E., Pülzl, H., Alkan-Olsson, J., & Frederiksen, P. (2011). Sustainability indicator development - science or political negotiation? *Ecological Indicators*, *11*(1), 61–70.

Reed, M. S., Fraser, E., & Dougill, A. J. (2006). An adaptive learning process for developing and applying sustainability indicators with local communities. *Ecological Economics*, *59*(4), 406–418.

Scheyvens, R., & Momsen, J. (2008). Tourism in small island states: From vulnerabilities to strengths. *Journal of Sustainable Tourism*, *16*(5), 491–510.

Spiro, M. E. (1986). Cultural relativism and the future of anthropology. *Cultural Anthropology*, *1*(3), 259–286.

Stumpf, T. S. (2016). Navigations: Enhancing qualitative hospitality and tourism research outcomes in Pacific Island Countries. *Asia-Pacific Journal of Innovation in Hospitality and Tourism*, *5*(2), 71–87.

Stumpf, T. S., & Reynolds, D. (2016). Institutional conformance and tourism performance: An efficiency analysis in developing Pacific Island countries. *Tourism Planning & Development*, *13*(4), 449–468.

Stumpf, T. S., & Swanger, N. (2017). Institutions and transaction costs in foreign-local hotel ventures: A grounded investigation in the developing Pacific. *Tourism Management*, *61*, 368–379.

Swedberg, R. (1994). Markets as social structures. In N. J. Smelser & R. Swedberg (Eds.), *The handbook of economic sociology* (pp. 255–282). Princeton, NJ: Princeton University Press.

Taylor, J. (2008). Indigenous peoples and indicators of well-being: Australian perspectives on United Nations global frameworks. *Social Indicators Research*, *87*(1), 111–126.

United Nations. (2016). Retrieved from <https://sustainabledevelopment.un.org/partnership/partners/?id=658>.

United Nations. (2017). Sustainable development goals 17 goals to transform our world. Retrieved from <http://www.un.org/sustainabledevelopment/sustainable-development-goals/#prettyPhoto>

UNWTO. (2018). *Yearbook of Tourism Statistics, Data 2012 – 2016, 2018 Edition*. Madrid, Spain: WorldTourism Organization.

Van Dijk, T. A. (2003). The discourse-knowledge interface. In G. Weiss & R. Wodak (Eds.), *Critical discourse analysis: Theory and interdisciplinarity* (pp. 85–109). New York, NY: Palgrave Macmillan.

Ward, R. G., & Kingdon, E. (1995). Land use and tenure: Some comparisons. In R. G. Ward & E. Kingdon (Eds.), *Land, custom, and practice in the South Pacific* (pp. 6–35). Cambridge: Cambridge University Press.

Weiss, G., & Wodak, R. (2003). Introduction: Theory, interdisciplinarity, and critical discourse analysis. In G. Weiss & R. Wodak (Eds.), *Critical discourse analysis: Theory and interdisciplinarity* (pp. 1–34). New York, NY: Palgrave Macmillan.

World Bank. (2016). *Doing business 2016: Economy profile: Federated States of Micronesia*. Washington, DC: The World Bank and the International Finance Corporation.

Colonizing space and commodifying place: tourism's violent geographies

Jennifer A. Devine

ABSTRACT
Across Latin America, non-traditional tourism sectors, like eco-tourism and cultural tourism, have grown dramatically in the last 30 years. Neoliberal ideologies, reforms, and infrastructure set the stage for growth, but neoliberalism alone does not explain why the tourism industry is an acute site of struggle over territory, identity and history. From Guatemala's Maya Biosphere Reserve in the heart of Central America's "Mayan World" tourism initiative, this article asks: Why is contemporary tourism such a powerful and pervasive site of contentious socio-spatial politics? It argues that neoliberal reforms have combined with tourism's dual violent practices of spatial colonization and the commodification of place to fuel this growth and politicization. Spatial colonization refers to the commodification of "nature" and practices of land dispossession that define capitalism's expansion into "underdeveloped" and "green" spaces. The industry also commodifies the culture, identity and experience of a place and its people as objects of tourist consumption, which infuses tourism's representational practices with immense, yet subtle power. This article further illustrates how tourism landscapes are carved out of practices of material and intangible dispossession, as well as practices of state territorialization. This violence often engenders diverse forms of resistance by "hosts" living in other people's playgrounds around the world.

Introduction

In Latin America, tourism is big business, but it has not always been that way. In the last 30 years, the region's tourism industry has grown dramatically, doubling in size since 1995 in terms of tourism receipts, number of tourists and tourism's contribution to GDP.[1] Neoliberal reforms, like the North American (NAFTA) and Central American Free Trade Agreements (DR-CAFTA), set the stage for this dramatic growth and enabled conditions for foreign direct investment and amended property rights to allow foreign ownership of land and tourism businesses. At the same time, structural adjustment policies reoriented economies away from economic protectionism and import substitution industrialization toward global economic integration based on ideologies of competitive advantage and free trade (Portes, 1997). Sustainable development and pro-poor development growth strategies emanating from the World Bank combined with these dynamics to identify non-traditional forms of tourism as a panacea to Latin America's economic and environmental crises (Mohan & Stokke, 2000).

Yet the dramatic growth in the non-traditional tourism industry in Latin America cannot be explained by the turn to neoliberalism alone. Simply focusing on how neoliberal reforms enabled this explosive growth does not explain why the sector is an acute site of struggle over territory, identity and history. This article asks: Why is contemporary tourism such a powerful and pervasive site of contentious socio-spatial politics? What characteristics of non-traditional tourism in the era of neoliberalism explain this growth and politicization?

This article argues that neoliberal economic reforms have combined with tourism's dual violent practices of spatial colonization and the commodification of place to fuel this growth. Drawing on examples from Guatemala's Maya Biosphere Reserve, this paper examines how eco-tourism development commodifies nature and expands capitalism's spatial frontier into "underdeveloped" and/or semi-proletarianized spaces. These dynamics in northern Guatemala further illustrate how these violent spatial processes are driven by neoliberal forms of accumulation by dispossession, and are tied to practices of state territorialization. In Guatemala, and around the world, capital's colonization of space through eco-tourism unfolds alongside militarized state efforts to render rebellious territories and people governable.

The second spatial practice defining tourism is the commodification of place. Unlike many industries, tourism sells the experience of a place and its people as its products. These experiential tourism products often commodify the identity of a place and its "hosts" in ways that fail to reflect the multiple and often competing narratives defining those places and identities. Rather, representations of people and places in cultural and eco-tourism are often essentialized and folkloric, reflect the preconceived and stereotypical imaginaries of tourists, and are sold and performed in tours, postcards, sex and everyday interactions (Salazar, 2012; Vich, 2006; Werry, 2011). When non-"hosts" own and manage cultural and eco-tourism projects, local residents are dispossessed of their ability to define their culture, identities, histories and landscapes. Eco-tourism's dual dynamics of spatial colonization and the commodification of place have combined with neoliberal reforms to fuel the industry's dramatic growth and state-sanctioned violence in "green" spaces, as well as engender diverse forms of resistance by "hosts" living in other people's playgrounds.

Despite the industry's inherent spatial dynamics of colonization and the commodification of place, tourism is not necessarily a site of exploitation. The article concludes by examining grassroots, Indigenous-owned tourism initiatives in Guatemala and Southern Africa to suggest that a key dynamic defining tourism's power relations, political possibilities and sustainability lies in who own the means of production; in other words, who decides which social relations defining culture and place are commodified and how.

This article draws on 2 years of ethnographic research in Guatemala from 2009 to 2011 that included 18 months of living and working in two Maya Biosphere villages where I participated in and observed activities related to tourism development, community forestry and Maya Biosphere governance. Additionally, oral histories and semi-structured interviews with 30 long-term residents and leaders in each village informed this work. Outside the reserve, I conducted an additional 40 interviews with state officials, military officers and conservationists, and attended multi-sector meetings regarding tourism development and Maya Biosphere governance. Since then, I have returned to Guatemala four times for two-week research trips (July 2013, August 2014, December 2015 and July 2016) and employed similar methods to inform this analysis. I further draw on Maya cultural tourism examples from the Guatemalan Highlands where I have observed the politics of tourism development since 2005.

Tourism and the colonization of space in the Mayan World

Eco-tourism is often defined as responsible travel to areas that conserves the environment and improves the welfare of the local people (Fennell, 2007; Honey, 2008). In contrast, I understand eco-tourism as a capitalist industry defined by tourists' travel and leisure practices that integrates people, places and resources into the global economy. The eco-tourism economy comes replete with normative ideas about what constitutes nature, who is an ideal tourist, and what types of activities are desirable or permissible in tourism spaces (Duffy, 2008, 2013; Gregory, 2006; Ojeda, 2012). I see eco-tourism as part of a broader set of neoliberal economic and political processes that render

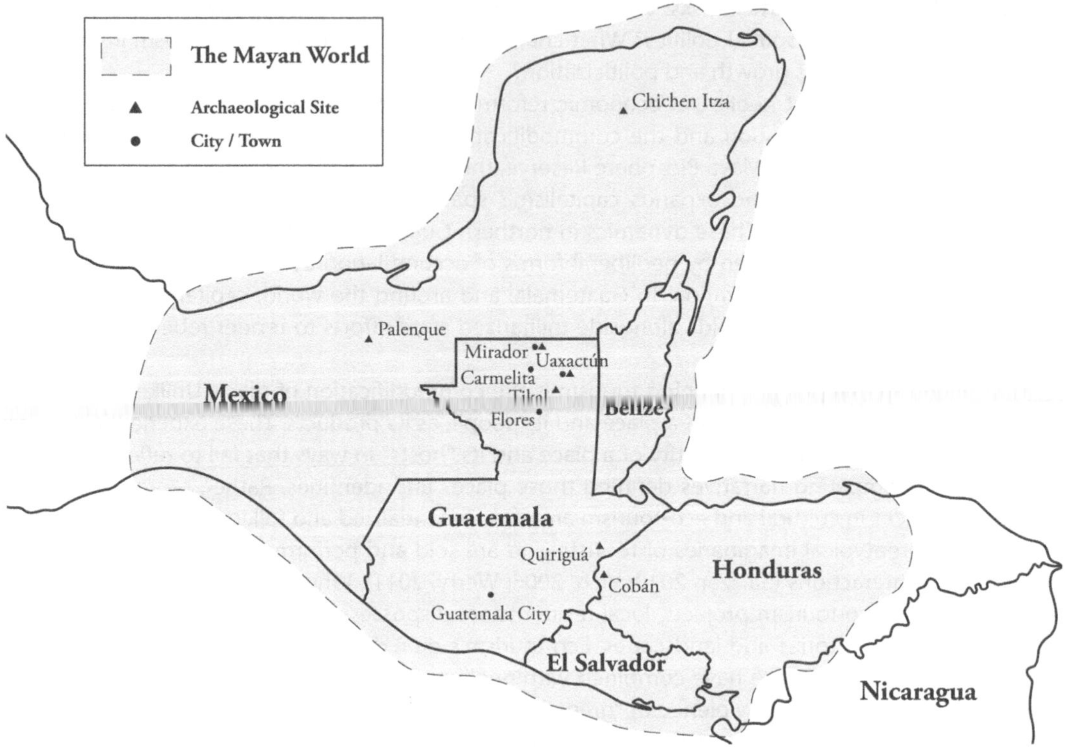

Figure 1. The "Mayan World" tourism region.

commonsense the idea that tourism development of green spaces and cultural places in the global south is the best, and perhaps only, sustainable development strategy (Duffy, 2013).

This neoliberal ideology and infrastructure took the form of branding and building the "Mayan World" as a global tourism destination in Central America. The project's roots date back to 1989 when *National Geographic* showcased Mexico's Yucatán Peninsula and the surrounding Guatemalan and Belizean lowlands as a newly christened geography, the "Mayan World" (see Figure 1) (Garrett, 1989). The popular piece segued into a multibillion dollar Central American marketing campaign and tourism industry stretching from the "Maya Riviera" beginning in Cancún to the forests of Guatemala's Maya Biosphere Reserve. As part of this marketing strategy, the state-run Guatemalan Tourism Commission (INGUAT) identified Guatemala as "The Heart of the Mayan World" in a national tourism campaign from 2000 to 2004, and then again from 2008 to the present.

In Guatemala, this Mayan World marketing and tourism development strategy was built on a human–ecological landscape reeling from centuries of state-sponsored Maya genocide, decades of civil war and a post-war period that has failed to produce peace. These living civil war geographies overlap in contradictory ways with the geographies and imaginaries of neoliberal eco-tourism development. The Maya Biosphere (see Figure 2) is home to hundreds of archaeology sites that include the UNESCO World Heritage Maya site of Tikal, which receives 200,000 tourists a year who contribute over $US 200 million annually to the national economy (GDT Tourism Consultants, 2010). Apart from Tikal, very few tourists visit the many other archaeology sites and exceptional ecological features characterizing the Maya Biosphere, due to lack of accessibility, security and information. Many state officials, tourism investors and conservationists lament the untapped eco-tourism potential of the Maya Biosphere and the Mirador archaeology site in particular. Mirador houses the oldest and largest

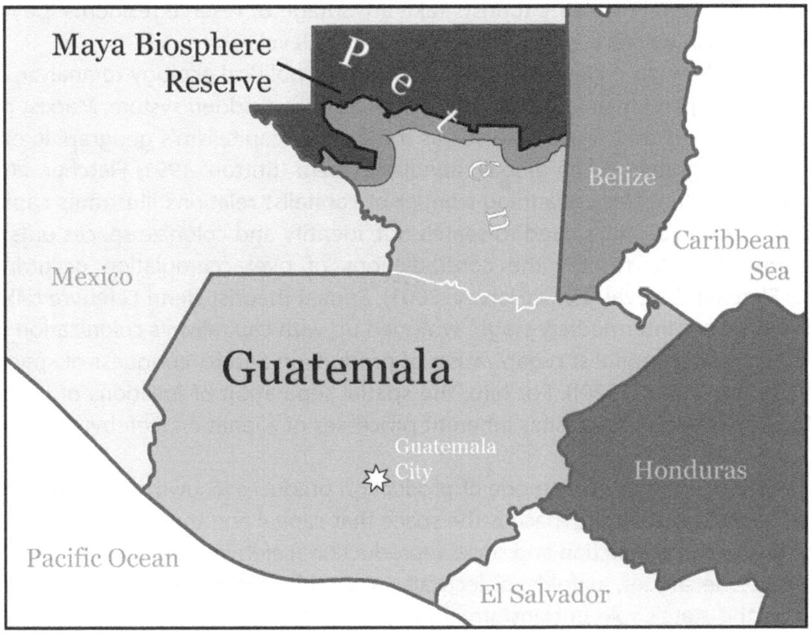

Figure 2. Maya Biosphere Reserve, Petén, Guatemala.

structures of the Pre-Columbian Maya, one of the largest pyramids in the world, and some of the earliest and well-preserved examples of Maya art and hieroglyphs (Hansen, 2012). Despite these exceptional features, the site receives only around 3000 visitors per year (GDT Tourism Consultants, 2010).

Maya Biosphere stakeholders forecast that Mirador's obscurity from the global tourism industry is coming to an end. Today, the Mirador archaeology site occupies center stage of several economic development and conservation initiatives that are working to convert the forests of northern Petén Guatemala into a global tourism destination, including a $39 million dollar World Bank loan (Inter-American Development Bank, 2006). Guatemalan President Alvaro Colom's (2008–2012) Four Jaguar eco-tourism project furthered this shared vision of turning the Maya Biosphere Reserve into the largest archaeological tourism park in Latin America.[2] At the same time, private-sector foundations run by Guatemalan elites are funneling millions of dollars into Mirador tourism development to promote and conserve the site as national patrimony and designate it as a UNESCO world heritage site (Global Heritage Fund [GHF], 2011). However, thousands of Maya Biosphere Reserve residents who make a living sustainably harvesting forest products and practicing subsistence agriculture fear that Mirador mega-projects and their elite interests threaten to destroy their homes, histories and livelihoods.

Whether non-traditional tourism development benefits people living and working in tourism spaces is highly contingent on historical and geographical contexts. More often than not, many critics argue that cultural and eco-tourism replicate colonial and neocolonial relations through tourism development and everyday tourism encounters. In these analyses, colonization refers to the physical presence of people from (former) colonial powers in (post) colonial spaces, and the connections between colonial and tourism routes/practices Craik (1994). Mowforth and Munt (2009), for example, examine the ways that non-traditional forms of tourism reinforce racial and colonial legacies of subservience and the aestheticization of poverty (p. 59). In her seminal novella, *A Small Place* (1988), Jamaica Kincaid (1988) poignantly illustrates the seamless connections between colonial violence and contemporary tourism's reproduction of racialized inequality in the Caribbean (see also Sheller, 2003). These critiques resonate loudly with the unequal exchanges characterizing many eco-tourism activities in Guatemala's Maya Biosphere, like the hike to Mirador that leaves from the village of

Carmelita in which relatively wealthy tourists take advantage of reserve residents' poverty and high unemployment to push workers' wages below subsistence level.

Here, however, I draw on critical political economy and political ecology to analyze the role tourism plays in expanding and maintaining capitalism as a conflict-ridden system. Marxist political economists have long argued that tourism serves as a means of capitalism's geographic expansion into spaces not (fully) integrated in the global capitalist system (Britton, 1991; Fletcher, 2011; Lefebvre, 1976, 2003; Harvey, 2003). This expanding frontier of capitalist relations illustrates capitalism's need for a "spatial fix", that is, capital's need to search out, identify and colonize spaces outside circuits of capitalist accumulation to resolve the contradictions of overaccumulation occurring elsewhere (Arrighi, 2006; Fletcher & Neves, 2012; Harvey, 2001). Spatial theorist Henri Lefebvre talks about tourism development as an "intermediary stage" wrapped up with capitalism's colonization of space, "the connection between the capitalist organization of production and its conquest of space as a whole" (1976, p. 84; see also 2003 [1970]). For him, the spatial separation of locations of leisure, work and social reproduction reflects capitalism's inherent processes of alienation (Lefebvre, 1991; MacCannell, 1976).

Lefebvre (1991) argues that every mode of production produces its own space, and "abstract space" is the space of capitalism. Abstract space is the space that capital and the state have "taken over" from other ways of organizing production and social reproduction (Lefebvre, 1991, p. 48). Thus, capital's colonization of space, he argues, unfolds dialectically alongside the production of "state space". This dynamic reflects the state's role in transforming previous socio-spatial formations into abstract space primed for capitalist development, and reproducing capitalism in the face of systemic crisis. Thus, for Lefebvre, the "spatial fix" is just one part of a broader process of transforming extant socio-spatial relations into abstract, state space. This violent process includes the attempted erasure of pre-existing socio-natural relations, histories and cultural landscapes through practices of privatization that render space equivalent and interchangeable, albeit differentiated, as part of a diversified menu of consumer commodities.[3] This process of producing abstract, state space is insatiably expansive (Brenner & Elden, 2009), and in the tourism industry manifests in violent practices of enclosure and land dispossession that separate local residents from land and livelihood (Duffy, 2013; Ojeda, 2012).

Herein lies an important nuance about capital's colonization of space via eco-tourism, cultural tourism and many other non-traditional forms of tourism. Resort style tourism in Latin America, particularly in Mexico and the Caribbean, has long served as a means for capitalist expansion into new spaces in order to exploit raw sources of materials and labor, as well as to provide relief for alienated workers. Non-traditional forms of tourism, however, have pushed the frontiers of commodification into new dimensions of nature and culture. This often takes the form of developers capitalizing on conservation projects by linking them with eco-tourism activities that include visits to archaeology sites, heritage sites, horse and elephant riding, and visits to "wilderness" areas for bird watching, hiking and hunting (Duffy & Moore, 2010). Along these lines, eco-tourism development of "green" or "natural" spaces can purportedly provide a way to pay for nature to "pay its own way" and incentivize conservation through foreign revenue generation (Büscher, Sullivan, Neves, Igoe, & Brockington, 2012; McAfee, 1999). Exemplifying this approach, the Foundation for Archaeological Research and Environmental Studies (FARES) financing archaeological research at the Maya Biosphere Mirador site describes: "The forest in this region is highly threatened, but [tourism] can provide new economic benefits for communities and the Republic of Guatemala through the establishment of world class archaeological parks and natural preserves" (n.d.).

The idea that forests and biodiversity are under threat from local populations has justified state militarization and securitization in the Maya Biosphere. In 2010, the Guatemalan State created the Green Battalion charged with protecting the natural and cultural patrimony of the Maya Biosphere. This mission has translated into the eviction of Maya Biosphere residents from the reserve for allegedly breaking conservation laws. The militarization of tourism and conservation spaces is not a phenomenon unique to northern Guatemala. These violent practices unfold around the world (Duffy, 2014) in places like Colombia (Ojeda, 2013), South Africa (Büscher, 2009), Mexico (Rocheleau, 2015) and Israel (Stein, 2008). These interlocking conservation and security narratives in Guatemala and elsewhere authorize new levels and forms of surveillance, intervention, and violence in the name of conservation and biodiversity (Bocarejo & Ojeda, 2016; Duffy, 2016; Lundstrom, 2014). Eco-tourism drives the militarization of space through an inherent logic of security embedded in the production of tourism spaces. Tourism necessitates security to convince the coveted global tourist that the wild, natural experiences they desire are not too wild, and not too dangerous (Devine, 2014). If a place is deemed "unsafe", people will not go there. This creates the impetus, the perceived necessity, of securing space as a condition of development. This profit-predicated-on-security equation puts armies, agencies and organizations into motion, and justifies the enormous amount of resources being poured into secure tourism spaces and breathes new life into old practices of colonization.

The case study of the Maya Biosphere further suggests that conservation violence is inseparable from practices of state territorialization and tourism development in the reserve. For years, the Maya Biosphere was a park on paper only, and poaching and trafficking of timber, fauna and Maya antiquities occurred with impunity (Finger-Stich, 2003). However, enforcing conservation law took on new meaning and form when President Alvaro Colom identified archaeological and eco-tourism development of the Maya Biosphere Reserve as one of his six presidential priorities. The terms "colonizing" and "war" are not just theoretical metaphors in this case, but are the vernacular terms employed by state agencies, conservationists and development practitioners who attempt to "colonize" the Petén and "recover governability" in the Maya Biosphere, in part, through tourism development and conservation efforts.

"Recovering governability" in Guatemala's Maya Biosphere

The Guatemalan government's project to develop the Petén and "recover governability" in the reserve through eco-tourism is simply the latest chapter of a long history of state efforts to "colonize" and integrate the region into the national fabric (Grandia, 2012; Schwartz, 1990). Northern Guatemala, home to the Maya Biosphere today, has a long frontier history of challenging and defining the limits of Guatemalan state sovereignty and territorial control (Schwartz, 1990). Spanish colonizers established the *Camino Royal* (Royal Road) through the area in the mid-sixteenth century, yet the Maya Itzá stronghold of Tayasal, the site of Petén's capital today, was the last Maya polity to be integrated into the Spanish colony in 1697. Individuals and political entities seeking to exert territorial control over Petén, its inhabitants and resources have long considered the region a frontier space (Grandia, 2012).

The national borders of three countries – Guatemala, Mexico and Belize – artificially divide Petén's lowland tropical geography that is united by history. Following independence in 1823, northern Petén was more closely tied to Belize and the Mexican Yucatán than Guatemala. Nearly 200 years later, the international borders themselves remain highly contested. The exact location of the Mexican–Guatemalan border is disputed by both states, and the Guatemalan state continues to accuse Mexican authorities of complicity in illegal logging by Mexican companies in Guatemalan territory. Guatemala's border disputes with Belize are even more contentious and date back more than 150 years. Guatemala recognized Belizean independence from Britain in 1991, but still claims half of Belize's current territory.

State-led "colonization" of Petén in the 1960s was tied to these international border disputes, but even more importantly, the brewing of civil war to the south in the rest of the country (Centeno Cordón, 1973). During Guatemala's "ten years of spring" (1944–1955), democratically elected President Jacobo Arbenz (1951–1954) initiated a land reform in 1952 that provided land for more than 100,000 landless peasants and their families (Jonas, 2000). Citing fears of communism, the CIA supported a military coup in 1954 to oust the reformist president, and the 1952 agrarian land reform was overturned (Jonas, 2000). The Guatemalan civil war (1960–1996) that followed culminated in the deaths or disappearances of 200,000 people and the displacement of a million more (Guatemala, Never Again!, 1999). The United Nations Mission for Historical Clarification determined that the

Guatemalan military was responsible for 93% of civilian deaths and committed acts of genocide and war crimes against the Maya (Guatemala, Never Again!, 1999).

As military governments sought an alternative to agrarian reform in the late 1950s, state officials looked north to Petén. Petén's low population density and large tracts of forest gave rise to an imaginary of the region as a vast open space ripe for settlement (Soza, 1970). Government officials and USAID identified what they called Petén's "colonization" as the answer to Guatemala's acute agrarian problems fueling social unrest in the rest of the country (Centeno Cordón, 1973). During the height of civil war violence in the 1980s, the state's project of colonization took a new territorial turn. The military defined the northern forests of Petén, today's Maya Biosphere, as a subversive space filled with guerilla brigades who used the forests and remoteness of the region to avoid capture and undertake guerilla warfare (Ybarra, 2012). During the peace negotiations (1985–1996), Guatemalan and global conservationists successfully campaigned for the creation of the Maya Biosphere in 1990, totaling 20% of the total national territory.

Following the Maya Biosphere's creation, Guatemalan authorities have been unable to stem the growth of illicit timber extraction, poaching, antiquities looting and drug trafficking activities in the reserve (Finger-Stich, 2003). Furthermore, the war's end left the issue of land inequality unresolved by failing to enact meaningful agrarian reform, and the national imaginary of Petén as an empty no man's land continued to draw landless peasants to "illegally" settle in the biosphere (Finger-Stich, 2003). It was not until eco-tourism development efforts in the form of the Four Jaguar initiative came to the fore of President Colom's presidential agenda that state forces re-galvanized efforts to enforce conservation law in the Maya Biosphere.

In late July of 2008, state officials, private sector investors and members of the press gathered at the National Palace of Culture to witness the unveiling of President Colom's much anticipated eco-tourism project. At the opening ceremony, an elaborate promotional video foretold the future of the Maya Biosphere in 2023 when an estimated 12 million people would visit the region, visiting dozens of archaeology sites and a new university specializing in biodiversity and Maya Studies. The video then quickly turns to issues of security:

> The creation and position of *Cuatro Balam* [Four Jaguar] at the world level demands the coordination of efforts at all levels. All of the parks that make it up are free of invaders, and traffickers of people, drugs, and animals… .[T] he rule of law prevails. There is judicial and administrative justice for land-owners and the state.

To bring this tourism/security vision to fruition, Colom created a Four Jaguar Special Presidential Commission in May of 2008. Following two years of the program floundering in Guatemalan City, the president moved the program to Petén and his team brought together 24 high-ranking individuals in Petén working across public, non-governmental and military sectors to form the Four Jaguar Regional Technical Committee. From September of 2010 to February 2011, the Technical Committee defined short- and long-term strategic objectives in the Maya Biosphere. Amid the diversity of initiatives proposed, the committee quickly identified "recovering governability" as the first priority of the Four Jaguar project.

Implemented by the army and other armed branches of the government, this mission has translated into the militarization of the reserve. While this process is detailed elsewhere (Devine, 2014), two dynamics are flagged here as illustrative of the ways eco-tourism investment in the reserve has occurred alongside what Lefebvre would call the production of state space. The first is the creation of the Green Battalion. In an elaborate ceremony in front of the national palace in 2010, President Colom introduced the Green Battalion to the nation and explained that six military detachments and 300 soldiers form the special forces that focus specifically on "recuperating in its totality the protected areas in the north of Petén" (Prensa Libre, 2010) and stopping the "depredation of natural resources, especially Maya ruins" (Molina, 2010). Government officials, non-profit and conservation workers and many Guatemalan citizens describe the reserve as ungovernable because the state does not have a monopoly on rule, order or violence. State agencies and armed forces face organized crime engaging in drug trafficking, money laundering, poaching and moving undocumented people across borders in the reserve, some of whom are trafficked. The Green Battalion partners with other armed forces, such as the Air Force, the National Civil Police and the Council on National Protected Areas, to carry out what are called "high-impact operations". These operations include removing cattle from the reserve, physically evicting residents identified as conservation threats, and patrolling the reserve for poachers and traffickers of drugs, humans, animals, antiquities, timber and arms. The newly established Green Battalion has amplified the military's territorial presence and surveillance of the region, yet "recovering governability" remains an elusive goal.

Second, the Four Jaguar project of "recovering governance" in the Maya Biosphere helped enable the construction of 14 new military outposts throughout the reserve, increasing the total number to 43 by 2012 (Consejo Nacional de Areas Protegidas and Wildlife Conservation Society, 2015). Multiple armed divisions of the Guatemalan state and military staff these military detachments, *Centros de Operación Conjunto* (Centers of Joint Operation) (COCs). These checkpoints exist at Maya Biosphere boundaries, in the structural bones of abandoned civil war outposts, and in the remains of villages whose residents the army evicted from the biosphere for breaking conservation law.

On the road to the village of Carmelita, the gateway to Mirador for hikers, tourists pass three COCs. The second one is the most memorable for most travelers; it marks the boundary and beginning of the multiple-use zone of the Maya Biosphere where residency and land use are restricted. Multiple forces of the Guatemalan military staff the site called the San Miguel COC, and every car that approaches the COC is stopped by a 40-foot-long metal bar lifted by a simple rope and pulley system. Stern men with accusatory glances approach the cars and ask drivers and sometimes passengers for their identification. At his discretion, an officer quizzes the travelers regarding their intentions and asks them what belongings they bring into the Maya Biosphere. After personal identification and possessions are inspected and dutifully noted, most tourists continue their trip unaware of how these practices of militarization are redefining the state's presence in the reserve and reviving living civil war legacies of counter-insurgency, like war time surveillance. Other COCs, like the *Nueva Esperanza* and *La Colorada* COCs, are located at the sites of former villages destroyed by state forces. The first was a community of landless Q'eqch'i peasants living in the reserve without land title, and the later was a community with legal residency who purportedly sold their lands to individuals linked to organized crime and thus broke conservation law. In both cases, the Green Battalion and other armed braches of the Guatemalan state evicted these residents for breaking conservation law and the state has built military detachments where these communities once stood. The complex dynamics underpinning insecurity in the Maya Biosphere are outside the scope of this article, and I do not intend to dismiss the contributions these armed forces make. They put their lives on the line to fight organized crime and drug trafficking organizations, and are often outgunned and outmanned. Nevertheless, these actions illustrate that the state is engaging in reserve evictions, surveillance and amplifying territorial control in ways that are reminiscent of the civil war era in the name of conservation and security (Ybarra, 2012), and eco-tourism development.

The dual practices of capital's spatial colonization and the production of state space have long histories in northern Petén. Today, however, eco-tourism development serves as the means and the end to re-establishing the state's presence, power and space at the frontiers of "ingovernability". Eco-tourism development of Mirador and the Maya Biosphere exemplifies tourism's twin dynamics of spatial colonization: the expansion of the capitalist frontier and the production of state space. By working to incorporate the "untapped" and "natural" resources of the Maya Biosphere in the global tourism economy, the Guatemalan state attempts to bring this seemingly ungovernable space into regulatory institutions and structures of contemporary green governance.

When Alvaro Colom left office in 2012, so did the Four Jaguar project. Despite the project's end and its limited success to increase tourism and security in the Maya Biosphere, militarization and efforts to "recover governability" in the reserve continue. While the potential for tourism development helps justify the resources spent to secure a tourism space, related practices of militarization often take on their own life once installed in the landscape as they articulate with other socio-spatial dynamics and living histories of civil war violence. Guatemala's recent history of genocidal violence and contemporary insecurity render these dynamics of tourism-enabled militarization poignantly acute and visible. However, these practices of tourism's spatial colonization are interconnected with cold war counter-insurgency operations and neoliberal development projects throughout the world, and manifest a variety of geographically specific ways, yet in familiar forms (see Gonzales, 2013; Lombard, 2012; Ojeda, 2013; Stein, 2008).

Commodification of place: Mirador in the "Heart of the Mayan World"

Tourism's violent geographies of spatial colonization parallel symbolic and epistemological forms of violence characterizing practices of commodification. Questions of commodification in tourism studies often analytically privilege questions of culture and the ways commodification shapes authenticity. Seminal analysis in this line of inquiry explores how tourism products and practices can be incorporated into cultures as dynamic and evolving entities over time (Cohen 1988), and how "hosts" engage in staging cultural authenticity to meet tourists' exoticized expectations (MacCannell, 1976). In an ethnically diverse Belizean community not far from the Maya Biosphere, Medina (2003) argues that residents living nearby an archaeological site began to revalorize Maya identity and culture following tourism growth, and use new channels, like anthropologist and epigraphers' publications, to access and revive Mayan traditions that have fallen out of quotidian usage. Village census data suggest this process has shaped villagers identities, for example, many villagers have changed their identity from *mestizo* (shared Spanish and Indigenous heritage) to Indigenous. In other words, commodification of culture in tourism may shape people's processes of identification as Indigenous subjects, as individuals and collectivities redefine their identities and cultures in ways that reflect their changing connections to people and places elsewhere (Massey, 1994).

Writing about the "Mayan World" initiative launched in 1992, Brown (1999) uses a geographical lens to argue that the tourism development of the Mayan World commodifies Maya landscapes in addition to Maya culture (see also Thomlinson & Getz, 1996). Brown argues that Maya working in tourism are in danger of recreating their cultural landscapes for exchange-value rather than use-value. Building on Brown's distinction between exchange-value and use-value in tourism landscapes, I want to emphasize how tourism's practices of commodification subordinate multiple dimensions of socio-spatial life to capitalist relations of accumulation (Greenwood, 1977). Commodification is related to objectification, which describes the ways social relations of labor are disguised as a thing, a commodity, with a monetary value that hides the social relations of labor exploitation that produced the product. Marx called this commodity fetishism, which he described as "a definite social relation between men…[that] assumes in their eyes, the fantastic form of a relation between things" (1978 [1887], pp. 301–302). Marx's analysis (1978 [1887]) and feminist theories (Bartky, 1990; MacKinnon, 1993) of objectification also explain how commodification of sexual or other interpersonal relations turn the seller of those services into an object to be bought and consumed and render invisible or irrelevant their own feelings, needs, desires and fears.

All practices of capitalist accumulation entail practices of commodification, objectification and commodity fetishism. All capitalist industries exploit and reproduce notions of race and gender and other forms of difference in relations of production (Massey, 1994; Wright, 2006). In that sense, tourism is similar to other service industries that hire women and people of color and pay them poorly, defining their work as unskilled precisely because it is socially defined as the work of women and people of color (Ehrenreich & Hochschild, 2003; Wright, 2006). At the same time, non-traditional forms of tourism are distinct from other industries in this regard. A consumer buying a t-shirt made in a sweatshop may be unaware of the exploitative and power-laden relations of production embedded in its fabric and swiftly sewn stitches. These relations are hidden and objectified in the t-shirt itself and by the temporal and physical distance between the production of the commodity and its final sale.

In contrast, when a tourist travels to the so-called Mayan World in search of an "authentic" cultural experience, the identity and history of a place and its hosts are the products that are bought, sold and consumed. I broaden my focus from culture to place to emphasize that the identity and history of places, not just people's cultures, are commodified in tourism development. Places are nodes in interconnected webs of extra local socio-spatial relations that articulate in a particular locality, and define its history, culture, racial diversity and physical form (Massey, 1994). Place is not a Cartesian point on a grid, but the geographic intersections and social relations and their consolidation in the landscape. In everyday eco-tourism development, people's homes, sacred spaces and livelihoods are commodified through acts of land dispossession, enclosure and privatization that occur in capitalist industries. While all industries exploit capital's geographical unevenness (Smith, 2008), tourism, unlike many industries, sells experiences as final consumable products (Duffy, 2015). This creates another dimension of the commodification of place in tourism representations and performances that delves into cultural landscapes and intangible heritage. David Harvey (2006) suggests that the commodification of intangible aspects of the social – history, heritage and culture – is a defining characteristic of neoliberal practices of ongoing accumulation by dispossession. He writes, "The commodification (through tourism) of cultural forms, histories, and intellectual creativity entails wholesale dispossessions" (pp. 35–36). While Harvey emphasizes the materiality of these practices, Diana Ojeda (under review) argues that dispossession is not a one-time event, does not even have to be an event at all and does not even require a "possession." Accumulation by dispossession in non-traditional tourism practices can occur when the state or other outside actors commodify Maya culture and landscapes, and thus, strip "hosts" or members of that group from the means of reproducing their multiple identities, landscapes and histories that define places-turned-destinations.

In Guatemala, one form this takes is when non-Indigenous tourism workers dress in Indigenous clothes called *traje* for cultural performances. In the tourism mecca of Panajachel on the majestic Lake Atitlán in the Guatemalan Highlands, for example, there is a four-star hotel and restaurant that employs *ladinas* (non-Indigenous women) to wear traditional Mayan *traje*. The *traje* is deeply infused with historical, geographical and cultural meaning, and it is often adorned with stitched images of animals, food, landmarks and other elements of Mayan cosmology (Hendrickson, 1995). Thirty years ago, women were killed for being identified as Indigenous, an identity marked, in part, by wearing the unmistakable *traje*. Today, non-Indigenous, private-sector tourism entrepreneurs cater to the desires of international tourists to experience Indigenous culture in Guatemala by having non-Indigenous women wear *traje*.

When tourists pay to consume cultural performances by non-Indigenous people wearing *traje* or buy *traje* as a souvenir, these items acquire an exchange value distinct from their "use-value" to the Maya. The "use-value" of *traje*, that is, its non-capitalist use in commodity form, is central to the reproduction of Maya culture, identity and cosmology (Hendrickson, 1995). Commodifying *traje* in tourism separates the Maya from the exclusive right to reproduce their culture through *traje*'s creation, exchange, performance and lived experience. As a commodity, an object, the cultural relations of *traje* production and use are rendered invisible. The commodification of culture in tourism enables a whole host of new meanings, practices and expressions to become associated with *traje* based on its commercial exchange in tourism settings that would not be possible otherwise.

In this and countless other examples, cultural tourism and eco-tourism development commodify spatial identity and history through processes of objectification that separate people living in or near tourism destinations from the ability to reproduce meanings and practices of place. Through accumulation by dispossession, "hosts" are divorced from the means of reproducing the identity, history and culture of the place-turned-destination. This practice of accumulation by dispossession reflects how tourism not only physically pushes the frontier of capitalist expansion into so-called green spaces (Benjaminsen & Bryceson, 2012; Büscher, 2009), but also into the social realm of the intangible – history, identity and culture (Girardo, 2011). Rather than questioning whether *ladinos* wearing *traje* is culturally authentic, "authenticity" might be rethought in terms of who defines, represents and performs subaltern cultures in societies defined by racism and other forms of domination.

What is being sold in tourism is not a diversity of identities and histories that define people and places, but an idealized abstraction that materializes in tourism-built environments, performances and marketing campaigns. The objectification of place in tourism produces an abstract spatial conceptualization of "the local" or "community" so easily identified in tourism narratives and practices. In this understanding of place defining tourism imaginaries, a rigid line is drawn around a tourism destination, like Mirador or the Maya Biosphere, which divorces a place from its interconnections with other sites defining and producing it. The history, experience, identity and culture sold in tourism promotions and tours are not reflective of the multiple and often times competing narratives continually redefining and reconstituting tourism destinations. Rather, they are singular, essentialized and typically folkloric tourism imaginaries that reveal the desires and ideologies of their creators (Salazar, 2012).

Commodifying Mirador: ancient space, natural place

Noel Salazar (2012) defines tourism imaginaries as socially constructed and transmitted representations of tourism destinations that interact with people's individual imaginations and everyday life experiences to become meaning-making and world-shaping devices (p. 864). Tourism imaginaries frame tourism destinations like Mirador and the Maya Biosphere as homogenous places, whose internal relations, living histories and global connections are invisible. In doing so, Mirador tourism development initiatives rewrite regional history through the erasure of contemporary subjects from the landscape. This practice as historical–geographical erasure works to clear the slate for national and global elites to articulate new place-based histories, land claims and notions of heritage out of malleable tourism landscapes in the making. Erasure here does not signify creating a blank slate, but the erasure of the complexity of place.

Mirador and Maya Biosphere Reserve tourism imaginaries are grounded in, and mutually reinforce, broader regional narratives about Guatemala as the "Heart of the Mayan World." Widely viewed TV programs, Hollywood celebrities, scientific publications, tourism marketing materials and public relation campaigns discursively construct the Mirador Basin and northern Petén as pristine wilderness home to ancient Mayan civilizations. Despite the vast diversity of actors involved in Mirador tourism development, their efforts coalesce around a shared tourism imaginary of Mirador and the surrounding region as an ancient space and natural place. The dominant tourism imaginary that Mirador ecotourism enthusiasts espouse paints a picture of a lost ancient city shrouded in pristine, yet endangered forest. For example, in the 2009 CNN Special "Mirador the Lost City", the reporter opens the program by setting the scene with an air of mystery and discovery: "From the air it looks like just jungle, but these forests in Guatemala hide an ancient secret: the city of Mirador often referred to as the Cradle of the Mayan Civilization." Similarly, the 2004 National Geographic Program "Dawn of the Maya" showcasing Mirador eerily commences the popular show by relating, "Deep in the jungles of Central America lie the ruins of a lost civilization ...".

A non-profit organization called the Foundation for Mayan Cultural and Natural Heritage (PACUNAM) promoting Mirador development describes the region as home to the "largest and earliest cities in the Western Hemisphere" found in the "last tract of virgin tropical rainforest left in Central America" (PACUNAM, n.d. a). Furthermore, the Global Heritage Fund's executive summary of the Mirador project concludes in bold blue oversized letters: "Described as the Cradle of the Maya Civilization, the Mirador Basin is the last tract of virgin rainforest remaining in Central America" (2011). Similarly, PACUNAM's new "Reino Kan" or "Snake Kingdom" marketing campaign also paints the picture of a lost city, awaiting (tourist) discovery. An invitation to a high-cache fundraiser in Los Angeles entices, "For 2000 years, the majesty and the mysteries of the Snake Kingdom have been protected by the forest...NOW YOU ARE INVITED TO REVEAL IT" (PACUNAM, n.d. b).

This "ancient space, natural place" tourism imaginary paints a false picture of remoteness, both spatially and temporally, and reproduces western notions of nature and wilderness as devoid of human occupation (Cronon, 1996). The region's settlement by *chicle* (chewing gum resin) workers and traders over a 100 years ago working for the US-based Wrigley's Chewing Gum Company is just one of the latest chapters of this forest's human history. Many descendants of these settlers today continue make a living as community forest concessionaires from sustainably harvesting forest products certified by the Forest Stewardship Council, like timber, a leafy palm called *xate* used in flower bouquets, and chicle (Schwartz, 1990). This living history of globalization at the turn of the twentieth century is not communicated in dominant tourism imaginaries defining Mirador as a place, but rendered invisible by the "ancient space, natural place" narrative. At the same time, a map or postcard speaks a thousand words, thousands more are silenced as geographies and places are left off the map. This dynamic could not be better encapsulated than in PACUNAM's conservation strategy: "Rewriting history to preserve the future" (PACUNAM, n.d. c).

Understanding tourism's practices of commodifying place is critical because the resulting tourism imaginary often ideologically underpins and justifies brutal acts of violence that include land and livelihood dispossession (Duffy, 2013; Ojeda, 2012) and sexual exploitation and assault (Enloe, 2000; Gregory, 2006). In the Maya Biosphere, the representation of Mirador and its surrounding forests as an "ancient space, natural place" in tourism imaginaries dovetails with legislative action to curtail practices of reserve resident community forestry neighboring the Mirador site. In 2002, following years of political activism, the archaeological team at Mirador successfully lobbied President Portillo to declare an area they identify as the Mirador Basin a special archaeological zone. Decree 129 superseded the boundaries and land uses of the Maya Biosphere, which included the rights of two forest concessions, Carmelita and Uaxactún, to sustainably manage the forest located with the newly legislated park by limiting land use within the Mirador Basin to scientific and tourism purposes, nullifying 66% of Carmelita's concession and 10% of Uaxactún's. The creation of the Mirador Basin archaeological zone legislated the land uses depicted in Mirador and Maya Biosphere tourism imaginaries. The lead archaeologist at Mirador opposes any type of forestry, however sustainable, in the region because he believes the forests' trees – through tourism, carbon credits and other conservation initiatives – are worth more standing than cut and sold. This argument fails to recognize the crucial role community forestry plays in preventing and fighting wildfires, denouncing poaching and trafficking, and providing jobs and economic earnings for thousands of people. Furthermore, the national parks in the Maya Biosphere, particularly the two largest parks in the west, have suffered much higher deforestation rates than areas managed by communities (Hodgdon, Hughell, Hugo Ramos, & McNab, 2015), suggesting the Guatemalan state is unable or unwilling to do the conservation work that communities are performing.

The battle over the overlapping territory of the community forest concessions and the Mirador Basin created by Decree 129 lasted over three years and reached Guatemala's Constitutional Court, the highest court in the land. In 2005, community foresters celebrated a land rights victory when Decree 129 was overturned, thus, restoring the land rights of Carmelita and Uaxactún's forest concessionaires and protecting the integrity of community forestry as a process. I write about this land struggle in depth elsewhere (Devine, in press), but marshal this example here to illustrate the connections between tourism imaginaries and material struggles for land rights and livelihoods.

Tourism is not the only industry where violent dynamics of spatial colonization and commodification unfold. Nonetheless, tourism remains outstanding in this regard because of the ways neoliberal reforms combined with the industry's spatial dynamics of colonization and the commodification of place to dramatically fuel immense tourism growth, and make tourism an often-violent terrain of struggle over territory identity and history. In contemporary conditions of late neoliberal capitalism, the growth of non-traditional forms of tourism reflects new processes of accumulation by dispossession enabled by the commodification of nature, culture and place in paradise. This takes the form of physical violence, like the eviction of Maya Biosphere residents from the reserve, or the enclosure and privatization of beaches and other resources for exclusive tourist use. It also occurs, however, on the intangible plane, at the level of phenomenological experiences such as the tour, the dance lesson, the meal and the show.

The power relations and political outcomes of tourism projects, however, are far from predetermined. Rather, tourism's violent spatial practices of colonization and commodification articulate with pre-existing historical geographies defining places around the world to politicize a whole host of issues that are at once familiar and at the same time different. They are familiar because many of the territorial and identity politics engendered by tourism development trigger similar concerns, but they are also unique because these concerns are embedded in the geographic and historical specificity of place.

Despite the industry's inherent spatial dynamics of colonization and the commodification of place, tourism is not necessarily a site of domination and oppression. In Guatemala and across the world, there are grassroots initiatives using tourism as a political platform to pursue socio-spatial justice, as well as find alternatives to neoliberal development and solutions to environmental and social crises. One such initiative is the New Horizon Cooperative in Santa Ana, Petén, Guatemala, founded by ex-guerilla combatants in 1998. "Solidarity tourism" is one of the four principal economic activities managed by the cooperative's 415 residents living on 900 hectares of collectively owned land. The other three collective projects include farm fishing tilapia, cattle ranching and reforesting 190 acres of the cooperative. New Horizon members chose the term solidarity tourism to distinguish their tourism approach from elite-owned and state-driven forms of cultural tourism sold in the "Mayan World". Solidarity tourism provides New Horizon members an opportunity to share their vision for Guatemalan society and their alternative sustainable development model with the world (Devine, 2016). Visitors tour the cooperative's collective projects, as well as visit "semi-collective" projects, like chicken-raising and food cultivation, owned and managed by smaller groups of residents. Solidarity tourists also receive lectures from ex-guerillas on Guatemalan politics and history, visit the cooperative's museum telling their "other history" of the civil war and spend a night in the forest learning to establish a guerilla camp. Despite the challenges facing the New Horizon Cooperative, nearly 20 years after its founding, the success of this social movement is evident in the quality of the resident's housing, health care and education (Devine, 2016). In contrast to the rest of the country, New Horizon residents do not have to pay tuition to attend high school, medical consultations are free, free day care is available for residents and members are eligible for low interest loans averaging 3% compared to 20% or higher in commercial banks.

In Botswana, South Africa and Namibia, Rachel Giraudo (2014) also finds important differences between cultural tourism projects that are run by the Indigenous ethnic group of the San rather than by (mostly white) outsiders. She argues that non-San tourism operators keep most of the profit and often sell a fantasy about the San as primitive, docile and distinct from other Africans, while they exploit the San as workers. When the San own and manage the tourism project, they still cater to tourists' romanticized expectations, but they have more control over what they choose to share with visitors about their heritage, and many use tourism to educate visitors about their history and hopes for the future in ways that counter prevailing stereotypes (Giraudo, 2014). Giraudo concludes "it is vitally important that communities have control in the decision-making process of creating a cultural tourism product" (2014, p. 6). Like New Horizon ex-guerillas, "these tourism encounters provide the San with a voice to the global community" (Giraudo, 2014, p. 6).

Part of New Horizon's mission through solidarity tourism is to share their successful sustainable development model with other organized communities in Guatemala and across the world. No longer fighting to overthrow the system through revolutionary warfare, New Horizon members are working within the system for transformative change on their own terms. Solidarity tourism has become a means through which the former guerillas continue to fight for their revolutionary ideals of collective land ownership, accountable and democratic governance, and more equitable access to health care and education in non-violent ways in a country where post-war "peace" has failed them.

Conclusion

Neoliberal reforms and development projects set the stage for the Latin American tourism boom since the 1980s, yet the industry's dramatic growth cannot be explained by the turn to neoliberalism alone. These reforms combined with tourism's dual practices of spatial colonization and the commodification of place to drive growth and make the industry a pervasive site of contentious politics over land and identity. Eco-tourism development physically colonizes spaces perceived as "green" or "natural" by extending the neoliberal frontier of capitalist relations into geographies outside of, or partially integrated into, global circuits of capital accumulation. The example of a state-led eco-tourism development in Guatemala's Maya Biosphere illustrates that capital's colonization of nature simultaneously functions to produce state territory in sovereignty's frontier.

Alongside enabling practices of spatial colonization by capital and state forces, non-traditional tourist industries also commodify place – its identity, history, culture and landscapes – as consumer products. As contemporary capitalism has persistently pursued new spaces and sites of commodification, eco-tourism and other forms of non-traditional tourism development have served as mechanisms extending relations of accumulation into nature, culture and into intangible social dimensions of place. These products take the physical form of private property, post cards and souvenirs, but they also take the form of commodified social relations and experiences, such as tours and cultural performances. The commodification of place and culture in tourism extends practices of accumulation by dispossession into the non-tangible realm of socio-spatial relations, and in doing so, divorces "hosts" from exclusively controlling the means of reproducing their identities, histories and cultures. This takes the form of the commodification of culture, for example, when Maya traditional clothing called *traje* is worn, made and sold by non-Maya in Guatemala, but it also takes the form of the commodification of place. Tourism imaginaries surrounding the Mirador archaeology site in the Maya Biosphere commodify and sell the site as an "ancient place" and "natural space" in ways that render invisible the lives and livelihoods of nearby residents. These imaginaries have material consequences. Mirador tourism development advocates successfully campaigned to bring their tourism vision to fruition in 2002 when the Guatemalan president restricted land use around the site to tourism and research only. This presidential decree dispossessed nearby residents of their land rights as community forest concessionaires, and was overturned after three years of popular and legal struggle.

While practices of spatial colonization and the commodification of place are inherent to state-led eco-tourism development efforts, tourism does not necessarily reproduce relations of domination and exploitation. The examples of New Horizon ex-guerilla solidarity tourism in Guatemala and San-owned cultural tourism in southern Africa suggest that a key dynamic defining tourism's power relations and political possibilities lies in who own the means of production, in other words, who decides which social relations defining culture and place are commodified and how. As these and other examples from around the world suggest, grassroots forms of tourism development are far from perfect, face immense power inequalities within them, and are still subject to the vulgarities of capitalism and its commodification of culture and place; the search for "the Just Destination" continues (Jamal & Camargo, 2014). Nonetheless, these examples of civil society organizations harnessing tourism's territorial and identity politics to forge alternative development paths and models suggest tourism can be sustainable under certain conditions, albeit not in the ways the architects of the "Mayan World" had imagined.

Notes

1. From 1995 to 2007, Latin American countries experienced a 49% increase in tourist arrivals, and a 61% growth in tourism receipts (expenditures by visitors) (WTTC, 2015a). In the last decade, tourism receipts increased again by 50% (IADB). In 2014, tourism contributed $US 417.4 billion to Latin America's GDP, making up 9.2% of the region's economy (WTTC, 2015a). In Guatemala, the figure is slightly lower, at 8.8% of total contribution to GDP (WTTC, 2015b).
2. The Four Jaguar project was called *Cuatro Balam* in Guatemala, which is a linguistic mix of Spanish and Mayan languages. The name draws on the project's focus on Maya archaeology, the Maya creation story of the universe, the *Popul Vuh,* and the four cardinal directions. Jaguars are endangered and endemic to the reserve, and are a symbol for many conservation efforts.
3. Thanks to John Stehlin for this clarification (personal communication, 14 June 2016).

Acknowledgments

A special thanks goes to Bernard Lane for his support and mentorship at the *Journal of Sustainable Tourism*. Thank you as well to Diana Ojeda, Kimberley Kinder, John Stehlin and three anonymous reviewers for your constructive feedback. All errors in fact or interpretation are mine.

Disclosure statement

No potential conflict of interest was reported by the author.

Funding

The Inter-American Foundation Grassroots Development Fellowship and the John L. Simpson Memorial Research Fellowship in International and Comparative Studies at the University of California, Berkeley, funded this research.

References

Arrighi, G. (2006). Spatial fixes, switching crises, and accumulation by dispossession. In C. Chase-Dunn (Ed.), *Global social change: Historical and comparative perspectives* (pp. 201–212). Baltimore, MD: John Hopkins University Press.

Bartky, S.-L. 1990, *Femininity and domination: Studies in the phenomenology of oppression*. New York, NY: Routledge.

Benjaminsen, T.A., & Bryceson, I. (2012). Conservation, green/blue grabbing and accumulation by dispossession in Tanzania. *Journal of Peasant Studies*, *39*(2), 335–355.

Bocarejo, D., & Ojeda, D. (2016). Violence and conservation: Beyond unintended consequences and unfortunate coincidences. *Geoforum*, *69*, 176–183.

Brenner, N. & Elden, S. (2009). Henri Lefebvre on state, space, territory. *International Political Sociology*, *3*(4), 353–377.

Britton, S. (1991). Tourism, capital, and place: Towards a critical geography of tourism. *Environment and planning D: Society and space*, *9*(4), 451–478.

Brown, D.F. (1999). Mayas and tourists in the Mayan World. *Human Organization*, *58*(3), 295–304.

Büscher, B. (2009). Letters of gold: Enabling primitive accumulation through neoliberal conservation. *Human Geography*, *2*(3), 91–93.

Büscher, B., Sullivan, S., Neves, K., Igoe, J., & Brockington, D. (2012). Towards a synthesized critique of neoliberal biodiversity conservation. *Capitalism Nature Socialism*, *23*(2), 4–30.

Centeno Cordón C. (1973). *Cooperatives de El Petén: Situación socioeconómicas* [Cooperatives in Petén: Socioeconomic situation]. Guatemala City: Editorial Universitaria, Universidad de San Carlos de Guatemala.

Cohen, E. (1988). Authenticity and commoditization in tourism. *Annals of Tourism Research*, *15*(3), 371–386.

Consejo Nacional de Areas Protegidas [National Council on Protected Areas] and Wildlife Conservation Society. (2015). *Monitoreo de la Gobernabilidad en la Reserva de la Biósfera Maya* [Monitoring governability in the Maya Biosphere Reserve]. San Benito: Government of Guatemala.

Craik, J. (1994). Peripheral pleasures: The peculiarities of post-colonial tourism. *Culture and Policy*, *6*(1), 21–31.

Cronon, W. (1996). The trouble with wilderness: Or, getting back to the wrong nature. *Environmental History*, *1*, 7–28.

Devine, J. (2014). Counterinsurgency ecotourism in Guatemala's Maya Biosphere Reserve. *Environment and Planning D: Society and Space*, *32*(6), 984–1001.

Devine, J. (2016). Politics of post-war tourism in Guatemala: Contested identities, histories, and futures. *L'Espace Politique* [Online], *28*. doi: 10.4000/espacepolitique.3723

Devine, J. (in press). Community forest concessionaires: Resisting green grabs and producing political subjects in Guatemala. *Journal of Peasant Studies*. doi: 10.1080/03066150.2016.1215305

Duffy, R. (2008). Neoliberalising nature: Global networks and ecotourism development in Madagascar. *Journal of Sustainable Tourism*, *16*(3), 327–344.

Duffy, R. (2013). *A trip too far: Ecotourism, politics, and exploitation*. New York, NY: Earthscan.

Duffy, R. (2014). Waging a war to save biodiversity: The rise of militarized conservation. *International Affairs, 90*(4), 819–834.
Duffy, R. (2016). War by conservation. *Geoforum, 69*, 238–248.
Duffy, R., & Moore, L. (2010). Neoliberalising nature? Elephant—back tourism in Thailand and Botswana. *Antipode, 42*(3), 742–766.
Ehrenreich, B., & Hochschild, A.R. (Eds.). (2003). *Global woman: Nannies, maids, and sex workers in the new economy*. New York, NY: Metropolitan Books.
Enloe, C.H. (2000). *Bananas, beaches and bases: Making feminist sense of international politics*. Berkeley: University of California Press.
Fennell, D.A. (2007). *Ecotourism*. New York, NY: Routledge.
Finger-Stich, A. (2003). Community concessions and certification in the Maya Biosphere Reserve. In E. Meidinger, C. Elliott, & G. Oesen (Eds.), *Social and political dimensions of forest certification*. Buffalo Legal Studies Research Paper Series, Paper No. 2015-007 (pp. 165–178). Buffalo: University at Buffalo.
Fletcher, R., (2011). Sustaining tourism, sustaining capitalism? The tourism industry's role in global capitalist expansion. *Tourism Geographies, 13*(3), 443–461.
Fletcher, R., & Neves, K. (2012). Contradictions in tourism: The promise and pitfalls of ecotourism as a manifold capitalist fix. *Environment and Society: Advances in Research, 3*(1), 60–77.
Foundation for Archaeological Research and Environmental Studies. (n.d.). Mirador Basin Project, the Cradle of Maya Civilization. Retrieved 11 August 2016 from http://www.miradorbasin.com/about/fares.php.
Garrett, W.E. (1989). La Ruta Maya. *National Geographic, 176*(4), 424–479.
GDT Tourism Consultants. (2010) *Plan de desarrollo del producto turístico para el cluster norte del area Cuatro Balam, Petén* [Development plan for tourism products in the northern cluster of the Four Jaguar area, Petén], Guatemala City: BID/FOMIN.
Giraudo, R.F. (2011). *Intangible heritage and tourism development at the Tsodilo World Heritage Site* (Unpublished doctoral dissertation). Berkeley: University of California.
Giraudo, R.F. (2014). San cultural tourism in Southern Africa. *Intellectual Property Issues in Cultural Heritage (IPinCH), 6*, 5–8.
Global Heritage Fund. (2011). Global Heritage Fund's Mirador project profiled in Smithsonian magazine cover story. Retrieved 11 August 2016 from http://globalheritagefund.org/index.php/news/global-heritage-funds-mirador-project-profiled-in-smithsonian-magazine-cover-story.
Gonzalez, V. (2013). *Securing paradise: Tourism and militarism in Hawai'i and the Philippines*. Durhan, NC: Duke University Press.
Grandia, L. (2012). *Enclosed: Conservation, cattle, and commerce among the Q'eqchi' Maya lowlanders*. Seattle: University of Washington Press.
Greenwood, D.J. (1977). Culture by the pound: An anthropological perspective on tourism as cultural commoditization. In V.L. Smith (Ed.), *Hosts and guests* (pp. 129–139). Philadelphia: University of Pennsylvania Press.
Gregory, S. (2006). *The devil behind the mirror: Globalization and politics in the Dominican Republic*. Berkeley: University of California Press.
Hansen, R. (2012). Kingship in the Cradle of Maya Civilization: The Mirador Basin. In M.A. Boxt, & B.D. Dillon (Eds.), *Fanning the sacred flame: Mesoamerican studies in honor of HB Nicholson* (pp. 139–172). Boulder: University Press of Colorado.
Harvey, D. (2001). Globalization and the spatial fix. *Geographische Revue, 2*(3), 23–31.
Harvey, D. (2003). *The new imperialism*. Oxford: Oxford University Press.
Harvey, D. (2006). Neo-liberalism as creative destruction. *Geografiska Annaler. Series B, Human Geography, 2*, 145–158.
Hendrickson, C.E. (1995). *Weaving identities: Construction of dress and self in a highland Guatemala town*. Austin: University of Texas Press.
Hodgdon, B.D., Hughell, D., Hugo Ramos, V., & McNab, R.B. (2015). 2000–2013 deforestation trends in the Maya Biosphere Reserve, Guatemala. Petén, Guatemala: Rainforest Alliance. Retrieved 11 August 2016 from http://www.rainforest-alliance.org/sites/default/files/publication/pdf/MBR-Deforestation_150213-2.pdf
Honey M, (2008). *Ecotourism and sustainable development: Who owns paradise?* Washington DC: Island Press.
Inter-American Development Bank. (2006). IDB approves $39 million loan for sustainable development program for the conservation of the Maya Biosphere Reserve. Retrieved 11 August 2016 from http://www.iadb.org/en/news/news-releases/2006-12-14/idb-approves-30-million-loan-for-sustainable-development-program-for-conservation-of-guatemalas-Maya-biosphere-reserve,3534.html
Jamal, T. & Camargo, B.A. (2014). Sustainable tourism, justice and an ethic of care: Toward the Just Destination. *Journal of Sustainable Tourism, 22*(1), 11–30.
Jonas, S. (2000). *Of centaurs and doves: Guatemala's peace process*. Boulder, CO: Westview Press.
Kincaid, J. (1988). *A small place*. New York, NY: Farrar, Straus and Giroux.
Lefebvre, H. (1976). *The survival of capitalism: Reproduction of the relations of production*. London: Allison & Busby.
Lefebvre, H. (1991). *The production of space* (D. Nicholson Smith, Trans.). Oxford: Blackwell
Lefebvre, H. (2003). The urban revolution. Minneapolis: University of Minnesota Press.

Lombard, L.N. (2012). Raiding sovereignty in Central African borderlands (Unpublished doctoral dissertation). Durham, NC: Duke University.
Lundstrom, E. (2014). Green militarization: Anti-poaching efforts and the spatial contours of Kruger National Park. *Annals of the Association of American Geographers, 104*(4), 816–832.
MacCannell, D. (1976). *The tourist: A new theory of the leisure class.* Berkeley: University of California Press.
MacKinnon, C.A. (1993). *Only words.* Cambridge, MA: Harvard University Press.
Marx, K. (1978 [1887]) Das Capital (S. Moore & E. Aveling, Trans.) In R.C. Tucker (Ed.), *The Marx-Engels Reader* (Vol. 1, 2nd ed.) New York, NY: W.W. Norton & Company.
Massey, D. (1994). *Space, place, and gender.* Minneapolis: University of Minnesota Press.
McAfee, K. (1999). Selling nature to save it? Biodiversity and green developmentalism. *Environment and planning D: Society and space, 17*(2), 133–154.
Medina, L. (2003). Commoditizing culture: Tourism and Maya Identity. *Annals of Tourism Research, 30*(2), 353–368.
Mohan, G., & Stokke, K. (2000). Participatory development and empowerment: The dangers of localism. *Third World Quarterly, 21*(2), 247–268.
Molina, C. (2010). Batallon Verde en la Zona Petenera: Ejército luchará contra narcos [Green Brigade in the Petén Zone: The army will fight against narcos]. Nuestro Diario. Retrieved 15 January 2011 from http://digital.nuestrodiario.com/Olive/ODE/NuestroDiario/Default.aspx?href=GND%2F2010%2F11%2F06&pageno=6&view=document
Mowforth, M, & Munt, I. (2009). *Tourism and sustainability: Development, globalisation and new tourism in the third world.* New York, NY: Routledge.
Ojeda, D. (2012). Green pretexts: Ecotourism, neoliberal conservation and land grabbing in Tayrona National Natural Park, Colombia. *Journal of Peasant Studies, 39*(2), 357–375
Ojeda, D. (2013). War and tourism: The banal geographies of security in Colombia's "Retaking". *Geopolitics, 18*(4), 759–778.
Ojeda, D. (under review). *Los paisajes del despojo: Propuestas para un análisis desde las reconfiguraciones socioespaciales* [Landscapes of dispossession: A suggested analysis from the implicated sociospatial reconfigurations]. *Revista Colombiana de Antropología* [Colombian Journal of Anthropology].
PACUNAM. (n.d. a), Fundación Patrimonio Cultural y Natural Maya. ¡Bienvenidos a Pacunam! *PACUNAM* [Welcome to Pacunam!]. Retrieved 19 December 2013 from http://pacunam.org/
PACUNAM. (n.d. b). Pacunam and Prince Albert II of Monaco Foundation – USA *Reino Kan* Fundraiser. [Invitation]. Guatemala: Pacunam.
PACUNAM. (n.d. c). *Mirador* [Brochure]. Guatemala: Pacunam.
Portes, A. (1997). Neoliberalism and the sociology of development: Emerging trends and unanticipated facts. *Population and Development Review, 23*(2), 229–259.
Prensa Libre. (2010). *Batallón Verde inicia operaciones para proteger Biósfera Maya* [Green Battalion initiates operations to protect the Maya Biosphere]. Retrieved 11 August 2016 from http://www.prensalibre.com/tema/batallon_verde_operaciones_proteccion_biosfera_maya.html
Guatemala, Never Again!. (1999). *Rocovery of historical memory project (REMHI), The official report of the human rights office, Archdiocese of Guatemala.* New York, NY: Orbis Books.
Rocheleau, D.E. (2015). Networked, rooted and territorial: green grabbing and resistance in Chiapas. *Journal of Peasant Studies, 42*(3–4), 695–723.
Salazar, N.B. (2012). Tourism imaginaries: A conceptual approach. *Annals of Tourism Research, 39*(2), 863–882.
Scheller, M.I. (2003). *Consuming the Caribbean.* London: Routledge.
Schwartz, N. (1990). *Forest society: A social history of Petén, Guatemala.* Philadelphia: University of Pennsylvania Press.
Smith, N. (2008). *Uneven development: Nature, capital, and the production of space.* Athens: University of Georgia Press.
Soza, J.M. (1970). Monografía del Departamento de El Petén [Monograph of the Department of Petén] (Vol. 1). Guatemala City: Ministerio de Educación.
Stein, R. (2008). *Itineraries in conflict: Israelis, Palestinians, and the political lives of tourism.* Durham, NC: Duke University Press.
Thomlinson, E., & Getz, D. (1996). The question of scale in ecotourism: Case study of two small Ecotour operators in the Mundo Maya Region of Central America. *Journal of Sustainable Tourism, 4*(4), 183–200.
Vich, V. (2006). La nación en venta: Bricheros, turismo y mercado en el Perú contemporáneo [The nation for sell: "Bricheros," tourism and the market in contemporary Peru]. In A. Ypeij & A. Zoomers (Eds.), *La Ruta Andina: Turismo y Desarollo Sostenible en Perú y Bolivia* (pp. 187–198). Quito: Abya Yala.
Werry, M. (2011). *The tourist state: Performing leisure, liberalism, and race in New Zealand.* Minneapolis: University of Minnesota Press.
World Travel and Tourism Council. (2015a). *Travel & tourism economic impact 2015: Latin America.* London: Author. Retrieved from file:///Users/Devine/Desktop/LatinAmerica2015.pdf
World Travel and Tourism Council. (2015b). *Travel and Tourism Economic Impact 2015: Guatemala.* London: Author. Retrieved 11 August 2016, from file:///Users/jendevine/Desktop/Guatemala2015_Economic%20Impact.pdf.
Wright, M.W. (2006). *Disposable women and other myths of global capitalism.* London: Routledge.
Ybarra, M. (2012). Taming the jungle, saving the Maya forest: Sedimented counterinsurgency practices in contemporary Guatemalan conservation. *Journal of Peasant Studies, 39*(2), 479–502.

Index

Page numbers in **bold** refer to tables and those in *italic* refer to figures.